# Radionuclides in the Food Chain

# Preface

The Symposium on Radionuclides in the Food Chain, sponsored by the International Life Sciences Institute in association with the International Institute for Applied Systems Analysis, was intended to bring together policymakers and other representatives of the food industry with radiation experts involved in measuring and assessing radioactivity in foodstuffs. The symposium was made timely by the problems arising from the nuclear reactor accident at Chernobyl, in the USSR, which brought out the lack of international agreement on guidance for responding to such radionuclide contamination of food and foodstuffs.

The presentations by the radiation experts covered the sources of radionuclides—natural radioactivity, fallout from nuclear weapons tests, routine releases from nuclear facilities, and various nuclear accidents. The speakers represented a broad distribution in both scientific disciplines and international geographic origin. They summarized the available data on measurements and indicated the current procedures for assessing radiation exposure. It was hoped that the food industry representatives would bring out the problems posed to industry and governments by the presence of radioactivity in food.

Because of the diverse sources of data and backgrounds of the speakers, there was not complete uniformity in the units used in the presentations. The conversions are not complex and are summarized here to allow the reader to have the data in the Système International (SI) units. For the amount of a radionuclide, 1 picocurie (pCi), or $10^{-12}$ curie, is equal to about 0.04 becquerels (Bq) in SI units. For absorbed dose of radiation in air or tissue, 100 rads is equal to 1 gray (Gy) in SI units. External radiation exposure has also been reported in terms of the roentgen (R), which can be considered to be numerically equal to the rad. To allow for the different biological effects of alpha, beta, and gamma irradiation, the dose equivalent replaces the absorbed dose in most of the data summaries presented here. The SI unit for dose equivalent is the sievert (Sv), which is equal to 100 rem. An absorbed dose of 1 Gy from beta or gamma radiation is 1 Sv, whereas for alpha radiation, 1 Gy = 20 Sv. Because of the magnitudes of the dose equivalents listed, they are usually reported in mSv, or thousandths of a sievert.

Basic guidance for addressing the accidental contamination of food exists in the recommendations of the International Commission on Radiological Protection and the several international organizations. The computation of concentrations permitted in food, referred to as derived intervention levels (DILs), involves many assumptions on the extent of contamination and dietary habits of a given population.

In response to the Chernobyl Nuclear Reactor accident, the levels and actions adopted by various countries, both in Europe and in the rest of the world, varied quite widely. This resulted in considerable concern by the public and political leaders as to the safety of the food supply. Certainly some of the response and concern was an overreaction and resulted in unnecessary actions and costs to society.

In response to this situation, the various international organizations (Commission of the European Communities, International Atomic Energy Agency, World Health Organization, Food and Agricultural Organization, and Nuclear Energy Agency) have initiated (or continued) efforts to develop guidance on acceptable levels of radionuclides in food (or DILs) resulting from accidental radionuclide contamination. As reported in various papers presented at this international conference, these efforts have resulted in different values for the DILs. Uncertainty as to the appropriate values still exists.

This conference has made it clear that there is an urgent need for harmonization of such levels (or DILs) to assure credibility and public safety worldwide and as a basis for the accommodation of world trade. The conference has highlighted the issues, and it now remains for the international organizations to agree on harmonized or generic DILs for radioactive contamination of food resulting from nuclear accidents and to establish effective procedures for dealing with radionuclides in the food chain.

It is of interest to note that in October 1988, some two and one half years subsequent to the Chernobyl Nuclear Reactor accident, problems are continuing to occur with regard to radionuclide contamination of certain foods in locations such as the United Kingdom, Europe, Scandinavia, Lapland, and the Middle East. The continuing strict controls are deemed warranted by national authorities to ensure the safety of foodstuffs for human consumption. This contamination resulted from fallout from the Chernobyl accident.

Atlanta, Georgia

*Melvin W. Carter*
*Editor-in-Chief*

Hoboken, New Jersey
Rockville, Maryland
Vienna, Austria

*John H. Harley*
*Gail D. Schmidt*
*Giovanni Silini*

# Contents

## Part III. Environmental Pathways Critical to Humans

## Part IV. Consequences of Radionuclide Release to Health, Safety, and the Environment

## Part V. Effects of Radionuclides in Food and Water Supplies

## Part VI. Risk Management of Food and Water Supplies

## Part VII. Development of Guidelines for Safety Evaluation of Food and Water After Nuclear Accidents

# Contributors

The complete affiliations for all authors are given as footnotes to the opening pages of their chapters. These page numbers are given in the list below.

*P. Linsalata, Ph.D.*, New York University Medical Center   217

*J.R. Lupien, B.Sc., M.Sc.*, Food and Agriculture Organization of the United Nations   389

*F. Luyckx*, Commission of the European Communities   436

*G.F. Meekings*, Ministry of Agriculture, Fisheries and Food   291

*D.P. Meyerhof*, Environmental Health Directorate   447

*N.Y. Novikova*, Ministry of Health   133

*R.J. Pentreath, B.Sc., Ph.D., D.Sc.*, Ministry of Agriculture, Fisheries and Food   99

*V.N. Petrov*, Institute of Applied Geophysics, USSR State Committee for Hydrometeorology   87, 285

*E.V. Petukhova*, Ministry of Health   133

*E.E. Pochin, M.A., M.D.*, National Radiological Protection Board   22

*C.R. Porter, M.S.*, US Environmental Protection Agency   302

*A.W. Randell, Ph.D.*, Food and Agriculture Organization of the United Nations   389

*R.J. Ronk, M.Sc.*, US Food and Drug Administration   409

*E.D. Rubery, M.B., Ch.B., Ph.D., F.R.C.R.*, Department of Health and Social Security   264

*K. Sankaranarayanan, Professor*, State University of Leiden   236

*G.D. Schmidt, M.Sc.* (retired), US Food and Drug Administration   365

*B. Segerståhl, Professor*, Research Institute of Northern Finland   10

*G. Silini, Ph.D.* (retired), United Nations Scientific Committee on the Effects of Atomic Radiation (UNSCEAR)   35

*W.K. Sinclair, Ph.D.*, National Council on Radiation Protection and Measurements   475

*E. Somers*, Environmental Health Directorate   447

*O. Svenson, Ph.D.*, University of Lund   453

*L.B. Sztanyik, M.D.*, Frédéric Joliot-Curie National Research Institute for Radiobiology and Radiohygiene   421

*P. Thompson, B.A.*, US Food and Drug Administration   409

*F.A. Tikhomirov, D.Sc.*, Moscow State University   136

*A.C. Upton, M.D.*, New York University Medical Center   217

*P.J. Waight, M.B., B.S., M.Sc.*, World Health Organization   381

*Y. Yamamoto, Ph.D.*, The Institute of Applied Energy   120

*M. Zifferero*, International Atomic Energy Agency   3

# Acronyms and Abbreviations

| | |
|---|---|
| AEC (US) | Atomic Energy Commission (superseded by the ERDA [DOE] and the NRC) |
| AECB (Canada) | Atomic Energy Control Board |
| AECL (Canada) | Atomic Energy of Canada, Limited |
| AERE (UK) | Atomic Energy Research Establishment |
| ARC (UK) | Agricultural Research Council |
| BEIR (US) | National Academy of Sciences "Biological Effects of Ionizing Radiation" Committee |
| BRH (US) | Bureau of Radiological Health (superseded by the CDRH) |
| BRMD (Canada) | Bureau of Radiation and Medical Devices |
| BUP (USSR) | Byelorussian-Ukrainian Polessye Area |
| CDC (US) | Centers for Disease Control |
| CDRH (US) | Center for Devices and Radiological Health |
| CEA (France) | Commisariat a l'Energie Atomique |
| CEC (Int.) | Commission of the European Communities |
| CEN (France) | Centre d'Etudes Nucleaires |
| CNEN (Italy) | Comitate Nazionale per l'Energia Nucleare |
| DOE (US) | Department of Energy |
| ECC (Int.) | Commission of the European Communities |
| EPA (US) | Environmental Protection Agency |
| ERDA (US) | Energy Research and Development Administration (superseded by the DOE) |
| FAO (Int.) | Food and Agriculture Organization of the United Nations |

| | |
|---|---|
| FDA (US) | Food and Drug Administration |
| GSF (Federal Republic of Germany) | Gesellschaft für Strahlen-und Umweltforschung |
| IAEA (Int.) | International Atomic Energy Agency |
| ICRP (Int.) | International Commission on Radiological Protection |
| ICRU (Int.) | International Commission on Radiation Units and Measurements |
| ILO (Int.) | International Labor Organization |
| IRPA (Professional) | International Radiation Protection Association |
| ISH (Federal Republic of Germany) | Institut für Strahlenhygiene des Bundesgesundheits-amtes |
| JAERI (Japan) | Japan Atomic Energy Research Institute |
| MAFF (UK) | Ministry for Agriculture, Fisheries and Food |
| MOH (USSR) | Ministry of Health |
| MRC (UK) | Medical Research Council |
| NAS (US) | National Academy of Sciences |
| NCRP (US) | National Council on Radiation Protection and Measurements |
| NEA (Int.) | Nuclear Energy Agency (of OECD) |
| NHW (Canada) | Ministry of National Health and Welfare |
| NIRS (Japan) | National Institute of Radiological Sciences |
| NPP (USSR) | Nuclear Power Plant |
| NRC (US) | Nuclear Regulatory Commission<br>Also National Research Council in Canada and in US |
| NRPB (UK) | National Radiological Protection Board |
| OECD (Int.) | Organization for Economic Co-operation and Development |
| RERF (Japan/US) | Radiation Effects Research Foundation |
| RPB (Canada) | Radiation Protection Bureau (superseded by the BRMD) |
| SCPRI (France) | Service Central de Protection contre les Rayonnements Ionisants |
| UKAEA (UK) | United Kingdom Atomic Energy Authority |
| UNEP (Int.) | UN Environmental Programme |

UNESCO (Int.)        UN Educational, Scientific and Cultural Organization

UNSCEAR (Int.)       UN Scientific Committee on the Effects of Atomic Radiation

USDA (US)            US Department of Agriculture

WHO (Int.)           World Health Organization

# Part I
# Introduction

# CHAPTER 1

# A Post-Chernobyl View

## M. Zifferero[1]

The nuclear accident at Chernobyl has, undoubtedly, been by far the most serious one in the 45-year history of nuclear reactor operation. Hopefully, that accident will forever remain so conspicuous.

Thanks to the prompt and exemplary gesture of willingness by the Soviet authorities to communicate and to cooperate through the International Atomic Energy Agency (IAEA) as the focal point, we have now a remarkably full and detailed technical report of the Chernobyl accident. This was made available at the "Post-Accident Review Meeting" held at the IAEA's Headquarters in August 1986.

Very briefly, the accident involved the release of some 50 MCi, or $2 \times 10^{18}$ Bq, of condensible radioactive fission and transuranium activation products. Approximately half this amount, of $\sim 10^{18}$ Bq, relates to radionuclides of possible significance to the human food chain, such as the isotopes of Sr, Ru, I, Cs, Ce, Np, and Pu (Table 1.1).

**Table 1.1.** Released radioactivity at Chernobyl affecting the food chain

| Radionuclide | Radioactive half-life | Emitted radiation | Radioactivity released (Bq) |
|---|---|---|---|
| $^{89}$Sr | 53 days | $\beta + \gamma$ | $8 \times 10^{16}$ |
| $^{90}$Sr | 28 years | $\beta$ | $8 \times 10^{15}$ |
| $^{103}$Ru | 40 days | $\beta + \gamma$ | $1.2 \times 10^{17}$ |
| $^{106}$Ru | 1 year | $\beta$ | $6 \times 10^{16}$ |
| $^{131}$I | 8 days | $\beta + \gamma$ | $2.6 \times 10^{17}$ |
| $^{134}$Cs | 2 years | $\gamma$ | $1.9 \times 10^{17}$ |
| $^{137}$Cs | 30 years | $\beta + \gamma$ | $3.8 \times 10^{16}$ |
| $^{141}$Ce | 32 days | $\beta + \gamma$ | $1 \times 10^{17}$ |
| $^{144}$Ce | 284 days | $\beta + \gamma$ | $1 \times 10^{17}$ |
| $^{239}$Np | 2.4 days | $\beta + \gamma$ | $4.2 \times 10^{15}$ |
| $^{238}$Pu, etc. | 13 years + | $\alpha$ | $5 \times 10^{15}$ |

[1] Deputy Director General, Department of Research and Isotopes, International Atomic Energy Agency, Wagramerstrasse 5, A-1400, Vienna, Austria.

The hot debris reached a height of about 1 km before "horizontal transport began." Most of the material was therefore expected to remain in the troposphere, to be deposited with the precipitation—as it was, indeed, observed.

It is believed that about half the emission of condensible products fell in an area extending to about 60 km from the accident site, while the remaining half was deposited—not uniformly, of course—over an area of Europe of some 10 million km$^2$ and far beyond. Measurements of ground deposits over the whole of Europe and beyond roughly confirm this pattern. Radioactive fallout was, in fact, detected over much of the northern hemisphere as far as Greenland to the north, the United States to the west, and Japan and China to the east. It is important, as well as reassuring, to note that in the context of environmental pollution, radionuclides can be detected unambiguously and with enormous sensitivity at levels far below those of any conceivable ecological or public health significance. A first evaluation of the Chernobyl impact on agriculture and forestry, undertaken by the Joint Food and Agriculture Organization (FAO)/ IAEA Division, concluded inter alia that it will be at least another year, that is, one complete cycle of sowing, cultivation, and harvest in Europe, before a relatively full and balanced appraisal will be possible.

Looking back over the 1½ years that have elapsed since the accident, the consequences can be summarized as follows: First, apart from the tragic loss of life, cases of serious radiation exposure, and community disruption within the critical 30-km zone around the Chernobyl power plant, public health across Europe was never exposed to serious danger or to the risk of significant consequences. This aspect of the accident and its implications for future reactor safety management, and radiological health protection have been summarized in nontechnical language by an IAEA publication titled "One Year After Chernobyl" [IAEA, Vienna, April 1987, Publication No. D3, pp 1–26].

A second major consequence was that the various health protective actions taken nationally frequently involved serious constraints on the use of exposed food crops, livestock production and dairying, food, and feed moving in international trade. These actions, in turn, sometimes caused psychological stress, worry about the future, considerable financial loss, and uncertainties of financial compensation among the affected and dependent communities. These consequences had been neither foreseen nor prepared for, either nationally or internationally.

A third and related consequence has been the clear demonstration of a new and urgent need for international communication, cooperation, and especially for harmonization of guidelines, terminology, and acceptable postaccident reference levels for radioactive contaminants of food and feed moving internationally.

It is easy to be wise in retrospect, but some of these problems evidently arose simply from different interpretations and implementation of existing internationally recognized principles. One, the so-called "ALARA" principle, stands for "doses should be As Low As Reasonably Achievable, economic and social considerations being taken into account" (ICRP, Publication No. 22). A second states "The establishment of . . . intervention levels is the responsibility of the competent authorities" (ICRP, Publication No. 40).

Both these principles are obviously reflected in the IAEA guidelines, but it should be remembered that they refer to normal operating conditions. Clearly, if these otherwise prefectly logical principles are applied unilaterally and nationally to food or feed moving internationally, then problems are likely to arise, and indeed they did. For example, post-Chernobyl limits applied for Cs levels in milk powder for import varied in different European countries by a factor of more than 100 (from 20 to 3,700 Bq/kg).

Turning now to the radioecological aspects of the Chernobyl accident—what are the scientific implications and needs for the future?

With regard to wildlife the idea has been confirmed that postaccident levels of radioactivity high enough to pose a direct threat to wildlife would represent such a horrendous threat to human health that the purely radioecological effects would not represent a primary concern. That is not to say that routine releases of radioactive wastes, for example, into inland water bodies, do not have to be carefully monitored for such effects. But even these would be relatively localized and, of course, would not represent any threat to local or public health.

It is in relation to the human food chain that radioecological studies are of greatest importance. A large number of radioelecological studies have been done since radioactive fallout started with the early atmospheric testing of nuclear weapons in the 1950s and 1960s. It is interesting to note that the total emissions from Chernobyl are believed to be less than those already accumulated in soils from those early tests. Recent data indicate that across Europe there still remains more residual $^{137}$Cs in soils from the weapons test fallout than has been added since Chernobyl.

An important difference between fallout from the atmospheric testing of nuclear weapons and that from Chernobyl must, however, be recognized. The weapons tests injected radioactive debris high into the stratosphere resulting in widespread and continuous deposition. Indeed, fallout of Pu isotopes and $^{137}$Cs over Europe and elsewhere was detectable right up until the Chernobyl accident itself. The emissions from Chernobyl, on the other hand, remained largely in the troposphere, resulting in single "spike" and "wet" depositions with rainfall. However, "dry" deposition was observed in some areas.

"Wet" or "dry" deposited radionuclides enter the food chain by direct interception of exposed crops, through the aquatic ecosystem of fisheries, or, later, by uptake or "transfer" from the soil through the root system. Likewise, grazing livestock take in radionuclides from contaminated pasture that then enter meat and dairy products. Obviously, a great deal can happen between initial deposit and dietary intake, quite apart from radioactive decay, livestock excretion, etc.

Many countries depend on imported food and feed and in this way may become exposed to the effects of distant transboundary fallout episode. Japan, for example, meets 35% of its food needs by importation. Another factor is that food processing beyond the "farm gate" may involve washing, cereal extraction for bread and flour manufacture, dilution or mixing with uncontaminated food constituents, concentration by drying as for milk powder, and the like.

Empirical equations and sophisticated computerized models have been developed and applied to the ecological and food chain behavior of radionuclides. These equations—soil-plant transfer coefficients, exposed crop interception factors, and integrating models—can and are being used for predicting various post-Chernobyl scenarios. One such relatively simple and cautious scenario is based on a single spike deposit of $^{137}$Cs at a level of 10,000 Bq/m$^2$ and indicates the dietary intake over a later postaccident year period when the radionuclide had become dispersed in the soil to the depth of the crop root zone. Radioactive decay was neglected and a "food basket" typical of "Central Europe" was assumed. According to this scenario, the combined intake of $^{137}$Cs per capita per year amounts to some 22,000 Bq (Table 1.2).

The data of this scenario indicate an average effective dose equivalent of less than 0.4 mSv per capita committed for that year of intake. This figure was obtained by applying the appropriate dose conversion factor; in this case $1.5 \times 10^{-8}$ Sv/Bq was assumed. I would like to note at this point that this value, 0.4 mSv per year, represents 20% of the annual dose equivalent received by an average human being from natural sources of radiation. Internal exposure due to the inhalation of Rn gas, that is, to a natural source, results actually in a much higher effective dose equivalent in large populated areas of Europe and North America.

Coming back to the scenario, the high contribution by fish is due to the high value of the water-fish bioconcentration factor assumed and to the neglect of the otherwise highly probable lowering of the concentration of $^{137}$Cs in the aquatic fish environment by particulate adsorption and sedimentation. However,

**Table 1.2.** Cautious scenario for average adult dietary intake of $^{137}$Cs in Central Europe, assuming an earlier spike deposit now equivalent to 10,000 Bq/m$^2$ dispersed down to crop root depth as a result of leaching, ploughing, etc

| Food basket commodity | Intake of $^{137}$Cs<br>Bq per capita per year |
|---|---|
| Green vegetables | 200 |
| Cereals | 160 |
| Potatoes | 100 |
| Cow's milk | 120 |
| Beef | 80 |
| Fruits | 20 |
| Pulses | 40 |
| Goat's milk | 60 |
| Lamb meat | 150 |
| Lake fish[a] | <20,000 |
| Drinking water | <1,600 |

[a] For fish, a water-fish bioconcentration factor of 1,000 was assumed. It is interesting to note that approximately 90% of this intake would originate from lake fish.

although it was once assumed that aquatic food chains would be a minor source of radionuclide dietary intake, post-Chernobyl experience indicated otherwise in some areas (e.g., in fish from some Swedish lakes). This suggests that the problem of inland fisheries exposed to fallout justifies further study.

It should also be noted that a different dose conversion factor might be appropriate for the protection of a particular "critical" population group, for example, infants.

In the same exercise it was estimated that a farm worker spending eight hours a day for six days of each week would receive as a result of external irradiation due to fallout a dose equivalent of 3% of the dose resulting from natural radiation background. These doses can be compared with the suggested international postaccident reference level of 1 mSv for subsequent postaccident years after the first one. It will be interesting to see how the emerging real-time observations compare with the various predictions and model forecasts for a given initial spike deposit.

A problem exposed by post-Chernobyl experience was the conflicting reports of the effectiveness of simply washing freshly harvested and contaminated fruit and vegetables. All food processing "beyond the farm gate," which may involve washing, milling, extraction, mixing, drying, etc., will affect the fraction of initial contamination persisting in the final plate of food or drink. In the post-Chernobyl publication, Safety Series No. 81, IAEA has indicated the effects of processing on fruit and vegetables.

The Chernobyl experience led to one, perhaps not surprising, conclusion: Local weather played a critical role in post-Chernobyl deposition and radiological behavior. Local weather, however, remains conspicuously immune to accurate forecasting!

Therefore, prompt notification of any future nuclear accident, and the mobilization of facilities for on-site monitoring of fallout and radionuclide movement into the food chain merit the highest priorities.

Recent experience has also raised questions and problems of scientific interest:

1. Could existing land-based radar rain-warning networks be used in the event of a serious future accident to enable farmers to move some of their livestock into temporary shelter with stored feed and water supplies?
2. Could exposed crops with high fallout interception factors be harvested after deposition to protect the soil and reduce future crop uptake?
3. Could more effective but simple washing methods be developed for removing recently deposited radionuclides from harvested fruits and vegetables, for example, by making available suitable salt mixtures for domestic use?

With the support of member states, IAEA took prompt action on the question of notification. Two new important international conventions are already in force: one on early notification of nuclear accidents involving transboundary releases and one on emergency assistance on request by a member state

Several of the specialized agencies of the United Nations are concerned with one or more aspects of the problems of accidental releases of radionuclides into the environment. The IAEA has a mandatory concern with all the peaceful applications of atomic energy worldwide, reactor safety, and with the application of safeguards for nonproliferation purposes. The IAEA also has extensive programs on radioactive waste management, radiological protection, radioecological research, as well as programs on the many applications of radioactivity in industry, medicine, agriculture, and exploration of natural resources.

The FAO of the United Nations has a mandatory concern with productivity and with the quality of all the products of agriculture, forestry, and fisheries worldwide. It is also concerned with the welfare of the dependent rural communities. In collaboration with the World Health Organization (WHO), and through the program of the Codex Alimentarius Commission, FAO is also concerned with the problems of chemical contaminants and additives in food, including radioactive contaminants, and, above all, with the problems of internationally acceptable limits for contaminants and additives. The WHO is concerned with the medical and public health aspects of radiation exposure and, reciprocally with FAO, also with the problems of food contaminants and additives. The World Meteorological Organization (WMO) is concerned with the role of climate and weather in the atmospheric transport of radioactive materials. The United Nations Environment Program is concerned with the possible effects of radioactive releases upon the quality of the environment.

The United Nations Scientific Committee on the Effects of Atomic Radiation (UNSCEAR), now also based in Vienna, has kept a role of watchdog on radiation levels worldwide and on the possible shorter—or longer—term effects on human health.

In addition to the UN agencies, several regional and nongovernmental organizations sustain important programs on nuclear safety, radioecology, and health protection. Extensive work in these areas is conducted inter alia by the Commission of the European Communities, the Nuclear Energy Agency of the Organization of Economic Cooperation and Development (OECD), and the International Commission on Radiological Protection (ICRP). Finally, of course, many countries with nuclear power programs support extensive research and development activities.

Clearly, there is considerable potential for duplication and possible confusion, as has been demonstrated by Chernobyl. The IAEA took prompt action on that matter and established an ad hoc Inter-Agency Committee to improve communication, cooperation, and coordination more than a year ago.

As a part of an expanded nuclear safety program that IAEA launched in the second half of 1986 following the detailed analysis of the Chernobyl accident, steps were taken toward the much needed harmonization of guidelines and terminology. In a recent publication, IAEA Safety Series No. 81, the development and application of "derived intervention levels" for food and environment following a radiological emergency is covered in detail.

The FAO also took prompt action in relation to the radioactive contamination of food moving in international trade. An international meeting of experts, which was convened in Rome in December 1987, recommended interim levels below which food could move freely in international trade. From its side, WHO, in cooperation with all other relevant international organizations, is developing numerical guidance for derived intervention levels based on public health consideration. The report has been published and distributed to all food control units of member states.

The FAO and IAEA are, together, in a unique position to address existing and possible future problems of radionuclides entering the food chain. This is due to the existence of a very active Joint FAO/IAEA Division in Vienna, which was created by formal agreement between the two organizations in 1964.

The Joint Division provides for bringing together, uniquely within the UN system, radiological health protection and radioecological expertise on the one hand, and agricultural production and protection expertise on the other hand. Sizable laboratory facilities operated by the IAEA at Seibersdorf, Austria, and in the Principality of Monaco assist the Joint Division in its work.

The IAEA and FAO plan jointly to improve training and preparedness for dealing with possible food and feed problems in the future. A new training and research program for "fallout radioactivity monitoring in environment and food" was initiated last year. This program is likely to be of considerable importance to developing countries that may currently lack an adequate infrastructure for the autonomous detection and monitoring of radioactive fallout over agriculture and fisheries but that could, nevertheless, be exposed as a result of a distant accident. Alternatively, they might wish simply to import and independently monitor food or feed from an affected area.

Finally, immediately after Chernobyl, the Joint FAO/IAEA Division undertook the preparation of two updating scientific reviews on the problems of radioactive fallout over agricultural, forestry, and fisheries ecosystems and on the movement of radionuclides into the food chain. These complementary reviews summarize and identify the nature of the problems, existing sources of information, countermeasures, research, and related needs for the future. They have been prepared with the nonspecialist reader in mind, and they attempt to integrate the immense range of available information and data that have accumulated during the last four decades.

To conclude, I have tried briefly to indicate the scale and nature of the problems of radionuclides entering the food chain as seen in the international context and with particular reference to plans for better preparedness in future. Hopefully, this preparedness will not be needed. But the signs are that nuclear power programs will continue and even extend in the immediate decades ahead. Let us try neither to forget the recent lessons nor to ignore the opportunities provided by that painful experience.

CHAPTER 2

# Structural Problems in Large-Scale Crisis-Management Systems*

## B. Segerståhl[1]

## Introduction

Recent events show that technology has an increasing ability to cause accidents that involve whole countries or continents in one way or another. The most recent, but certainly not the last example of this, is the accident at the Chernobyl nuclear power plant.

Large-scale industrial accidents are serious in their immediate effects and disturbing in their long-trerm consequences. As Lagadec (1) states:

The very large-scale event causes sudden immersion in a universe quite different from that of the "conventional" emergency. Enormous and unexpected difficulties that defeat or wrong-foot the operational arrangements in force; agonizing and paralyzing uncertainty; a critical phase that goes on and on and is therefore wearing on mechanisms, men, and organizations, and an extraordinary increase in the number of people involved: these are some of the features of the post-accident dynamics following a major accident. The logic is scaled up from that of the "ordinary" accident to that of the crisis. Disproportion, hypercomplexity, and strongly destabilizing tendencies are the hallmarks of the crisis phenomenon which we must now learn to understand better and bring under control.

Traditional organizational procedures are not adequate in handling large-scale emergencies. Too many organizations are involved; no common command structure can be established; common goals are vague or nonexistent; the time frame involved is too large; economic, political, and social conflicts emerge; and the dynamics and structure of the event as a whole are not understood.

Our purpose here is to draw a preliminary sketch of the systemic structure of a large-scale crisis-management system. We do not give a recipe for a management system design. That would be futile, as all new disasters are unique. We try, however, to point out a few basic facts about the structure and interactions in a typical, more or less ad hoc, crisis-management system.

* Keynote Address, Conference on Radionuclides in the Food Chain, Vienna, Austria, November 2–5, 1987.
[1] Director, Research Institute of Northern Finland, 90570 Oulu, Finland.

## Dimensions of the System

We start with a few definitions to indicate how the system can be structured and analyzed. There are many dimensions to the problem and no commonly accepted terminology exists. The terminology used in this chapter is not in any way definite. It is, however, an effort to create internal consistency in this presentation. We will look at three different dimensions of the system—time, type, and actors.

### Time

The event that is the object of our investigation is called a **disaster.** Over time a disaster goes through different stages:

precatastrophe,
catastrophe,
crisis,
reconstruction, and
monitoring.

The first stage—precatastrophe—is often neglected in crisis-management systems. The reason is simple: You never know about a catastrophe before you are in the middle of it. One common fact should, however, be noted. In many cases the transfer from normal operating conditions to a catastrophe is not direct. The prelude to a disaster is often a period of maintenance, testing, changes in operating procedures, or other activities that cannot be considered as an integral part of the steady-state operation of the plant. In Chernobyl the accident was triggered by badly planned tests and ignorance of safety requirements. In a recent train accident in Sweden the accident was preceded by maintenance of the safety and signal system.

To use an analogy from the military environment, instead of having either normal operating conditions (complete peace) or a catastrophe (war), a military organization uses a sequence of alert levels. This is routinely done at airports when a plane has to land after having reported abnormal operating conditions. A system of alert levels automatically introduces a system of checks and clearances for nonroutine activities in an industrial plant. This type of precatastrophe alertness can eliminate the accident in some cases. When an accident occurs, the preparedness would be better and, consequently, the damages should be smaller.

Precatastrophe conditions are easy to identify when an accident with possible but avoidable catastrophic consequences has occurred. One well-known case is the Mississauga evacuation in Canada. The accident itself never reached the proportions of a catastrophe, but the evacuation of more than 220,000 people took on all the characteristics of a crisis.

The catastrophe stage ranges over anything from 10 seconds to several days. The border line to the next stage is fluid. One way to identify a transition from catastrophe to crisis is to judge whether there is any immediate unavoidable danger for loss of life caused directly by the original catastrophe. Another indicator showing that the catastrophe has mutated into a crisis is that outside agencies and organizations become involved and the geographical dimensions of the event have grown beyond what can be controlled by plant management. In the case of Chernobyl, the crisis continued less than three months and was preceded by a catastrophe period ranging over five to 15 days.

Reconstruction is, on one hand, technical and economic reconstruction of damaged property and structures and, on the other hand, rehabilitation of destroyed credibility, including introduction of new regulations and standards. Part of the reconstruction after Chernobyl (excluding local work) was the passing of two international conventions prepared by the International Atomic Energy Agency (IAEA). These conventions define procedures and guidelines for early notification and emergency assistance in the event of a nuclear accident or radiological emergency—specifically, in the event of "a release of radioactive material which occurs or is likely to occur and has resulted or may result in an international transboundary release that could be of radiological safety significance" (2). Draft agreements on the conventions were signed in September 1986. The convention on early notification entered into force on October 27, 1986, and the convention on emergency assistance entered into force on February 26, 1987.

## Type

There are many ways to classify technological risks. One of the more ambitious efforts to generate a taxonomy was done by Hohenemser and his colleagues (3) at Clark University. Risk taxonomies, which order risks according to causes of hypothetical accidents, are not suitable for classification of disasters. For this purpose we suggest a classification that puts most of the emphasis on the physical and chemical characteristics of the catastrophe itself. One of several possible classifications according to type is

   contamination of large areas,
   river and lake pollution,
   gas releases, and
   explosions and fires.

It should be noted that a particular disaster can exhibit, simultaneously or sequentially, characteristics from several of these types. In addition, the characteristics of a specific disaster are different, depending on the geographical position of your point of reference. In Chernobyl the catastrophe was an explosion and fire. In other European countries the crisis emerged as a consequence of large-scale contamination.

The main reason for including a typology as one dimension of a crisis-management system is that it puts the system on a fast learning curve. A complete classification system has to include several stages of refinement. After an indication that gases have been released, a need arises to determine whether the gases are heavy or light and what the concentrations and toxicity levels are. This information guides protective measures for the crew working on the site of the catastrophe. Combined with meteorological data, the information gives a basis for the team that has to make the most difficult decisions of all in an emergency. Should an evacuation be ordered? And if so: when, how, and what scope?

## Actors

A minor or "routine" emergency is taken care of by the plant or company being affected. In a major disaster it is inevitable that several different societal actors become involved. The main actors are

plant and company staff;
fire departments, police, military, and hospitals;
industry and insurance companies;
government and local agencies;
regulatory agencies;
the press and television; and
the political system and general public.

To organize coordination and communication between these actors is a formidable task. In many cases the structures are semirandom and nondeterministic. It is natural that nobody is completely in control in a large-scale crisis-management system. The control function is with the plant staff during the first stage of the catastrophe. This control will be transferred as soon as fire departments and police are on site. After a crisis has replaced the initial catastrophe, events can occur rather randomly, depending on the power structure in society as a whole and on the regulatory framework. The importance of a well-designed regulatory framework is not in its ability to provide exact rules but in its ability to create predictable patterns for interaction among different actors in the crisis-management system.

## System Structure

The general structure of the system is shown in Fig. 2.1. The system is driven by the case-specific information available. The structural behavior of the system is influenced by earlier experience, the general knowledge base available, and by the regulatory framework within which the activity takes place. As a vertical structure the system is straightforward:

identification,
assessment,
decisions,

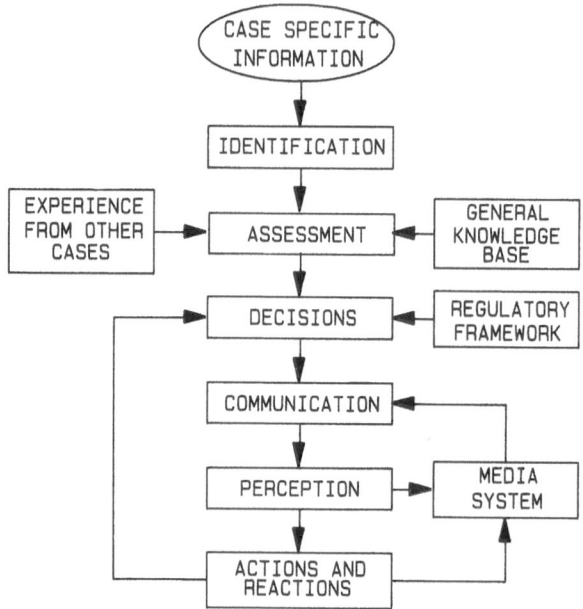

**Fig. 2.1.** Crisis-management system.

communication,
perception, and
actions and reactions.

It is, however, clear that the system behaves in a nonhierarchical manner. Connections between levels are created, bypassed, and destroyed, depending on the dynamic changes in the overall situation. The situation is further complicated by the fact that analysis and management of a large-scale disaster cannot be based on the assumption that only one system is in operation. In cases where several countries are affected, each country has its own system. The connections between national systems is one of the main problems for efforts to manage a disaster efficiently. These connections between parallel systems exist mainly on the communication level. It is obvious that coordinated connections on several levels might be preferable in an ideal case. In reality temporary, and often informal, links between national systems are created for coordination of decisions and actions.

## Regulatory Framework

Every disaster is managed within a national and international regulatory framework. This framework determines, to a large extent, the dynamics of the system's behavior. The priorities, responsibilities, and actions within the system are not predetermined and fixed.

The national framework for risk regulation varies from country to country. In some cases it is strongly centralized, whereas in other countries a substantial amount of regulatory power is delegated to the regional and local level.

The policy principles for regulation range from those requiring very little theory, data, and analysis, and offering very little flexibility, to those at the opposite extreme requiring substantial research efforts and almost continuous interaction between policy-setting institutions and research organizations.

Lester Lave (4) has suggested four criteria that could be used to assess and compare frameworks for regulation. The first is comprehensiveness. The second criterion is the intellectual foundation required for each framework. The third is the resources required to implement the framework. The fourth criterion is felicitousness. The world changes so rapidly that an agency rarely gets to second-order priority issues. The most important issues must be treated first, and they must be raised in an easily comprehended fashion. If the issues are posed in a confused or obscure manner, the decision is likely to be made on an ad hoc basis. The clarity of the framework is more important than its comprehensiveness.

The real question is to what extent a policy framework can support decision makers. Must all effects be quantified accurately and all valuations agreed upon? If complete quantification is required by a framework, but for practical reasons this is not possible, a need arises during a crisis to switch to a less demanding framework. This, in turn, leads to inconsistencies and credibility problems that easily backfire as communications disasters. A policy framework for decision making during a catastrophe and crisis has to be robust, simple, and flexible enough to survive unchanged throughout the entire emergency as the provider of stability and consistency in an otherwise often chaotic and confrontational process.

It is clear that this planning has to rely on consistent and well-known national and international regulatory frameworks in order to be efficient. A preliminary analysis of the situation in western Europe after the Chernobyl accident shows that correct, not too strong and not too weak, protective measures were taken in only a few European nations. Particular problems included belated publication of data and slow institution of food-chain controls. For example, in Switzerland food-chain controls were not put into effect until 2 months after the accident, and in France, exposure data were withheld for months after the accident. In West Germany a great deal of confusion accompanied the announcement of a variety of protective action standards by different government units (5).

One of the major outcomes of the Chernobyl accident was the creation of two international conventions for early notification and mutual assistance in cases of severe nuclear accidents (2). The notification convention effectively establishes an "early warning" system for all nuclear accidents holding potential transboundary consequences, requiring states to report the accident's time, location, radiation releases, and other data essential for assessing the situation. Notification can be made either directly to affected states or through the IAEA. Reporting is mandatory for any nuclear accident involving civilian or military nuclear facilities or materials, with only the exception of nuclear weapons or nuclear weapons testing. The second convention sets out an international framework to facilitate prompt assistance by requiring states to notify the IAEA of available experts, equipment, and other materials. The IAEA will serve as a focal point for coordination by channeling information, supporting efforts, and providing its available services.

This is a typical example of how major accidents influence and change the international regulatory framework. The same type of regulatory process was started in countries belonging to the European Community after the accident in Seveso, Italy, almost exactly 10 years earlier. No loss of life has been reported as a consequence of the Seveso accident. Mortality data from 1975 to 1981 do not indicate that the accident appreciably altered the specific mortality rates by sex, age, and cause in the area around Seveso (6). The impact of this accident on public policy relating to chemical safety in Europe and possibly worldwide has nevertheless been substantial. It had a considerable impact on both government and public opinion in a number of countries and convinced them of the need for stricter control of hazardous chemicals throughout the whole life cycle. The introduction of new legislation concerning chemical safety was considerably accelerated (7). Examples of this are the following Commission of European Communities (EEC) directives that have had a major impact on improving chemical safety in Europe:

1. The EEC Council Directive to prevent major accidents that might result from certain industrial activities and to limit their consequences for man and environment (EEC Directive 82/501).
2. The sixth amendment of the EEC Council Directive (67/548) on approximation of laws, regulations, and administrative provisions relating to the classification, packaging, and labeling of dangerous substances (EEC Directive 79/831).
3. The EEC Directive on supervision and control of transfrontier shipment of hazardous waste.

## Identification

Identification of the characteristics of a major disaster ranges from the immediate and trivial to the vague and confusing. The first indication that a nuclear accident might have occurred came in western Europe when increased radiation levels were measured at the end of April 1986. At that time there was no clear information available from the USSR on what the source for this radiation might be. The identification process proceeds in parallel with emergency actions. Its main purpose is to serve as input for the assessment process that, in turn, creates a base for decision making. In the case of a nuclear accident the identification process is, to a large extent, based on measurements. In addition, atmospheric transport models can be used to design an optimal monitoring strategy.

It became very clear after the Chernobyl accident that many countries in Europe were unprepared for monitoring nuclear contamination on the scale required. Two problems dominated: a need for immediate information, which led to an overload on measurement systems and laboratories, and unforeseen problems with large variations in local contamination levels. The effect of rain on fallout was much higher than expected and led to large variations in local contamination levels. A lot of confusion and distrust emerged as the system was unable to cover these aspects of the situation adequately.

From a management point of view, two requirements are important for the identification process. National and international consistency have to be ensured. Varying standards and very uneven quality of instruments and professional skills in different countries have led to repeated arguments concerning real radiation levels in different countries. The second requirement is that the relevant substances are monitored and measured. In the case of the recent fallout in Europe, this was no problem. From past experience it was known that Cs and I were the important substances to monitor.

## Assessment

We will not discuss the scientific aspects of risk assessment or the methodological problems and technical priorities, but concentrate on the problems that arise when this activity enters the emergency management process.

Experts are in a very difficult position with respect to their credibility during an emergency, real or imagined, as has emerged in the last years. As pointed out by Krohn and Weingart (5), the expert has to represent a scientific "consensus" that has to be presented in the political context as fixed, irreversible, and final. This is the expert's chief contribution to the transformation of probabilistic knowledge into knowledge for political decision making.

The situation is contradictory and clearly impossible. How can an assessment be fixed, irreversible, and final when it has to be based on information that is vague, contradictory, and constantly changing. It is not possible to escape this dilemma with the knowledge we have today of the overall crisis-management system. A serious effort is needed to create procedures that would enable us to harmonize the entire process.

## Decisions

Decisions are based on a projection of information from the assessment process on the regulatory framework within which the crisis-management system operates. There is not much freedom to evaluate alternatives if the regulatory framework is rigid. This makes the decision-making process simple. The main problem encountered is the fact that no rigid framework can predict all possible future disasters.

A rigid framework therefore means that decisions are either completely predictable and consistent with standards or almost completely unpredictable, based on ad hoc judgment. An alternative to making ad hoc decisions when a real situation falls outside the predicted pattern is to simply delay all decisions. A policy of "wait and see" is applied. This makes sense in some cases. It has been said that a safe process can be recognized by the fact that when something goes wrong there is still time for the operator to have a cup of coffee before any decisions have to be made. Completely unforeseen decision-making situations are not desirable for the simple reason that confusion and inconsistency are generated all over the crisis-management system. In addition, there are no clear views of who is allowed to make the rules for decision making and to apply them simultaneously.

The only viable alternative to a rigid set of rules is the use of clearly defined procedures for decision making. It is impossible to make lists of all possible decisions for all situations. It is, however, possible to establish procedures for decision making in such a way that emergency patterns are identified and decisions are made based on patterns and procedures instead of by matching information and regulations on a one-to-one basis. This flexibile approach to decision making in emergencies will certainly benefit from different types of knowledge-based information systems with abilities to deduct patterns from confusing information and assessments and with instant ability to connect the present situation to previous similar events.

Resource allocation is one of the critical tasks during the early stage of a disaster. In the future this specific task will, without doubt, be supported by computerized advisory systems that can take into account case-specific problems like a road being blocked by a population being evacuated or nonavailability of staff and management, which have been foreseen in standard emergency plans.

## Communication

The communication process and its deficiencies are at the core of much of the confusion and of many of the conflicts that emerge during and after a catastrophe. Let us first separate between "professional communication" and "public communication." Professional communication takes place between organizations and units collaborating in the rescue and salvage operations during a catastrophe and crisis. This type of communication follows established rules and procedures. It can be predefined, organized, and rehearsed. This does not ensure that the professional communication system functions properly in all situations. Incompatible equipment, confusion with radio-frequency usage, overloaded switchboards that prevent telephone communication, conflicting terminology, and many other problems create serious communication problems within the professional system.

The real quagmire is in the field of communication with the general public. This communication process changes with time and is dependent on local circumstances and culture. The public broadcasting systems can be included in the professional communication system during the initial stage—the catastrophe—of a disaster. Instructions to the public and unedited news bulletins from the management center are transmitted without distortion. This means in many cases that the messages are also transmitted without clarification.

The dialogue between crisis-management staff and the public can break down within a few minutes when the mode of communication reverts to the traditional style where the media are expected to be the watchdogs and deliverers of the "real truth." To this is added the confusion generated by the use of "concerned scientists" and other sources who see as their main objective to correct the disinformation spread to the public by the officials in charge.

Credibility and consistency are the two main requirements for communication with the public. It is often impossible to fulfill these requirements completely. Information becomes available gradually and leads to reassessment of the situation. As a consequence, a need will arise to upgrade or downgrade the level of alertness required by the public.

Harry Otway (8) has made an important point in differentiating between two kinds of communication. The first is intended to persuade people to accept policies or technologies and the risk they imply; in essence, it encourages passive compliance with the intentions of those providing the information. It is fundamentally manipulative. The second tells people how they can avoid or mitigate risks, or gives information that helps them to form their own opinions; it supports the needs of the audience rather than those of the communicator. The difference is subtle, perhaps sometimes more a question of intent than content, but with this choice between a technocratic or a democratic path, risk analysis stands at a crossroads.

After the Chernobyl accident, situations were observed where the intent of communication fell into the second category—efforts to help the public form their own opinions—but owing to several factors and specific circumstances the communications were interpreted as manipulation and turned out to be counterproductive. After statements by specialists on television that there was no need to take iodine pills, pharmacies were crowded with people buying these pills. The technical and professional implementation of communication strategies is still far from adequate and often looks like a sequence of trials and errors.

The public was at the center of what is sometimes called the "Chernobyl experiment." Assumptions about its behavior, however vague and unfounded, were also the basis for all measures planned and implemented after the accident. These assumptions should now be tested and conclusions drawn.

## Perception

The way people perceive, order, and react to risks is often a mystery both to scientists trained in the natural sciences and to decision makers with a professional involvement in the control and management of a crisis. Whereas scientists rely on risk assessment to evaluate hazards and risks in an objective way, the majority of people rely on intuitive, subjective risk judgments. We use "subjective" and "objective" not as a way of indicating any superiority of one compared with the other, but as a way to stress the point that we are dealing with two different views of reality and of the relationship between scientific abstractions and real societal phenomena.

Research on risk perception is shedding more light on what people mean when they say that something is risky and on what factors and processes underlie those perceptions. The basic motivation for this work lies in the fact that both for regulation related to health and safety and for crisis management, a better understanding of the risk perception process is needed.

Risk analysis describes the impact of an event in terms of direct or indirect, short- or long-term harm to the population. Risk perception seems to be more concerned with what Slovic (9) calls the signal potential of an event. The signal potential of an event, and thus its potential social impact, appears to be systematically related to the characteristics of the hazard and the location of the event within a factor space defined by one factor labeled "dread risk" and one factor labeled "unknown risk."

One important implication of the signal concept is that effort and expense beyond levels indicated by a cost-benefit analysis might be warranted to reduce the possibility of a "high-signal accident." Another important implication is that a reduction in the potential impact of an accident does not influence its position in the signal-factor space. If the wrong conclusions are drawn from these two implications, we are in danger of promoting policies that lead to very low accident probabilities, and the impact of one of these very improbable accidents could be enormous.

The important implication for the crisis-management framework into which we put risk perception is that communication is a translator between the "assessment space" and "perception space." If we learn how to do this translation, many of today's communication and credibility problems could be, if not eliminated, at least reduced to manageable proportions.

## Actions and Reactions

Actions and reactions occur as an integral part of the decision-making process when decisions are made and actions taken in the same organization. These actions are not being discussed here. Problems arise when a decision maker has little or no control over the organizations or groups that should act (or refrain from acting) in accordance with his decisions. The general public cannot easily be forced to act according to decisions unless these decisions are perceived as being in the best interest of the individual person. If an evacuation is ordered, the implementation depends, to a large extent, on the level of credibility transmitted through communications to the public. Unpredictable and counterproductive actions can follow on instructions from the authorites if there is a lack of credibility. The earlier mentioned example of pharmacies crowded by requests for iodine pills is a typical example of a situation where a reaction instead of a nonaction followed a specific communication.

In extreme cases the reactions generate perturbations that affect the whole crisis-management system. Staff changes and transfers of responsibility can be initiated through a political process driven more by demands from the public than by a need to minimize the effects of the disaster. All this boils down to the earlier mentioned need for consistency, clarity, and credibility in communication.

# Concluding Remarks

Our environment is heavily polluted by chemical compounds with a high potential for generating cancer. Compared with these health hazards, radionuclides in the food chain are not a high-priority issue in risk management, not even after the Chernobyl accident. Put into its proper societal context, the problem caused by radionuclides gains importance enormously. It has already been demonstrated that there is a potential for enormous financial losses, extended and traumatic social conflicts, and widespread anxiety and insecurity in the population as a consequence of radioactive contamination—real or imagined.

Much is known about the different components of a crisis-management system involved. Existing bodies of knowledge are nevertheless isolated entities. Very little is known about the dynamic behavior of the system as a whole. Crisis-management systems are activated on an ad hoc basis by pulling agencies and other societal agents into the structure without anybody being able to preplan the overall efficiency of the system.

More work has to be done on analyzing action patterns that can guide the structuring process the next time a large-scale international disaster has to be managed. We intentionally use the term "action patterns" to indicate the futility of efforts to rely on extensive preplanning. Success is more a matter of generating correct thinking habits than a matter of detailed planning.

# References

1. Lagadec P (1986) From Sevesco to Mexico and Bhopal: Learning to cope with crises. In Kleindorfer PR, Kunreuther HC (eds) Insuring and managing hazardous risks: From Seveso to Bhopal and beyond. Springer-Verlag, New York, pp 13–46
2. International Atomic Energy Agency (1986) IAEA Newsbrief, Vol 1, No 1, October 1986
3. Hohenemser C, Kates RW, Slovic P (1983) The nature of technological hazard, Science 220:378–384
4. Lave L (1984) Eight frameworks for regulation. In Ricci PF, Sagan LA, Whipple CG (eds) Technological risk assessment. Martinus Nijhoff Publishers, pp 169–189
5. Krohn W, Weingart P (1987) Nuclear power as a social experiment—European political "fall out" from the Chernobyl meltdown. Sci Tech Human Value 12 (2):52–58
6. Apricena M, Ghioldi R, Pentoni R, Stagnaro E, Vercelli M, Santi L, Dorigotti G, Meazza L (1983) Mortality study in the Seveso area, 1975–1981, Report to the Lombardy Region Authority (Special Office for Seveso, Lombardy Region, Milan).
7. Pocchiari F, Silano V, Zapponi G (1986) The Seveso accident and its aftermath. In Kleindorfer PR, Kunreuther HC (eds) Insuring and managing hazardous risks: From Seveso to Bhopal and beyond. Springer-Verlag, New York, pp 60–78
8. Otway H (1986) Experts, risk communication and democracy. Commission of the European Communities, Joint Research Center, Ispra Establishment, November 1986.
9. Slovic (1987) Perception of risk. Science 236:280–285

CHAPTER 3

# Links in the Transmission of Radionuclides Through Food Chains

E. E. Pochin[1]

## Introduction

The transmission of radionuclides into and through the human food chain has been studied in some detail during the last 30 years. Initially the occasion for these studies was the environmental contamination resulting from atmospheric testing of nuclear weapons, particularly those of high yield, which caused large discharges of fission products into the stratosphere, with subsequent worldwide deposition of various moderately long-lived radionuclides of biological importance. The first report of the United Nations Scientific Committee on the Effects of Atomic Radiation (UNSCEAR), published in 1958, reviewed the evidence on the transmission of radioisotopes of Cs and Sr to human diets from such tests, and the significance of [131]I and other radionuclides in early fallout (1). The same report mentioned the "high concentrations of radon and of its decay products [that had] been observed in ill-ventilated rooms of masonry buildings in certain areas." It also noted, prophetically, that "radioactive contamination of man's environment [might] arise from radioactive waste disposal and accidents involving dispersion of radioactivity" adding that "at the present time, the radiation doses from these two sources are negligible, but in the future they might become appreciable."

In the 30 years since this report was published, increasing attention has been given to the observed or predicted transmission—through food chains—of radionuclides released through waste disposal and from accidents and, more recently, to the exposures resulting from inhalation of Rn daughter products. As a result, we now have extensive information, to be reviewed in these proceedings, on the complex linkages between becquerels of activity released to the environment in many different ways and the sieverts of dose equivalent likely to be delivered to the various body tissues of populations on diets contaminated by such releases.

The quantitative analysis of the successive links in these chains has involved

[1] National Radiological Protection Board, Chilton, Didcot, Oxfordshire, OX11, ORQ, England.

research in many different scientific disciplines, particularly in meteorology, oceanography, soil chemistry, and animal husbandry. Results from this research have been combined with knowledge about the passage of solutes through different geological formations, the dietary practices of many populations and at different ages, and the transfer functions governing different modes of intake of a range of radionuclides into the body plus the doses delivered to body organs at different ages as a result of such intakes.

It has been critically important that even such a wide-ranging inquiry be comprehensive and that no major route be overlooked by which radionuclides released into the environment might return to humans in significant amounts.

## Meteorological Evidence

At present radionuclides are most likely to enter the terrestrial food chain following their initial discharge into the atmosphere. As a result, many aspects of atmospheric distribution, retention, and deposition of injected activity have been examined in detail, during the periods of heavy atmospheric testing of nuclear weapons in the 1950s and 1960s and, more recently, by modeling the potential effects of reactor accidents or by measuring the results of actual discharges.

These two modes of release differ greatly in the amount of heat generated at the time of release and therefore in the height to which the released radioactive materials are carried. In large atmospheric weapon tests, the greater part of the fission products formed were carried into the stratosphere at heights greater than 10 to 15 km. Since no precipitation forms at these heights, they remained in the stratosphere for average periods of one to two years before sedimenting through the tropopause into lower levels of the atmosphere.

Once they reached the troposphere, however, they were rapidly deposited in rain or other forms of precipitation. By this time radionuclides of short half-life such as $^{131}I$ had decayed, but longer-lived fission products such as $^{137}Cs$ and $^{90}Sr$ became widely distributed (at least over the hemisphere into which the release had occurred, stratospheric air transfer over the equator being slow), causing very low doses to very large populations (2).

In accidental releases from nuclear plants, on the other hand, radionuclides remain within the lower troposphere, and this was the case even at Chernobyl, despite the extensive reactor fires and explosions that characterized this accident and carried much of the discharged material initially to a height of about 1 km. In any such releases, therefore, the contents of the radioactive cloud are rapidly depleted by fallout of the fresh fission products onto the ground, causing relatively higher doses to much smaller population groups than in the case of stratospheric discharges, which cause worldwide fallout.

The local sites and amounts of deposition are determined by the movement of air masses at different relevant altitudes, the occurrence of rainfall through air at these altitudes, and dry deposition of radionuclides suspended in the cloud, the speed of the last being influenced largely by the size of any particulate material with which fractions of the nuclear debris may be associated.

The predicted patterns of local and more distant deposition of released material therefore depend—both in position and timing as well as in amount—on a range of familiar meteorological factors, such as wind direction at different altitudes, turbulence, inversions, convection of air masses, and precipitation through them at, and following, the time of release.

## Agricultural Factors

Radionuclides deposited on land may enter human food chains in the following ways: (1) by direct deposition onto the leaves or exposed parts of plants that are eaten by humans or other animals, (2) by persistence in layers of soil from which they are taken up into growing plants through their roots, (3) by resuspension as dust from the soil or from other exposed surfaces, (4) by being washed from the surface or deeper ground layers into water sources that are used ultimately for human or animal drinking or for irrigation, or as media from which fish or other food sources are drawn. The importance of each of these potential sources of dietary contamination depends on the amounts of different materials used in regional types of diet and, in the case of short-lived radionuclides, on the interval between their deposition and their passage through soil, or through animal species, into food to be consumed.

In the case of foliar deposition, variable proportions of the deposited material may be removed from vegetables before human consumption by washing or removing the outer leaves, although some fraction will have been incorporated into the tissues of the growing plant by foliar absorption (3). Foliar deposition also is important in determining the early contamination, particularly with $^{131}$I, of milk from cattle or other species that have grazed outdoors following fresh fallout onto their pastures. The typically short interval (a few days) between grazing and appearance of $^{131}$I in the milk, and the selective concentration of I from blood into milk by mammary glands, can result in high activities in fresh milk. The 8-day half-life of $^{131}$I, however, implies that its activity in butter and other storable dairy products falls by more than an order of magnitude during each month of storage.

After cessation of a release, the level of $^{131}$I in fresh milk will fall at least as rapidly as the deposited activity decreases. For longer-lived radionuclides, however, the situation is different, since few such materials are selectively concentrated in milk, and their persistence in the soil can result in their continuing to enter the food chain over long periods by root uptake into growing plants, in addition to foliar deposition and absorption initially. Isotopes of Cs become widely distributed through plant and animal tissues in broadly the same way as is K. The relatively abundant fission products $^{137}$Cs and $^{134}$Cs, which remain largely in the top few centimeters of soil, may therefore be taken up into growing plants during successive seasons and cause prolonged contamination of animal feeds and human diets (4). They may thus cause moderate but continuing contamination of human diets comparable to a more intense but much briefer activity due to $^{131}$I.

The dose commitments due to radiocesium deposition, however, depend on soil chemistry—the retention of Cs in the upper soil profile being greater in soils of high clay content than in those rich in K or organic humus, from which Cs is more rapidly leached. In some circumstances the form of plant cover is also important, particularly in the case of arctic lichens of which the growth cycle is very long, so that $^{137}$Cs may remain in plant tissues onto which it was deposited for up to 15 years (5). This prolonged contamination, combined with the large areas grazed daily by reindeer or caribou, can result in much higher $^{137}$Cs levels in the tissues of these animals than in the meat of other animals subjected to the same levels of initial pasture contamination.

Soil characteristics are thus important in relation to both the physical and the chemical states of deposited radionuclides (3). The latter may be critical, in regard to

1. their deposition in soluble, insoluble, or particulate form, as affecting particularly their leaching by water; and
2. their fixation by soil constituents and their "dilution" by the very much larger masses of the same elements present in stable form, or of chemically analogous elements, as of Cs by K, or of Sr isotopes in Ca-rich soils.

For these reasons it has been valuable to maintain maps of high- or low-Ca soil types as affecting the $^{90}$Sr content of milk supplies. This expedient can usefully supplement the now conventional maps of normal land usage, so that the likely implication of a given discharge from any site under specified conditions of wind direction and turbulence can be read out rapidly from a prepared program in the event of any particular release.

Geochemistry also has an important influence on the speed or the extent to which different materials, after having been leached from soil or washed from other surfaces, are carried through soil or rock formations into the water table and then into possible sources of drinking water, or of fish or other food growth. Some radioelements, particularly if in relatively insoluble forms, are found to travel from the soil to the stream at considerably lower rates than the water itself travels (2), owing to periods of retention in clay or other formations. As a result, substantial radioactive decay may have occurred in some cases before the water, or food materials grown in it, are likely to be consumed.

## Transmission Through Animal Tissues

Radionuclides present in the feed or water supply of an animal become distributed through its body tissues at a rate, and in a way, that depends mainly on the chemical nature of the material ingested. Just as the speed at which a short-lived radioisotope of I appears in the milk is important in the early stages of an environmental contamination, so the turnover rate of the longer-lived Cs isotopes is relevant to the longer-term activity of animal muscle and other body tissues that are used in human diet. It is important to know this mean turnover for different incorporated radioelements to assess not only how long the contamination of meat will lag behind the corresponding contamination of the animals' diet, but also the value of transferring animals from highly contaminated pasture to pasture that is less contaminated, for example, because rain did not fall on it during passage of the radioactive cloud. Recent evidence (6) has indicated a rather rapid loss of radiocesium from sheep when transferred from hill pastures on which heavy rainfall and contamination had occurred following the Chernobyl release, to neighboring low land and relatively uncontaminated pasture.

This speed of turnover will certainly vary with the chemical nature of the radionuclides involved, with a faster turnover of the widely distributed Cs resembling that of the metabolically similar K and with a slower turnover of Sr isotopes, which, like Ca, become incorporated and retained in bone. Since under most conditions of atmospheric release, however, much greater body doses are likely to result from Cs than from Sr ingestion, a survey of this subject is of practical importance. Many other aspects of long-term dietary contamination were rather fully examined and reported in the early publications of UNSCEAR (7). Since the fallout resulting from atmospheric nuclear tests was much more uniform on a regional scale than that from any brief period of discharge at low level from a single source, the possibility of reducing dietary contamination by local movements of animals to uncontaminated pasture for brief periods before use as food did not have equal practical value for these different situations. Also, the predictably long continued fallout from the stratospheric reservoir offered decreased incentive to study the possibilities of transfer to any uncontaminated feed.

## Waterborne Contamination

Much useful work has been done on the entry of radionuclides into food chains after their discharge into lakes or rivers, or their entry into seas and oceans. Both short-term and long-term distributions need attention here.

In the short-term—the first few years after the arrival of radioactive material into lakes or seas—the concentration ratios of different radionuclides are important, as between sea or fresh water and the edible tissues not only of fish, mollusks, and crustacea but also of any locally edible forms of aquatic plants (8). In the animal forms, the concentration of Cs isotopes is of major importance, although Sr nuclides may be significant when whole small fish, including their bones, are eaten. It is valuable that the dosimetric significance of dietary contamination by radioactive Cs can be directly assessed by whole-body counting methods (9). Thus, in 1977, during an inquiry into the development of nuclear fuel-reprocessing facilities at Sellafield in the United Kingdom, 17 people living at or near Windscale volunteered to have such counts made after a period in which the radioactivity of their normal consumption of locally caught fish and crustacea was measured by duplicate sampling of their diet (10).

Similar whole-body counting is also reported as having been of value in assessing the doses due to radioactive Cs in terrestrially derived diets in the Soviet Union following the Chernobyl release. This possibility of a direct validation of the size of a major contribution to dose from food-chain contaminations is facilitated by the ease and accuracy with which the body content of radiocesiums can be determined and by the relatively long half-time of retention of $^{137}$Cs in humans—about 100 days for the major fraction of the intake by adults (11), with shorter half-time in children. Some assessment can therefore be made of the intake over the preceding few months in a way that is not possible for earlier intakes of $^{131}$I.

In the short-term also, the assessment of dietary contamination requires taking account of any releases of radionuclides into water supplies that are used either directly for drinking or for irrigation of land used for food production. It is likely, too, that resuspension of material deposited on beaches may be significant in some circumstances; windborne spray from coastal waters also may contribute to this activity, as suggested by increased activity of some narrow zones of coastal land (12).

In the much longer term of centuries, significant contributions to population doses may be made at a very low annual dose rate from $^{14}$C and even longer-lived radionuclides such as $^{129}$I and, for some aquatic foods, from transuranic elements. The aquatic distribution tends to be important in these cases, since mobile fractions of their total deposition on land are likely to be leached into oceans of these time scales.

Despite the lack of detailed knowledge of the rate of mixing of water layers at different ocean depths, the ultimate distribution of very long-lived elements within the biosphere may be simplified by knowledge of the distribution of the corresponding stable elements (where stable forms exist) and by assuming that distribution equilibrium has been reached in various parts of the environment and that global circulation models may be constructed on this basis.

Estimation of the fraction of a released radionuclide that will ultimately be present in the terrestrial biosphere, however, depends on knowledge of the fraction present in oceans and oceanic deposits. In the case of $^{129}$I, the latter fraction has been estimated on the basis of an estimated oceanic water mass and the stable I content of sampled water layers (13).

For C, however, the regular, and apparently constant rate at which $^{14}$C is formed by the action of cosmic radiation allows a direct estimate of the $^{14}$C concentration in living tissues and hence a dose estimate for a given release of this radionuclide (14).

The dose commitments that may result during millennia from the presence of low activities of these long-lived radionuclides have a certain unreality, however, when they add so little to the normal life-long exposure of any one generation. The contributions to committed doses from dietary contaminations in any one generation are likely to come mainly from radionuclides of only moderately long half-life, such as those of Cs, Ru, Sr, and Zr, and from $^3$H; and in some circumstances from $^{131}$I and actinides.

## Dietary Information

The variety of routes by which radionuclides released into the environment are known to enter the human diet and the range of scientific disciplines that have been enlisted in their study indicate the importance of the studies that have been made on food chains. It will be a valuable function of this conference, in its papers and in its discussions, to review the variety of possible modes of release and the special or local forms of food intake that might need more detailed investigation.

It is also important that the normal composition of diets used in different regions be known and documented, with their seasonal variations, the localities from which different main constituents are drawn, and the typical intervals between their harvesting and their consumption. The major variations of these diets, particularly in infants and in children, should also be known.

There are several reasons why it is important that a database be collected, even if only in broad outline, on regional patterns and variations in dietary composition, particularly

1. to facilitate quick action in emergency;
2. to ensure appropriate continuing action in applying sound guidelines in control of food supplies; and
3. to establish, and explain, a proper balance between the detriment from banning or restricting a food supply and that from continuing it without restriction.

In an emergency it is necessary to know what kinds of food are likely to cause the highest radiation doses in the early stages of an accidental release so that correct immediate action can be taken in the light of simple, early measurements and in knowledge of the type of release involved.

As more detail emerges on the radionuclide composition of environmental and dietary contamination, it will be important also to review what local or regional modifications should be made to the general intervention levels for foods that have been applied earlier on less complete information. Efficient radiation protection will necessarily depend on a compromise between widespread application of the same intervention levels in all regions and the more administratively, and presentationally, complex procedures involved in applying special restrictions in regions using markedly different dietary sources. Moreover, in principle at least, any restriction—whether local or general—should take account of the total difficulties, detriments, and risks that would result from the restriction, as well as of the radiation risks of continuing an unrestricted use of the dietary components in question.

# Doses and Risks from Ingested Radionuclides

The risks from any given dietary (and water) intake of radionuclides can be assessed with reasonable confidence. With knowledge of the concentration of different radionuclides in an article of diet, and of the average amount of that food or liquid consumed daily, data are available to estimate the amounts of each radionuclide that will be absorbed from the gut and retained in the body or in body organs. The lengths of retention in these organs, and hence the radiation doses delivered to them, have been assessed—both for adults (15) and for children (e.g., at age 10) and infants (age 1) (16). In the case of some elements, absorption and retention may depend on the chemical forms in which the radionuclides are present in the food, but approximate estimates can be made of the total doses delivered to body organs from continued use of a dietary component at a given level and type of contamination. The possible biological effects of using this food can then be assessed in the light of evidence from human epidemiological surveys of the risks of causing cancer or other effects by whole-body or organ irradiation (17,18), or of data on genetic damage in other species (18,19).

It is clearly important to make such objective assessments of the doses and hence of the potential risks that would be avoided by banning or restricting the use of certain classes of food at specified levels and types of radionuclide concentration. "Reference levels" of dose can then be proposed, below which action is likely to be inappropriate, but above which intervention ought to be taken or at least seriously considered (20). The reference level of dose adopted might be expressed in terms of that from prolonged use of a continuingly contaminated food or, more simply and predictably, that from each year's use. In the latter case, the level of dose adopted could be considered in the general context of the recommendation of the International Commission on Radiological Protection (21) that doses to members of the public (in addition to those from natural sources and from medical exposures) should be restricted to "a lifetime average annual dose of 1 millisievert" with "a subsidiary dose limit of 5 mSv in a year for some years" provided that the lifetime average limit is not exceeded.

It seems important that objective estimates of the safety or risk of continued use of a food should be taken into account when decisions are made on the numerical value of "reference levels" of dose and of corresponding "derived reference levels" of radionuclide concentrations in particular foods. Implications of this subject merit discussion, since our detailed and quantitative review of transmission of radionuclides through food chains should lead to a quantitative application of the results of this review.

It must be recognized, however, that we need more information, discussion, and evaluation of the difficulties and risks of restricting the use of any article of diet, particularly of the way in which these difficulties vary in different regions and with different patterns of diet. It is foolish to examine in detail the size of the risks we wish to avoid without any estimate of the risks that we would incur by avoiding them. Certainly it may be difficult to assess the size of the latter risks, to compare the risks of radiation exposure with the temporary inconveniences of mobilizing alternative food sources, to assess the occasional risks of sudden changes in an infant's diet from cow's milk to some powdered alternative, or to assess the effects of abrupt dietary changes throughout the Arctic territories.

At least it is important that "derived reference levels" for foods should be set with the intention of minimizing total risk and detriment and that the lowering of a reference level does not necessarily achieve a greater total safety.

What is clear, however, is that if the detriments and probable risks of banning the use of a major food material differ greatly in some particular circumstances from what they are in general, the use of the same reference levels in such circumstances might, and logically would, represent less efficient rather than more efficient protection.

## References

1. UNSCEAR (United Nations Scientific Committee on the Effects of Atomic Radiation) (1958) 1958 report of the General Assembly, Annex D. United Nations, New York
2. UNSCEAR (1972) 1972 report to the General Assembly. Ionizing radiation: levels and effects. Vol 1, Levels, Annex A. United Nations, New York
3. Russell RS (ed) (1966) Radioactivity and human diet. Pergamon Press, Oxford
4. UNSCEAR (1977) 1977 report to the General Assembly, Sources and effects of ionizing radiation, Annex C. United Nations, New York
5. Hanson WC (1967) Cesium 137 in Alaskan lichens, caribou and Esquimos. Health Phys 13:383–389
6. Howard BJ, Beresford NA, Burrow L, Shaw PV, Curtis EJC (1987) A comparison of caesium 137 and 134 activity in sheep remaining on upland areas contaminated by Chernobyl fallout with those removed to less active lowland pastures. J Soc Radiol Protect 7 (2):71–73
7. UNSCEAR (1969) 1969 report to the General Assembly, Annex A. United Nations, New York
8. Hunt GJ (1986) Radioactivity in surface and coastal waters of the British Isles, 1985. Aquatic environment monitoring report No. 15. Ministry of Agriculture, Fisheries and Food, Directorate of Fisheries Research, Lowestoft
9. Fry FA, Sumerling TJ (1984) Measurements of caesium 137 in residents of Seascale and its environs. NRPB-R172. Report of the National Radiological Protection Board, Chilton, Didcot, Oxon, England
10. Parker Mr Justice (1978) The Windscale Inquiry; Report to the Secretary of State for the Environment, Vol 1. London, HMSO
11. UNSCEAR (1964) 1964 report to the General Assembly, Annex F. United Nations, New York
12. Stather JW, Wrixon AD, Simmonds JR (1984) The risks of leukaemia and other cancers in Seascale from radiation exposure. NRPB-R171. Report of the National Radiological Protection Board, Chilton, Didcot, Oxon, England. Strather JW, Dionian J, Brown J, Fell TP, Muirhead CR (1986) Appendum to Report R171. London, HMSO

13. UNSCEAR (1982) Report to the General Assembly, Ionizing Radiation: sources and biological effects, Annex F. United Nations, New York
14. UNSCEAR (1977) Report to the General Assembly, Sources and effects of ionizing radiation, Annexes C and D. United Nations, New York
15. Limits of intakes of radionuclides by workers. International Commission on Radiological Protection, Publication 30, Annals of the ICRP 2 (¾) 1979, 4 (¾) 1980 and 6 (⅔) 1981; with Supplements at Annals of the ICRP 3 1979, 5 1981, 7 1982 and 8 (⅓) 1982
16. Kendall GM, Kennedy BW, Greenhalgh JR, Adams N, Fell TP (1987) Committed doses to selected organs and committed effective doses due to intakes of radionuclides, NRPB-GS7. Report of the National Radiological Protection Board, Chilton, Didcot, Oxon, England
17. Report of the National Institutes of Health ad hoc working group to develop radio-epidemiological tables (1985) NIH Publication No 85-2748, Washington, DC
18. UNSCEAR (1977) Report to the General Assembly, Sources and effects of ionizing radiation, Annexes G and H. United Nations, New York
19. UNSCEAR (1986) Report to the General Assembly. Genetic and somatic effects of ionizing radiation, Annex A, United Nations, New York
20 Radiation protection in the European Community (15 December 1986) Evaluation and suggestions by a Committee of high level independent scientists. Directorate-General, Science, Research and Development, paper 1/UH4, Brussels
21. Statement from the 1985 Paris meeting of the International Commission on Radiological Protection (1985) Annals of the ICRP *15* (3) i–ii

# Part II
# Fundamental Information

CHAPTER 4

# Biological Effects of Ionizing Radiation*

## G. Silini[1]

## Introduction

Ionizing radiation affects human tissues by depositing energy as it passes through
the body. During the tenth of a trillionth of a second after radiation hits an
atom in a tissue, an electron is stripped from this atom. This process is called
*ionization*. As the electron is negatively charged, the remainder of the atom,
which was previously neutral, becomes positively charged. Both the electrons
and the ionized atom are very unstable, and during the next tenth of a trillionth
of a second, they undergo a complex chain of physicochemical reactions, creating
new molecules, including particularly reactive ones known as "free radicals."
These, in turn, during the next millionth of a second may interact among themselves or with other molecules and, through processes not yet fully understood,
may produce chemical changes in some molecules that are biologically important
for the cells, such as those carrying the essential information for their function.
It is through these changes that biological effects occur over any time span,
from a few seconds to many decades after irradiation. These changes may kill
cells outright or alter them in ways that will lead in the course of time to
cancer or to genetic effects.

   As the effects of radiation are related to the energy absorbed in tissues, it is
important to measure this energy in order to precisely correlate energy deposited
and effects. To this end, one defines the quantity absorbed dose, that is, the
energy absorbed per unit mass. This is, however, a very simplistic way to
study radiation-related phenomena. There are various other ways of expressing
the dose, such as the dose equivalent, the effective dose equivalent, the collective
effective dose equivalent, and the collective effective dose equivalent commitment. These other quantities are suitable for defining the dose received by an
individual or a whole population, now or in the future, and they are used for
special purposes in order to assess the doses with an increasing degree of precision.

---

* Opinions expressed in this chapter do not necessarily reflect the views of UNSCEAR.
[1] United Nations Scientific Committee on the Effects of Atomic Radiation (UNSCEAR)
Vienna International Centre, P.O. Box 500 A-1400 Vienna, Austria.

All these quantities must, of course, be measured: to this end, one defines a system of units such as the becquerel (Bq), the gray (Gy), or the sievert (Sv). Their precise definition is beyond the scope of this presentation, but their knowledge is essential to the specification of dose-effect relationships, particularly at the lowest doses of interest for practical purposes.

There are two ways in which the body may absorb radiation energy. The first is when the radiation source is located outside the body as, for example, when one takes an x-ray picture at the dentist: In this case we speak of external irradiation. The second is when the source of radiation is within the body, such as, for example, when one takes up a radionuclide through breathing or eating: In this case we speak of internal irradiation. Although in these two instances the radiation source is differently located with respect to the body, the result is always the same: irradiation leading to energy deposition in cells and tissues. The effects are, by and large, the same, for the same amount of energy absorbed in a given organ or tissue.

Two major classes of effects may be distinguished in radiation biology: the hereditary and the somatic ones. To the first class belong effects that are induced by radiation in the germinal cells of the exposed individuals (i.e., the cells that we pass on to our descendants). These effects are then transmitted to the offspring, to appear, sooner or later, in future generations. To the other class belong the effects induced on the somatic cell lines: They will appear on the exposed individuals themselves and therefore, by definition, within the lifetime of the individuals. I will address these two classes in turn.

## Hereditary Effects

On account of their mode of transmission, hereditary effects may be divided ideally (although in practice this distinction is not always so clear) into dominant— those appearing in the first-generation offspring—and recessive—those appearing in later generations. On account of their mechanisms of induction, these effects are (always ideally) either of the gene type, affecting one single gene, or of the chromosome type, affecting more profoundly the subcellular structures, called the chromosomes, on which the genes are ordered. This latter case may include the so-called numerical aberrations involving loss or additions of entire chromosomes, or the structural aberrations, changes in the makeup of single chromosomes, involving losses or additions of pieces of chromosomes. Many genetic characters are not determined by simple genes, but by the interplay of many genes; they are the so-called polygenic characters. About these we know very little, but we have no reason to suspect, so far, that radiation may affect them more than it affects single genes; the evidence is, in fact, that the polygenic characters may be less easily affected than the monogenic ones.

It is important to remember that there is a difference between alterations induced by radiation at the subcellular level on the genetic material and clinical effects brought about by such alterations, which are manifested under the form of hereditary diseases. Clinically, these conditions may be fairly trivial (imagine, for example, a condition such as color blindness) or very severe (Down's syndrome, for example). One should also remember that a sizable proportion of people are carriers of such genetic diseases, even in the absence of irradiation above the normal background, and that there is no way to recognize these "natural" conditions from the radiation-induced ones. Equally important is the notion that the correlation between genetic effects and genetic detriment is rather loose, in the sense that relatively minor genetic effects compatible with life may be the cause of more harm and suffering than the biologically disruptive effects causing the death of the affected individual early in its development. From the point of view of radiation protection, it is the detriment that matters, rather than the genetic alteration at its origin. Among the detrimental conditions, the most relevant are those that bring about the most severe diseases, rather than the clinically mild ones that cause relatively less suffering and are compatible with the life and reproduction of the affected individuals.

Up to the present time, there is no direct evidence of radiation-induced hereditary effects in man, although there is plenty of such evidence in other species and there is no reason to think that man may be an exception to the rule. Therefore, our estimates of risk for man are drawn from the data obtained on experimental animals, with suitable corrections to adapt them to man, on the basis of our knowledge of human genetics. Although such interspecies extrapolation is not very satisfactory, it is nonetheless the only possibility, in view of the lack of direct human data. It is unlikely, at least in the short term, that we may abandon this practice of extrapolating between species.

One way of extrapolating is through the so-called direct methods. With these one aims at an assessment of the risk by the multiplication of three factors, namely, the number of genes at which mutations can occur, the mutation rate per unit dose, and the radiation dose. Table 4.1 shows the currently accepted estimates of genetic diseases, obtained according to the direct methods. They are expressed as the number of genetically abnormal children born in the first generation after irradiation, per 0.01 Gy of sparsely ionizing low-dose-rate irradiation. They are drawn from the 1986 UNSCEAR report (1) and are essentially the same as those arrived at in the 1982 report.

Alternatively, extrapolation may be carried out indirectly by the so-called doubling-dose method. This method assumes that a given dose of radiation

**Table 4.1.** Risk of induction of genetic damage in man per 0.01 Gy at low dose rates of sparsely ionizing radiation according to the direct method[a]

| Genetic damage | Expected number of genetically abnormal children in the first generation after irradiation of 1 million people | |
| --- | --- | --- |
| | Males | Females |
| Mutations having dominant effects | ~10 to ~20 | 0 to ~9 |
| Recessive mutations | 0 | 0 |
| Unbalanced products of reciprocal translocations | ~1 to ~15 | 0 to ~5 |

[a] Reprinted with permission from ref. 1.

delivered over one generation will double the natural incidence of genetic diseases, leaving unchanged the types of such diseases expressed naturally by the population. Thus, with the doubling-dose method, the risk is related to, and expressed as a fraction of, the spontaneous prevalence of gene and chromosomal disorders, as well as disorders of more complex etiology. Table 4.2 shows the risk estimates derived by UNSCEAR according to the doubling-dose method, expressed as the number of diseased children after 0.01 Gy per generation of sparsely ionizing low-dose-rate irradiation. Table 4.2 also shows separately the number of genetically abnormal children that would be expected to appear in a population of 1 million people either in the first generation after receiving that dose, or at equilibrium, that is, a long time after the dose of 0.01 Gy per generation added to the natural background. These numbers have been derived in 1986 under the assumption that the doubling dose is 1 Gy. They should be compared with the "natural" incidence of the same diseases, shown in the left column of Table 4.2.

These estimates are also essentially the same as those derived in 1982, except that the disorders of complex etiology are no longer included in the estimates. The UNSCEAR had calculated in 1982 that the incidence of these diseases would be of the order of 5 and 50 cases per 0.01 Gy in the first generation and at equilibrium, respectively, against a total of about 90,000 cases per generation in the nonirradiated population. These estimates had been based on some early data obtained on individuals up to 21 years of age. If one takes, however, more recent data on the incidence of disorders of complex etiology in people up to 70 years of age, one finds that the natural incidence of these conditions is much higher because they become manifest mostly in old age.

These new findings will have to be validated by additional evidence, and there will need to be discussion on other points, namely, (1) whether the doubling dose of 1 Gy, based on clear-cut genetic end points, is also valid for disorders of complex etiology: and (2) whether the assumption that the true mutational component of these conditions of 5%, as assumed in 1977 and 1982, is realistic for these disorders. In the absence of firm information, particularly on the

**Table 4.2.** Estimated effect of 0.01 Gy per generation of low dose or low-dose-rate sparsely ionizing irradiation on a population of 1 million live-born, according to the doubling-dose method[a]

| Disease classification | Current incidence per million | Effect of 0.01 Gy per generation | |
|---|---|---|---|
| | | First generation | Equilibrium |
| Autosomal dominant and | | | |
| X-linked diseases | 10,000 | 15 | 100 |
| Autosomal recessive diseases | 2,500 | Slight | Slow increase |
| Chromosomal diseases due to | | | |
| Structural anomalies | 400 | 2.4 | 4 |
| Numerical anomalies | 3,400 | Probably very small | Probably very small |

[a] Reprinted with permission from ref. 1.

mechanisms of maintenance of these disorders in the population, UNSCEAR is not in a position to provide meaningful risk estimates. Another exercise that UNSCEAR is pursuing is an attempt to refine the estimates by going from the incidence of the most severe conditions to new estimates, taking into account not only the incidence but also the severity of genetic diseases. This may be accomplished by calculating the years of life lost as a result of genetic diseases of different severity. The hope is to obtain new estimates, more meaningful in terms of the individual, familial, and societal burden imposed by such conditions. Such a refinement is not yet ready, however, for immediate use.

# Somatic Effects

According to the definition given previously, somatic effects are caused by radiation affecting the cells of the somatic lines. These effects will, therefore, be expressed within the lifetime of the exposed individuals. There are two classes of such effects: those that appear rather early after high doses of radiation and those that become manifest a long time after small doses. The former effects are conventionally known as nonstochastic; the latter as stochastic. There are many differences between these two classes in addition to those already mentioned. Stochastic effects, as the name suggests, are induced at random in an irradiated population; they are brought about by malignant transformation of one or very few irradiated cells in the body; nonstochastic effects, on the contrary, appear when many cells in the body are killed and the functions supported by such cells are severely disrupted.

## Nonstochastic Effects

According to a well-known definition, both the incidence and the severity of nonstochastic effects increase when the dose of radiation increases. These effects are deterministic in nature in the sense that on the basis of empirical knowledge, one may reasonably predict their occurrence and degree in a person irradiated at a given dose.

The clinical form and severity of the nonstochastic effects are very different in the case of partial-body or of total-body irradiation. When the whole body is exposed, the early damage takes the form of the so-called radiation syndromes, of which the hemopoietic, the gastrointestinal, and the cerebral are paradigmatic clinical manifestations between a few and a few tens of gray of acute sparsely ionizing radiation.

For partial-body irradiation, early effects take the form of tissue damage, particularly of the so-called self-renewing tissues (bone marrow, skin and skin adnexa, gastrointestinal lining, testis, lens of the eye) but also of other radiosensitive tissues and organs (ovary, lung, kidney, central nervous system). The net result of this damage is a disappearance of the parenchymal cells, giving rise to a loss of specific functions of the irradiated tissues. Even after repair and repopulation, some damage may still persist long after treatment; it generally consists in a relative loss of parenchymal cells (and of the related functions) and in a relative increase of the connective tissue component, giving rise to tissue fibrosis.

These are the effects that became apparent at Chernobyl soon after the recent accident and these were responsible, together with other damages from fire or steam, for the death of about 30 people. Such effects would not be expected to appear far away from the site of the accident because the dose absorbed under these conditions would be too low.

The 1982 UNSCEAR report (2) systematically examined the information available in man and other mammals on irradiation of single organs and tissues by external and internal sources. It discussed the nature of the early and late nonstochastic damage in these tissues, the levels of dose at which specific effects become apparent, the relationships between dose and the probability of nonstochastic damage, and the influence of time and radiation quality. The conclusions of that study may be summarized as follows:

1. Nonstochastic effects on tissues are generally characterized by nonlinear relationships with dose and by a minimum dose of radiation that must be administered to produce the effect. This is called the threshold dose.
2. The concept of threshold is a difficult one to define and must be discussed in relation to each tissue and each effect because it depends, to a large extent, on the sensitivity of the measuring technique. For example, sophisticated ultrastructural analyses may show subtle morphological changes in some tissues at doses much lower than those needed to produce a loss of functional capacity.
3. There is a need to distinguish between thresholds for effects that have no apparent deleterious consequences and are easily repaired and thresholds for damage having a clear pathological connotation. The second is more relevant to radiation protection.
4. Since most information in man is derived from radiotherapy experience, allowance must be made for the fact that the degree of damage that may be acceptable when treating a patient (often carrying a malignant tumor) is clearly higher than for the same damage assumed for planning of radiation protection. It seems obvious that the threshold doses must be appropriately scaled down for protection purposes.
5. The type of radiation in relation to its capacity for ionizing is an important variable in the production of nonstochastic damage. For the same absorbed energy, different radiations may produce more or less effect. It is important to study the effectiveness of these radiations, and differences must be taken into account when assessing dose equivalent.
6. The time over which a dose is delivered is also very important. The results of long-term irradiations are the most valuable for establishing dose threshold for planning of radiation protection because chronic exposures are much more frequent and important than acute ones.
7. Finally, and most importantly, the mechanisms of production of nonstochastic effects make it unlikely that the thresholds might be abolished at low doses

and dose rates. The presence of the threshold is the one characteristic that clearly differentiates the nonstochastic from the stochastic effects and that justifies the goal of radiation protection to avoid nonstochastic damage completely at sufficiently low doses and dose rates.

A special class of nonstochastic effects is that induced by irradiation of mammals in utero, a subject that was treated extensively in the 1977 and 1986 UNSCEAR reports (1,3).

The effects induced by irradiating the developing mammal have peculiar features related essentially to the fact that the product of conception is an organism where tissues are not functioning in a steady state, but are continuously differentiating and growing. Since development follows accurately planned pathways that are closely correlated to and dependent on each other, irradiation of a developing organism may lead to profound alterations or even to disruption of the complex sequence of events that end in the production of a mature animal. The variety of the consequences is extreme; they include effects leading to intrauterine death of the animal, effects that are visible at birth, or effects that have a long-term expression in the extrauterine life. The clinical severity of these lesions is also variable: Some are incompatible with life before or after birth; others are perfectly compatible with a fairly normal life and reproduction. There is little correlation between the degree of biological damage and its individual, familial, or societal importance. For example, the most damaging lesions may result in failure to implant, an effect that may go clinically undetected, whereas the less damaging conditions allowing an individual to survive may actually allow only a poor quality of life.

Conventionally, lethal effects, malformations, and growth disturbances are the major effects to be seen, with ample scope for combinations of the various classes. Roughly, the lethal effects in utero are characteristic of the preimplantation period; malformations result from embryo irradiation during the phase of major organogenesis; growth disturbances arise from irradiation of the fetal stages. Difficulties in the classification of the effects could reflect upon uncertainties in the estimates of induction. Radiation is by no means the only agent causing these effects, particularly the malformative ones, because other chemical toxic substances may induce lesions of the same kind. They are, moreover, similar to those appearing in a seemingly spontaneous fashion in about 6% of the live-borns in the human species.

Data on animals represent by far the greatest amount of the information available. From them it appears possible to derive generalized conclusions about the main characteristics of developmental damage. These are thought to be applicable in principle to man, at least qualitatively, since there is no reason to believe that the response of man may substantially depart from that of other mammals when one allows for the characteristics of each species.

In contrast with the relative abundance and precision of observations in experimental animals, data in man are rather scanty and inadequate, owing to the lack of many important pieces of information. For example, dose estimates are approximate and often indirect and variable, even in the best-documented series. The test samples are invariably too small, particularly at the low doses, to draw very precise estimates of induction. It is sometimes difficult to refer the radiobiological effects to well-matched and homogeneous control groups. There are uncertainties as to the time of exposure and the best series are for irradiation during the second or third trimester of pregnancy, although the most sensitive periods in humans are between 8 and 15 and 15 and 25 weeks from fertilization.

The best data in humans are those derived from children exposed in utero at the time of the explosions in Hiroshima and Nagasaki. The most common malformations found in these individuals are in the cerebral cortex, resulting in severe mental retardation, sometimes accompanied by small head size. However, other effects could also be induced in man. Taking all evidence into account, UNSCEAR has attempted to derive quantitative risk estimates for a number of radiation-related effects in utero, such as mortality and induction of malformations, mental retardation, tumors, and leukemia. Under a number of qualifying assumptions, it is possible to conclude that for the small doses likely to be encountered in practice, the overall risk of these harmful conditions is no more than 0.002 for the live-born at 0.01 Gy. This should be compared with the natural incidence of malformations in nonirradiated individuals, which is of the order of 0.06 in the human species.

## Stochastic Effects

This term covers the induction of leukemia and tumors, which are by far the most difficult and uncertain subjects in the field of somatic effects. For the biologist, the difficulties are related to the doubts still existing as to the nature and pathogenesis of all neoplastic lesions, which reflect heavily on our capacity to correctly interpret the role of radiation in the biological process. For the radiation protectionist, the difficulties are imprecision of the risk estimates for tumor induction on which the entire system of radiation protection is founded. These fields were extensively covered in the 1977 and 1986 reports of the UNSCEAR (1,3).

As to the radiobiological aspects, the main conclusion is that generalizations about tumors are totally unwarranted because the range of conditions under which different tumor types are induced (by radiation or otherwise) is very broad. Different forms of tumors recognize different etiological factors (viruses, somatic mutations) interacting with radiation in a variety of pathogenetic mechanisms, with interplaying subcellular, cellular, tissue, and humoral factors. Some of these mechanisms are gradually being uncovered, but until better knowledge of the phenomenon of tumor induction as a biological process becomes available, the role of radiation in inducing tumors will not become clearer. It should be stressed that further information on mechanisms is not only the logical answer to the most fundamental questions, but the necessary step for the development of precise models of tumor induction for the use of radiation protection. Short of this knowledge, research on tumor induction by radiation must remain phenomenological and descriptive.

The massive amount of data accumulated and still being produced in experimental animals has significantly facilitated the projection of results from one to another species by allowing some cautious generalization of mechanisms. It has also allowed the extrapolation of general trends in radiation carcinogenesis in respect, for example, to the effects of radiation quality, dose rate, and dose fractionation. However, up to now, species- and strain-specificities have not allowed the extrapolation of quantitative rates of induction for specific tumors, which is an essential step in order to arrive at precise estimates of risk. For these, human data will probably continue to be used, despite the shortcomings of many of the series available.

Turning now to the human data, the following brief discussion of these shortcomings is appropriate:

1. The first and most important of them is that human data are only available for relatively high acute doses, while the low doses and dose rates are the conditions of exposure of most practical interest. It is therefore necessary to extrapolate the risk to these latter modalities of exposure. This is a difficult exercise, since each estimate applies strictly to the exposure conditions under which it is derived, and extrapolation to other conditions requires some knowledge (and not only some assumption) of the precise form of the dose-effect relationships.

2. Dosimetry in epidemiological series is sometimes unclear, often imprecise, and never satisfactory. The dose rate, the pattern of dose fractionation, and the variability of doses between different individuals are frequently unknown for external irradiation. The same is true for the dose distribution in body organs after incorporation of radionuclides. Of course, indirect inferences, comparisons of organ sensitivities, and model calculations may be used to derive values of the risk coefficients, but the lack of direct dosimetric data adds to the uncertainties of these values.

3. The observation period is rarely long enough to ensure that all potentially induced tumors have, in fact, appeared, in consideration of the long latency periods and their variability. This fact requires some correction to account for cancers expected to appear before the extinction of the observed population.

4. The size of the exposed population is almost invariably too small to ensure statistical significance of the observation, considering the high natural incidence of tumors that are not related to radiation and indistinguishable from the radiation-induced ones. This produces large statistical errors in the risk estimates.

5. The only way to show that radiation does induce tumors is through appropriate comparisons with a control population similar in all respects to the test sample, except for the radiation received. However, this condition is seldom achieved. For example, in irradiated patients (ankylosing spondylitis, hyperthyroidism) it is difficult to be sure that radiation, rather than the disease

itself requiring treatment, is responsible for the extra tumors detected. In atomic bomb survivors the establishment of appropriate control groups has always been difficult for a variety of reasons (living conditions, differences in sex and age distributions, immediate mortality at the time of explosion).

Given all these reservations, it is no wonder if some people regard the derivation of risk estimates more as an art than as science. Whatever it may be, from a careful analysis of all existing data, UNSCEAR attempted in 1977 to derive estimates of tumor induction per unit dose, which, as far as I know, were quite well received. They could be summarized as follows:

1. The number of fatal malignancies induced by sparsely ionizing radiation in the range of about 1 Gy could be of the order of $10^{-2}$ per sievert, whereas the number appropriate at the much lower levels of occupational or environmental exposure could be substantially lower.
2. The risk of inducing nonfatal malignancies would probably be of the same order of magnitude.
3. Figures twice as high might be applicable to malignancies induced by exposure of the fetus in utero. They are included in the risk estimates for prenatal irradiation.

The most recent reports of UNSCEAR have endorsed these figures provisionally, while recognizing that they may need some revision. Such need is dictated by the accumulation of new information and by a reevaluation of the findings in the survivors of the atomic bombs in Hiroshima and Nagasaki. As a result of some new calculations, the neutron component of the total dose now appears to be substantially lower in both cities, whereas the gamma component is increased substantially at Hiroshima and reduced slightly at Nagasaki. These necessary readjustments may result in some changes of the relevant risk estimates to an extent that is now difficult to predict. It is not expected that new risk estimates will become available before next year.

# References

1. United Nations (1986) Genetic and somatic effects of ionizing radiation. Report of the United Nations Scientific Commitee on the Effects of Atomic Radiation. Official Record of the General Assembly, Forty-first Session, Supplement No. 16 (A/41/16). New York
2. United Nations (1982) Ionizing radiation: Sources and biological effects. Report of the United Nations Scientific Committee on the Effects of Atomic Radiation. Official Record of the General Assembly, Thirty-seventh Session, Supplement No. 45 (A/37/45). New York
3. United Nations (1977) Sources and effects of ionizing radiation. Report of the United Nations Scientific Committee on the Effects of Atomic Radiation. Official Record of the General Assembly, Thirty-second Session, Supplement No. 40 (A/32/40). New York

CHAPTER 5

# Assessment of Dose From Man-Made Sources

W. Jacobi[1]

## Introductory Remarks

The intensity of ionizing radiation and the activity of radionuclides in our environment can be measured with rather simple methods. During the last decades, large surveys have been carried out to investigate the population exposure from natural and man-made sources. In particular, the observance of regulations limiting radiation exposure from nuclear plants required the establishment of comprehensive monitoring programs. Thus, for purposes of radiation protection, exposure to ionizing radiation can be estimated with sufficient accuracy. This is particularly valid for operational discharges from nuclear power reactors. The results have been summarized and critically reviewed in the reports of United Nations Scientific Committee on the Effects of Atomic Radiation (UNSCEAR) (1,2), together with an analysis of the biological effects of ionizing radiation.

In this chapter an overview of the dose assessment from man-made sources is given, with emphasis on the contributions from operational activity releases from nuclear and conventional power plants.

## General Objectives and Dose Quantities

The primary objective for judgment of the exposure from a given source is the estimation of the spatial and temporal dose distribution to the affected population. For purposes of radiation protection, the following exposure quantities are of main importance:

1. the maximum dose to individuals or to critical groups;
2. the mean individual dose or the mean per caput dose, averaged over the exposed population; and

[1] Gesellschaft für Strahlen und Umweltforschung (GSF), Institut für Strahlenschutz, Ingolstädter Landstr. 1, D-8042 Munich-Neuherberg, FRG.

3. the collective dose or dose commitment from the considered source to the local, regional, and global populations, taking into account the size of these populations.

With respect to the dose to individuals, the annual dose and the total dose cumulated over their whole lifetime are of main interest. In this chapter all dose values are expressed in terms of the effective dose equivalent.

The collective dose is usually applied to characterize the total radiological impact of the considered source to the population. It is given by the number of individuals in the considered population multiplied by the mean per caput dose (effective dose equivalent); the unit of the collective dose is 1 person-sievert (person-Sv). This quantity enables a relationship with the total detriment of health, which may be caused by the radiation-induced risk of carcinogenic and hereditary effects in the considered population.

## Current Population Exposure: Overall Situation

It seems appropriate to obtain an overview of current radiation exposure of populations from different sources. Figure 5.1 shows the estimated mean annual per caput dose from natural sources, from medical exposure, and from all other man-made sources expressed in terms of the effective dose equivalent. It refers to the years before the Chernobyl accident.

The available data indicate for the populations in most countries a mean natural radiation exposure of 2 to 2.5 mSv per year. About one half of this value is attributed to the inhalation of short-lived radon decay products in domestic

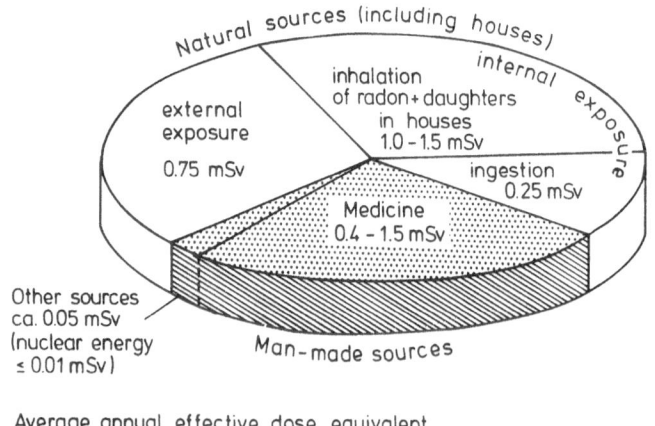

**Fig. 5.1.** Estimated mean values of the annual effective dose equivalent from natural and man-made sources.

houses, which leads to a rather high alpha dose equivalent to the bronchial epithelium. About 0.25 mSv per year, or about 10% of the total natural exposure, is caused by the ingestion of natural radionuclides $^{40}$K, $^{226}$Ra, $^{210}$Pb, and $^{210}$Po with food.

By far the greatest dose contribution from man-made sources comes from medical exposure, particularly from x-ray diagnostics. It differs strongly with the types and frequencies of such examinations. The mean annual effective dose equivalent of 0.4 to 1.5 mSv, given in Fig. 5.1, has been suggested for countries with high radiological health care (2). There is some indication that the mean per caput dose from diagnostic procedures in these countries is decreasing now. This reduction is mainly due to the application of alternative nonradiation methods such as, for example, the use of ultrasound in mammography.

Compared with natural and medical exposure, the dose contribution from all other man-made sources is much less and will normally be lower than 0.1 mSv per year. This corresponds to less than 5% of the total mean dose of 2.5 to 4.0 mSv from all sources, which should be expected in these countries (see Fig. 5.1). As outlined later on, the dose contribution from the operational, normal discharges of nuclear power plants is less than 0.01 mSv per year.

It has to be emphasized that the values in Fig. 5.1 refer to the mean per caput dose. The actual dose to individual members of the public varies over a large range. This is particularly valid with respect to the natural and medical radiation exposure. For example, measurements of Rn in houses indicate that the attributable inhalation dose covers a range of more than three orders of magnitude (2,3).

On the other hand, the variation range of the dose to operational discharges from nuclear power plants is restricted by authorized limits. As a consequence, also, the maximum dose to members of the public from these sources is normally considerably lower than the mean dose from natural sources. These dose limits for individual members of the public living near the site of these plants lead to a limitation of the collective dose to the local and regional population.

## Exposures From Nuclear Power Production

The monitoring systems in nuclear facilities, particularly in nuclear power reactors, provide quantitative data on the operational releases of specific radionuclides with effluent air and water. On the basis of these data, the attributable radiation exposure near the plant site and the total collective dose to the population can be evaluated. For this purpose, simplified radioecological models for chronic releases are applied, taking into account site-specific parameters. Reference models of this type are described in the reports of UNSCEAR (1,2). Taking into account the assumptions made, most of these models lead to exposure estimates that are on the safe side. In the following, some general issues of the exposure analysis for the production of nuclear power are outlined. Results are shown as normalized values of electric energy produced per year [GW(el)a].

## Nuclear Power Reactors

The main source of environmental exposure from nuclear reactors comes from the atmospheric discharges. For example, in Fig. 5.2 the normalized discharges from nuclear power reactors in the Federal Republic of Germany during the period from 1970 to 1984 are plotted, as taken from the monitoring measurements of the different plants (4). Five groups of radionuclides are specified:

radioactive noble gases (krypton, xenon);
tritium, mainly released as tritiated water (HTO);
$^{14}C$;
radioiodine, mainly $^{131}I$; and
radionuclides released in particulate form.

The results indicate a rather strong decrease of the normalized discharges, particularly of short-lived noble gases, I, and particulates with time. This finding characterizes the technological improvements, especially of the hold-back systems.

For dose assessments, the physical decay data and the different behavior of the radionuclides in the environment plus their metabolic distribution in the human body after incorporation have to be taken into account. With respect to noble gases, only the dose contribution by external gamma- and beta-radiation is of importance because the solubility of Kr and Xe in body tissues is rather small.

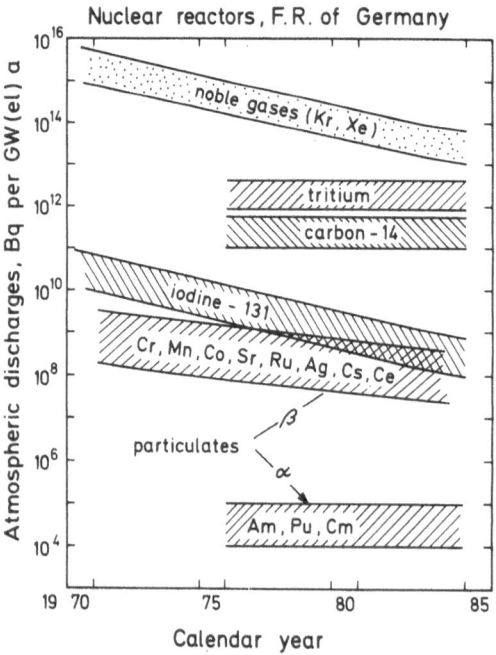

**Fig. 5.2.** Annual mean values of the normalized atmospheric discharges of radionuclides from nuclear reactors in the Federal Republic of Germany, 1970–1984. Produced using data from ref. 4.

Local exposure is mainly due to external radiation from noble gases and from deposited activity of I and particulates, and the internal exposure by $^{131}$I; the latter leads to a preferential irradiation of the thyroid. For the collective dose to the regional population the released activity of long-lived gases and vapors becomes the dominant source, particularly $^{14}$C and $^3$H.

Estimated average values of the collective dose commitment per GW(el)a electric energy produced are listed in Table 5.1. They are based on the compilation of worldwide data on the operational releases from nuclear reactors and the environmental models described in the report of UNSCEAR (2). It results in a total value of about 4 person-Sv per GW(el)a for the local and regional population from which about 70% is attributed to the ingestion with food, mainly of $^{14}$C.

## Total Nuclear Fuel Cycle

In the UNSCEAR report (2) estimates of the population exposure from the total nuclear fuel cycle have been made, taking into account the operational releases from uranium mining and milling, fuel fabrication, nuclear reactors, and reprocessing plants. Figure 5.3 shows a graphical presentation of the resulting collective (effective) dose (equivalent) commitment, normalized to 1 GWa of electric energy produced, as a function of the integration period of the collective dose.

For the operational releases from the total fuel cycle, a collective dose commit-

**Table 5.1.** Collective effective dose equivalent commitment (S) to the local and regional populations by operational releases from nuclear reactors: estimated average values per GWa electric energy produced[a]

| Radionuclide | S, in person-Sv per GW(el)a | |
|---|---|---|
| Noble gases | 0.6 | |
| Tritium | 0.5 | total |
| $^{14}$C | 2.8 | 4 |
| I[b] | 0.1 | |
| Particulates (Co,Sr,Ru,Cs) | 0.1 | |

Relative contribution to total dose:

| | |
|---|---|
| External radiation ($\gamma,\beta$) | 20% |
| Ingestion with food | 70% |
| Inhalation | 10% |

[a] Reprinted with permission of ref. 2.
[b] The corresponding dose value to the thyroid is about a factor of 30 higher.

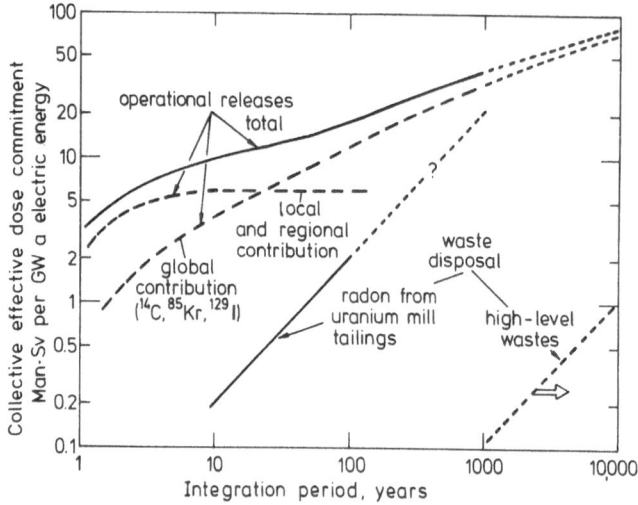

**Fig. 5.3.** Estimated effective collective dose equivalent commitments from the nuclear fuel cycle, normalized to 1 GWa electric energy produced. Produced using data from ref. 2.

ment of about 6 person-Sv per GW(el)a results from the local and regional contribution. It is expected, however, that on a long-term scale, the global contribution from long-lived gases and vapors, particularly from [14]C, might be considerably higher; for an integration period of 1,000 years, a value of about 30 person-Sv per GW(el)a from this contribution has been estimated. Taking into account a world nuclear energy production in 1985 of about 100 GW(el)a, a total collective dose commitment of about 4,000 person-Sv over 1,000 years should be expected from the current annual production of nuclear energy.

In Fig. 5.3, the collective dose commitment from the disposal of radioactive wastes is also plotted, as derived from estimates in the reports of INFCE (5) and UNSCEAR (2). On a long-term scale, particularly, the release of [222]Rn from U mill tailings might be of importance. However, the extrapolation of this contribution involves large uncertainties. In any case this contribution is very small compared with the value of about $5 \times 10^6$ person-Sv per year, which is delivered by Rn and its daughters in our natural environment.

## Radiation Exposure From Conventional Power Plants

The combustion of fossil fuels, particularly of coal, leads to a release of natural radionuclides in the atmosphere. As an example, Fig. 5.4 shows the mean values of the specific activity in the coal and fly ash, which we have measured at two modern coal-fired power plants in the Federal Republic of Germany (5,6).

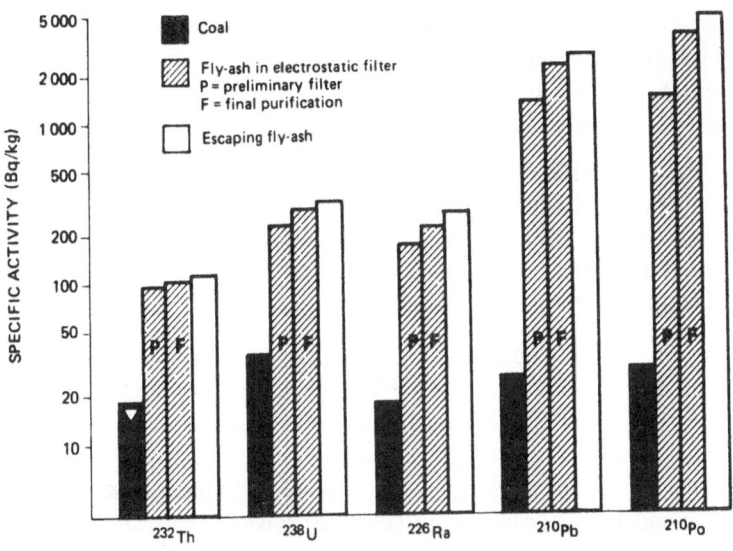

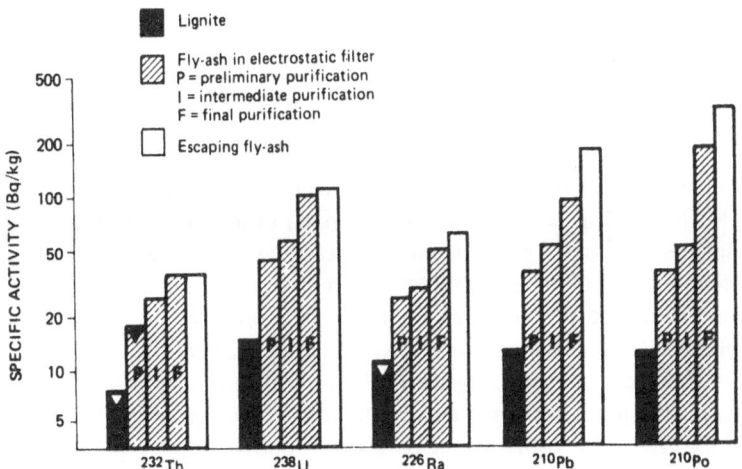

**Fig. 5.4.** Measured mean values of specific activity in the coal, in the fly ash deposited in the electrostatic filter stages, and in the escaping fly ash of two different types of power plants in the Federal Republic of Germany.

Both types of plants were equipped with multistage electrostatic precipitation systems. As shown in Fig. 5.4, the specific activity in the fly ash increases from stage to stage and reaches the highest value in the escaping fly ash. This observation indicates that the enrichment factor increases with decreasing particle size. Compared with the inserted coal, the enrichment factor for the escaping fly ash is about 5 to 10 for $^{232}$Th, $^{238}$U, $^{226}$Ra, and 50 to 150 for $^{210}$Pb and $^{210}$Po in the case of the plant fired with hard coal (upper graph). At the lignite power plant (lower graph), the enrichment factors are considerably lower owing to the lower furnace temperature.

Taking into account the released mass of fly ash, the normalized discharges from these plants (hard coal power plant/lignite power plant) are as follow:

$^{238}$U to $^{226}$Ra, $5/1 \times 10^8$ Bq per nuclide per GW(el)a;
$^{232}$Th to $^{224}$Ra, $2/0.4 \times 10^8$ Bq per nuclide per GW(el)a;
$^{222}$Rn, $600/700 \times 10^8$ Bq per nuclide per GW(el)a;
$^{210}$Pb, $40/2.2 \times 10^8$ Bq per nuclide per GW(el)a;
$^{210}$Po, $80/4.3 \times 10^8$ Bq per nuclide per GW(el)a.

Excluding Rn, the total normalized atmospheric discharges of natural radionuclides in particulate form from the coal-fired plant are about one order of magnitude higher than the normalized particulate discharges of artificial radionuclides from a modern nuclear power plant (see Fig. 5.2); in the case of the lignite power plant the values are comparable. It should be noted, however, that the natural radionuclides released from coal-fired power plants are partly alpha-emitters.

The resulting radiation exposure to the local population in the environment of these coal-fired power plants is very small, in the order of 0.0001 to 0.001 mSv per year. The main dose contributions are obtained from inhaled Th and from the ingestion of $^{210}$Pb and $^{210}$Po (6).

UNSCEAR has started, on a worldwide basis, an evaluation of the collective dose commitment by the operational releases of natural radionuclides from the different types of electricity-producing power plants. The preliminary results, which are based on rather crude models, are listed in Table 5.2. For comparison, the value for nuclear power plants is also given (Table 5.1). In general, it can be concluded that for the same amount of electric energy produced, the collective dose commitment of operational activity releases from nuclear power plants and from coal-fired power plants is of the same order of magnitude. For oil-fired plants and plants using natural gas, the normalized collective dose commitment might average about a factor of 10 or 100 lower, respectively. Taking into account a total world production of 1,000 GW(el)a per year in 1985, a total collective dose commitment of about 3,000 person-Sv per year to the world population should be expected (see Table 5.2). This value, which does not include the contribution from other by-products and from the attributed occupational exposure, is about a factor 3,000 lower than the mean annual population exposure from natural sources (see Fig. 5.1).

In this context it is of interest that the increase of $CO_2$ in the atmosphere from the combustion of fossil fuels leads to a decrease of the $^{14}$C/C-ratio in the biosphere ("Suess effect"). Because the radiation exposure from $^{14}$C is proportional to this ratio, the combustion of fossil fuels causes a decrease of the radiation exposure from $^{14}$C in our environment. Rough estimates yield a negative collective dose commitment due to this effect of about 50 person-Sv

**Table 5.2.** Collective effective dose equivalent commitment[a] (S) per year of operation by atmospheric activity releases from electricity producing power plants, estimated from UNSCEAR data

| Type of plants | World energy production, 1985 GW(el)a | S in person-Sv | |
|---|---|---|---|
| | | per GW(el)a | Total in 1985 |
| Coal-fired | | | |
| "old" plants | 400 | 6 | 2,400 |
| "modern" plants | 200 | 1 | 200 |
| Oil-fired plants | 200 | 0.5 | 100 |
| Nuclear reactors | 100 | 4 | 400 |
| Natural gas | 100 | 0.03 | 3 |
| Geothermal energy | 2 | 5 | 10 |
| Total (1985) | 1,000 | | ca. 3,000 |

[a] Taking into account a world population of 5 billion and an integration period of 1,000 years.

per GWa electric energy production by combustion of fossil fuels, distributed over hundreds of years.

## Contributions From Other Man-Made Sources

In addition, humans are exposed to a great variety of other man-made sources. Although the individual doses from most of these sources are very low, their wide application leads to a considerable collective dose. For the population exposure, the combustion of coal in domestic houses, the use of coal ash in building materials, and the application of by-products from the phosphate industry are of main importance; to the last belong particularly the use of phosphogypsum as building material and the use of phosphate fertilizers.

The population exposure from these sources can be only roughly estimated. Preliminary values for the global collective dose commitments per year of practice are listed in Table 5.3. Further studies are necessary to improve these estimates. The relatively high values resulting for the domestic combustion of coal and the use of coal ash and phosphogypsum in building materials especially involve large uncertainties and have to be reexamined.

With respect to the transfer through food chains, the use of phosphate fertilizers is of interest. In the Federal Republic of Germany, for example, nearly 7 million tons of phosphate fertilizer are spread over agricultural areas every year. This mass contains about $2 \times 10^{12}$ Bq $^{238}$U, $1.3 \times 10^{12}$ Bq $^{226}$Ra, $1 \times 10^{12}$ Bq $^{232}$Th, and about $2 \times 10^{13}$ Bq $^{40}$K. Taking into account an average consumption of 50 g/m$^2$ soil, this yields a surface-related input of about 14 Bq/m$^2$ $^{238}$U, 10 Bq/m$^2$ $^{226}$Ra, and 7 Bq/m$^2$ $^{232}$Th. As the specific activity of these radionuclides in the fertilizer is about five to ten times higher than in normal soils, this type of artificial manuring leads to an enhancement of the natural radiation exposure.

On a worldwide basis a collective dose commitment in the order of $10^4$ person-Sv is expected from this enhanced exposure. The main part of this total value is attributed to the contamination of foodstuffs.

Altogether, the enhanced natural radioactivity in our environment from these industrial processes and their by-products might cause a collective dose commitment to the world population of about $1-2 \times 10^5$ person-Sv per year of practice (see Table 5.3). Taking into account a world population of 5 billion, this corresponds to a mean per caput dose of about 0.02 to 0.04 mSv per year. It should be recognized that only a small fraction of this additional exposure is attributed to the operational releases of natural radionuclides from coal- and oil-fired power plants.

## Summary and Conclusions

As shown in Fig. 5.1, the medical exposure from diagnostic procedures yields by far the largest contribution to the population exposure from man-made radiation sources. This is particularly valid for the population in countries with high radiological health care. Contrary to most other man-made sources, it has to be recognized that the patient himself gains or should gain the benefit of such diagnostic examinations. Thus, risk and benefit are received by the same person.

The second most important dose contribution comes from the enhanced natural radioactivity resulting from industrial activities and from the artificial radioactivity in consumer products. Rough estimates indicate a mean per caput dose from these sources of about 0.02 to 0.05 mSv per year, corresponding to about 1% to 2% of the mean natural radiation exposure. This value is about a factor of two higher than the present dose contribution by the global fallout from nuclear weapon tests. Of main importance are the dose contributions caused by by-products and wastes from the phosphate industry. Compared with these, the

**Table 5.3.** Estimates of the collective effective dose equivalent commitment[a] (S) from man-made sources due to enhanced natural radioactivity

| Source, practice | S, in person-Sv per year of practice |
|---|---|
| Coal combustion | |
| in power plants | 3,000 |
| in domestic houses | 2,000–40,000 |
| Use of coal ash in building materials | 40,000 |
| Use of phosphogypsum as building material | 100,000 (?) |
| Use of phosphate fertilizers | 10,000 |
| Total ca. | 100,000–200,000 |

[a] Referring to a world population of 5 billion.

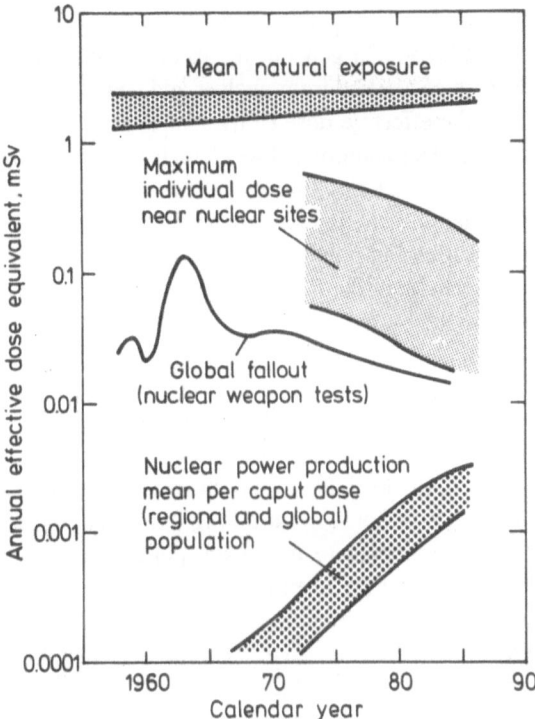

**Fig. 5.5.** Trends with time of the mean annual effective dose equivalent from different sources of radiation. Produced using data from ref. 2.

dose from the enhancement of the natural radioactivity in our environment resulting from operational atmospheric releases of coal- and oil-fired power plants is rather small. Due to the "Suess effect" the combustion of fossil fuels leads to a decrease of the global radiation exposure from $^{14}C$.

In compliance with the recommendations of the International Commission on Radiological Protection rather restrictive regulations have been issued by the responsible authorities in most countries for the monitoring and limitation of activity discharges from nuclear power plants. Owing to these precautions, the doses to the local, regional, and global population resulting from operational releases of nuclear power facilities are small compared with the natural radiation exposure and its individual variation.

The estimated trends with time of the doses from nuclear power production until 1985 are shown in Fig. 5.5. They refer to the operational releases under normal conditions. For comparison, in this figure the mean annual doses from natural radiation sources and from the global fallout of nuclear weapon tests are given. The trends in time are characterized by two features: first, the development of improved techniques for the reduction of discharges, and, second, the increase of nuclear power production. Owing to the first item, the maximum dose to individuals and the collective dose to the local population are lower for modern nuclear reactors. In general, the estimates indicate that the normal operational discharges of activity from nuclear and coal-fired power plants lead to similar values of the effective dose equivalent to the local population in the vicinity of these plants. This additional dose is small compared with the individual variation of the dose from natural sources.

It has been suggested that the indoor exposure to radon decay products might have been somewhat lower in previous years than nowadays. This is indicated in Fig. 5.5 by the slight rise of the lower boundary of the band, which represents the natural exposure.

On the other hand, the increase of nuclear power production during the last decades has led to an increase of the attributable collective dose or mean per caput dose, respectively, in the regional and global populations. The estimates yield, for 1985, a mean annual dose of about 0.002 to 0.005 mSv, which corresponds to about 0.1% to 0.2% of the mean population dose from natural sources, including radon in houses. On a long-term scale the main fraction of the collective dose commitment to the global population from nuclear power production will be caused by the release of $^{14}C$.

The evaluation of the dose commitment from accidental releases, particularly from the Chernobyl accident, is beyond the scope of this chapter. The experiences from this accident have shown the difficulties of dose assessments in the case of major reactor accidents and the problems of decision making on optimum intervention strategies and on the setting of primary and secondary intervention levels.

For these purposes, dynamic radioecological models are required that enable an early prediction of the potential dose contributions from relevant exposure pathways as a function of time. For the predictive estimation of the contamination of foodstuffs, such models have to take into account the spatial and temporal distribution of the deposited activity, the vegetational growth at the time of deposition and afterwards, the specific types of soils and their agricultural use in the affected area, and the spatial distribution of the foodstuffs produced in these areas.

In recent papers we have outlined the general capabilities of such dynamic models and have demonstrated, as an example, some typical results of the ECOSYS-model, which we have developed in the Institute for Radiation Protection in the Federal Republic of Germany (7,8). This model also considers external exposure. In areas with higher contamination, the external gamma-radiation from deposited radionuclides can yield a significantly higher contribution to the long-term dose than the ingestion of contaminated food.

Preliminary, and still incomplete, dynamic radioecological models have been prepared in several countries. They have reached a state of the art such that further validation studies and improvements can be expected to provide a very useful tool for the emergency management after reactor accidents.

# References

1. UNSCEAR (1977) Sources and Effects of Ionizing Radiation. United Nations Publication No. E.77.IX.1, New York
2. UNSCEAR (1982) Ionizing Radiations: Sources and Biological Effects. United Nations Publication No. E.82.IX.8, New York
3. International Commission on Radiological Protection (ICRP) (1987) Lung cancer risk from indoor exposure to radon daughters. ICRP Publication 50, Pergamon Press, Oxford
4. Bundesminister für Umwelt, Naturschutz und Reaktorsicherheit (BMU) (1986) Umweltradioaktivität und Strahlenbelastung. Jahresbericht 1984. Bonn, ISSN 0533-9456
5. Chatterjee B, Hötzl H, Rosner G, et al (1980) Untersuchungen über die Emission von Radionukliden aus Kohlekraftwerken. Analysenverfahren und Messergebnisse für ein Steinkohle- und ein Braunkohlekraftwerk. GSF-Report S-617
6. Jacobi W (1981) Umweltradioaktivität und Strahlenexposition durch radioaktive Emissionen von Kohlekraftwerken. GSF-Report S-760
7. Jacobi W, Paretzke HG, Müller H (1987) Applicability of dynamic models to establishing derived intervention levels for foodstuffs. International scientific seminar on foodstuffs intervention levels following a nuclear accident. Luxembourg, Apr 27–30, 1987, p 39–57 in CEC Report EUR 11232, Commission of the European Communities, Bruxelles, 1987
8. Jacobi W (1988) Environmental radioactivity and man: The 1988 Sievert Lecture. Health Phys 55. To be published

CHAPTER 6

# Naturally Occurring Sources of Radioactive Contamination

J. H. Harley[1]

## Introduction

Radiation from natural sources gives more than 80% of the total exposure received by the average member of a population. A portion of this exposure comes from dietary intake, and we can obtain some perspective on this portion by looking at a breakdown of natural background radiation.

In the United States, the National Council on Radiation Protection and Measurements (NCRP) has prepared a report (1) on population exposure to natural radiation in the United States and Canada. Some of this material will be used here, since it should apply equally well to average exposures in many countries. Other reviews are included in earlier reports (2–5).

Table 6.1 shows the annual effective dose equivalent delivered from the various natural sources. The largest contribution (2,000 µSv) is from inhaled radionuclides, mostly the short-lived decay products of $^{222}$Rn. The next largest, radionuclides in the body (400 µSv), is our particular interest at this symposium and Table 6.2 shows the annual dose equivalent to soft tissues and to bone surfaces from specific radionuclides. These doses are almost entirely derived from measured concentrations of the radionuclides in autopsy specimens.

Ideally, we should have data representing the average intake of the important radionuclides and the distribution of values around the average. Actually, the data are extremely limited, partly because the contribution from food is only about 10% of the total effective dose equivalent from all sources of natural radiation and partly because both sampling and analysis of the dietary contribution is difficult. This report will attempt to summarize our knowledge of the dietary intake of the important natural radionuclides and to give some of the background information necessary to the understanding of the values found.

## The Human Food Chain

Human dietary composition is quite variable from place to place and even from individual to individual. This chapter emphasizes average intakes of natural

---

[1] Consultant, P.O. Box M-268, Hoboken, NJ 07030, USA.

**Table 6.1.** Annual effective dose equivalent from natural background in the United States (1)

| Source of radiation | Effective dose equivalent ($\mu$Sv) |
|---|---|
| Cosmic radiation | 270 |
| Cosmogenic radionuclides | 10 |
| Terrestrial radiation | 280 |
| Inhaled radionuclides | 2,000 |
| Radionuclides in the body | 400 |
| Rounded Total | 3,000 |

radionuclides for population groups and their geographic variability. Data on special high background areas are only mentioned briefly.

The natural radionuclides entering the food chain are contained in the soil, therefore, soil content is a prime source of geographic variability. Uptake varies with plant species; thus, the intake of different vegetable foods is the second source of variability. Finally, the amount of animal-derived foods consumed also modifies the radionuclide intake.

Most of the radionuclides of interest enter the food chain from the soil by plant root uptake. In some cases the radionuclide is formed in the plant by decay after the parent has been taken up. Thorium isotopes, for example, are poorly absorbed but $^{224}$Th can appear in the body following uptake and radioactive decay of $^{228}$Ra.

One specific case of natural radionuclide contamination is not entirely root uptake. Gaseous $^{222}$Rn escapes from soil into the atmosphere where it decays. The long-lived member of the $^{222}$Rn decay chain is $^{210}$Pb, which exists in the atmosphere attached to the ambient aerosol. The aerosol deposits on plant surfaces and some of this $^{210}$Pb is absorbed. In addition, the $^{210}$Pb deposited on the

**Table 6.2.** Annual dose equivalent to selected tissues from natural radionuclides in the body for the United States (1)

| Radionuclides | Soft tissue ($\mu$Sv) | Bone surfaces ($\mu$Sv) |
|---|---|---|
| $^{14}$C | 10 | 10 |
| $^{40}$K | 180 | 140 |
| $^{87}$Rb | <10 | 10 |
| $^{238}$U–$^{230}$Th | <10 | 10 |
| $^{226}$Ra | <10 | 90 |
| $^{210}$Pb–$^{210}$Po | 140 | 700 |
| $^{228}$Ra–$^{224}$Ra | <10 | 120 |
| Rounded totals | 350 | 1,100 |

**Table 6.3.** Transfer factors for $^{226}$Ra and $^{210}$Pb in the food chain (10)

| Radionuclide | Dry soil to | | Wet forage to | |
|---|---|---|---|---|
| | Vegetables | Forage | Meat | Milk |
| $^{226}$Ra | $2 \times 10^{-2}$ | $9 \times 10^{-2}$ | $5 \times 10^{-3}$ | $8 \times 10^{-3}$ |
| $^{210}$Pb | $1 \times 10^{-2}$ | $2 \times 10^{-1}$ | $8 \times 10^{-3}$ | $2 \times 10^{-3}$ |

Transfer factors are dimensionless but usually expressed as radioactivity per unit weight of product divided by activity per unit weight of precursor. Soil concentration is in terms of dry weight, others in terms of fresh wet weight.

soil surface with the aerosol is readily soluble and can enter plants through the roots. The degree to which these two processes control the $^{210}$Pb content of plants is still controversial.

There have been few systematic studies attempting to define the transfer of natural radionuclides through the food chain to man. Tracy et al. (6) measured the uptake of several types of vegetation for $^{238}$U, $^{226}$Ra, and $^{210}$Pb. The respective transfer factors were $8 \times 10^{-5}$, $1 \times 10^{-3}$, and $4 \times 10^{-4}$ in transferring from dry soil to the total plant. Frindik (7) reported a much higher uptake for U with a factor of $1 \times 10^{-3}$ for cereals and vegetables, and Schreckhise and Cline (8) list values of $2 \times 10^{-3}$ and $4 \times 10^{-3}$ for barley and peas, respectively. Morishima et al. (9) found a range of $10^{-5}$ to $10^{-3}$ for different species, so it would appear that there is a considerable range of uptake, at least for U.

The transfer from forage to animals provides additional discrimination against the natural radionuclides. McDowell-Boyer et al. (10) surveyed the literature on $^{226}$Ra and $^{210}$Pb and provided estimates of the transfer factors for several stages in the chain. These are shown in Table 6.3. When these factors are applied to average soil concentrations, they overestimate the levels actually measured for fruits and vegetables, milk and meat by severalfold. This could result from lack of correspondence of the concentrations in the soils, plants, and animals or from problems in estimating transfer factors.

Water is potentially a highly variable and significant source of radionuclide ingestion. This could be very significant if the more efficient uptake of $^{226}$Ra from water as postulated by Lucas (11) is correct and if it extends to the other radionuclides of interest.

A number of the dietary estimates reported here also give a water intake, and these are generally not more than 10% of the diet. This may be due to the low concentrations in the surface water supplies used by most large population groups. It appears that concentrations are always higher in ground water supplies and that the most highly contaminated supplies are used by smaller groups.

## Dietary Intake Surveys

Investigators have been interested in the dietary intake of natural radionuclides for more than 50 years. Typical references are Burkser et al. (12) and the more recent Turner et al. (13). In general, however, the effort devoted to such work has been minimal. The required measurements on soils, plants, animals, and humans had no programmatic model until the intensive work on fallout from nuclear weapons began in the late 1950s.

The sampling effort involved in measuring dietary intakes is considerable. In the "market-basket" approach, national or local statistics on food consumption are required, the food items must be collected from sources used by the population, and each item must be prepared according to custom before submission for analysis. In the "parallel sample" approach, duplicate quantities of each dietary item selected by members of the study population are combined for the sample. Institutional sampling, combining suitable portions of food prepared in bulk, lies between the two.

There is another approach to estimating total dietary intake for a particular radionuclide that has been used. If we assume intake/output equilibrium, a measurement of total daily excretion can be considered equal to the daily intake. Some of these data are included in the tables.

The modern, urbanized population tends to have a diet that comes from a number of regions in the country or even in the world. This is true of fruits and vegetables in the United States, where Central and South America, New Zealand, and the Caribbean countries now seem to supply more than Florida and California, the former centers. There are relatively few subsistence farmers, and national diets tend to become uniform in radioactive content.

The dietary contributors most likely to be local are milk, water, and seasonal fresh vegetables. Unlike our experience with fallout radionuclides, milk is usually a minor source of natural alpha-radioactivity. Surface water supplies also tend to be low but, ground water is often high in the more soluble radionuclides, $^{238}U$, $^{226}Ra$, and $^{222}Rn$.

In using the data collected here, a careful scientist should have some reservations. The total quantities of food stated to be consumed vary by a factor of about three, and it is not obvious that caloric intake is sufficient in some cases.

## Measured Dietary Intakes

The dietary intakes of the natural radionuclides are summarized here from data that are readily available and pertinent. Isotopes of some major body elements contribute significant doses, but the most interesting dietary radionuclides are members of the U and Th series.

### Major Elements

Three of the major elements of the body have naturally radioactive isotopes, $^{14}C$, $^{3}H$, and $^{40}K$. These make up a fixed fraction of the corresponding stable elements, although the radiocarbon and tritium fractions have been disturbed by contributions from nuclear weapons testing. The amount of radioactivity from each of these radionuclides in the body depends on the percentage of the element in the body and on the specific activity of the radionuclide in the environment. The same holds for the various foodstuffs, and the body concentrations are not controllable.

The largest contributor to annual radiation dose equivalent is $^{40}$K, with the $^{14}$C and $^3$H giving only about 10 and 1 μSv per year, respectively, to soft tissues. The daily intake of $^{40}$K is about 350 mBq per day but, regardless of intake, healthy subjects maintain $^{40}$K body contents of about 400 Bq, giving an annual dose equivalent to soft tissues of almost 200 μSv.

## Series Radionuclides

The radionuclides that may vary from place to place are the members of the U and Th series. In reviewing the literature, there is appreciable material on $^{238}$U, $^{226}$Ra, and the $^{210}$Pb-$^{210}$Po pair in the U series and very little on members of the Th series. This is in line with the relative contributions of the radionuclides to internal dose. The major data tables cover the dietary intake of the U series members, while the Th series is noted in the text.

## Uranium

The two isotopes of U in the series are $^{238}$U and $^{234}$U. These undergo some natural separation in geologic processes, and the $^{238}$U is usually in slight excess in soil, while the lighter isotope is in excess in many waters and in plants. Uranium compounds in nature are often quite soluble, and root uptake is distributed well throughout the plant.

Dietary intake data are limited to a few countries as shown in Table 6.4. The values reported here are not intended to include drinking water, but it is not always clear in the literature if this is true. Intake with drinking water can

**Table 6.4.** Daily dietary intake of $^{238}$U in various countries

| Country | Region | mBq/d | Reference |
|---|---|---|---|
| Europe | | | |
|   France | | 12 | UNSCEAR (3) |
|   West Germany | | 30 | Frindik (15) |
|   USSR | Moscow | 45[a] | Drutman (16) |
|   UK | | 12 | Hamilton (17) |
| Asia | | | |
|   Japan | Sapporo, Kyoto | 18 | Nozaki (18) |
| | Okayama | 11–60[b] | Masuda (19, 20) |
| North America | | | |
|   United States | Chicago | 17 | Welford (21) |
| | New York (1963) | 16 | Welford (21) |
| | (1978) | 15 | Fisenne (22) |
| | San Francisco | 16 | Welford (21) |

[a] From excretion measurements.
[b] Higher values were in U mining areas.

**Table 6.5.** Daily dietary intake of $^{226}$Ra in various countries

| Country | Region | mBq/d | Reference |
|---------|--------|-------|-----------|
| Europe | | | |
| Belgium | | 44 | Smeets (23) |
| Czechoslovakia | Bohemia | 110 | Truelle (24) |
| France | | 37–52 | Gahinet (25) |
| | Paris | 41 | UNSCEAR (3) |
| Italy | 7 Cities | 30 | Mastinu (26) |
| | Varese | 52 | deBortoli (27) |
| Netherlands | | 74 | Smeets (23) |
| Poland | | 22 | Pietrzak-Flis (28) |
| West Germany | | 110 | Muth (29) |
| UK | | 44 | Smith (30) |
| | | 30 | Smith-Briggs (31) |
| Asia | | | |
| China | Nanchang | 85 | Ye (32) |
| | Hupeh | 33–89 | Ye (32) |
| India | Bombay | 24 | Chabra (33) |
| | Kerala | 95 | Chabra (33) |
| | Kerala | 140 | Lalit (34) |
| Japan | Tokyo | 40 | Kametani (35) |
| North America | | | |
| US | Est. mean | 52 | Holtzman (36) |
| | 10 Cities | 67 | Michelson (37) |
| | Chicago | 27 | Spencer (38) |
| | New York (1966) | 63 | Fisenne (39) |
| | (1971) | 59 | Morse (40) |
| | (1978) | 51 | Fisenne (22) |
| | Puerto Rico | 25 | Hallden (41) |
| South America | | | |
| Argentina | | 26 | Beninson (42) |
| Brazil | Rio de Janeiro | 110 | Lobao (43) |

be considerable and has been reported, in some cases (14), to be equal to that from diet. Intake by inhalation is negligible, so that the body content is dependent on the diet plus drinking water.

## Radium

Radium is considered to be rather insoluble in the environment but this may be balanced by the tendency for Ra to be on the surface of soil grains rather than in the particles. Radium concentration does vary within the plant, with leaf > root > stem and seeds and fruit showing little uptake. Animal foods are low, with milk exhibiting a high discrimination against Ra as compared with its alkaline earth analog, Ca.

There are more data on $^{226}$Ra intake than on any other radionuclide, probably for historical considerations. The estimates for different countries are shown in Table 6.5. Drinking water makes a smaller contribution to intake but Lucas

(11) has reported that Ra in water is more available for uptake by the body than that from diet. Once again, the intake by inhalation is negligible.

The dietary intake values lie within a rather narrow range—perhaps a factor of 3 for normal background areas. Figure 6.1 shows the distribution of values for $^{226}$Ra in human bone (44) and the range seems to be a factor of 10. This is not a real discrepancy, since all the reported dietary intakes are from countries at the center of the bone curve. The extreme bone values, however, should be caused by corresponding differences in intake.

## $^{210}$Pb and $^{210}$Po

The behavior of this pair of radionuclides adds some complexity to our considerations. First, a large fraction of the $^{210}$Pb in the environment has been formed in the atmosphere following the decay of $^{222}$Rn. This gives higher concentrations in surface soils (45) and also provides a route for surface contamination to contribute to the concentration in plants. Second, the 138-day half-life of $^{210}$Po means that measured values in the environment can result from either transfer of the $^{210}$Po itself or by ingrowth from the $^{210}$Pb already present. Lead and Po are both quite soluble in the environment. As in the case of U, distribution is uniform throughout the plant.

The data on dietary intake of the two radionuclides are shown in Table 6.6. Takata et al. (59) and Kametani et al. (35) have reported a high intake for Japan, stated to result from high concentrations in local seafood. The data are not included here, since it is not clear that they are representative of the country.

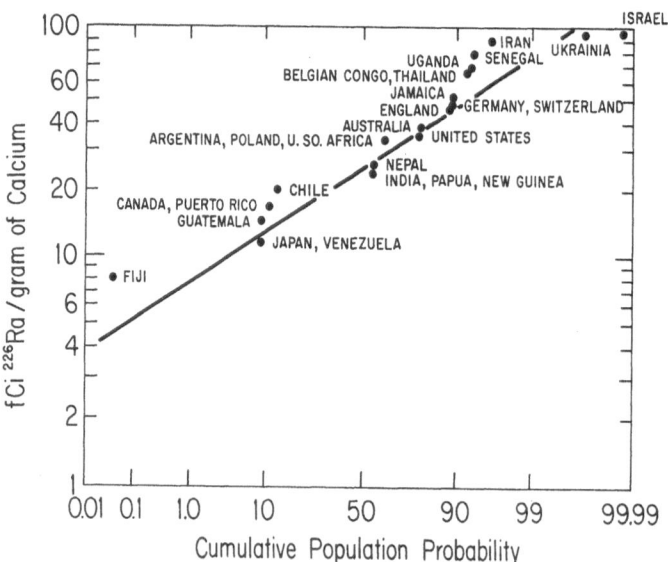

**Fig. 6.1.** Cumulative population frequency distribution of measured $^{226}$Ra/g Ca in human bone ash (44). Note that 1 fCi = 0.037 mBq. (Reproduced from the journal *Health Physics* by permission of the Health Physics Society.)

**Table 6.6.** Daily dietary intake of $^{210}$Pb and $^{210}$Po in various countries

| Country | mBq/d $^{210}$Pb | mBq/d $^{210}$Po | Reference |
|---------|------------------|------------------|-----------|
| Europe | | | |
| Bulgaria | 60–78 | | Keslev (46) |
| France | 50 | | Servant (47) |
| Italy | 110 | 110 | Clemente (48) |
| West Germany | 170 | | Gloebel (49) |
| USSR | 230 | 150 | Ladinskaya (50) |
| | | 140 | Yermolaeva (51) |
| UK | | 120 | Hill (52) |
| | 82 | 78 | Smith-Briggs (53) |
| | 45 | | Chamberlain (54) |
| Asia | | | |
| India | | 56 | Khandekar (55) |
| North America | | | |
| US-Est. mean | 52 | 60 | Holtzman (36) |
| 6 Cities | 62 | | Magno (56) |
| Chicago | 46 | 60 | Spencer (57) |
| New York | 44 | | Morse (40) |
| | 44 | | Bogen (58) |
| South America | | | |
| Argentina | | 48 | UNSCEAR (3) |

Considering that the radiation dose from this source is larger than for $^{226}$Ra (1), it is surprising that more attention has not been paid to its measurement. Once again, the chemical and physical properties of these radionuclides add complications. The intake by inhalation is not negligible, and it is believed that the uptake to blood from the lung is more efficient than from the gastrointestinal tract. In addition, smokers receive an added contribution from $^{210}$Pb and $^{210}$Po in cigarettes and other tobacco products (60).

Table 6.7 shows the relative contribution of several classes of foods to the intake of the U series radionuclides. The data represent a Western diet—actually New York City (22,40). Eastern diets with lower meat intake, with or without

**Table 6.7.** Daily intake of some natural radionuclides in the U series for New York City (22, 40)

| Food class | Intake (g/d) | $^{238}$U (mBq) | $^{226}$Ra (mBq) | $^{210}$Pb (mBq) |
|------------|--------------|-----------------|------------------|------------------|
| Cereals | 260 | 4 | 22 | 15 |
| Meat, fish, eggs | 340 | 6 | 6 | 6 |
| Milk products | 550 | 0.4 | 3 | 6 |
| Fruits | 270 | 0.4 | 10 | 5 |
| Vegetables | 330 | 4 | 10 | 12 |
| Water | 1,400 | 1 | 1 | 2 |
| Totals | | 16 | 52 | 46 |

**Table 6.8.** Daily intake of natural $^{226}$Ra for Bombay (33)

| Food class | Intake (g/d) | $^{226}$Ra (mBq) |
|---|---|---|
| Cereals | 550 | 16 |
| Meat, fish, eggs | 80 | 2 |
| Milk products | 250 | 1 |
| Fruits | 150 | 2 |
| Vegetables | 30 | 5 |
| Water | 2,500 | 2 |
| Miscellaneous | 150 | 1 |
| Total | | 29 |

substitution by fish, show a different distribution. One example is taken from the report by Chabra (33) for India and is shown in Table 6.8.

**Thorium Series**

The initial radionuclides of the two series, $^{238}$U and $^{232}$Th, have the same global radioactivity, yet the final quantities transferred to man are quite different. The important radionuclides from the Th series are the Th isotopes $^{232}$Th and $^{228}$Th and the Ra isotope $^{228}$Ra. Thorium tends to be very insoluble in the environment and transfers poorly to foods and to man. The $^{228}$Ra behaves chemically like the other Ra isotopes and is taken up quite readily by plants. The dietary intake of the three radionuclides are reported as 4 mBq per day for each of the two Th isotopes and 35 mBq per day for the $^{228}$Ra for New York City (22). Linsalata et al. (61) used fecal analysis to estimate a 3 mBq per day intake for $^{232}$Th in New York State, and Petrow et al (62) found a mean intake of about 40 mBq per day of $^{228}$Ra for New York City, Chicago, and San Francisco. Frindik (15) reported 6 mBq per day for $^{232}$Th and 46 mBq per day for $^{228}$Th in West German diets.

There are a few measurements for high background areas: Mistry et al. (63) reported a $^{228}$Ra intake of 6,000 mBq per day for the Kerala district of India and Penna-Franca et al. (64) a range of 4,000 to 9,000 mBq per day for high

**Table 6.9.** Daily intake of natural thorium radionuclides for New York City (22)

| Food class | Intake (g/d) | $^{232}$Th (mBq) | $^{230}$Th (mBq) |
|---|---|---|---|
| Cereals | 260 | 0.7 | 2 |
| Meat, fish, eggs | 340 | 0.6 | 0.8 |
| Milk products | 550 | 0.2 | 0.2 |
| Fruits | 270 | 0.05 | 0.1 |
| Vegetables | 330 | 2.7 | 3.1 |
| Water | 1,400 | 0.1 | 0.2 |
| Rounded totals | | 4 | 6 |

background areas in Brazil. These contrast with the Lalit and Ramachandran (34) value of 130 mBq per day of $^{228}$Th for Bombay and Penna-Franca et al.'s (64) value of 110 mBq per day for Rio de Janeiro.

Table 6.9 shows the contributions of several classes of foods to dietary Th intake (22) for comparison with Table 6.7 for the U series. The intakes with diet and drinking water are not well known for a broad range of areas with normal backgrounds. This may not be of importance for the diet, but additional valid data for $^{228}$Ra in drinking water are needed.

## Summary

This chapter has attempted to summarize our knowledge on the dietary intake of several natural radionuclides. It is obvious that most of the many studies in the field are rather unsystematic and do not allow us to assemble a coherent picture. Much of the problem is common to all studies involving humans as compared with laboratory animals. This is compounded by the difficulty in making high-quality measurement of low concentrations of alpha-emitting radionuclides in large samples of foods.

Daily dietary intakes are of the order of 20 mBq per day for $^{238}$U, 60 mBq per day for $^{226}$Ra, 80 mBq for $^{210}$Pb, 5 mBq for $^{232}$Th, and 40 mBq for $^{228}$Ra. Water ingestion from urban surface supplies would generally add a few percent to these intakes, but local water intakes can be much higher.

A current study in the United States estimates the total effective dose equivalent rate as about 3,600 μSv per year. About 3,000 μSv per year, or more than 80%, comes from natural sources, with 2,000 μSv per year being the dose from inhaled radon decay products. The other 1,000 μSv per year from natural sources includes the 400 μSv per year from radionuclides contained in the body, which is our interest in this particular chapter. So, about 10% of the radiation exposure from natural sources could come from the diet.

In assessing dietary contributions to dose, it is more instructive to look at organ dose and this presentation has stressed the average dose rate to soft tissues, bone surfaces, and bone marrow. If we continue our subtraction to remove the doses from the major body elements, C and K, we are left with organ dose rates of about 150 μSv per year to soft tissues, 900 μSv per year to bone surfaces, and 180 μSv per year to marrow. These are the doses that vary with dietary intake of the major ingested radionuclides, $^{226}$Ra, $^{210}$Pb, and $^{228}$Ra.

The radiation doses delivered are moderate and mostly cannot be significantly reduced by remedial action. Certain high background areas have shown exposures from individual radionuclides to be several times the average values given here, but the increase in total radiation exposure caused by ingestion is unlikely to exceed a factor of 2. Such a change should not produce any observable health effect. This has the ancillary effect of making these populations unlikely candidates for epidemiological studies.

# References

1. NCRP (1987) Exposure to the Population in the United States and Canada from Natural Background Radiation. National Council on Radiation Protection and Measurements Report No. 94. NCRP, Bethesda, MD
2. NCRP (1984) Exposures from the Uranium Series with Emphasis on Radon and its Daughters. National Council on Radiation Protection and Measurements Report No. 77. NCRP, Bethesda, MD
3. UNSCEAR (1972) United Nations Scientific Committee on the Effects of Atomic Radiation. Ionizing Radiation: Levels and Effects. United Nations, New York
4. UNSCEAR (1977) United Nations Scientific Committee on the Effects of Ionizing Radiation. Sources and Effects of Ionizing Radiation. United Nations, New York
5. UNSCEAR (1982) United Nations Scientific Committee on the Effects of Ionizing Radiation. Ionizing Radiation: Sources and Biological Effects. United Nations, New York
6. Tracy BL, Prantl FA, Quinn JM (1983) Transfer of Ra-226, Pb-210 and uranium from soil to garden produce—Assessment of risk. Health Phys 44:469–477
7. Frindik O (1986) Uranium contents in soils, plants and foods. Landwirtsch Forsch 39:75–86
8. Schreckhise RG, Cline JF (1980) Uptake and distribution of U-232 in peas and barley. Health Phys 38:341–343
9. Morishima H, Koga T, Kawai H, Honda Y, Katsurayama K (1977) Studies on the movement and distribution of uranium in the environments—Distribution of uranium in agricultural products. J Radiat Res (Japan) 18:139–150
10. McDowell-Boyer LM, Watson AP, Travis CC (1979) Review and Recommendations of Dose Conversion Factors and Environmental Transport Parameters for Pb-210 and Ra-226. US Nuclear Regulatory Commission Report NUREG/CR-0574. NTIS, Springfield, VA
11. Lucas HF (1960) Correlation of the natural radioactivity of the human body to that of its environment: uptake and retention of Ra-226 from food and water. In Argonne National Laboratory Radiological Physics Division Semi-Annual Report, July-December 1960, ANL-6297. ANL, Argonne, IL, p 55
12. Burkser E, Schapiro M, Bronstein K (1929) Radium content of some foodstuffs. Biochem Z 211:323–325
13. Turner RC, Radley JM, Mayneord WV (1958) The naturally occurring alpha-ray activity of foods. Health Phys 1:268–275
14. Drury JS, Reynolds S, Owen PT, Ross RH, Ensminger JT (1981) Uranium in US Surface, Ground and Domestic Waters. US EPA Report EPA-570/9-81-001. NTIS, Springfield, VA
15. Frindik O (1983) Uranium in diet. In Schelenz R (ed) Essential and Toxic Food Constituents in the Daily Total Diet. Bundesforschungsanstalt fuer Ernaehrung-Berichte BFE-R-83-02. BFE, Karlsruhe, pp 301–306
16. Drutman RD, Mordasheva VV (1985) Natural uranium content in human organs and excreta. Gigiena i Sanit. Translation for Oak Ridge National Laboratory, ORNL/TR-86/27 7:61–64
17. Hamilton EI (1972) The concentration of uranium in man and his diet. Health Phys 22:149–153
18. Nozaki T, Ichikawa M, Sasuga T, Inarida M (1970) Neutron activation analysis of uranium in human bone, drinking water and daily diet. J Radioanal Chem 6:33–40

19. Masuda K (1971a) Intake and urinary excretion of uranium in non-occupationally exposed persons: II. Uranium in the daily diet. Jpn J Hyg 26:438–441
20. Masuda K (1971b) Intake and urinary excretion of uranium in non-occupationally exposed persons: IV Discussions on dietary intake and urinary excretion of uranium. Jpn J Hyg 26:447–450
21. Welford GA, Baird R (1967) Uranium levels in human diet and biological materials. Health Phys 13:1321–1324
22. Fisenne IM, Perry PM, Decker KM, Keller HW (1987) The daily intake of U-234, 235, 238, Th-228, 230, 232, and Ra-226, 228 by New York City Residents. Health Phys 43:357–363
23. Smeets J, van der Stricht E (1970) Comparison of the radioactive contamination of the total diet of adolescents in the Community. Report EUR-3945 (pt 2). European Communities, Luxembourg
24. Truelle MA (1977) Content of Ra-226 in selected food produced in the South Bohemian Region (cited by Holtzman, Ref. 36). Cesk Hyg 22:141–146
25. Gahinet ME, Remy ML, Moroni JP, Pellerin P (1969) Study of radioactivity in total diet in schools. In Symposium-Environmental Contamination by Radioactive Materials STI/PUB/226. IAEA, Vienna, pp 357–472
26. Mastinu GG, Santaroni GP (1980) Radium-226 levels in Italian drinking waters and foods. In Gesell TF, Lowder WM (eds) Natural Radiation Environment III, CONF-780422. NTIS, Springfield, VA, pp 810–825
27. deBortoli M, Gaglione P (1972) Ra-226 in environmental materials and foods. Health Phys 22:43–48
28. Pietrzak-Flis Z (1972) Ra-226 in Polish diet and foodstuffs. Nukleonika 17:227–231
29. Muth H, Rajewsky B, Handtke HJ, Aurand K (1960) The normal radium content and the Ra-226/Ca ratio of various foods, drinking water and different organs and tissues of the human body. Health Phys 2:239–245
30. Smith KA, Watson PG (1964) Radium-226 in diet in the United Kingdom in 1963. In Annual Report 1963–1964, Agricultural Research Council Radiobiological Laboratory Report 12. ARCRL, Wantage, England, p 79
31. Smith-Briggs JL, Bradley EJ (1984) Measurement of natural radionuclides in UK diet. Sci Total Environ 35:431–440
32. Ye C (1984) Estimation of intake and distribution of Ra-226 in bone of inhabitants of Nanchang. Radiation Protection, English translation provided by the author. 4:430
33. Chabra AS (1966) Radium-226 in food and man in Bombay and Kerala State (India). Br J Radiol 39:141–146
34. Lalit BY, Ramachandran TV (1980) Natural radioactivity in Indian foodstuffs. In Gesell TF, Lowder WM (eds) Natural Radiation Environment III, CONF-780422. NTIS, Springfield, VA, pp 800–809
35. Kametani K, Ikebuchi H, Matsumura T, Kawakami H (1981) Ra-226 and Pb-210 concentrations in foodstuffs. Radioisotopes 30:681–683
36. Holtzman RB (1980) Normal dietary levels of Ra-226, Ra-228, Pb-210 and Po-210 for man. In Gesell TF, Lowder WM (eds) Natural Radiation Environment III, CONF-780422. NTIS, Springfield, VA, pp 755–782
37. Michelson I, Thompson JC Jr, Hess BW, Comar CL (1962) Radioactivity in total diet. J Nutr 78:371–383
38. Spencer H, Kramer L, Samachson J, Fisenne IM, Harley NH (1973) Intake and

excretion patterns of naturally occurring Ra-226 in humans. Radiat Res 56:354–369

39. Fisenne IM, Keller HW (1970) Radium-226 in the diet of two U.S. cities. In Health and Safety Laboratory Report HASL-224. NTIS, Springfield, VA, pp I-2

40. Morse RS, Welford GA (1971) Dietary intake of lead-210. Health Phys 21:53–55

41. Hallden NA, Harley JH (1964) Radium-226 in diet and human bone from San Juan, Puerto Rico. Nature 204:240–241

42. Beninson D, Beninson AMde, Menossi C (1972) Ra-226 in Man. USAEC Report NP-19358. USAEC, Washington, DC

43. Lobao N, Penna-Franca E (1973) Radium-226 in diet and human bones in the State of Rio de Janiero (cited in Holtzman, Ref. 36). An Acad Bras Ciene 45:489–495

44. Fisenne IM, Keller HW, Harley NH (1981) Worldwide measurement of Ra-226 in human bone: Estimate of skeletal alpha dose. Health Phys 40:163–171

45. Fisenne IM, Welford GA, Perry PM, Baird R, Keller HW (1978) Distributional U-238, 234, Ra-226 and Po-210 in soil. Environ Int 1:245–246

46. Keslev D, Novakova E, Boyadzhiev A, Kerteva A (1975) Contents of polonium-210 in food products of Bulgaria (cited in Holtzman, Ref. 36)

47. Servant J, Delapart M (1981) Blood lead and lead-210 origins in residents of Toulouse. Health Phys 41:483–487

48. Clemente GF, Renzetti A, Santori G, Breuer F (1980) Assessment of polonium-210 exposure for the Italian population. In Radiation Protection: A Systematic Approach to Safety. Pergamon Press, Oxford, pp 1091–1094

49. Gloebel B, Muth H, Oberhausen E (1966) Intake and excretion of the natural radionuclides Pb-210 and Po-210 by humans (cited by Holtzman, Ref. 36) Strahlentherapie 131:218–226

50. Ladinskaya LA, Parfenov YD, Popov DK, Fedorova AV (1973) Lead-210 and polonium-210 content in air, water, foodstuffs and the human body. Arch Environ Health 27:254–258

51. Yermolayeva-Makovskaya AP, Pertsov LA, Popov DK (1969) Polonium-210 in the human body and in the environment UNSCEAR Document A/AC. 82/9/L.1260. In USAEC Health and Safety Laboratory Translation. HASL, New York, pp 163–170

52. Hill CR (1965) Polonium-210 in man. Nature 208:423–428

53. Smith-Briggs JL, Bradley EJ, Potter MD (1986) The ratio of lead-210 to polonium-210 in UK diet. Sci Total Environ 54:127–133

54. Chamberlain AC (1983) Fallout of lead and uptake by crops. Atmos Environ 17:693–706

55. Khandekar RN (1977) Polonium-210 in Bombay diet. Health Phys 33:148–150

56. Magno PJ, Groulx PT, Apidianakis JC (1970) Lead-210 in air and total diets in the United States during 1966. Health Phys 18:383–388

57. Spencer H, Holtzman RB, Kramer L, Ilcewicz FH (1977) Metabolic balances of Pb-210 and Po-210 at natural levels. Radiat Res 69:166–84

58. Bogen DC, Welford GA, Morse RS (1976) General population exposure of stable lead and Pb-210 to residents of New York City. Health Phys 30:359

59. Takata N, Watanabe H, Schikawa R (1968) Lead-210 content in foodstuffs and its dietary intake in Japan. J Radiat Res (Japan) 9:29–34

60. Holtzman RB, Ilcewicz FH (1966) Lead-210 and polonium-210 in tissues of cigarette smokers. Science 153:1259

61. Linsalata P, Eisenbud M, Penna-Franca E (1986) Ingestion estimates of Th and the light rare earth elements based on measurements of human feces. Health Phys 50:163–167
62. Petrow HG, Schiessle WJ, Cover A (1965) Dietary intake of radium-228. In Radioactivity Studies, USAEC Report NYO-3086-1, New York Operations Office. NTIS, Springfield, VA, pp 1–10
63. Mistry KB, Bharathan KG, Gopal-Ayengar AR (1970) Radioactivity in the diet of population of the Kerala Coast, including monazite-bearing high radiation areas. Health Phys 19:535–542
64. Penna-Franca E, Fiszman M, Lobao N (1970) Radioactivity in the diet in high background areas of Brazil. Health Phys 19:657–662

CHAPTER 7

# International Recommendations on Radiation Protection

## B. Lindell[1]

## Introduction

International recommendations on radiation protection have been issued since
1928, when the organization that is now called the International Commission
on Radiological Protection (ICRP) was founded at the 2nd International Congress
of Radiology in Stockholm. The early recommendations were limited to radiation
workers, but in the 1950s ICRP began to recommend a system of dose limitation
also for members of the public.

The ICRP recommendations for protection in controllable situations against
radiation exposures from normal uses of radiation and radioactive substances
are now in worldwide use and have been largely accepted by international
organizations such as the International Atomic Energy Agency (IAEA), the
International Labour Organization (ILO), the UN Food and Agricultural Organization (FAO), and the World Health Organization (WHO), and by most regional
and national bodies with responsibility for radiation protection, for example,
the Commission of the European Communities (CEC), and the Organization
for Economic Cooperation and Development/Nuclear Energy Agency (OECD/
NEA).

With the threat of major accidents in nuclear power plants and subsequent
environmental contamination, these bodies are also faced with the obligation
to advise on an appropriate emergency preparedness. For this purpose some,
but not all, of the principles recommended by ICRP for protection in the normal
uses of radiation sources may be applied. This chapter presents these principles
as a background for later discussions of the additional aspects that are characteristic for remedial intervention in emergency situations.

---

[1] National Institute of Radiation Protection, S-10401 Stockholm, Sweden.

**Fig. 7.1.** The scientific background needed for radiation protection is provided by ICRP and UNSCEAR. The recommendations of ICRP form the basis of the safety standards of other international organizations.

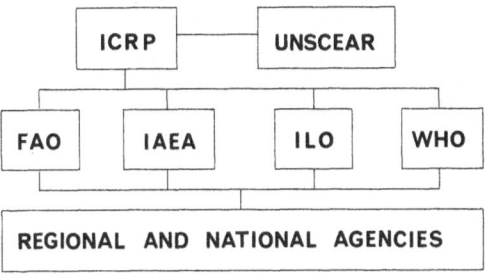

## International Organizations

The scientific knowledge needed to assess radiation exposures and radiation risks is evaluated and published by the United Nations Scientific Committee on the Effects of Atomic Radiation (UNSCEAR), a body that was created by the UN General Assembly in 1955. UNSCEAR and ICRP have close contacts and may be considered as the two scientific bodies that provide the necessary background information for radiation protection (Fig. 7.1). Both of these organizations have surprisingly small staffs—only one full-time scientific secretary each—but rely upon great contributions from scientific members and their institutions; UNSCEAR also makes short-term use of consultants.

The international organizations, IAEA, ILO, and WHO, together with OECD/NEA, have based their joint Basic Safety Standards (1) on the ICRP recommendations, thus influencing recommendations of other organizations as well as national regulations.

## Recommendations of ICRP for Normal Operations

### ICRP Publication 26

The main recommendations of ICRP have been published in ICRP Publication 26 (2) in 1977. These recommendations are supplemented by a number of committee and task group reports and have been slightly modified by a series of ICRP statements in 1978 (3), 1980 (4), 1983 (5), 1984 (6), 1985 (7), and 1987 (8).

### Biological Assumptions

The biological effects of radiation exposures have been described in Chatper 4. A short summary is given here merely to explain the assumptions behind current protection recommendations.

A simplified picture of the biological effects of ionizing radiation may be related to radiation-induced changes in the reproductive functions of irradiated cells. If cells fail to reproduce, there will be a loss of cells at the first attempt at cell division. If a sufficient number of cells are affected, for which a high radiation dose is needed, tissues and organs will be noticeably harmed. These effects are called *nonstochastic* effects because there is no major randomness in their clinical appearance; if the dose is high enough, they are certain to appear, but if the dose is below some threshold value, they are extremely unlikely. The threshold doses in the case of short-time exposures are usually higher than 1 Gy from beta- or gamma-radiation (one could also say 1 Sv, although this is strictly not appropriate at high doses).

In contrast, *stochastic* effects, which include cancer and hereditary harm, seem to appear at random. They are caused when the change in the cells' reproductive system is not grave enough to prevent cell division into new, viable cells, but will initiate functions that may lead to cancer if a somatic cell is affected or hereditatry harm if a germ cell is affected. For the stochastic effects, therefore, there are no threshold values of doses below which there is complete safety; the "risk," in the meaning of probability of harm, increases with the dose.

The dose-response relationship at low radiation doses is not known because the risk is so low that it is difficult to distinguish the radiation-induced harm from naturally occurring harm of the same type. However, it is usually assumed that the risk increases in proportion to the *effective dose equivalent,* a quantity that sums up radiation doses in all body organs and tissues after weighting for differences in their sensitivities. The proportionality coefficient is believed to be of the order of 1%/Sv for the risk of death from cancer and about the same for severe hereditary effects in future generations. Obviously, the risk will depend on age and sex, if for no other reason because the stochastic effects may not have time or possibility to develop in old persons.

A third class of effect, which is not easily classified as either stochastic or nonstochastic, is developmental harm to individuals who were exposed in utero. The dominant risk in this case is believed to be mental retardation, for which the risk coefficient has been assessed to be as high as 40%/Sv if the exposure occurred during weeks 8 to 15 of pregnancy. It is yet not known whether there is a threshold dose for this effect.

## Implications for Protection

In the case of nonstochastic effects, the magnitude of the risk increment $\Delta R$, caused by a small dose increment $\Delta D$, will depend on both doses previously received and doses to be expected in the future. However, if the dose-response relationship is linear (Fig. 7.2), the risk increment will always be the same, independent of other dose contributions. This means that the consequence of a particular exposure, e.g., caused by the consumption of a contaminated foodstuff, can be evaluated without regard to other exposures. This is not the case for nonstochastic effects where it is the total dose from all exposures that determines whether a dose threshold will be exceeded or not. Failure to comprehend this difference between stochastic and nonstochastic effects is the source of much misunderstanding.

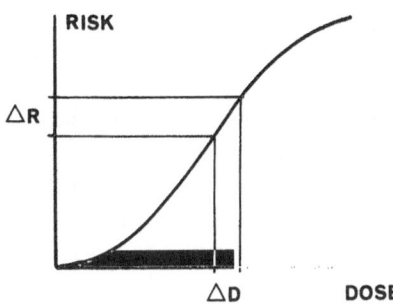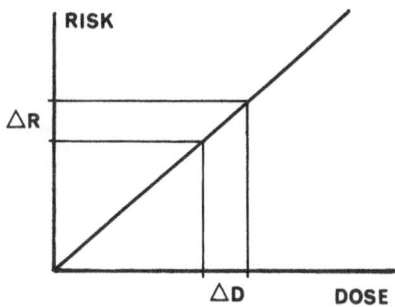

**Fig. 7.2.** If the relationship between risk (e.g., probability of cancer) and radiation dose is linear within the dose region of practical interest, a small dose increment $\Delta D$ will always cause one and the same risk increment $\Delta R$, regardless of previous doses.

## Basic Principles

The ICRP has recommended the application of the following three basic principles for radiation protection (2):

1. The practice causing radiation exposure must be justified, i.e., its introduction must produce a positive net benefit;
2. there should be limits for the individual doses and hence the individual risks for those who are exposed to radiation; and
3. even below these dose limits, all exposures should be kept as low as reasonably achievable, economic and social factors being taken into account.

The third principle is usually referred to as "optimization of protection." This can be done by various methods, one of which is cost-benefit analysis. The acronym "ALARA" ("as low as reasonably achievable") is sometimes used, but since some authors use "ALARA" for procedures that are not consistent with the ICRP principle, the acronym is not used by ICRP.

Values for the dose limits are recommended by ICRP, but the limits only apply to doses from controllable sources. The dose limits, therefore, are not relevant in the case of accidents.

## Dose Limits for Normal Operations

### Basic Dose Limits

With the current principles for dose limitation, there is no risk of nonstochastic effects except in cases of accidents. The purpose of dose limits is therefore mainly to limit the probability of stochastic effect.

In the absence of threshold values, there is no obvious level for a dose limit since any dose will cause some risk. If optimization is obtained by means of cost-benefit analysis, the collective doses will be kept as low as reasonably achievable, and this often reduces the maximum doses. The dose limits may be seen as indicating the lower end of a region of unacceptable doses. Doses above the dose limits are never acceptable; doses below the limits are acceptable only if they are as low as reasonably achievable.

Since "acceptable" is a subjective judgment, the limits may be seen as a fairly arbitrary convention for limiting risks from one group of risk sources, namely, controllable man-made radiation sources. There are other groups of sources of risks of stochastic harm in life, each subject to its own limitation procedure. As yet, however, there is no attempt to limit the aggregate risk from all sources of stochastic risk.

The main dose limits recommended by ICRP are expressed as limits for the annual effective dose equivalent. For workers the limit is 50 mSv per year; for the public it is 1 mSv per year, with the acceptance of 5 mSv for some years, provided that the average annual dose over a lifetime does not exceed the principal limit of 1 mSv.

## Source-Related Upper Bounds

Particularly in the case of the public, each individual may be exposed by radiation from a number of different sources, e.g., consumer products and various nuclear installations causing environmental contamination. Since there is no practicable way of directly controlling the exposure of individul members of the public, dose restrictions must be achieved by restrictions at the source, such as release limits for radioactive substances into the environment. These cannot be derived from the full dose limit since several sources together might then cause a total dose exceeding the limit. For source control, there must therefore be source-related limits, often called source-related upper bounds, being only some fraction of the basic dose limit. It is the responsibility of competent national authorities to set such upper bounds.

## Secondary Limits and Derived Limits

The source-related upper bounds may be basically expressed as dose limits, but may also be expressed as the annual limits of activity intake, which correspond to the source-related upper bounds. The annual limit of intake (ALI) that corresponds to the full basic dose limit is called a secondary limit. The limits that are actually applied are on other quantities, such as air concentrations, shield thicknesses, or quantities of radioactive material released per specified time intervals. Such limits are called derived limits if they relate to the basic dose limit and derived upper bounds if they relate to the source upper bound.

## Authorized Limits

The limits actually set by national authorities or by the management of a particular installation are called authorized limits. They should reflect the situation when the protection is optimized, but they must never exceed levels that denote the source-related upper bound.

## Optimization of Protection

### Lower Than the Limits

Since no radiation dose can be said to have absolutely no risk, there could be a demand to reduce doses further from whatever level has already been obtained. In practice, a zero dose can never be reached nor be proven against the background of radiation doses from cosmic rays and from unavoidable natural sources of radiation in the environment. Therefore, there must be a practical limitation of how far the dose reduction should be pursued. This is expressed by the ICRP principle of optimization of protection.

### Detriment and Collective Dose

The ICRP has defined the term "detriment" to mean the mathematical expectation of the amount of harm from a radiation exposure, taking into account not only the probability but also the severity of the various harmful effects. The detriment in terms of stochastic and nonstochastic biological effects is sometimes referred to as the "objective health detriment." In addition, there may be economic consequences and nontangible effects such as concern and anxiety.

On the assumption of proportionality between risk and radiation dose, the objective health detriment is proportional both to the number of exposed persons and to their average dose, and therefore also to the product of these two quantities, the *collective dose,* measured in man sievert (man Sv).

### Cost-Benefit Analysis

In cost-benefit analysis, the monetary cost of radiation protection is compared with the "cost" of the radiation detriment. For this to be possible, the detriment, or at least the collective dose, must also be expressed in monetary terms. The "cost" of a unit collective dose is usually denoted $\alpha$. The value of $\alpha$ is often taken to be about US $10,000 per man Sv.

Translating the collective dose into monetary terms implies that it is considered possible to pay a certain amount of money to avoid a case of premature death in a statistical sense. This is not equivalent to "putting a value on human life." It only means that if all lifesaving is based on this limitation, the maximum number of lives will be saved that can be saved for the amount of money that society is willing to set aside for stochastic lifesaving. The optimum protection is obtained when the sum of the protection cost and the "cost" of the collective dose is at a minimum (Fig. 7.3).

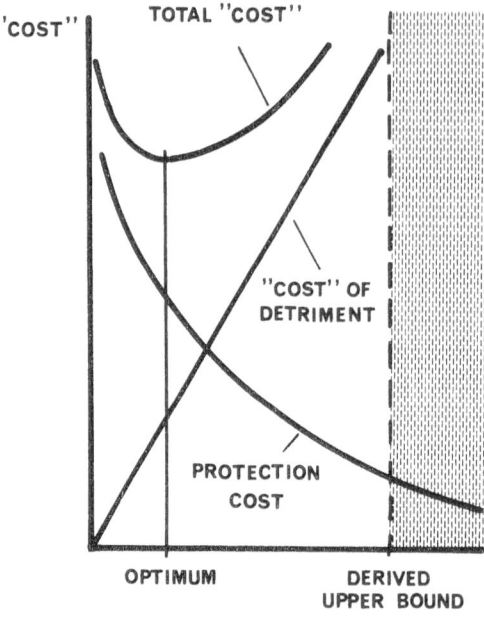

**Fig. 7.3.** If the optimization assessment is carried out by means of a cost-benefit analysis, optimum protection is provided when the sum of the protection cost and the "cost" of the radiation detriment has a minimum. The source-related upper bound for individual dose provides a boundary condition for the optimization result. In the diagram, the example illustrates the principle when the method of protection is limitation of activity released into the environment.

## Noncontrollable Sources

### Accidents

The post facto condition after an accident is only one example of an exposure situation that is a "fait accompli" and can no longer be influenced by source control but only by remedial action. Another example is our natural radiation environment.

### Natural Radiation

Most natural radiation, e.g., cosmic rays and radiation from naturally occurring radionuclides in the environment and in our own bodies, cannot be appreciably influenced by any reasonably practicable human action. Doses from such types of radiation, therefore, are exempted from the ICRP dose limits (doses from medical examinations and treatments are also exempted because it is considered to be the responsibility of the physician to decide what is in the best interest of each patient).

### Enhanced Natural Radiation

In recent years the high lung doses from radioactive daughter products of $^{222}$Rn in dwellings have caused considerable concern. The doses are sometimes higher, and even much higher, than the normal ICRP dose limits for members of the public. In the choice, for *new* houses, between including this source under the normal dose limits (which would then have to be raised since they were never intended to include any natural radiation) or recommending a special source-related upper bound, the ICRP preferred the second alternative (9).

The situation in *existing* houses is of particular interest in relation to accidental situations since, as in accidents, the doses can only be controlled by remedial actions. Neither the ICRP dose limit nor the recommended upper bound for new houses is applicable in this case. Remedial action has often not been taken even when the annual effective dose equivalent to occupants has approached 50 mSv, although the ICRP has suggested an intervention level of about 20 mSv per year.

# Remedial Actions

For the situation after an accident, the normal dose limits do not apply because the doses caused by accidents are only controllable by remedial action. Instead, recommendations on *intervention levels* are needed.

## ICRP Publication 40

In 1984 ICRP published recommendations on protection of the public in the event of major radiation accidents (ICRP Publication 40, ref. 10). In that document the basic principles of remedial action are described and recommendations on radiation doses that should be avoided by various types of remedial actions are given. These recommendations have been supplemented by documents issued by WHO in 1984 (11) and IAEA in 1985 (12).

### Objectives of Remedial Actions

In ICRP Publication 40, the following three principles are given for planning intervention in the event of an accident:

1. Serious *nonstochastic* effects should be avoided by preventing doses from reaching threshold values for such effects;
2. the probability of *stochastic* effects should be limited by countermeasures that achieve a positive net benefit to the individuals involved; and
3. the overall incidence of stochastic effects should be limited by reducing the collective dose as much as reasonably achievable.

The first two of these principles relate to the risk of *individuals* and correspond to the basic protection principle of individual dose *limitation*. The third principle relates to the total detriment and calls for optimization of *collective* dose *reduction*.

## Intervention Levels

### No Universal Intervention Level

It is clear from the first two of the quoted principles, says ICRP in its Publication 40, that it is the level of individual dose that is the quantity of importance in deciding on the introduction of countermeasures. As in the case of dose limits for normal operations, there is some level of dose that should be avoided even if it is not cost-effective in the same sense as optimization of protection. If it is expected that this dose level would be exceeded unless some remedial action is taken, the situation will require such action.

The intervention level has nothing to do with the normal dose limit for members of the public. Since the intervention must improve the overall situation for those at risk, the appropriate intervention level will depend on the types of actions that are available. For example, there may be little penalty in avoiding some contaminated food if alternative supplies are available. However, if this is not the case, it is obviously not advisable to starve rather than to accept the moderate radiation risk from doses somewhat higher than the normal dose limits. There is, therefore, no universal value for the intervention level of projected individual dose.

### Intervention Levels Recommended by ICRP

Although it is difficult to decide on intervention levels before the actual emergency situation and before the available countermeasures are known, ICRP has ventured to recommend ranges of values that might be considered for particular types of remedial actions. These recommendations are given in ICRP Publication 40 (10). They are expressed both as effective dose equivalents and as dose equivalents in organs receiving the highest doses (see Table 7.1).

## Optimization of Protection

Remedial actions may be taken not only to prevent reaching a certain unacceptable individual dose but also to reduce the expected collective dose "as far as reason-

**Table 7.1.** Ranges of intervention levels recommended by ICRP[a]

| Countermeasures | Effective dose (mSv) | | Organ dose (mSv) | |
|---|---|---|---|---|
| | Lower | Upper | Lower | Upper |
| Early phase | (Short-term doses) | | | |
| Sheltering and stable iodine administration | 5 | 50 | 50 | 500 |
| Evacuation | 50 | 500 | 500 | 5,000 |
| Intermediate phase | (First-year doses) | | | |
| Control of foodstuffs | 5 | 50 | 50 | 500 |
| Relocation | 50 | 500 | (no values given) | |

[a] Data from ref. 10.

ably achievable." Such reduction may or may not reduce individual doses to levels below whatever intervention level that would have been judged appropriate if only avoidance of unacceptable individual risks had been taken into account.

## Nonaction Levels

An emergency situation in which the highest individual doses are below the lower value in the range of intervention levels recommended by ICRP and in which, in addition, further dose reduction is not cost-effective should not require any intervention if the optimization assessment is carried out by means of cost-benefit methods. If some dose-reduction were found to be cost-effective, the cost-benefit assessment would indicate dose levels below which it would not be reasonable to go. These may be called nonaction levels.

General agreements on nonaction levels for various postulated emergency situations have been considered helpful, and such levels have sometimes been calculated by equating the "cost" of the corresponding dose reduction (using commonly applied values for $\alpha$) with the production cost of the particular foodstuff. Experience after the Chernobyl accident, however, indicates that more factors than this cost must be considered in the optimization procedure and that the simple cost-benefit analysis may not suffice for the purpose. Nonaction as well as remedial action for control of foodstuffs or advice about their wholesomeness may also be considered at quite different levels for intervention against import, trade, and private consumption.

## Derived Levels

Regardless of whether the purpose is to avoid high individual doses or to eliminate a certain amount of collective dose, two parameters determine the result: the activity concentration ($C_i$) in various foodstuffs ($i$), and the individual consumption ($Q_i$) of these foodstuffs over some period of time, for example, one year. If the values (average and/or maximum) of $Q_i$ are known, therefore, the necessary restriction of $C_i$ (average and/or maximum) can be calculated.

In the practical application, the restriction of $C_i$ for the various foodstuffs, or for groups of foodstuffs, will have the character of limits, for example, for import or sale. This creates a difficult information problem, since any "limit" is perceived by the public as a borderline between complete safety and imminent danger. It is not easy to explain that applied action levels related to the activity concentration are merely a means to achieve the appropriate protection by limiting the total intake and not a primary objective. To make this more understandable, authorities sometimes also present their objective to limit, for example, the annual intake, by stating that it is the product of $Q_i$ and $C_i$ to which a derived intervention level should primarily apply. In the case of households that grow their own vegetables and have direct access to berries, mushrooms, fish, and game meat, advice on the limitation of the total activity intake, in combination with information on activity concentrations, may be the only means of protection.

Under normal circumstances, radiation protection is governed by a system of dose limitation that is widely accepted over the world and that has been developed over many decades. The normal application of this system usually does not cause surprises, nor does it have many repercussions outside the radiological field. In contrast, some situations for which this system was not originally intended, for example, emergency situations and high radon levels in dwellings, extend into fields in which the principles of radiation protection were never applied and in which the consequences of their application may well cause surprises and unforeseen repercussions. In particular, any intervention with regard to control of foodstuffs may involve consequences and considerations that are not familiar to those who advise on radiation protection. This makes it prudent in such cases to exercise great caution in applying the usual methods of radiation protection. For example, it may be inappropriate merely to use crude cost-benefit analysis for the purposes of optimization of protection.

## Calculation of Derived Levels

To calculate derived levels from annual intakes, it is necessary to know the effective dose equivalent or organ dose per unit intake of the various readionuclides that may contribute significantly to the exposure. If several nuclides contribute significantly, the intervention level for each will have to be reduced by taking the exposure from the others into account.

Different assumptions by various authorities on the dose per unit intake caused some confusion after the Chernobyl accident. Efforts are therefore being made by a number of organizations to derive agreed dose coefficients. One set of numbers was published by the IAEA in 1986 (13).

## References

1. International Atomic Energy Agency, International Labour Organisation, OECD Nuclear Energy Agency, and World Health Organisation (1982) Basic Safety Standards for Radiation Protection. International Atomic Energy Agency Safety Series No 9, 1982 Edition. IAEA Vienna
2. International Commission on Radiological Protection (1977) Recommendations of the International Commission on Radiological Protection. ICRP Publication 26, Annals of the ICRP Vol 1 No 3. Pergamon Press, Oxford
3. International Commission on Radiological Protection (1978) Statement from the 1978 Stockholm Meeting of the ICRP. Annals of the ICRP Vol 2 No 1. Pergamon Press, Oxford
4. International Commission on Radiological Protection (1980) Statement and Recommendations of the 1980 Brighton Meeting of the ICRP. Annals of the ICRP Vol 4 No 3/4. Pergamon Press, Oxford
5. International Commission on Radiological Protection (1984) Statement from the 1983 Washington Meeting of the ICRP. Annals of the ICRP Vol 14 No 1. Pergamon Press, Oxford
6. International Commission on Radiological Protection (1984) Statement from the

1984 Stockholm Meeting of the ICRP. Annals of the ICRP Vol 14 No 2. Pergamon Press, Oxford

7. International Commission on Radiological Protection (1985) Statement from the 1985 Paris Meeting of the ICRP. Annals of the ICRP Vol 15 No 3. Pergamon Press, Oxford

8. International Commission on Radiological Protection (1988) Statement from the 1987 Washington Meeting of the ICRP. ICRP Publication 51, Annals of the ICRP Vol 17 No 2/3. Pergamon Press, Oxford

9. International Commission on Radiological Protection (1984) Principles for limiting the exposure of the public to natural sources of radiation. ICRP Publication 39, Annals of the ICRP Vol 14 No 1. Pergamon Press, Oxford

10. International Commission on Radiological Protection (1984) Protection of the public in the event of major radiation accidents: Principles for planning. ICRP Publication 40, Annals of the ICRP Vol 14 No 2. Pergamon Press, Oxford

11. World Health Organization (1984) Nuclear power: Accidental releases—principles of public health action. WHO Regional Publications, European Series No 16. WHO Regional Office for Europe, Copenhagen

12. International Atomic Energy Agency (1985) Principles for establishing intervention levels for the protection of the public in the event of a nuclear accident or radiological emergency. IAEA Safety Series No 72. IAEA, Vienna

13. International Atomic Energy Agency (1986) Derived intervention levels for application in controlling radiation doses to the public in the event of a nuclear accident or radiological emergency. IAEA Safety Series No 81. IAEA, Vienna

# Part III
# Environmental Pathways Critical to Humans

# CHAPTER 8

# Airborne Contamination

Yu. A. Izrael[1] and V. N. Petrov[1]

## Atmospheric Transport and Dynamics of Radioactive Product Releases

Data on radioactive releases and contamination obtained for the first months after the Chernobyl accident were given in the paper presented by the Soviet delegation to the IAEA expert meeting in August 1986 (1). For the period after the accident, much has been done in specifying the pattern of radioactive contamination in the areas of close-in and remote fallout.

The most intensive radioactive release from the reactor zone was observed to the north-westward and north-eastward for the first two to three days after the accident (1). The plume height on April 27, based on aircraft data, exceeded 1,200 m. The maximum radiation level near the nuclear power plant (NPP) was recorded at a height of 600 m. On subsequent days the plume height did not exceed 200 to 400 m.

Additional meteorological information was acquired for the period of principal release of radioactive products. Balloon data on wind direction and velocity at the airports of Kiev (Juljany, Borispol), Mozyr, Gomel, and Chernigov as well as radiosonde data at Kiev from April 26 to May 1, 1986, were used as input weather information to the model. Mean wind directions and velocities within the layer from the ground surface to a specific height were calculated based on primary observation data under a specially developed program. The calculated values of mean wind velocity and direction in the layers at 0 to 500 m and 0 to 1,000 m are given in Fig. 8.1 for the observation period of five days after the accident. The data were used to calculate the airborne particles in the layers at 0 to 1,000 m and 0 to 500 m.

The analysis of meteorological data on the wind direction (Fig. 8.1) as well as the 925 mb trajectory shows that during five days (April 26 to May 1), the airborne particle transport direction in the layer from the ground surface to 1 to 1.5 km changed by 360°, making a full circle.

[1] USSR State Committee for Hydrometeorology, 12 Pavlik Morozov, 123376 Moscow, USSR.

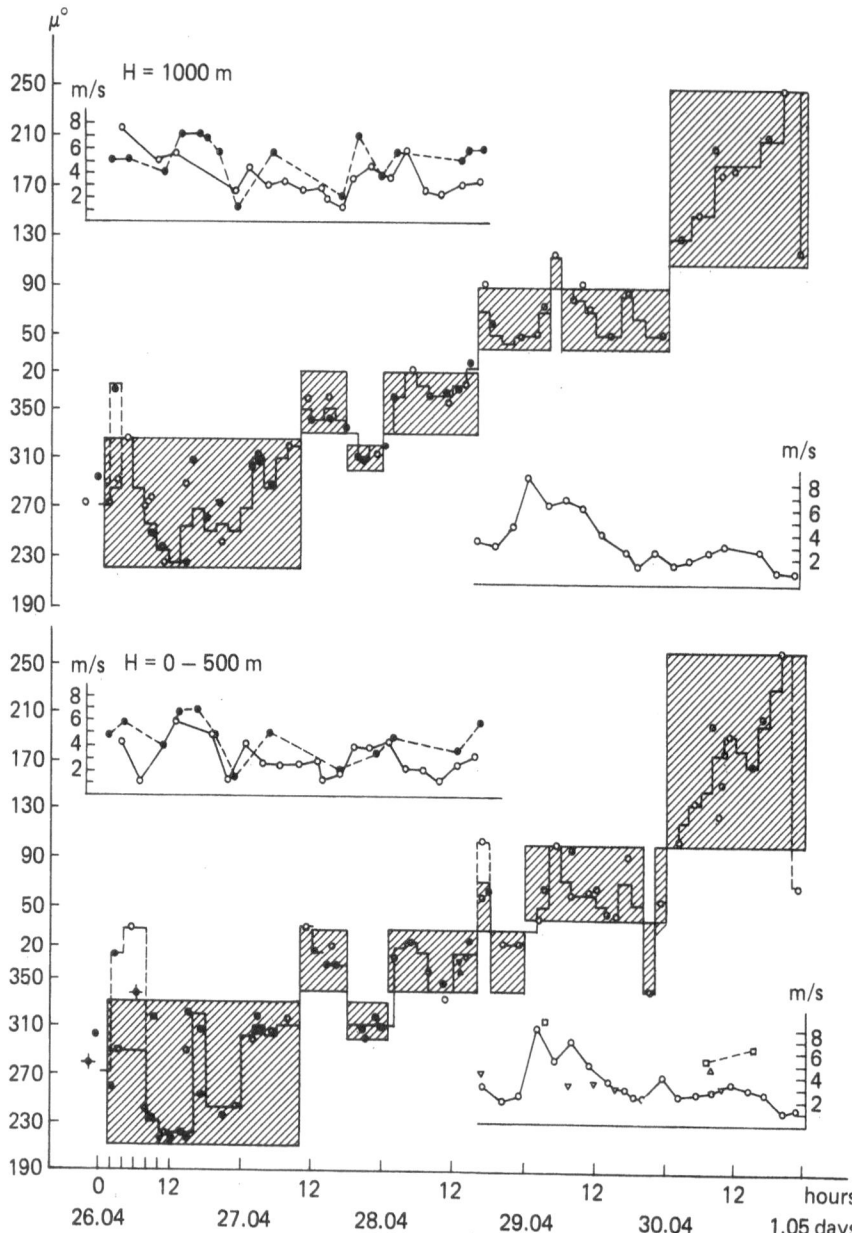

**Fig. 8.1.** Mean values in layer 0–500 m and 0–1,000 m and wind direction and speed values from 26.04 to 01.05.86 in the region adjacent to Chernobyl atomic power station. + Kiev (radio probe), ○ Kiev, airport: ● Borispol, △ Mozyr, □ Gomel, ▽ Chernigov.

Specific meteorological peculiarities of the radioactive product transport from the reactor zone, in comparison with the radiation fallout distribution pattern, allowed us to obtain additional characteristics of the release dynamics.

Data on the diurnal radioactive release from the reactor zone into the atmosphere are given in reference (1). It is well approximated during the first four to five days by

$$Q(t) = 0.32 \exp(-0.28t), \text{ where } t = 0.1 \ldots 4 \text{ days.}$$

Actually, the radioactive fallout in the close-in area ceased during the first four to five days. The daily radioactive release into the atmosphere, approximated by the given exponential function, includes the complete spectrum of radioactive particles that caused local, regional and global fallout.

The hatched sectors (in degrees) of Fig. 8.1 indicate the mean wind direction in the layer from the ground surface up to 500 and 1,000 m, where particles were transported and fell out from the plume at different time intervals after the accident. These sectors (230° to 320°, 320° to 20°, 20° to 90°, 90° to 220°) were drawn in the close-in fallout map, where the integral gamma-radioactivity was established in mR $\times$ km²/h for April 29, 1986. Total gamma-radioactivity of the close-in fallout was at that time $4.4 \times 10^4$ mR $\times$ km²/h ($2.8 \times 10^6$ Ci) (1). Hourly estimated gamma-radioactivity released into the atmosphere and deposited as close-in fallout is shown in Fig. 8.2 as histograms. The dashed

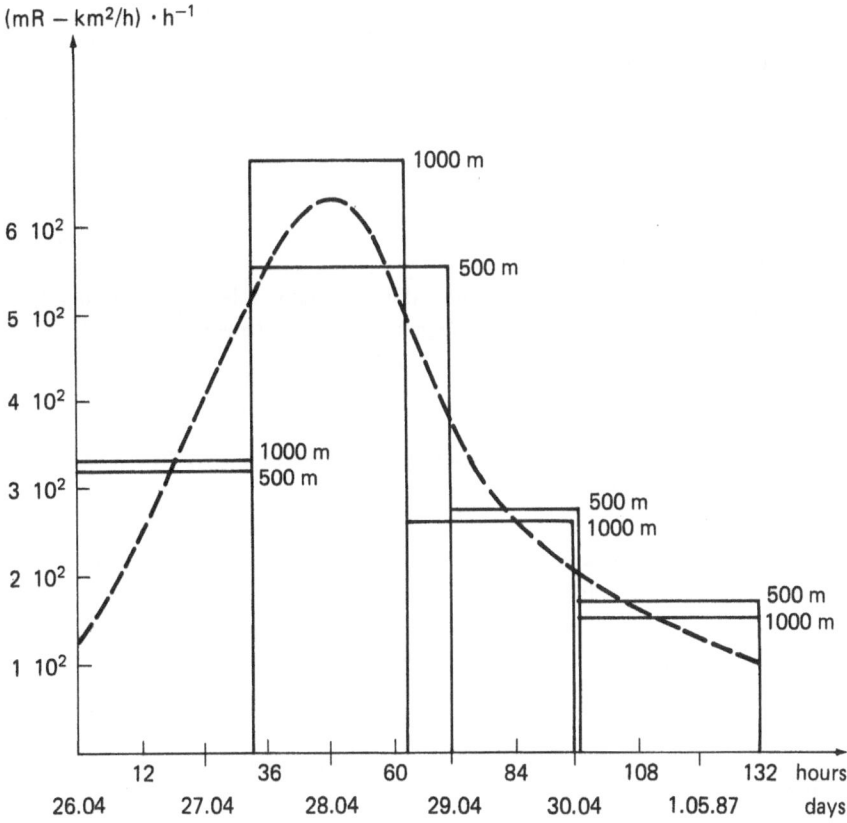

**Fig. 8.2.** Hourly emission of gamma-radioactive substances into the atmosphere fallen out in the close-in zone.

**Table 8.1.** Relative distribution of radioactive release in the atmosphere for the first five days related to the total radioactivity that entered the atmosphere or was deposited in the close-in fallout for the first five days

| Data | Relative release, ref. (1) | Relative release in this chapter | | | | | |
|---|---|---|---|---|---|---|---|
| | | Total radioactivity | | | Individual radionuclides | | |
| | | t. = 1,000 m | t. = 500 m | Mean value | $^{144}Ce$ | $^{137}Cs$ | $^{131}I$ |
| 26.04.86 | 0.32 | 0.17 | 0.17 | 0.17 | 0.19 | 0.1 | 0.09 |
| 27.04 | 0.24 | 0.29 | 0.25 | 0.27 | 0.28 | 0.3 | 0.31 |
| 28.04 | 0.19 | 0.29 | 0.29 | 0.29 | 0.3 | 0.4 | 0.42 |
| 29.04 | 0.14 | 0.14 | 0.15 | 0.14 | 0.12 | 0.18 | 0.15 |
| 30.04 | 0.11 | 0.11 | 0.14 | 0.13 | 0.11 | 0.02 | 0.03 |

curve indicates the relative variation in the hourly release rate from April 26 to May 1, 1986.

Similarly, based on the relationship between radioactive fallout density of a given nuclide (Ci/km²) and the dose rate (mR/h) for different sectors (west, north, south) given in reference (3), the individual radionuclide releases into the atmosphere were computed on the subsequent days after the accident. Table 8.1 gives computation results of the relative daily release of the total gamma-radioactivity and individual radionuclides into the atmosphere, related to the total radioactivity that entered the atmosphere (1) or deposited in the close-in fallout for the first five days.

Approximately the same change in the relative radioactive release was obtained from the sectors of the meteorological trajectories and the radiation distribution given in Fig. 8.3.

One can say for certain, then, that the maximum release of radionuclides $^{131}I$, $^{137}Cs$, and $^{144}Ce$ into the atmosphere, which resulted in the radioactive contamination of the USSR territory, occurred on April 27 and 28, 1986.

## The Atmospheric Radioactive Contamination and Fallout Over the USSR Territory

The radionuclide composition of the atmospheric aerosol was determined through air sampling by various filters for subsequent laboratory analysis by semiconductor gamma-spectroscopy and radiochemistry methods. The radionuclide composition is a wide range of debris radionuclides with a characteristic peculiarity: the atmospheric radioactive products are rich in I, Cs, and to some extent, in Ru radioisotopes. Depending on the meteorological conditions, the radioactive products entered the surface atmospheric layer at different times from various points.

The atmospheric fallout is monitored within the Goscomhydromet network with the help of trays exposed during 24 hours. The data obtained indicate

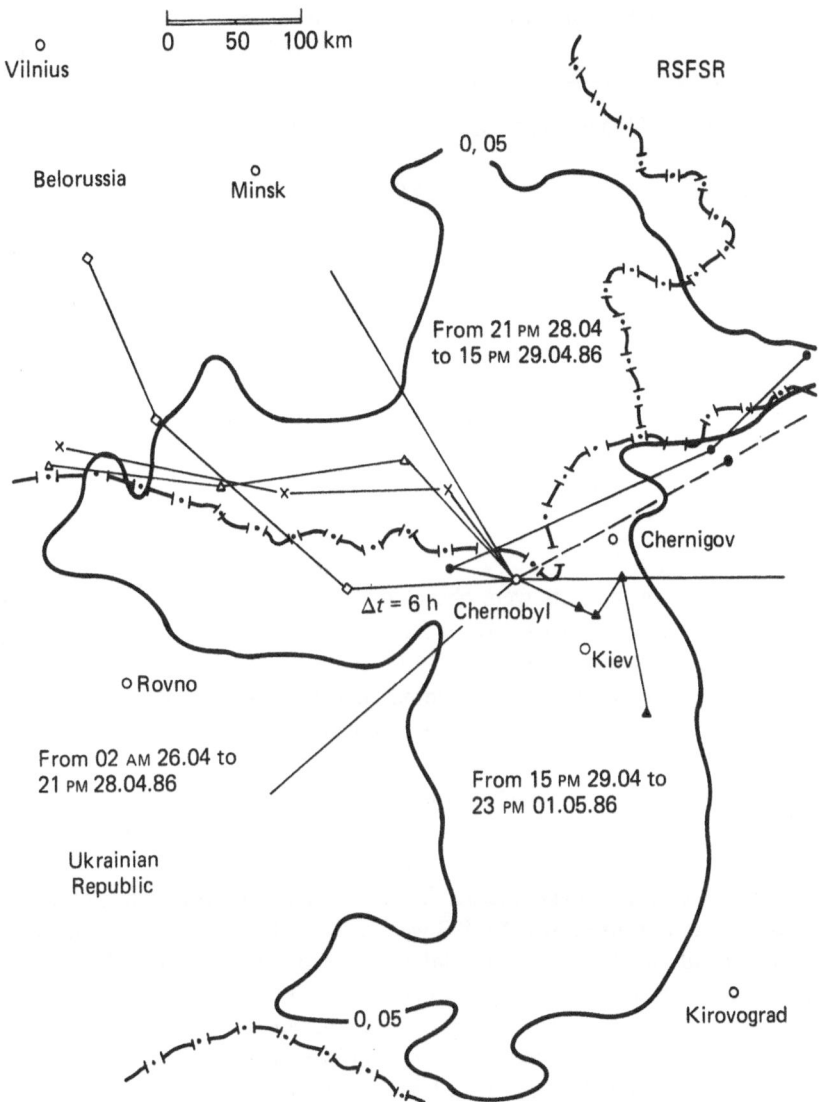

**Fig. 8.3.** Gamma-field distribution with the USSR territory according to the isopleth of dose rate 0.05 mR/h on June 10, 1986, and sectors of particle trajectory distributions at the 925 mb level (x—x from 3 AM 26.04.86, ◇—◇ from 15 AM 26.04.86, △—△ from 3 AM 27.04.86, ●—● from 15 PM 27.04.86, — — — from 3 AM 29.04.86, ▲—▲ from 15 PM 29.04.86).

that atmospheric and earth's surface contamination is irregular. For a short period of time, the concentration values exceeded those observed on the previous day by two to three orders of magnitude. In the next period the radioisotope concentration levels in the air and their fallout amounted to several percent of the maximum value. The arrival of maximum is different for various points.

**Table 8.2.** Maximum radionuclide concentrations in the surface air (Bq/m$^3$)

| | $^{95}$Zr | $^{103}$Ru | $^{131}$I | $^{132}$Te | $^{134}$Cs | $^{137}$Cs | $^{140}$Ba | $^{141}$Ce | $^{144}$Ce | $^{90}$Sr |
|---|---|---|---|---|---|---|---|---|---|---|
| Minsk | | | | | | | | | | |
| April 28–29, 1986 | 2.94 | 15.5 | 317 | 73.5 | 47.2 | 93.0 | 27.1 | | | |
| Vilnius | | | | | | | | | | |
| April 28–29, 1986 | 3.0 | 17.0 | 27.0 | 69.0 | 3.8 | 6.0 | 2.4 | 6.0 | | |
| Baryshevka | | | | | | | | | | |
| April 30–May 1, 1986 | 24.5 | 24.5 | 303 | 3310 | 51 | 77 | 225 | 26 | 26 | |
| Kiev | | | | | | | | | | |
| May 1–2, 1986 | 12 | 80 | 340 | 160 | 50 | 100 | 160 | 20 | 30 | |
| | Maximum radionuclide fallout, (kBq/m$^2$d) | | | | | | | | | |
| Minsk | 1.27 | 2.1 | 24 | 12 | 0.58 | 0.37 | 1.3 | 0.14 | | 0.02 |
| Vilnius | 0.19 | 0.6 | 14.4 | — | 0.22 | 0.32 | 0.054 | | | |
| Baryshevka | 45 | 56 | 330 | 54 | 0.7 | 7.7 | 64 | 32 | 45 | 1.0 |
| Kiev | 22 | 89 | 381 | 211 | 6.0 | 12 | 12 | 26 | 37 | 5.6 |

Estimated from the radioactive fallout.

The radioactive products were observed in the northwest in Minsk on April 27, in the north in Gomel on April 28, and in the south in Baryshevka and Kiev on April 30 and May 1.

It is of interest to estimate the external and internal doses through the food chains and inhalation from the atmospheric contamination and radioactive fallout in different places.

Table 8.2 shows maximum radionuclide concentrations in the surface air and radioactive fallout when radioactive air masses passed Minsk, Vilnius, Baryshevka, and Kiev, located in different areas of the USSR European territory.

The effective external and internal dose equivalents for the first year of human exposure in points given in Table 8.3 were estimated from the integrated values of radionuclide concentrations and fallout of certain radionuclides for the ten-day period after the accident. The coefficients for estimating the inhalation doses are taken from reference (4) for open terrains. The coefficient for the expected dose from a unit density of contamination by $^{131}$I is taken from the UN Scientific Committee on the Effects of Atomic Radiation (UNSCEAR, 1982). When estimating the dose of ingested $^{137}$Cs, a mean ration was taken

**Table 8.3.** Effective dose equivalents of external and internal radiation

| Input pathway | Minsk | | Vilnius | | Baryshevka | | Kiev | |
|---|---|---|---|---|---|---|---|---|
| | rem | % | rem | % | rem | % | rem | % |
| Inhalation | $1.3 \times 10^{-2}$ | 25 | $2.9 \times 10^{-3}$ | 20.0 | $1.5 \times 10^{-2}$ | 2.7 | $2 \times 10^{-2}$ | 2.5 |
| Ingestion | $1.5 \times 10^{-2}$ | 29 | $5.4 \times 10^{-3}$ | 37.2 | $1.7 \times 10^{-1}$ | 30 | $2.5 \times 10^{-1}$ | 31.3 |
| Gamma radiation from fallout | $2.4 \times 10^{-2}$ | 46 | $6.2 \times 10^{-2}$ | 42.8 | $3.8 \times 10^{-1}$ | 67.3 | $5.3 \times 10^{-1}$ | 66.2 |

for the European territory of the USSR with a half-removal time of Cs from soil equal to 7 years.

The estimate of effective dose equivalent from ingestion for the first year is given for [131]I and [137]Cs. Gamma-radiation from fallout makes the greatest contribution to the effective dose equivalent. It should be noted that under low densities of radioactive fallout, the importance of inhalation input increases, although its general contribution to the dose is insignificant.

## Amount of Radioactive Products in Close-In and Remote Fallout Over the USSR European Territory

Measurements of gamma-fields were regularly performed in the close-in zones of the Chernobyl NPP in the period after the accident. Figure 8.4 represents the map of radioactivity distribution over the area for May 1, 1987.

Using data on the gamma-dose-rates distribution over the area at different times after the accident, we have estimated the total amount of radioactive products deposited in the close-in fallout and its change with time due to radioactive decay and other factors.

The total amount of gamma-products in the close-in fallout decreased by

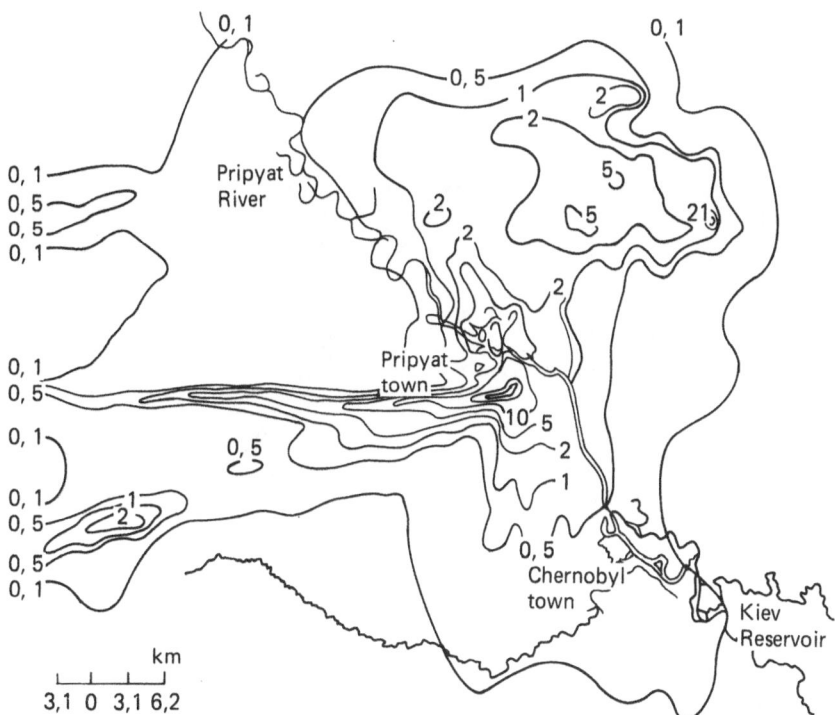

**Fig. 8.4.** Gamma-field distribution (mR/h), May 1, 1987.

some 56 times in the year after the accident and was $2.7 \times 10^6$ (R/h)m² ($5 \times 10^5$Ci) for May 1, 1987.

Fig. 8.4 shows areas included in the dose rate isolines for May 1, 1987. They are R = 1.0 mR/h − 500 km²; R = 2.0 mR/h − 280 km²; R = 5.0 mR/h − 70 km²; R = 10 mR/h − 20 km²; R = 20 mR/h − 8.0 km²; R = 50 mR/h − 3.0 km². A year after the accident, the areas included in the above dose rate isolines decreased 50 to 150 times. Figure 8.5 shows variations in the areas (R) of close-in radioactive fallout depending on dose rates that limit these areas. Aerogamma survey data are reduced to one and the same time, namely, May 29, 1986. Inside the area limited by the isolines, the dose rate is equal to or higher than the specified one. As it follows from Fig. 8.5, the results obtained from different surveys are rather close.

Investigations of the isotope composition in the radioactive fallout did not

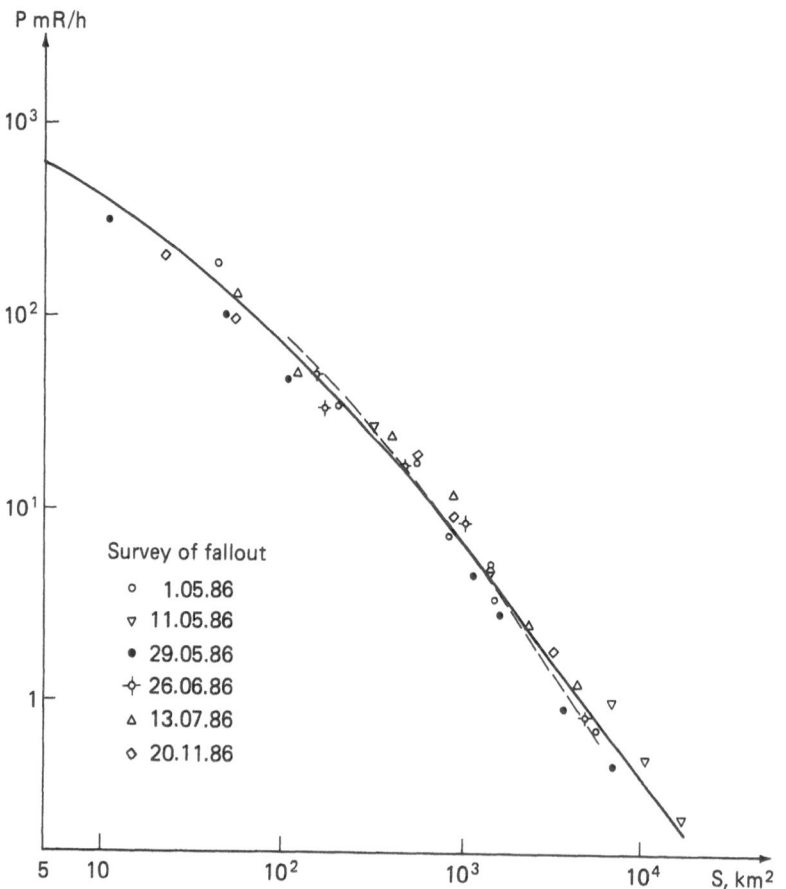

**Fig. 8.5.** Dependence of radioactive fallout area in the local zone on dose-rate isolevel (data are given for May 29, 1986), ------- calculated value, ——— measured mean value.

show any essential radioisotope fractionation at distances from 15 to 30 km. However, at greater distances, especially in the northern fallout, significant enrichment in $^{137}$Cs (some ten times and more) was observed. Contamination by Cs is of a spotty character, which is determined both by the release dynamics and weather conditions for spreading the radioactive products. The aerial survey and sample analysis show the amount of $^{137}$Cs deposited in the close-in fallout as equal to approximately 0.2 MCi.

Repeated aerogamma-survey and aerospectral survey of the USSR territory was carried out in May 1987. Distribution of the gamma-field over the USSR territory (Fig. 8.3) by the dose rate isopleth 0.05 mR/h taken for June 10, 1986 (3) cannot be presented as a closed isopleth even in 1987; gamma-field is presented as many separate unclosed spots. Data from aerogamma-survey in late May 1987 make it possible to estimate the total amount of gamma-radioactive materials deposited over the USSR territory beyond the close-in fallouts as (6 − 9) × $10^6$(R/h)m$^2$ as compared with 1.2 × $10^8$ (R/h)m$^2$ in the first ten days of June 1986 (3). The total of close-in and remote fallout amounts to (9 − 12) × $10^6$(R/h)m$^2$ or about 4.0% of the theoretical total of the radioactivity energy rate in the reactor for the given time. According to specific data from the spectral gamma-survey, the amount of $^{134}$Cs and $^{137}$Cs over the USSR territory is about 0.9 MCi. The total $^{137}$Cs amount in close-in and remote zones over the USSR territory is estimated as 1.1 MCi or approximately 13% of the produced $^{137}$Cs.

## Radioactive Contamination of Rivers and Reservoirs

Radioactive releases into the atmosphere along with contamination of soil and vegetation cover led to the contamination of surface waters. Radioactive contamination of water bodies and streams occurs both due to the direct aerosol deposition onto water surfaces and to radionuclides discharged with contaminated waters from catchment areas.

A peculiarity of airborne contamination is the practically simultaneous contamination of the hydrographic network over large territories. As a result, many of the water masses may be significantly contaminated in the initial period. Then the washing out of contaminated territories and bed sediments becomes decisive in the contamination process of river and water bodies.

The example of the river Pripyat shows that levels of total beta-activity in water due to aerosol fallout were 100 times higher than in subsequent months when fallout had ceased.

Table 8.4 gives the concentrations of Cs isotopes in water samples from the Kiev water reservoir as well as from the rivers Pripyat and Dnieper from July 1986 up to May 1987. The table shows that Cs concentrations in water decreased more than 20 times during this period. A very sharp decrease of Cs concentration was observed in the summer and autumn of 1986. Water samples taken in autumn showed that up to 70% of the Cs radioactivity was sorbed by suspended

**Table 8.4.** Mean Cs concentrations in water samples from the Kiev water reservoir and rivers Pripyat and Dnieper taken in July 1986 to May 1987 ($10^{-11}$Ci/L)

| Water subject | July 1986 | | October 1986 | | April–May 1987 | |
|---|---|---|---|---|---|---|
| | $^{137}$Cs | $^{134}$Cs | $^{137}$Cs | $^{134}$Cs | $^{137}$Cs | $^{134}$Cs |
| Kiev water reservoir | (20 + 50) | (10 + 20) | (1 + 3) | (0.5 + 1.5) | (0.4 + 1.2) | (0.2 + 0.6) |
| River Pripyat (below Chernobyl) | (40 + 50) | (15 + 25) | (2 + 5) | (1 + 2) | (2 + 5) | (1 + 2) |
| River Dnieper (St. Ter- emtsy) | (1 + 1.4) | (0.4 + 0.6) | (0.5 + 0.6) | (0.2 + 0.3) | (0.4 + 0.6) | (0.2 + 0.3) |

particles. Thus, the suspension in water and resulted radioactivity increase are related to hydrometeorological conditions in water bodies under study. The most striking relation was observed in the Kiev water reservoir where suspensions from the rivers Pripyat and Dnieper are discharged.

In autumn, the water reservoir showed increased roughness and, as a result, a second water contamination occurred, owing to wind roiling of the upper level of contaminated silts.

In the winter of 1986 and 1987, under conditions of stable frosty weather without any essential thaws, radioactivity of the Dnieper cascade waters changed insignificantly.

In late winter the total water beta-activity in the Kiev–Kremenchug water reservoirs was $(1–2) \times 10^{-10}$Ci/L, and in the Dnieper–Kakhovka water reservoir it approximated $(1–2) \times 10^{-11}$Ci/L. $^{90}$Sr and $^{137}$Cs made a significant input to water contamination.

The spring flood of 1987 developed smoothly and slowly without sharp high-water waves. As a result, no essential increase in radionuclide concentrations was obvserved in rivers and reservoirs. Table 8.4 shows that Cs concentrations in samples taken in the period from late April to May 1987 remained at the level of those taken in autumn or decreased slightly.

Table 8.5 presents values of $^{137}$Cs average concentrations measured in autumn

**Table 8.5.** Seasonal concentrations of Cs isotopes in the main Dnieper tributaries over the Belorussian territory ($10^{-11}$Ci/L)

| River | Autumn 1986 | Spring 1987 |
|---|---|---|
| | $^{137}$Cs | $^{137}$Cs |
| River Dnieper (the area between the towns Zhlobin and Lvov) | (0.5 + 0.8) | (0.5 + 0.8) |
| River Pripyat in the region of the town Mozyr | (0.5 + 1) | (1 + 1.4) |

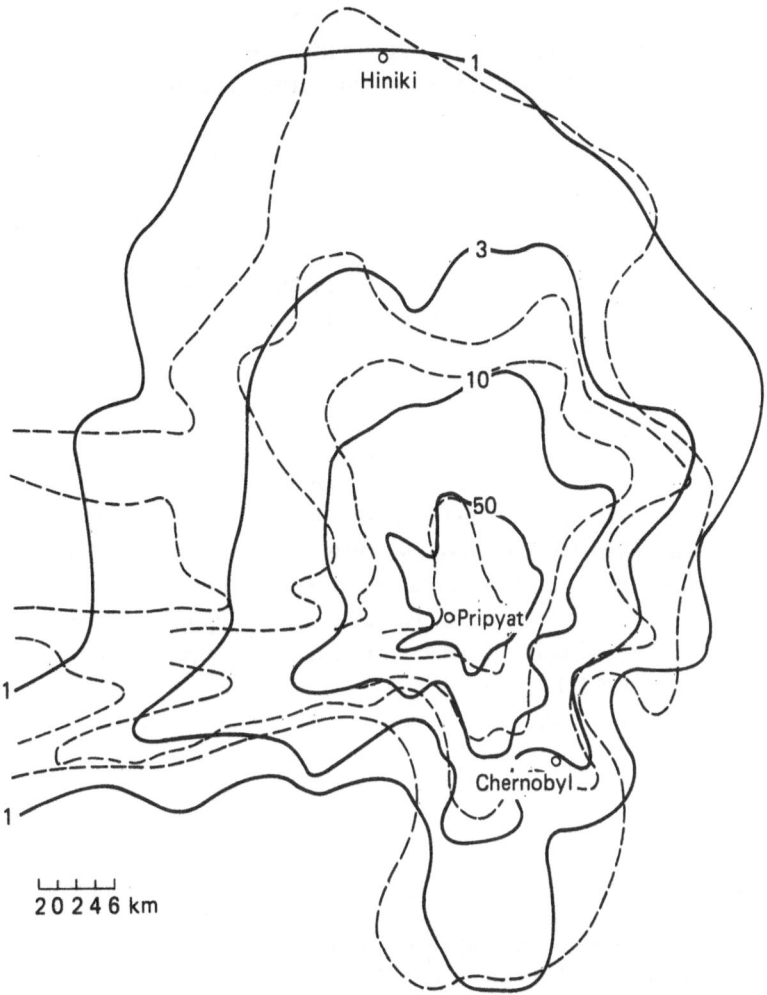

**Fig. 8.6.** Calculated and actual distribution of radiation levels at the ground surface mR/h on May 29, 1986.

1986 and spring 1987 in the rivers Dnieper and Pripyat running through the Belorussian territory. The table shows that all the given concentrations are much lower than maximum permissible concentration.

## Simulation of Radioactive Fallout in the Close-In Zone of the Chernobyl NPP

The model is a mathematical description of radioactive fallout products released into the atmosphere that resulted in close-in radioactive contamination of the terrain (2).

The release of products from the damaged block being raised to the effective height (H) is represented as a continuous lifted point source of nonuniform aerosols. The radioactivity distribution with particle size follows a lognormal law.

The fallout from this source is determined by the kinematics of the particle sedimentation in the time-variable wind field, taking account of their horizontal diffusion. The particle transport trajectories are calculated from the temporal distribution of wind direction and velocity in the layers $H = 0\text{-}1{,}000$ m or $H = 500$ m (Fig. 8.1), depending on the initial source height.

The radioactive release dynamics used in the estimations are presented in Fig. 8.2.

Values $\zeta$ and $\sigma$ were determined by the numerical experiment to obtain the best coincidence with a real curve drawn in Fig. 8.5 from the aerogamma survey data. The dashed curve shows the calculated relationship between S and P with distribution parameters $\zeta = 50$ $\mu$m and $\sigma = 0.25$ (Fig. 8.5).

Figure 8.6 shows maps of radioactivity distribution in the close-in fallout from the Chernobyl NPP on May 29, 1986, calculated by a model with the above parameters of activity distribution on particle sizes and assumed initial data for height, dynamics of radioactive release, and temporal wind-field characteristics. The dashed curve indicates actual radiation level distributions from aerogamma surveys (1).

The initial height of the radioactive plume used in calculations was assumed to be 1,000 m up to 17.00 h Moscow time on April 28, 1986; then it was assumed as 500 m. The wind direction was taken from the histogram (Fig. 8.1) every two hours.

## References

1. Information for IAEA on the accident of Chernobyl APP and its consequences (1986) Atomnaya energia Vol 61, No 5, November 1986
2. Izrael Yu A, Petrov VN, Severov DA (1987) Modelling of the close-in fallout from Chernobyl atomic power plant accident. Meteorolog gidrolog No 7, 1987
3. Izrael Yu A, Petrov VN, *et al*. (1987) Radioactive contamination of natural environmental media over the area of Chernobyl atomic power plant accident. Meteorolog gidrolog No 2, 1987
4. Permissible emission of radioactive and chemical substances to the atmosphere (1985) Moscow, Energoatomizdat

CHAPTER 9

# Radionuclides in the Aquatic Environment

R. J. Pentreath[1]

## Introduction

The aquatic environment is immense. Not only do the oceans cover 70% of the Earth's surface but, with an average depth of about 4 km, they contain some $1.37 \times 10^{21}$ L of water plus dissolved salts. Fresh waters are much smaller, accounting for no more than about 3% of the water on the planet; about $3.6 \times 10^{16}$ L per year enters the oceans as continental runoff (1). The aquatic environment has also received the greatest input of radionuclides from the atmospheric testing of nuclear weapons and continues to receive low levels of radioactive wastes that are discharged, under authorization, from the nuclear industries. As a source of food, however, the aquatic environment is a small contributor, compared with the terrestrial environment, in terms of total food production; it may, nevertheless, provide a large fraction of the diet of some individuals and of certain local populations.

Radionuclides become widely distributed in seas, lakes, and rivers, depending on their source, their chemical behavior, and the nature of the receiving environment. Their accumulation into aquatic food chains, and particularly into pathways leading to humans, can be very complex. Food chains can be long, members of food chains can be highly mobile, and most aquatic fauna derive some fraction of their body burdens by direct accumulation from the medium as well as from ingestion. Simplifying assumptions are thus usually made in order to relate the concentrations of radionuclides in aquatic organisms to that of the ambient water. Such an approach is often taken for predictive or generalized modeling purposes; for dose assessments relating to current practices, however, particularly where the doses received are likely to be substantial fractions of the dose limit, it is more usual to base the assessments on site-specific environmental monitoring and habit-survey data.

It is not generally appreciated that because of naturally occurring radionuclides in aquatic environments, high consumers of seafoods can receive a substantial

[1] Ministry of Agriculture, Fisheries and Food, Directorate of Fisheries Research, Fisheries Laboratory, Lowestoft, Suffolk NR33 OHT, England.

fraction of their background radiation dose from this source. Dose rates received by the public from authorized discharges vary considerably from site to site and can be shown to be highly dependent on the composition of the diet and the radionuclides present. In terms of annual collective dose commitment, however, discharges from nuclear industries are a small contributor compared with that arising from naturally occurring radionuclides in the aquatic environment, and this in turn is small compared with other natural sources in the terrestrial environment.

## Sources and Quantities of Radionuclides in Aquatic Environments

### Naturally Occurring Radionuclides

The oceans of the world consist of a single body of water that contains dissolved salts plus suspended solids. The major dissolved constituents arise from both the input of weathering products from the continental land masses and from biological activity. The major ionic constituents include chloride, Na, sulfate, Mg, Ca, K, bicarbonate, bromide, borate, and Sr. These dissolved solids collectively constitute the "salinity" of the oceans, which varies from 33 to 38 g/kg in areas away from the coast. A very large number of other elements and their compounds also exist at very low concentrations, together with a great variety of dissolved organic compounds. By and large, the oceans are more or less in equilibrium with respect to the input and output of elements via runoff from land, exchange with the atmosphere, and removal via particulate materials that descend through the water column and subsequently deposit onto the sea floor. As to be expected, therefore, they also contain the major naturally occurring radionuclides of primordial and cosmogenic origin.

**Table 9.1.** Naturally occurring radionuclides in sea water: Typical concentrations and approximate inventories

| Radionuclide | Concentration (Bq/L) | Approximate inventory (TBq) |
|---|---|---|
| Primordial | | |
| $^{40}$K | 12 | $1.6 \times 10^{10}$ |
| $^{87}$Rb | 0.110 | $1.5 \times 10^{8}$ |
| $^{234}$U | 0.046 | $6.3 \times 10^{7}$ |
| $^{238}$U | 0.040 | $5.5 \times 10^{7}$ |
| $^{226}$Ra | 0.0035 | $4.8 \times 10^{6}$ |
| $^{210}$Pb | 0.0030 | $4.1 \times 10^{6}$ |
| $^{235}$U | 0.0022 | $3.0 \times 10^{6}$ |
| $^{210}$Po | 0.0020 | $2.7 \times 10^{6}$ |
| Cosmogenic | | |
| $^{14}$C | 0.006[a] | $8.0 \times 10^{6}$ |
| $^{3}$H | 0.1[a] | $8.5 \times 10^{5}$ |

[a] Surface-water values.

The most abundant nuclides in sea water are the ubiquitous $^{40}$K and $^{87}$Rb. Taking the concentrations as given in Table 9.1, the total oceanic inventories of these two nuclides amount to some $1.6 \times 10^{10}$ and $1.5 \times 10^8$ TBq, respectively. Of the other primordial nuclides, because the concentration of dissolved U— at about 3.3 μg/L—is relatively constant throughout the oceans, the U isotopic content is fairly easily calculated, as also indicated in Table 9.1. Estimating the total inventories of the other nuclides, however, is far more complex because their concentrations vary. Thorium is essentially insoluble, and its nuclides are removed from the water column adsorbed to particulate material. Protactinium is also insoluble in sea water, and the isotopes of both Pb and Po also adsorb to particulate materials. Radium, on the other hand, is somewhat more soluble, and Rn readily diffuses from one medium to another. The environmental distributions of the various nuclides in the natural series therefore depend on the chemical behavior of the nuclides, their immediate precursors in the decay chain, their rates of decay (half-lives), and their sources of input. Concentrations in the marine environment can therefore vary considerably between nearshore and offshore, between the sea surface and close to the sea-bed, and between coastal waters and the deep ocean. This, in turn, leads to many states of disequilibrium between various pairs of nuclides, a phenomenon that is widely used to determine the rates of oceanographic processes (1,2). The values given for $^{226}$Ra, $^{210}$Pb, and $^{210}$Po in Table 9.1 are thus only rough approximations, based largely on data from Broecker and Peng (1).

Of the major cosmogenic radionuclides, as to be expected, the greatest fraction ($\sim$ 65%) of $^3$H resides in the oceans; although this is somewhat different from the distribution of water on the planet, the difference being due primarily to its half-life of 12.4 years. Because of radioactive decay, its concentration also decreases with depth. The other major cosmogenic radionuclide in the sea is $^{14}$C, more than 90% of its global inventory being contained in deep oceanic waters (3).

Marine sediments also contain a considerable quantity of naturally occurring radionuclides. Inventories of such nuclides are, of course, dependent upon the depth assumed. It is sufficient to note that such sediments have relatively high concentrations of particular radionuclides and that these are frequently higher in the deep sea than in coastal waters. Red clay deposits, in particular, may contain concentrations of up to 2,000 Bq/kg of $^{230}$Th, 1,500 Bq/kg of $^{226}$Ra, and about 100 Bq/kg of $^{232}$Th (4).

Despite these variations, the seas as a medium are relatively homogeneous with regard to their major chemical constituents. Fresh waters are quite different. The dissolved constitutents of ground waters vary considerably, being dependent upon the rocks and soils through which they flow. The quantity of dissolved carbonates and bicarbonates, which determines water "hardness", considerably affects the quantities of many elements in solution and their rates of removal by co-precipitation. The presence of organic materials also determines the quantities present in the dissolved phase. Added to which, fresh waters can vary greatly in pH and Eh, may be static or flowing, may experience considerable variations in temperature, and are subject to extreme seasonal changes in biological and geochemical activity. Large bodies of fresh water, such as the Great Lakes, are relatively stable and have the characteristics of many small seas; but otherwise it is difficult to generalize.

River waters usually contain less U than sea water, dissolved concentrations typically lying in the range of 0.3 to 0.6 $\mu$g/L (5). Concentrations of Ra and Rn also vary greatly. When river waters reach estuaries, their dissolved Fe and Mn precipitate as oxyhydroxides during mixing with sea water, and dissolved organic matter may also precipitate to some extent. Both of these processes can result in the co-precipitation of other elements. Little removal of U occurs, but Th and Pa are rapidly scavenged. In contrast, the effective concentration of $^{226}$Ra may increase in estuarine conditions as a result of desorption from suspended particulate matter in river water as it crosses the salinity gradient (5).

The nuclides of particular interest are those of Rn—$^{222}$Rn and $^{220}$Rn. Radon concentrations in ground water are related to the Ra concentration of the rock through which it percolates, the rate of percolation, the porosity of the rock, and a number of other factors. The lowest concentrations occur in waters from rocks such as limestone and sandstone; the highest occur in waters percolating or collecting in granites and sedimentary rocks that have experienced considerable metamorphic change. Concentrations of up to $10^3$ Bq/L of $^{222}$Rn have been determined in ground waters in Maine, USA (3). The principal radiological interest in such concentrations are the doses received by ingestion of water and inhalation of the Rn daughters in gas rather than via their accumulation in freshwater foodstuffs.

## Artificial Radionuclides

Artificial radionuclides have entered aquatic environments on a worldwide scale from the detonation of nuclear weapons in the atmosphere plus at least five underwater tests. At local and regional levels, the authorized discharges of liquid wastes from the nuclear industries have considerably increased ambient concentrations in certain areas, and other sources include the disposal of packaged wastes in the deep sea, lost military hardware and, most recently, a power reactor accident (6).

Input from fallout has been by far the major source (Table 9.2), but the nuclides have been widely dispersed. In contrast, radionuclides discharged by the nuclear industries can result in an increase in concentrations within confined water masses, and their rates of discharge are thus controlled by taking such factors into account. Power reactors discharge only small quantities of radionuclides under normal operation, the isotopic composition being dependent upon the type of reactor and the operating and effluent management practices at the site. Approximate rates of release of liquid effluents from power reactors, normalized to annual energy generated, are given in Table 9.3. The dominant nuclides in these effluents are those of Cs, Co, and I, in addition to $^3$H.

The main sources arise from the two ends of the nuclear fuel cycle: the mining and milling of U and the reprocessing of spent fuel. Taking these in

**Table 9.2.** Introduction and inventories (to 1983) of a number of long-lived radionuclides arising from weapons fallout and SNAP-9A, a space satellite[a]

| Radionuclide | Deposited inventory (TBq) | |
| --- | --- | --- |
| | Northern hemisphere | Southern hemisphere |
| $^{3}$H | $6.0 \times 10^{7}$ | $1.5 \times 10^{7}$ |
| $^{90}$Sr | $3.0 \times 10^{5}$ | $9.5 \times 10^{4}$ |
| $^{137}$Cs | $4.7 \times 10^{5}$ | $1.5 \times 10^{5}$ |
| $^{238}$Pu[b] | $3.2 \times 10^{2}$ | $4.8 \times 10^{2}$ |
| $^{239 + 240}$Pu | $1.0 \times 10^{4}$ | $2.6 \times 10^{3}$ |
| $^{241}$Pu | $5.0 \times 10^{4}$ | $1.2 \times 10^{4}$ |
| $^{241}$Am | $2.8 \times 10^{3}$ | $7.0 \times 10^{2}$ |

[a] Reprinted with permission from ref. 9.
[b] Includes Systems for Nuclear Auxilliary Power.

reverse order, the largest commercial reprocessing plants are all situated on the coast and discharge routinely into coastal waters. The two best-known plants are those of Sellafield in the United Kingdom and La Hague in France. The approximate cumulative inventories of a number of longer-lived nuclides discharged from these two sites are given in Table 9.4 (6).

Another source to the marine environment worth mentioning is the disposal of low-level packaged waste into the deep sea. Such wastes have been dumped by a number of countries, at various locations, since 1947 (7,8). Over the last two decades the majority of wastes has been disposed of in waters of about 4-km depth. The best characterized operations are those involving the Nuclear Energy Agency (NEA) dump site in the northeast Atlantic Ocean (9). Approximately $42 \times 10^{3}$ TBq of mixed radionuclides have been dumped in this area, of which a large fraction, $23 \times 10^{3}$ TBq, has been $^{3}$H.

Other less well-characterized sources are those resulting from accidents. One of these, about 1 TBq of $^{239/240}$Pu arising from a B-52 bomber that crashed in 1968 off Thule, Greenland, has been particularly well studied (10,11).

There is one major reprocessing facility that discharges into freshwater—Marcoule, situated on the banks of the River Rhone, in France. Its discharge

**Table 9.3.** Normalized release rates of radionuclides from different types of reactors between 1975 and 1979[a]

| Reactor type | Release rate, TBq/GW(el)a | |
| --- | --- | --- |
| | $^{3}$H | Other nuclides |
| PWR | 38 | 0.18 |
| BWR | 1.4 | 0.29 |
| GCR | 25 | 4.77 |
| HWR | 350 | 0.47 |

PWR = pressurized water reactor, BWR = boiling water reactor, GCR = gas cooled reactor, and HWR = heavy water reactor.
[a] Reprinted with permission from ref. 3.

**Table 9.4.** Approximate introduction and inventories (TBq) of a number of radionuclides arising from Sellafield and La Hague[a]

| Radionuclide | Sellafield (to 1986) | | La Hague (to 1985) | |
|---|---|---|---|---|
| | Total | Cumulative inventory | Total | Cumulative inventory |
| $^3$H | 25,230 | 16,735 | 10,190 | 8,470 |
| $^{106}$Ru | 27,545 | 285 | 4,905 | 800 |
| $^{137}$Cs | 41,080 | 33,480 | 940 | 760 |
| $^{90}$Sr | 6,120 | 4,690 | 755 | 675 |
| $^{239/240}$Pu | 680 | 680 | — | — |
| $^{238}$Pu + $^{239/240}$Pu | — | — | 3 | 3 |
| $^{241}$Am | 535 | 815[b] | — | — |
| $^{125}$Sb | 170[c] | 65[c] | 1,000 | 795 |
| $^{99}$Tc | 305[c] | 305[c] | — | — |

[a] Calculated largely from data reported in Stather et al. (25) and Calmet and Guegueniat (26).
[b] Includes grow-in from $^{241}$Pu.
[c] Since 1978.

rates are very much lower than those of the marine sites (3). Other potential sources that could lead to locally enhanced levels of radionuclides in freshwater environments are the effluents from mining and milling operations. Liquid wastes can arise from mine drainage and process-feed water and from the leaching of tailings from mines and mills. Tailings are usually held in various impoundments from which the water is either allowed to evaporate or is treated with Ba chloride to co-precipitate the Ra. As a result, the dissolved quantities of $^{226}$Ra arising in fresh waters downstream of such plants are usually less than 0.4 Bq/L (3).

## Distribution and Behavior of Radionuclides in Aquatic Environments

The gross distribution of radionuclides in aquatic environments is essentially dependent on the extent to which they may become associated with particulate materials, the quantity of particulate materials present, and the physical nature of the receiving water mass. In very simple terms, the relationship between the quantity of a radionuclide associated with particulate material, relative to that in solution, can be represented as a distribution coefficient ($K_d$). As a dimensionless term this can be expressed as Bq/kg particulate/Bq/kg filtrate, assuming the two quantities to be in equilibrium. Radionuclides that are relatively soluble, and thus have low ($< 10^3$) $K_d$ values, include $^3$H and those of Sr, Tc, Sb, Cs, and U. Radionuclides with relatively high $K_d$ values include those of Pb, Th, the transuranium nuclides (Np, Pu, Am and Cm), and most of the transition elements. The precise values, however, can vary considerably in relation to the valence state of the nuclide in solution. A generalized list of $K_d$ values for coastal water and deep-sea sediments is given in IAEA (12). If it is assumed that such values are relatively constant, then the fraction of the nuclide in solution (filtrate) will be dependent on both its $K_d$ and the quantity of suspended material present.

Mathematical models can employ far more complicated expressions, which include rates of adsorption and desorption rather than equilibrium values. Estimates are made of rates of removal of radionuclides attached to particles to the sediment, the rates of mixing of sediments, and the diffusion of radionuclides in sediments. In the water column, advective and diffusive processes transport radionuclides away from their source of input, and a plethora of general, site-specific, reigonal, and global models are used to calculate their distributions in space and time. For freshwater situations, particularly rivers, the modeling can become very difficult because not only does the water flow downstream in a complex manner but the bed sediments may also flow. And estuaries, again, have many unique features that render attempts to represent them in mathematical terms a rather complicated process. The transport of radionuclides in ground waters can be described to various depths of complexity, from those models that consider transport by convection—and in which velocity is assumed to be both unidirectional and constant—to those that allow for water movement in all directions by using vector analysis techniques.

## The Accumulation of Radionuclides by Aquatic Organisms and Food Pathways Leading to Human Exposure

### Radionuclide Accumulation

Both aquatic plants and animals accumulate the majority of elements to concentrations greater than those of the ambient water, even though there is not always an apparent metabolic requirement for them. To some extent this may be due to a limited ability in regulating the essential elements plus an inability to differentiate specifically between chemically similar elements. The blood of marine animals is more or less on an osmotic par with sea water, but fresh water is considerably lacking in the monovalent elements that dominate in sea water. Many animals do manage to maintain their blood and tissue concentrations of important monovalent ions in conditions of reduced salinity, however, as do freshwater fauna; the tissue concentrations of these elements in such fauna are thus much increased relative to the ambient water.

Many elements may be accumulated onto biological surfaces simply because of the physical and chemical properties of those surfaces. For those that have such an affinity, the passive adsorption of radionuclides may account for a large fraction of the body burden in organisms that have a high surface-to-volume ratio, such as many planktonic species. Benthic algae also tend to accumulate many radionuclides in this way because of the adsorptive properties of their coating of extracellular polysaccharides, which protects them from physical damage and dessication in the intertidal environment.

The absorption of elements involves active or passive transport across membranes. Active transport and direct exchange with water occur to maintain homeostatic concentrations of a number of elements in tissues. The degree of absorption of an element across the gut wall differs considerably from one phylum to another. Many invertebrate species can absorb radionuclides quite well across the intestinal epithelium, even if they are attached to particles, whereas vertebrates may not. Thus uptake from food at any one link in a food chain depends not only on the type of food and the concentration of the nuclide in it, but also the ability of the animal to absorb it.

Nevertheless, one would expect that nuclides that adsorb onto sedimentary materials would also adsorb onto algae and that they would also be found in filter-feeding animals or adsorbed onto the exoskeletons of crustaceans. To some extent this is so, but there are important exceptions. Iodine, for example, is quite soluble in sea water but is highly accumulated by benthic algae, particularly phaeophycean (brown) species. Technetium behaves in a similar manner, although this element has a remarkable ability to be highly accumulated by some species of both plants and animals but not by others in the same phylum (13). Strontium, too, is a fairly soluble element; it is not highly accumulated by most algae, but because of its similarity to Ca, it does occur in calcareous algae, in the hard tissues of invertebrates, and in vertebrate bone. Cesium and K are also similarly accumulated at tissue level in a wide variety of plants and animals.

It is impossible to summarize briefly the various ways, means, and rates at which different types of aquatic organisms accumulate different nuclides. By whatever means—active or passive, direct from water or via the food—it is usually found that under more or less equilibrium conditions, the concentration of an element in the organism, relative to the water, can be represented as a concentration factor (CF). The values are usually related to filtered water—to eliminate variability in the data resulting from different quantities of particulate matter in the water. Concentration factors are widely used in radiological assessment modeling, and a representative set of values for marine seafoods is given in IAEA (12). A much more limited data set is available for freshwater species (14). It has generally been found that because the tissue concentrations of many elements remain fairly constant, the apparent CF value increases as the water concentrations decrease. This also occurs with chemically similar elements, and thus the expected concentrations of $^{137}$Cs and $^{90}$Sr in freshwater fish flesh can, to some extent, be predicted because of their inverse relationship with the respective concentrations of K and Ca in the ambient water (15).

It should again be stressed the CF data only apply to what are assumed to be equilibrium conditions. When conditions are not in equilibrium, dynamic models are used that represent the rates at which radionuclides are taken up and excreted, usually over periods of days to months. Unlike the terrestrial environment, food chains in aquatic environments can be both long and tortuous, to the extent that they are more usually termed food "webs." This in itself makes modeling complicated, although it is sometimes necessary to employ such a method, as in the assessment of possible "short-circuit" pathways from the deep sea to humans (9).

## Food Pathways to Humans

Fish are not the only foodstuffs that are taken from the sea and fresh waters for human consumption; a wide variety of invertebrates—particularly mollusks and crustaceans—are consumed, together with more unusual forms such as echinoderms, amphibians, and even plankton in the form of krill. Many species of algae, too, are harvested either for direct consumption or for conversion into alginates and other compounds that are used in the manufacture of a number of foods. And even the fish may not be eaten directly, but used as fish meal, which is fed to terrestrial livestock. Benthic algae may be used as fertilizers in the production of crops for human consumption. Farm animals may graze in the intertidal zone. In addition, for freshwater pathways, not only is water directly imbibed, but it is used for the irrigation of crops. Radionuclides can also be inhaled from water vapor and from gases that arise from water. And, indeed, it should not be overlooked that exposure also results from external sources, although these pathways are not the subject of this chapter.

The potential pathways back to humans are thus considerably varied. One important feature in assessing dose to humans is that of making allowance for the fact that radionuclide concentrations in foodstuffs differ greatly from one tissue to another and that different portions of marine animals are consumed. People do not usually eat mollusk shells, or fish bone, although some desorption of nuclides from them may occur when the food is cooked. The values given in tabulations of concentrations factors should therefore be representative of that part of an animal that is eaten, or allowance made in some way for the fraction in a typical diet that is eaten whole. Such allowances were made in the IAEA tabulations (12).

The fact that radionuclides differ considerably from one tissue to another, often by many orders of magnitude, also has an effect on the extent to which statements are made with regard to the oft-asked question of whether or not certain radionuclides are "accumulated," "concentrated," or "biomagnified" along a food chain. Many invertebrates, particularly planktonic organisms, worms, the soft tissues of mollusks such as winkles, mussels, and so on, are usually analzyed whole. Thus, although these samples contain a mixture of tissues and organs—the gut and its contents, gonads, muscle tissue, and the equivalents of such organs as liver and kidney—only a single representative value is obtained. In contrast, other fauna, such as crustaceans and fish, are often examined selectively, with only the edible portions—muscle and gonad in fish, for example—being analyzed. Because of this variability in the database, quite misleading conclusions may be drawn. In general, the tissue with the lowest concentrations of many of the divalent and multivalent cations, apart from blood plasma, is the muscle. The tissues with the highest concentrations are usually the liver, kidney, or other organs serving similar functions of storage or excretion. In comparing concentrations along a food chain, therefore, the conclusions depend on what values are used. Comparisons are usually made between a selected organ or tissue of one animal, with the whole-body concentration plus gut contents of its food. If the predator's liver were analyzed, the concentration might well be higher than that of the prey; if the predator's muscle tissue were analyzed, the concentration might be lower than that of the prey. When predator and prey are compared on a whole-body basis, however, the general tendency is for a decrease to occur in most elements at the higher trophic levels because muscle tissue not only has relatively low concentrations but accounts for a larger fraction ($>$ 50% in fish) of the body weight of top predators. There are some notable exceptions: Hg is one of them and Cs another. These are both concentrated in muscle tissues to levels not greatly different from those of other organs—although, in the case of Hg, not in the same chemical form—and both are excreted slowly relative to their rates of intake. Of course, this does not negate the fact that high concentrations of chemicals, relative to food, can occur in specific organs and adversely affect the animal. This is most unlikely to occur as a result of the routine discharges of radionuclides into the environment, the discharges being very carefully controlled and monitored, as discussed in the next section.

## Doses to Man From Radionuclides in Aquatic Foodstuffs

The discharges of radionuclides to aquatic environments are controlled at national level through a system of authorization, inspection, monitoring, and assessment to ensure that the level of radiation protection of the public is optimized and that the dose limits, as recommended by the International Commission on Radiological Protection (ICRP), are not exceeded. In addition, there are international regulations and agreements that limit the quantities of radioactive wastes that can be dumped at sea (16,17).

In assessing dose to humans via the consumption of foodstuffs, it is therefore necessary to consider the dose to individuals and the collective dose arising from a practice. The former is to ensure that the dose rates received are within the nationally set limits and the latter in relation to optimization requirements. But it is also instructive to compare doses to man from discharged radionuclides with those arising from naturally occurring radionuclides in the marine environment because both contribute to the total dose received.

### Doses to Individuals

In order to assess doses to individuals it is usually necessary to have basic information on the annual quantities of seafoods consumed and on the concentrations of nuclides in them. Assessments may be made on the basis of predictive models or from data gained from surveys of the food consumed by groups of

**Table 9.5.** Dose per unit intake values and assumed concentrations of the principal naturally occurring radionuclides in marine foodstuffs

| Radionuclide | Dose per unit intake (Sv/Bq) | Concentration (Bq/kg wet) | | | |
|---|---|---|---|---|---|
| | | Fish[a] | Crustaceans[b] | Mollusks | Algae |
| $^{14}C$ | $5.6 \times 10^{-10}$ | 15 | 15 | 15 | 15 |
| $^{40}K$ | $5.1 \times 10^{-9}$ | 100 | 100 | 100 | 500[e] |
| $^{87}Rb$ | $1.3 \times 10^{-9}$ | 1 | 1 | 1 | 1 |
| $^{210}Po$ | $4.3 \times 10^{-7}$ | 1.5 | 25 | 50[c] | 4[f] |
| $^{210}Pb$ | $1.4 \times 10^{-6}$ | 0.04 | 0.2 | 3[c] | 0.2[b] |
| $^{226}Ra$ | $3.0 \times 10^{-7}$ | 0.1 | 0.02 | 0.3[d] | 0.02[b] |
| $^{234}U$ | $7.0 \times 10^{-8}$ | 0.012 | 0.12 | 0.3[d] | 2.5[g] |
| $^{238}U$ | $6.3 \times 10^{-8}$ | 0.011 | 0.11 | 0.27[d] | 2.2[g] |

[a] Based largely on data in Pentreath (27) and Pentreath et al. (28).
[b] Values adjusted, relative to fish, on the basis of CF data in IAEA (1985), except for $^{210}Po$ (23).
[c] Based on data in Bangera and Patel (23, 29, 30).
[d] Based on data in Bangera and Patel (23).
[e] Based on data in Thompson et al. (31).
[f] Based on data in Hodge et al. (32).
[g] Based on data in Holm and Persson (33).

people and the direct radiochemical analysis of the foodstuffs. Food obtained from the aquatic environment forms the major fraction of the diet of relatively small numbers of people, although they may be the majority of inhabitants of some communities. A maximizing daily diet of seafoods used in dose assessment models for the sea disposal of packaged wastes consists of the following: fish, 600 g; crustaceans, mollusks and seaweed, 100 g each; plankton, 3 g; and, for good measure, fish caught at depths > 4000 m, 60 g. Assuming a diet consisting of surface-caught fish only, plus crustaceans, mollusks, and seaweed (all consumed fresh), together with the values for dose-per-unit intake and concentrations given in Table 9.5, this rate of intake would result in an annual dose of about 2 mSv per year from naturally occurring radionuclides. The overwhelming contributor to the dose rate is $^{210}Po$ (Fig. 9.1a) from the molluscan components of the diet (Fig. 9.1b). Only a small fraction arises from $^{40}K$, and it should be noted that this dose would be received in any case because K is under homeostatic control within the human body, and the $^{40}K/K$ quotient is the same in all materials. It is not, therefore, an "extra" dose to the individual. It should also be noted that the $^{210}Po$ and $^{226}Ra$ concentrations given in Table 9.5 could probably vary by about an order of magnitude either way from the values used (18). In frozen and stored foods, the $^{210}Po$ would also decay to lower concentrations.

As far as freshwater pathways are concerned, there is little on which to base comparative estimates, but they are probably of similar magnitude. The principal interest in radionuclides in fresh waters has centered largely around concentrations of Rn, particularly in relation to "balneology," the use of spas and spa waters in the alleviation of illness. Resultant effective dose equivalents

Dose to critical groups—marine pathways
(a) Contribution by nuclide

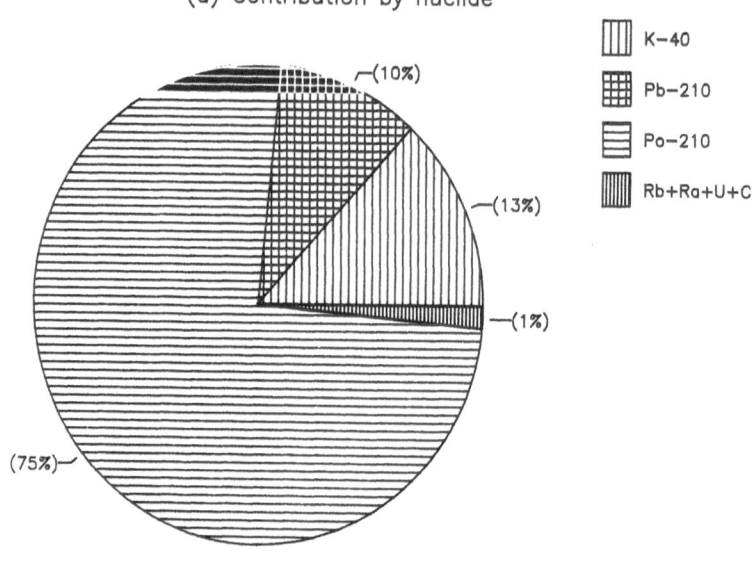

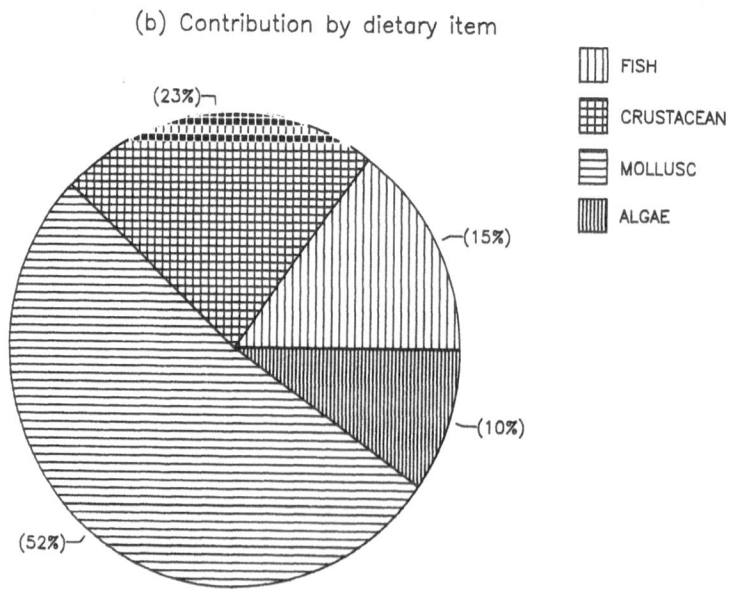

**Fig. 9.1.** Estimated relative contributions of (a) different naturally occurring radionuclides and (b) different dietary items to the annual dose rate to critical groups consuming 600 g of fish and 100 g of crustaceans, mollusks, and algae per day.

in the range of 8 to 300 mSv are believed to be received from such sources (3).

Assessments of dose to individuals resulting from authorized discharges from nuclear industries are usually based on direct monitoring data of seafoods and local consumption rates. The best documented assessments are those relating to the discharges from the reprocessing plant at Sellafield in the United Kingdom, and these are worth examining, briefly, as an example of the interplay between discharge rates, consumption rates, dietary changes, and doses received. The means by which doses are received by members of the public as a result of the Sellafield discharges have been under constant review since the plant first began (19,20). Over the last few years the rates of discharge have fallen greatly as new plant has been brought into operation; discharges of some of the dominant radionuclides are shown in Figs. 9.2a, b. Regular habit surveys of the local population by the Ministry of Agriculture, Fisheries and Food [MAFF (UK)], one of the regulatory authorities, have shown that there are two groups of high seafood eaters in the area. One of these, consumers associated with commercial fishing operations, is currently assumed to eat 225 g of fish, 50 g of crustaceans, and 40 g of mollusks a day (21). But a more important group (in a local fishing community) is one that, although only eating 100 g of fish, 18 g of crustaceans, and 18 g of mollusks per day, receives a higher dose rate because the food consumed is obtained closer to the point of discharge. The total seafood consumption rates of both groups have remained fairly constant in recent years (Fig. 9.3a), and the concentrations of radionuclides in foodstuffs, as determined from monitoring programs, have decreased more or less in parallel with the decline in discharges (22). But the estimated dose to the local fishing community has varied considerably (Fig. 9.3b), although it has always been below the ICRP-recommended limits. Fish consumption is the principal pathway for [134]Cs and [137]Cs, and, as can be seen, the contribution from these nuclides has fallen steadily. In contrast, although the consumption of molluscan shellfish— primarily winkles—has increased slightly and then decreased, its effect upon the dose received has been considerable. This is because shellfish consumption is currently the principal pathway for the transfer of [106]Ru and the transuranium nuclides back to man.

The dose rates received by the public in general are far less than these values. A typical fish-eating consumer, at about 40 g per day, would receive only 0.02 mSv per year from fish landed at the fishing ports close to Sellafield. And in the United Kingdom as a whole, dose rates received by local critical groups as a result of discharges to the sea from nuclear power stations are of the order of a few μSv per year (21).

It is worth noting that the group consisting of the local fishing community at Sellafield and that associated with commercial fishing operations could be receiving up to 0.2 and 0.5 mSv per year, respectively, from [210]Po alone in these foods (23), and the typical 40 g per day fish eater would receive a total of about 0.02 mSv per year from naturally occurring nuclides. It is also worth

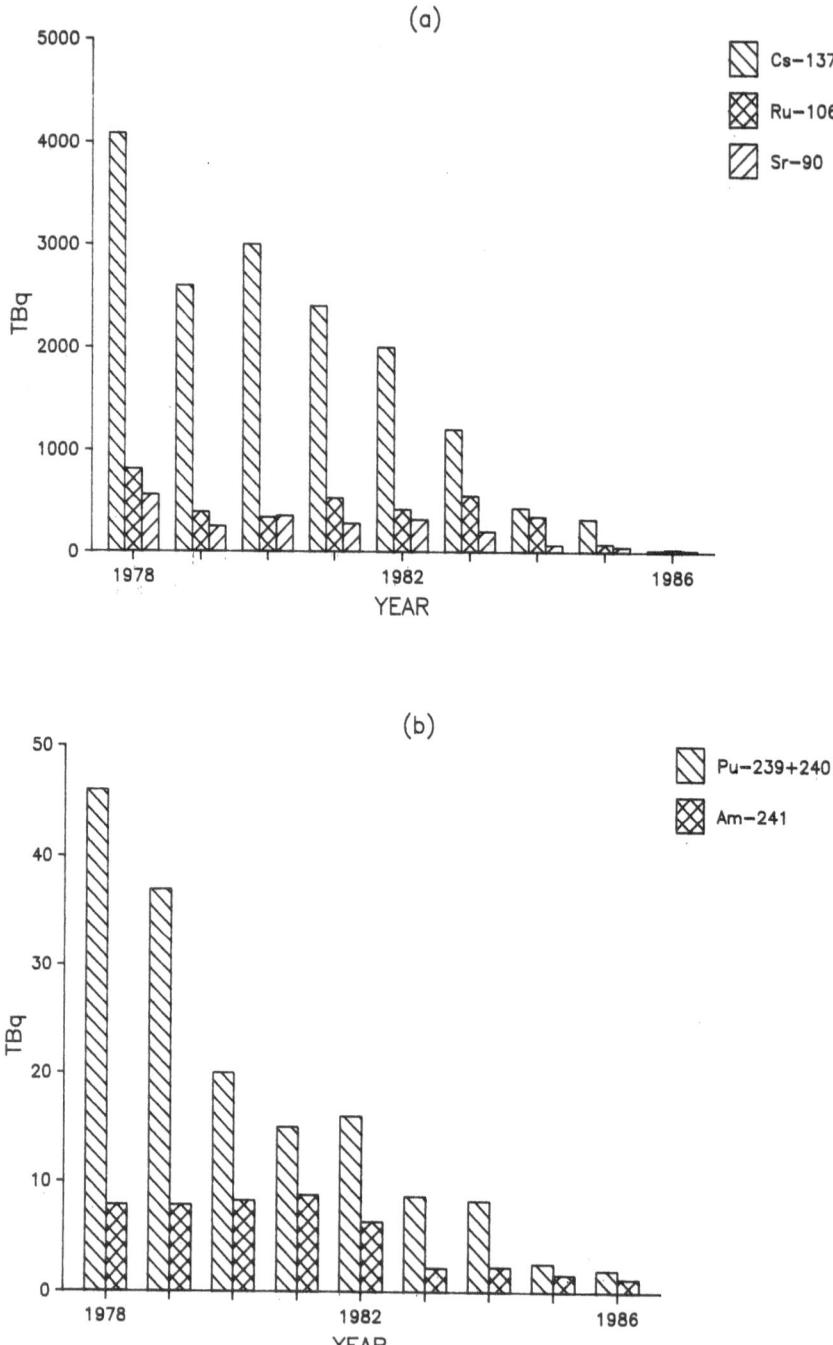

**Fig. 9.2.** Recent discharges to sea from the British Nuclear Fuels Limited plant at Sellafield (UK) of (a) some fission product and (b) some transuranium nuclides.

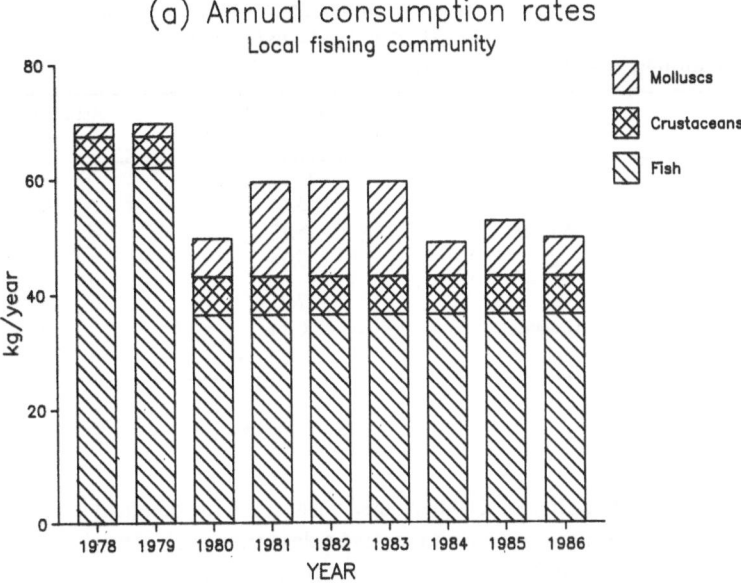

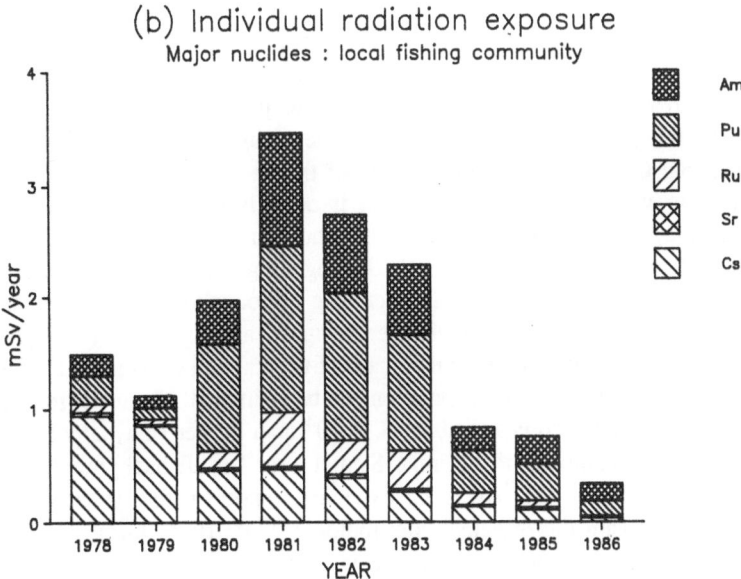

**Fig. 9.3.** (a) Consumption rates of and (b) dose rates to consumers of seafoods in the local fishing community near Sellafield who have constituted the critical group. (A gut transfer factor of 0.0005 was used for Pu.)

**Table 9.6.** Quantities of foodstuffs landed in marine and freshwater fishing areas of the world—mean values for the period 1978 to 1984—(34) together with assumed fraction that is ultimately eaten by man (9)

| Area | Foodstuff | Quantity landed (Mt/y) | Assumed fraction eaten |
|------|-----------|------------------------|------------------------|
| Marine | Fish | 58.27 | 0.5 |
| | Crustaceans | 3.04 | 0.35 |
| | Mollusks | 5.13 | 0.15 |
| | Algae | 3.24 | 0.1 |
| Freshwater | Fish | 7.77 | 0.5 |
| | Crustaceans | 0.14 | 0.35 |
| | Mollusks | 0.27 | 0.15 |

noting that for the sea disposal of packaged wastes, the hypothetical avid seafood eaters, who were estimated above to receive a dose from such naturally occurring nuclides of about 2 mSv per year, would be unlikely ever to receive more than a maximum of some $10^{-4}$ mSv per year from the radionuclides dumped at sea to date (9).

## Collective Dose Rates

Collective dose-rate estimates are based on radiometric analyses of seafoods and data derived from annual catch-rate statistics for appropriate areas of sea water or fresh water. Such catch-rate data are published for the world as a whole by the FAO and at a more regional level by such organizations as the International Council for the Exploration of the Sea (ICES). It is necessary to adjust these values for the fraction that is actually consumed; for mollusks, for example, the data refer to tonnes of whole animals landed, and most of this weight consists of shells, not flesh. Taking the global picture first, assuming that the concentrations of naturally occurring radionuclides in marine samples (Table 9.5) apply equally to freshwater samples, and using the landings data given in Table 9.6, together with estimates of the fraction consumed, the annual collective dose to the world's population can be estimated. The total, for marine freshwater pathways, comes to about $8 \times 10^4$ man Sv per year, approximately 1% of the total estimate of $10^7$ man Sv per year for all background sources made by UNSCEAR (3). As to be expected, marine pathways predominate (> 90%) and again the principal contributor is $^{210}$Po (Fig. 9.4a), but fish are the most important single dietary item (Fig. 9.4b).

In comparison, the annual collective dose arising from the Sellafield discharges—mainly Cs nuclides via fish consumption—is at present (21) about 140 man Sv per year (Fig. 9.5). The maximum that is likely to arise from sea dumping of packaged wastes is about 3 man Sv per year.

# Collective dose rate—aquatic pathways

## (a) Contribution by nuclide

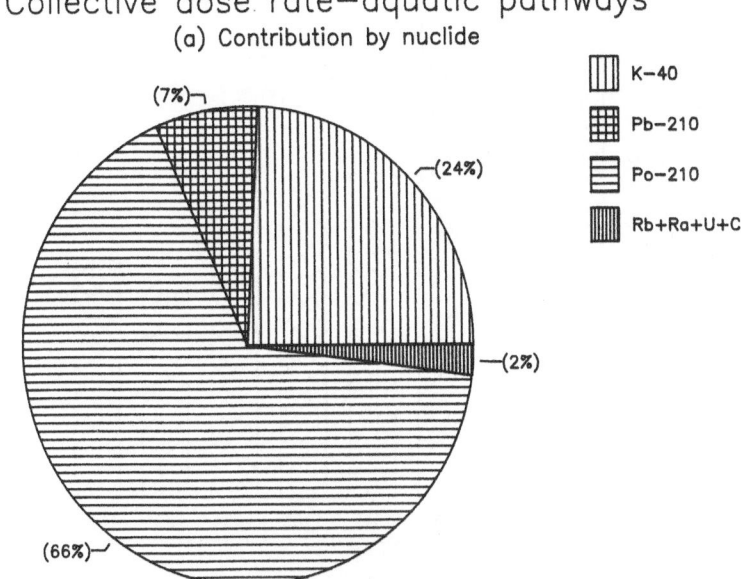

## (b) Contribution by dietary item

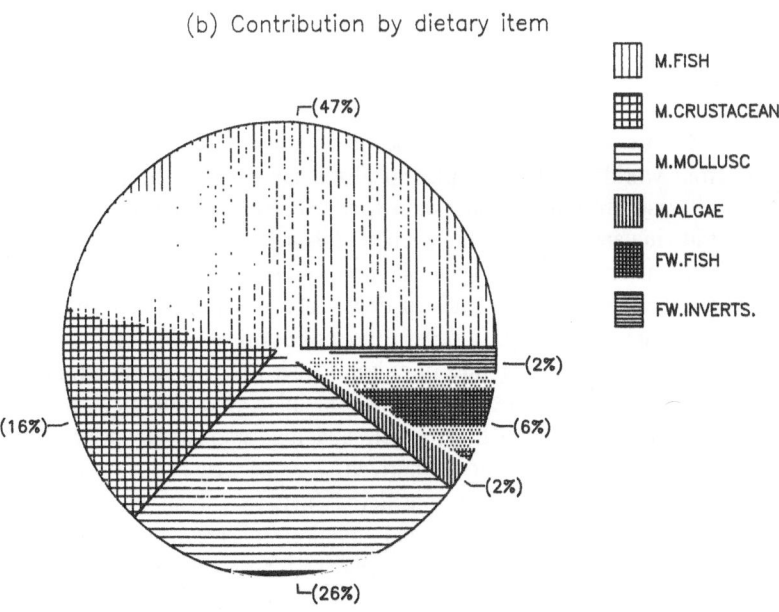

**Fig. 9.4.** Estimated relative contributions of (a) different naturally occurring radionuclides and (b) different dietary items to the annual collective dose from all aquatic food pathways (M = marine; FW = freshwater; inverts = invertebrates).

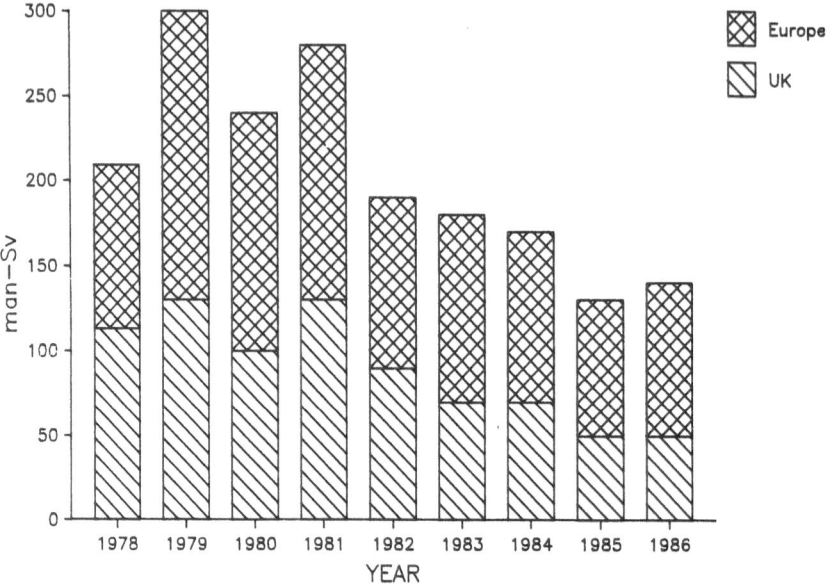

**Fig. 9.5.** Annual collective dose arising from the BNFL Sellafield (UK) discharges in recent years.

## Discussion

Aquatic environments contain a considerable quantity of naturally occurring radionuclides to which has been added a number of artificial radionuclides arising from worldwide fallout, plus localized inputs from the authorized discharges of waste from the nuclear industries. Because the quantities of seafood eaten worldwide are not large, the average dose rate resulting from any source, via seafood consumption, is very small. For high seafood consumers, however, there is a potential for higher rates of dose; such consumers may therefore be the subject of special surveys in order to identify critical pathways, and critical groups, relating to localized inputs of radionuclides (19,24). Regular monitoring and assessment programs ensure that the principles of radiological protection, as recommended by the ICRP, are met, and it is upon these principles that discharges to the environment are authorized. Any increase in dose to individuals arising from such discharges is considered in terms of an increase in risk, irrespective of the dose from background radiation. It is estimated by UNSCEAR (3) that the annual effective dose equivalent, for persons living in areas of "normal" background, is about 2 mSv per year. About one third of this is due to external sources. Of the internal sources, inhalation of $^{222}$Rn accounts for about 0.8 mSv per year. Thus it is likely that high seafood eaters could receive a higher-than-average background dose rate. This need not be of any concern to the public because the dose rates are very low; but it does highlight the difficulties inherent in attempting to relate epidemiological data to local consumers of seafoods in the vicinities of nuclear establishments. And in terms of collective dose rates, that arising from naturally occurring radionuclides is so large, and variable, that estimates from the input of radioactive wastes are probably no more than a fraction of the standard error associated with it. Further work on this subject would thus be of value.

# References

1. Broecker WS, Peng T-H (1982) Tracers in the sea. Eldigio Press, Columbia University, New York
2. Ivanovich M, Harmon RS (eds) (1982) Uranium series disequilibrium: applications to environmental problems. Oxford University Press, Oxford
3. UNSCEAR (United Nations Scientific Committee on the Effects of Atomic Radiation) (1982) Report to the General Assembly, with annexes. United Nations, New York
4. Woodhead DS, Pentreath RJ (1983) A provisional assessment of radiation regimes in deep ocean environments. In: Park PK, Kester DR, Duedall IW, Ketchum BH (eds) Wastes in the ocean, Volume 3: radioactive wastes and the ocean. John Wiley & Sons, New York, p 133
5. Scott MR (1982) The chemistry of U- and Th-series nuclides in rivers. In: Ivanovich M, Harmon RS (eds) Uranium series disequilibrium: applications to environmental problems. Oxford University Press, Oxford, p 181
6. Pentreath RJ (1988) Sources of artificial radionuclides in the marine environment. In: Radioactivity and oceanography. Elsevier London, in press
7. Hagen AA (1983) History of low-level radioactive waste disposal in the sea. In: Park PK, Kester DR, Duedall IW, Ketchum BH (eds) Wastes in the Ocean, Volume 3: radioactive wastes and the ocean. John Wiley & Sons, New York, p 47
8. Preston A (1983) Deep sea disposal of radioactive wastes. In: Park PK, Kester DR, Duedall IW, Ketchum BH (eds) Wastes in the ocean, Volume 3: radioactive wastes and the ocean. John Wiley & Sons, New York, p 107
9. NEA (Nuclear Energy Agency) (1985) Review of the continued suitability of the dumping site for radioactive waste in the North-East Atlantic. OECD, Paris
10. Aarkrog A (1977) Environmental behaviour of plutonium accidentally released at Thule, Greenland. Health Phys 32:271–284
11. Aarkrog A, Dahlgaard H, Nilsson K (1984) Further studies of plutonium and americium at Thule, Greenland. Health Phys 46:29–44
12. IAEA (International Atomic Energy Agency) (1985) Sediment $K_d$s and concentration factors for radionuclides in the marine environment. Tech Rep Series No 247, IAEA, Vienna
13. Beasley TM, Lorz HV (1986) A review of the biological and geochemical behavior of technetium in the marine environment. In: Desmet G, Myttenaere C (eds) Technetium in the environment. Elsevier, London, p 197
14. IAEA (International Atomic Energy Agency) (1982) Generic models and parameters for assessing the environmental transfer of radionuclides from routine releases: exposures of critical groups. Safety Series No 57, IAEA, Vienna
15. Preston A, Jefferies DF, Dutton JWR (1967) The concentrations of caesium-137 and strontium-90 in the flesh of brown trout taken from rivers and lakes in the

British Isles between 1961 and 1966: the variables determining the concentrations and their use in radiological assessments. Water Res 1:475–496

16. IAEA (International Atomic Energy Agency) (1986) Definition and recommendations for the convention on the prevention of marine pollution by dumping of wastes and other matter, 1972. Safety Series No 78, IAEA, Vienna

17. IMCO (Inter-Governmental Maritime Consultative Organization) (1976) Inter-governmental conference on the convention on the dumping of wastes at sea (London, 30 Oct–13 Nov 1971). Final act of the conference. IMCO Publications No 76.14E

18. Woodhead DS (1982) The natural radiation environment of marine organisms and aspects of the human food chain. J Soc Radiol Protect 2(4):18–25

19. Leonard DRP, Hunt GJ (1985) A study of fish and shellfish consumers near Sellafield: assessment of the critical groups including consideration of children. J Soc Radiol Protect 5(3):129–139

20. Pentreath RJ (1985) Radioactive discharges from Sellafield (UK). In: Behaviour of radionuclides released into coastal waters. IAEA-TECDOC-329, IAEA, Vienna, p 67

21. Hunt GJ (1987) Radioactivity in surface and coastal waters of the British Isles, 1986. Aquat Environ Monit Rep, No 18, MAFF Direct Fish Res, Lowestoft, UK

22. Hunt GJ (1986) Time-dependent estimates of dose to the critical group of fish and shellfish consumers near Sellafield. J Soc Radiol Protect 6(3):125–130

23. Pentreath, RJ, Allington, DJ (1988) Dose to man from the consumption of marine seafoods: a comparison of the naturally-occurring $^{210}$Po with artificially-produced radionuclides. Proc Int Rad Protect Assoc 7, Sydney, Vol 3. Pergamon, pp 1582–1585

24. Sumiya M, Ohmono Y (1980) Dietary survey around nuclear site in the Tokai area of Japan and their radiological significance to the relevant population. In: Marine radioecology, Nuclear Energy Agency, OECD, Paris, p 349

25. Stather JW, Dionian J, Brown J, Fell TP, Muirhead CR (1986) The risks of leukaemia and other cancers in Seascale from radiation exposure. National Radiological Protection Board, Chilton, NRPB-R171 Addendum

26. Calmet D, Guegueniat P (1985) Les rejets d'effluents liquides radioactifs du centre de traitement des combustibles irradies de la Hague (France) et l'evolution radiologique du domaine marin In: Behaviour of radionuclides released into coastal waters. IAEA-TECDOC-329, IAEA, Vienna, p 111

27. Pentreath RJ (1977) Radionuclides in marine fish. Oceanogr Mar Biol Ann Rev 15:365–460

28. Pentreath RJ, Lovett MB, Harvey BR, Ibbett RD (1979) Alpha-emitting nuclides in commercial fish species caught in the vicinity of Windscale, United Kingdom, and their radiological significance to man. In: Biological implications of radionuclides released from nuclear industries, Vol 11. (STI-PUB-522), IAEA, Vienna, p 227

29. Bangera VS, Patel B (1984) Natural radionuclides in sediment and in arcid clam (*Anadara granosa* L.) and gobiid mudskipper (*Boleophthalmus boddaerti* Cuv. and Va.). Ind J Mar Sci 13:5–9

30. McDonald P, Fowler SW, Heyraud M, Baxter MS (1986) Polonium-210 in mussels and its implications for environmental alpha-autoradiography. J. Environ Radioact 3:293–303

31. Thompson N, Cross JE, Miller RM, Day JP (1982) Alpha and gamma radioactivity in *Fucus vesiculosus* from the Irish Sea. Environ Pollut B 3:11–19

32. Hodge VF, Hoffman FL, Folsom TR (1974) Rapid accumulation of plutonium and polonium on giant brown algae. Health Phys 27:29–35
33. Holm E, Persson BRR (1980) Behaviour of natural (Th, U) and artificial (Pu, Am) actinides in coastal waters. In: Marine radioecology. Nuclear Energy Agency, OECD, Paris, p 237
34. FAO (Food and Agriculture Organization of the United Nations) (1968) Yearbook of fishery statistics, Vol 58, FAO, Rome

# CHAPTER 10

# Soil-Borne Radionuclides

Y. Yamamoto[1]

## Introduction

The transfer of man-made radioactive materials from soil to man through the food chain by the intake of vegetables, fruits, meats, and milk continues for a long time. In a UN Scientific Committee on the Effects of Atomic Radiation (UNSCEAR) report (1), the transfer coefficient from fallout radionuclide deposition to diet was precisely described. However, an evaluation of pathways of soil-borne radionuclides that are related to the food chain is not simple. The transfer rate of radionuclides in soil depends on many factors, such as types of soil, kinds of crops, cultivation conditions, air temperature, humidity, and so on. Therefore, data obtained in one place sometimes cannot be applied to other places.

Pathways are mutually tangled and, moreover, the behavior of radionuclides in soil and transfer mechanisms to crops are very complicated. Reflecting these reasons, the current knowledge is still limited.

Defining soil is not easy because of its wide variety, ranging from soil whose components are almost sand to soil composed mostly from organic materials.

Radionuclides existing in soil (solid phase) in adsorbed form have a significant effect on their transfer to crops. The concentration of radionuclides in soil solution is related to the following parameters:

1. Particle size distribution of soil and kinds of content of clay minerals and silt. The function of clay minerals is somewhat similar to organic materials.
2. Content of organic materials. Many organic materials show strong adsorptivity to metal elements. Thus, they hold these elements in them and result in the increase of uptake by crops, but at the same time transforming these elements into barely adsorbable forms.
3. The pH value. The solubility of heavy metal elements increases with the decrease in pH values, and such elements are thus more easily absorbed by roots.

[1] The Institute of Applied Energy, Shinbashi SY Bldg. 14-2 Nishishinbashi 1-chome, Minato-ku, Tokyo, 105, Japan.

4. Redox potential.
5. Voidage, aggregate conditions, density, and soil moisture.
6. Cation ion-exchange capacity. The form and quantity of oxide precipitates of Fe, Al, and Mn, which act as adsorbers of anions and have functions similar to those of organic materials and clay minerals, play important roles.
7. Coexisting elements, especially metal elements.

Possible pathways of soil-borne radionuclides in food chains are

1. intake of crops;
2. intake of meat, milk, and poultry;
3. drinking contaminated well water, river water, etc.; and
4. intake of fishes, shellfishes, and seaweeds that are taken from hydrosphere contaminated by soil-borne radionuclides.

Radionuclides that were of great concern and studied in the past were radioiodine (mainly short-lived radioiodine), radiocesium, and Sr. The behavior of transuranic (TRU) nuclides in soil also was studied. Doses from other nuclides in normal discharge from nuclear facilities are evaluated and monitored. In addition to these, we now must consider the behavior of many other nuclides in soil, as it relates to nuclear accidents.

# Radionuclides in Soil

## Sources of Man-Made Radionuclides in Soil

Radionuclides enter soil from the following sources:

1. The normal operations of nuclear facilities, such as nuclear power stations, reprocessing plants, and so on. Nuclides that are considered when regarding the atmospheric discharge from nuclear facilities include, other than inert gases, volatile $^{129}I$, $^{131}I$, $^{133}I$, and particulate $^{51}Cr$, $^{54}Mn$, $^{59}Fe$, $^{58}Co$, $^{60}Co$, $^{89}Sr$, $^{90}Sr$, $^{134}Cs$, $^{137}Cs$, and $\alpha$-emitting nuclides (tritium is excluded).
2. Accidental release from nuclear facilities. The largest accidental release experienced in the past was that from the Chernobyl accident in April 1986. Almost all the kinds of radionuclides in nuclear fuel were detected over a very wide area. Other than rare gases, detection of I and Cs, Nb, Mo, Tc, Ru, Ag, Sb, Te, Ba, Ce, Sr, and TRU elements were reported in Europe.
3. Atmospheric atomic bomb tests. Long-lived radionuclides that resulted from the atmospheric atomic bomb tests in the 1950s and 1960s still remain in soil.
4. Land disposal of radioactive wastes. Several kinds of radionuclides arising from the peaceful uses of nuclear energy are finally disposed of as wastes either near ground surface or in deep geologic formations. In these cases, however, environmental effects are thoroughly investigated and checked before site selection and disposal.

5. Other sources. $^{14}C$ in pesticides is one example. The effects of tritium and $^{14}C$ are not treated in certain dose-estimation models.

Deposition to the ground surface due to 1, 2, and 3 depends on meteorological conditions. Rainfall plays an especially important role. This fact is widely known from the radiation survey after the atomic bomb explosions in Hiroshima and Nagasaki (Japan) and also from the Chernobyl accident. Release mode of type 1 is expressed as a time-dependent quantity and of types 2 and 3 as a discrete quantity.

## Behavior of Radionuclides in Soil

Rainwater plays an important role in the migration of radionuclides in soil. Nuclides, deposited on the soil surface from the atmosphere by dry/wet deposition, resuspension or irrigation, migrate downward with rain water or irrigation water through soil (vadose zone), repeating the adsorption/absorption and desorption processes. The migration process depends on many factors, such as the kind of nuclide and the physical and chemical conditions of soil. Thus, the process is a very complex combination of physical, chemical, and biochemical processes.

Upland soil is usually well ploughed and mixed. As a result, the radionuclide concentration at rhizosphere depth can actually be assumed to be uniform. However, in the case of undisturbed soil such as pasture, concentration varies with depth and, in many cases, radionuclides migrate along the roots. The simplest case is the percolation through the sandy layer. Figure 10.1 shows the distribution profile of the migration of some radionuclides through a sand-filled column (2). This is a very simple experiment in which physical adsorption is the only factor retarding migration. However, the profile shows a rough trend of migration behavior of each nuclide in soil. Radionuclides migrate through the vadose zone before reaching the saturated zone and are then carried, usually horizontally, by ground water to the hydrosphere.

## Soil Models Used for the Food Chain Calculation

Radionuclides migrated downward in soil have a respective concentration profile along the depth, and their concentration at a point in consideration changes with time owing to various causes such as runoff and radioactive decay.

Almost all compartment models currently used assume a uniform concentration in the rhizosphere. This is the case for upland soil, which is well ploughed and well mixed. To the contrary, in the case of pasture, the concentration of radionuclides cannot be assumed to be uniform. Simmonds et al. (3) showed a compartment model for such a case.

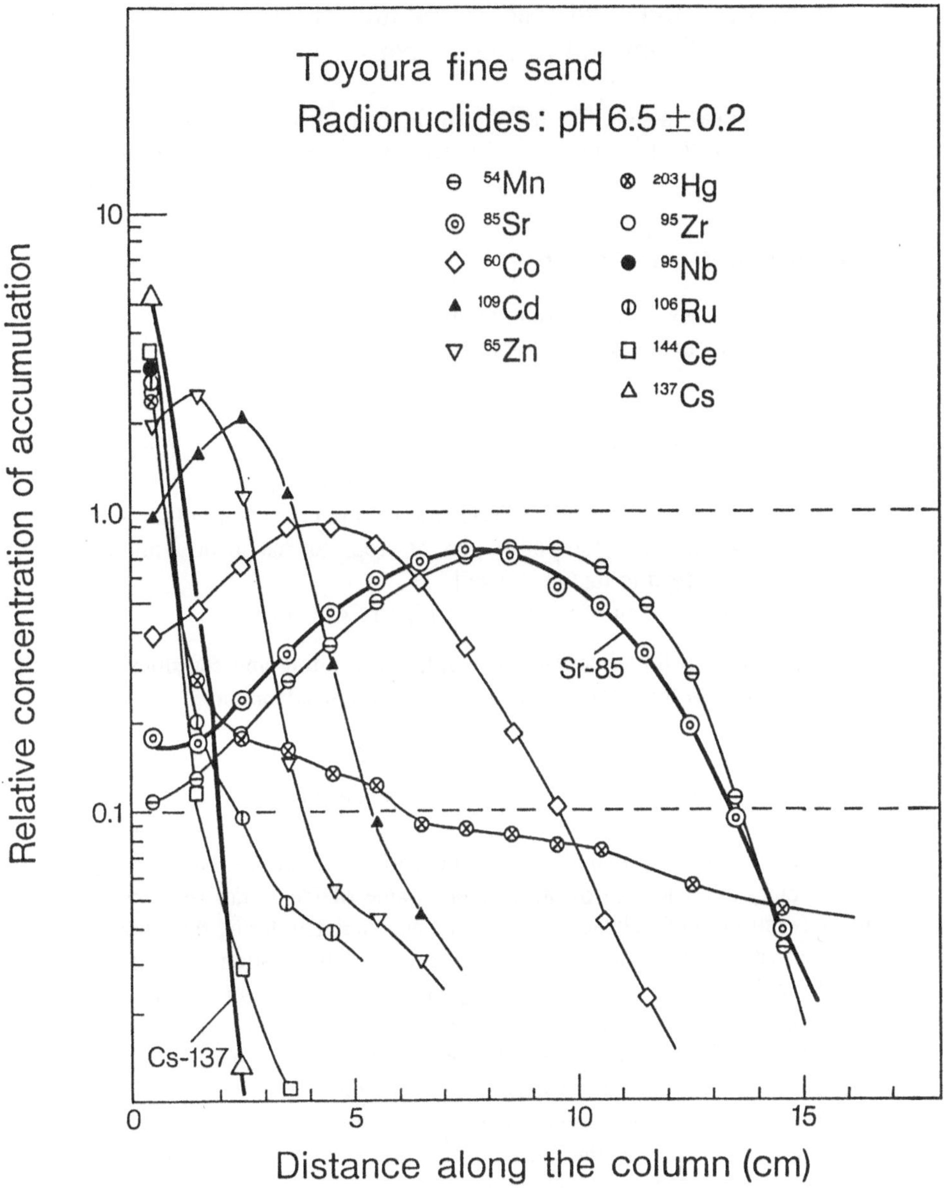

**Fig. 10.1.** Concentration distribution of some radionuclides in a sand-filled column experiment. (Reprinted with permission from ref. 2.)

# Mathematical Models for the Estimation of Transfer of Radionuclides From Soil to Field Crops

Most mathematical models for the uptake of radionuclides to date are based on the distribution coefficient concept. The most widely used is the concentration factor method.

## Concentration Factor Model

This method assumes an equilibrium relationship between soil and field crops/pasture grass. The expression of this model is given by

$$C_{v,i} = (CF)_{v,i} \cdot C_{s,i} \tag{10.1}$$

$C_{v,i}$ = the concentration of nuclide $i$ in the edible part of crops (wet basis) or in the livestock feed (wet or dry bases) (Bq/g).

$(CF)_{v,i}$ = the concentration factor of a nuclide $i$. This is the concentration ratio of a nuclide $i$ in the edible part of crops to that in dry soil, or in livestock feed to that in dry soil.

$C_{s,i}$ = the concentration of radionuclide $i$ in dry soil (Bq/g).

When the hold-up time interval between harvest/cutting and consumption of food (or dry feed to livestock), $t_h$ (d) is taken into consideration.

$$C_{v,i} = (CF)_{v,i} \cdot C_{s,i} \exp(-\lambda_i t_h) \tag{10.2}$$

where $\lambda_i$ is the radioactive decay constant of nuclide $i(d^{-1})$

In this model the definition of concentration factor (CF) is clear and simple. For this reason, this expression is convenient for practical use. But in spite of its simplicity, the selection of an adequate value of CF is not so easy. It is defined as the ratio of radioactivity in crops accumulated during their life span to radioactivity in soil. Studies on the concentration distribution of some metal elements existing at a very low concentration in soil show log-normal frequency distribution, and their normal range often is two orders of magnitude (4). Observed data of the concentration factors are obtained under various experimental conditions. Soils used for one observation differ from other experiments. Likewise the cultivation period and methods are diverse. Chemical forms of radionuclides have significant effects on the concentration factor. In the case of $^{131}I^-$ and $^{131}IO_3^-$, for instance, their accumulation in leaves of komatsuna (*Brassica rapa var pervidis*, a kind of leafy vegetable) by water culture shows distinct differences (Fig. 10.2) (5).

Other factors, such as climate, cultivation method, and concentration of radionuclide in soil solution, affect CF values. The value also varies with time from seeding to harvesting of the crop. Therefore, when we use this model, cautious selection of a CF value becomes necessary. Usually, a conservative value may be selected, but an overestimation of greater than ten times would not be desirable. Among the various factors above, soil properties and conditions

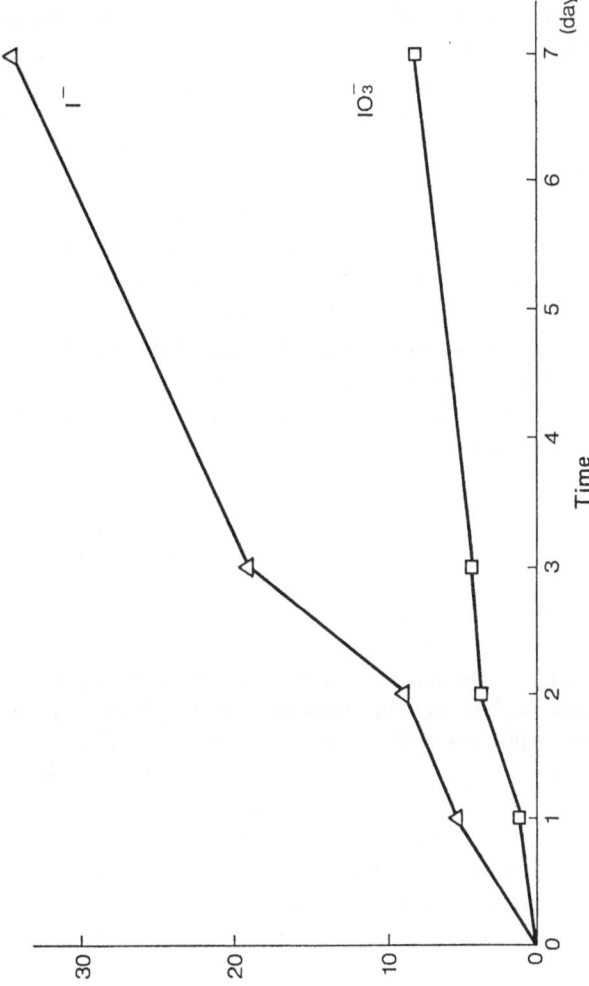

**Fig. 10.2.** Accumulation of I (I and iodate) in shoots of komatsuna (*Brassica rapa var pervidis*) as a function of time. (Reprinted with permission from ref. 5.)

have the most significant effects. Values of CF for many kinds of radionuclides and field crops are reported in many sources (6–10).

## Model Based on the Availability of Radionuclides in Soil

In the concentration factor method, the denominator of CF is the total radionuclide concentration in soil. Because all the nuclides do not always exist in a form transferable to roots in soil, a modified model that utilizes the concentration in absorbable form in soil as a denominator seems reasonable. When field crops take up radionuclides in soil solution together with water, some parts of radionuclides existing in the soil (solid phase) in the adsorbed form move into the solution and transfer to the roots.

Soil consists of three phases, that is, solid, liquid, and air, and radionuclides are contained in the solid and liquid phases. Radionuclides deposited on the ground surface and then migrating downward are contained in soil solution, and a part of them is fixed in the solid phase. After some period, an equilibrium is established between these two phases. The dissolved nuclides in the soil solution are in an available form and, when a part of them transfers to crops, deficient nuclides are supplied from the solid phase to the soil solution.

The model is written as

$$C_{v,i} = (TR)^*_{v,i} \cdot C^*_{s,i} \qquad (10.3)$$

where $(TR)^*_{v,i}$ is the transfer factor of the nuclide $i$ in available form in soil to edible part of crops or pasture (Bq/g/Bq/g dry soil), and $C^*_{s,i}$ is the concentration of the radionuclide $i$ in available form in unit weight of dry soil (Bq/g dry soil).

Several methods are reported to measure the availability of an element in soil. The chemical method uses an extraction agent, which is selected from (a) weak or strong acids, (b) neutral salts and acids, (c) some buffer solutions, (d) chelating agents, (e) redox agents, or (f) water itself. Mitsui et al. (11) recommended 1mol/L ammonium acetate as an extraction agent for $^{90}$Sr and $^{137}$Cs when they studied the transfer of these nuclides deposited in soil from the fallout of nuclear weapons tests to crops. Other methods reported are (a) the so-called "specific activity method," which utilizes radioactive isotopes for the measurement, and (b) a method that is based on the thermodynamic equilibrium concept.

As to the "specific activity method," some data on $^{90}$Sr are reported (2), but data for other radionuclides seem to be lacking.

The approach used in the availability method seems more reasonable than the CF method. But for the availability method to become more practical to use, quick and easier measuring procedures need to be developed. Among the several measuring methods mentioned above, chemical extraction procedures seem appropriate. However, even by this method, as the extraction agents and extraction conditions differ from case to case, its practical use seems, at present, not feasible. Therefore, it can be said that there still remains many unknown problems to be solved for the practical use of the method based on the availability concept for the pathway evaluation.

## Modified Model Based on the Distribution Coefficient Concept

This model is developed by van Dorp and his co-workers (13), based on the assumption that the field crops absorb radionuclides in soil solution together with water, and an equilibrium is assumed between the radionuclides adsorbed in solid phase and those existing in liquid phase.

A distribution coefficient, $K_d$, defined as a ratio of concentration of radionuclides in soil to that in soil solution under equilibrium, is given as

$$K_d = \frac{\text{concentration of radionuclides in solid phase of soil (Bq/g)}}{\text{concentration of radionuclides in soil solution (Bq/mL)}}$$

$$= \frac{C_t - C_s L \theta}{C_s L \rho} \tag{10.4}$$

or

$$C_s = \frac{C_t}{L(\theta + \rho K_d)} \tag{10.5}$$

where $C_s$ is the concentration of radionuclides in soil solution (Bq/mL), $C_t$ is the amount of radionuclides contained in soil of unit area (Bq/cm$^2$), $L$ is the depth of rhizosphere (cm), $\theta$ is the water content of soil (ml/cm$^3$), and $\rho$ is the soil density (g/cm$^3$).

Van Dorp et al. (12) proposed the following equation to give the concentration of radionuclide $i$, ($C_{vi}$), in the edible part of a crop, (Bq/g) as,

$$C_{vi} = F \cdot S \cdot \frac{P_t}{P_{ep}} \cdot (TC) \cdot \frac{C_t}{L(\theta + \rho K_d)} \tag{10.6}$$

When a field crop absorbs a radionuclide from soil solution, the transpiration of soil solution per unit weight of crop, $(TC)$ (mL/g), and selectivity of water to radionuclide, $S$, have an important meaning, and these two quantities are taken into consideration in the above equation. The transferred fraction of a radionuclide to the edible part, $F$, is also in the equation. $P_t$ and $P_{ep}$ in eq. (10.6) are the yearly production of a crop per unit area (g/cm$^2$/y) and the yearly production of edible part of a crop per unit area (g/cm$^2$/y), respectively.

Many parameters are contained in the equation, but knowledge of $(TC)$, $P_t$ and $P_{ep}$ are accumulated to some extent, and the measurement of $\theta$ and $\rho$ is a relatively simple procedure. Therefore, when the values of $F$, $S$ and $K_d$ are available, eq. (10.6) will give better information. Of these three parameters, $F$ may be obtained by a tracer experiment or from the field data of stable isotopes, but data on $S$ are scarce. $K_d$ is the most important parameter in this model, and whether the relationship shown in eq. (10.4) always holds may need further study.

**Table 10.1.** Concentration factor and distribution coefficients of some stable elements

| Element | CF | | | $K_d$ | | |
|---|---|---|---|---|---|---|
| | CF (NRC) (7) | Range of reported data (10) | Reference (14) | Reference (15) | Reference (16) |
| Sr | $1.7/10^2$ | $1.6/10^3 \sim 1.7/10^0$ | 20 | $10 \sim 700$ | — |
| Ru | $5.0/10^2$ | $4.8/10^5 \sim 1.4/10^1$ | — | — | $90 \sim 800$ |
| I | $2.0/10^2$ | $2.0/10^4 \sim 1.2/10^1$ | 0 | — | — |
| Cs | $1.0/10^2$ | $1.5/10^5 \sim 5.9/10^2$ | 200 | $200 \sim 8,000$ | $90 \sim 500$ |
| Ce | $2.5/10^3$ | $4.6/10^6 \sim 1.8/10^2$ | — | — | $500 \sim 2,000$ |

As a whole, the model proposed by van Dorp et al. is attractive, but more research will be necessary to use it for a pathway evaluation as a part of the total calculation model.

In Table 10.1, some figures are cited from the literature.

## Others

A model based on the physiological approach is given by Barber (7). This dynamic model is derived to predict nutrient uptake, combining equations describing nutrient influx and equations describing plant growth. The resultant equation having 11 parameters was verified by Owa (18) and presumed to be useful for the radionuclide transfer from soil to plant.

## A Summary of Models for Estimation of the Transfer of Radionuclides From Soil to Man in Food Chains

Several compartmental terrestrial food-chain models have been reported in the literature. These models are composed of many submodels like those reviewed earlier. They include models for soil to agricultural products, soil to pasture grass and then to livestock products.

Models for the transfer of radionuclides to pasture grass are different from those of soil to field crops because the distribution of radionuclides in soil is different. A model for the latter case is given by Haywood et al. (19).

In the northern European countries, the stock-farming pasturage is an important means of livelihood. In contrast, agriculture, such as cultivation for the production of vegetables, grapes, and fruit culture, becomes an important means of livelihood in the southern European countries. The models for such a case will be those based on assumptions for well-mixed soils.

Several overall models that contain such submodels already exist for the purpose of prediction of the dose to man. Comparison of different models is also reported (20).

In Asian countries rice is a staple food, and its cultivation, especially paddy field, occupies a wide area. A calculation by the Japan Atomic Energy Research Institute shows that when $^{129}$I is discharged from a 100-m stack of a reprocessing plant, ingestion of rice gives the highest thyroid dose compared with the other pathways.

A compartment model for the transfer of $^{129}$I to man in the paddy field rice pathway is shown in Fig. 10.3 (21). Water cultivation of rice in the paddy field is complex, and the model considers every cultivation stage including the atmospheric deposition. The stages considered are the following:

1. *Noncultivation period* (extending from October of the preceding year to March). Water is not present on the field surface during this period. Directly deposited and accumulated $^{129}$I from the air on the field appears in the water, with the remaining $^{129}$I in the root of the rice plant when water is introduced to the paddy field during March and April.
2. *Growing period.* The growing period after the transplantation; $^{129}$I is absorbed from the water to the root, and the deficient $^{129}$I in water is supplied from soil together with that from deposition.
3. *Ear growing period.* The growth of stalks and leaves almost stops in August

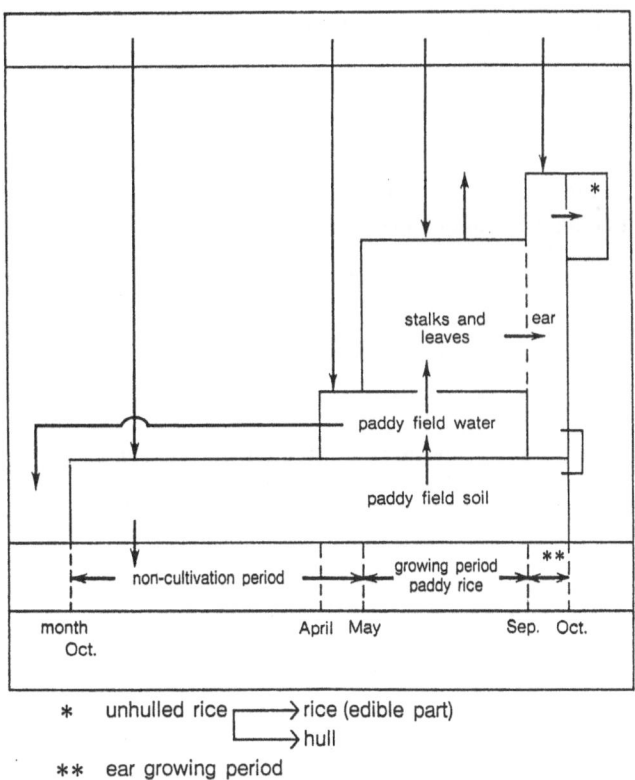

**Fig. 10.3.** Rice paddy-field pathway model. (Reprinted with permission of ref. 21).

and heading time begins. In the late heading period, extending from September to the early October, the ear grows but the transfer of $^{129}I$ via the root almost stops and only the accumulated $^{129}I$ in stalks and leaves moves to the ear by translocation. In the course of calculation, parameters such as evaporation from the paddy field, transpiration from the leaves, and percolation through soil must also be taken into consideration.

## Calculation Methods

### Concentration Factor (CF) Method

In this calculation method, the transfer of radionuclides between two connected compartments is unilateral. The calculation is straightforward and outlined in detail in International Commission on Radiological Protection (ICRP) Publication 29 (1).

### Systems Analysis (SA) Method and the Transfer Coefficient

This method is useful to delineate the course of time-dependent phenomena. To use this method, the transfer coefficients connecting two adjacent compartments becomes necessary, that is, the forward transfer coefficient $k_f$ and the backward transfer coefficient $k_b$. Usually, the observed data are the net transportation, and the values of $k_f$ and $k_b$ are not known.

If the transfer between the compartments is represented by the inventory change of radionuclides per unit time in each compartment, the amount of decreased nuclide $i$ in compartment $I$ is denoted by

$$k_f \cdot Q_I \qquad (i).$$

This is equal to the increase of $i$ in compartment $I$, and similarly the decrease of $i$ in compartment $I$ is denoted as

$$K_b \cdot Q_{II} \qquad (i),$$

where $Q_I$ and $Q_{II}$ are the inventory of the radionuclide $i$ in each compartment. Under the equilibrium conditions,

$$k_f \cdot Q_I = k_b \cdot Q_{II} \qquad (i). \tag{10.7}$$

If the total mass of each compartment is represented as $M_I$ and $M_{II}$, the concentration factor CF is written as

$$CF = (Q_I/M_I)/(Q_{II}/M_{II}) \qquad (i). \tag{10.8}$$

From this relationship,

$$k_f/k_b = CF\,(M_{II}/M_I) \qquad (i). \tag{10.9}$$

By this equation the ratio of $k_f$ and $k_b$ can be obtained from the value of CF. The mechanisms of radioactive nuclide transfer from soil solution to the roots of a plant (or vice versa) is complex, but Simmonds et al. (3) assumed that equilibrium is established very rapidly between these two phases and chose $k_b$ as 1.0/s.

When the transfer coefficient is to be obtained from the above relationships, the CF values must be given as realistically as possible and at least for every stage of cultivation.

## Conclusions

Several models are now used to estimate the transfer of radionuclides from soils to plants. The CF method used to calculate the transfer of radioactive nuclides from one compartment to the connecting compartments is relatively simple and quick, and many CF data have been reported. Therefore, this method can be used most effectively to find out what nuclide is critical in food chains and what pathway gives the greatest effect on dose to man. Expertise is indispensable in the selection of relevant parameter values.

In the selection of the value of CF, it must be understood that soil parameters have, among others, significant effects on its value.

In actual dose estimation, the dilution effect of contaminated crops or hay by noncontaminated materials has, many times, a more significant effect than the selection of the proper CF value. But this is not within the scope of the behavior of "soil-borne radionuclides" and is not discussed here.

When more precise information is required, after the critical nuclide(s) are known, use of a model that expresses the transfer of radionuclides more realistically, based on the scientific principles, is desirable. Such a model seems to be in the research stage, however, and the data for the use of these models are not fully available. Therefore, further research in this field is anticipated.

The distribution of man-made radionuclides on land is by no means uniform, therefore, depending on the purpose of the calculations, an adequate model and calculation method should be selected.

## References

1. Annals of the ICRP. ICRP Publication 29 (1978) Radionuclide release into the environment: Assessment of doses to man
2. Fukui M (1979) Sorption of radionuclides on a soil. Proceedings of the Seminar on Assessment of Radiation Dose due to Terrestrial Contamination by Radionuclides organized by National Institute of Radiological Sciences, NIRS-M-31:67–80
3. Simmonds JR, Jones JA, Linsley GS (1979) A general model for the transfer of radioactive materials in terrestrial food chains. National Radiological Protection Board, NRPB-R89
4. Yamazaki S (1987) Transfer factor. Proceedings of the Seminar on Transfer of Radionuclides to Agricultural Products organized by National Institute of Radiological Sciences, NIRS-M-65:25–30
5. Muramatsu Y, Christoffers D, Ohmomo YN (1983) Influence of Chemical Forms on Iodine Uptake by Plant. J Radiat Res 24, pp 326–338

6.  Menzel RG (1965) Soil-plant relationships of radioactive elements. Health Phys p 1325
7.  USNRC Regulatory Guide (1977) 1.109
8.  IAEA Safety Series No 57 (1982) Generic Models and Parameters for Assessing the Environmental Transfer of Radionuclides from Routine Releases: Exposures of Critical Groups pp 62–64
9.  International Union of Radioecologists (IUR) (1982) Report on a workshop on the measurement of soil to plant transfer factors of radionuclides, Parts I, II
10. Ng, YC, Colsher CS, Thompson SE (1982) Soil-to-plant concentration factors for radiological assessments, NUREG/CR-2975
12. Kobayashi H (1984) Environmental Radioactivity. Saiki M (ed). Soft Science Inc Tokyo pp 189–199
13. van Dorp F, Eleveld R, Frissel MJ (1979) A new approach for soil-plant transfer calculations. In Biological Implication of Radionuclides Released from Nuclear industries, vol II. IAEA Publication STI/PUB/522, IAEA, Vienna pp 399–406
14. Denham DH, Baker DA, Corley JP, Soldat JK (1974) Radiological evaluation for advanced waste management studies, BNWL-1764
15. Inoue Y, Morisawa S (1976) Distribution coefficient $K_d$ of radionuclide between sample soil and water. J Atom Energy Soc Japan p 524
16. Watabe T (1987) Application of soil $K_d$s to prediction models of the soil-to-plant transfer of radionuclides. Proceedings of the Seminar on Transfer of Radionuclides to Agricultural Products organized by National Institute of Radiological Sciences Chiba City, Japan: NIRS-M-65:63
17. Baber SA (1984) Soil Nutrient Bioavailability—A Mechanistic Approach. Wiley-Interscience, pp 114–135
18. Owa N (1987) Mathematical models for predicting nutrient transfer from soil to plant by distribution coefficient of nutrient between soil solid and solution. Proceedings for the Seminar on transfer of Radionuclides to Agricultural Products organized by National Institute of Radiological Sciences: NIRS-M-65:54–60
19. Haywood SM, Linsley GS, Simmonds JR (1980) The development of models for the transfer of Cs-137 and Sr-90 in the pasture-cow-milk pathway using fallout data. National Radiological Protection Board, NRPB-R110
20. Hoffman FO, Bergström V, Gyllander C, Wilkens A-B (1984) Comparison of predictions from internationally recognized assessment models for the transfer of selected radionuclides through terrestrial food chains. Nucl Safety p 533
21. Iijima T (1987) Compartment model for migration of I-129 to rice plant. Proceedings of the Seminar on Transfer of Radionuclides to Agricultural Products organized by National Institute of Radiological Sciences, NIRS-M-65:71–82

# CHAPTER 11

# Effect of Local Conditions on Coefficient of Radionuclide Transfer Through Food Chains

R. M. Barkhudarov,[1] V. A. Knizhnikov,[1] N. Y. Novikova,[1] E. V. Petukhova[1]

This chapter presents the results of studies carried out for years in some regions of the Soviet Union that have a relatively high migration of global $^{137}$Cs to vegetation. This is especially characteristic of wooded lowlands where soils are, for the most part, made of peat bog and grassy-podzol sand varieties. An increased $^{137}$Cs migration to vegetation is found to result primarily from lack (or low content) of clay minerals, for example, hydromica, in these soils. Further contribution is made by a limited content of mobile K and a high water content. For these soils, the $^{137}$Cs transfer coefficient is ten times higher than the average for the country. As a result, the $^{137}$Cs content of local foods, particularly of milk, is 10 to 100-fold higher.

In the late 1960s to the early 1970s, monitoring of the global radioactive fallout over the USSR territory revealed an increased $^{137}$Cs concentration in local foods of some regions. It could not be associated with a higher $^{137}$Cs content of soil since the ground contamination level in these areas was typical of middle latitudes: 3 to 4 GBq/km$^2$. The phenomenon was observed mostly in wooded lowlands along the southern boundary of the forest belt from Poland to the Urals, with noticeable prominence in the Ukrainian and Belorussian Polessye (marshy woodlands). The soils here are mainly grassy-podzol and peat bog varying in swamping and gleization. These are characterized by high acidity and a low content of exchangeable forms of phosphate, Ca, K, and Na due to considerable soil moistening and, in some cases, intensive washing out of the soil cover, which causes transport of exchangeable bases to the underlying layers.

The $^{137}$Cs migration to vegetation is found to be a maximum in virgin lands generally used as pastures (Table 11.1).

Table 11.1 shows that the higher accumulation coefficients are observed for peat bog and grassy-podzol sand soils, especially in the Byelorussian-Ukranian Polessye (BUP). For sand loam and loam soils, migration is six to ten times lower. Abnormally high coefficients of $^{137}$Cs accumulation in pasture vegetation are due to mechanical and mineral composition of soils, a low content of K

---

[1] Institute of Biophysics, Ministry of Health, Moscow, USSR.

**Table 11.1.** $^{137}$Cs Transfer from soil to pasture vegetation

| Soil | Belorussian–Ukrainian Polessye | | | Meshchyora Lowland | | |
|---|---|---|---|---|---|---|
| | K content of soil[a] (mg/kg) | $^{137}$Cs content of vegetation (Bq/kg) | Accumulation coefficient | K content of soil[a] (mg/kg) | $^{137}$Cs content of vegetation (Bq/kg) | Accumulation coefficient |
| Peat bog and podzol sand varieties | 63 | 88 ± 30 | 4.5 | 110 | 35 ± 28 | 1.2 |
| Grassy-podzol sand loam and loam varieties | 120 | 17 ± 12 | 0.5 | 180 | 3 ± 1 | 0.2 |

[a] Forms of stable potassium.

available for plants and a high moisture content of the active root zone. With regard to mechanical composition, light soils belong to the group of bound and loose sands; heavier soils are represented by sand loam and light loam. The main soil-forming minerals include quartz and feldspar. The small silt fraction (0.001 mm) amounts to 0.5% to 0.6% in the BUP soils and 2.3% to 7.8% in the Meshchyora Lowland (ML), its bulk being composed of organic matter (in BUP) or clay minerals, sometimes mixed with organic matter (in ML).

As evidenced by our studies, it is the mineral part of the soil, namely, availability and qualitative composition of clay minerals of which hydromica is the principal one, that is responsible for $^{137}$Cs transfer to vegetation in the area under consideration.

The amount of K, particularly of its mobile forms, contained in soil is a most important factor for $^{137}$Cs migration to vegetation. For the above soils the K content is not high: 0.97% to 0.22% (1,2). Table 11.1 gives the content of mobile K, exchangeable and acid-soluble forms included. As seen from the Table, the $^{137}$Cs content of vegetation is in inverse relation to the mobile K content of soil, a fact already reported in the literature (3,4). Excess moistening, other conditions being equal, is known to decrease $^{137}$Cs fixation by soils and increase the cation migration velocity (5,6). This agrees with the data obtained in our studies (Table 11.2).

**Table 11.2.** $^{137}$Cs Content of vegetation as influenced by soil characteristics

| Soil | Ground water level from surface (m) | $^{137}$Cs content of grass (Bq/kg) | K content of soil[a] (mg/kg) | C content of organic matter (%) |
|---|---|---|---|---|
| Humus, peat, gley, sand | 0.5 or less | 180 | 22 | 4.5 |
| | 1.0 or more | 50 | 27 | 1.7 |
| Grassy-podzol, gley | 0.5 or less | 130 | 37 | 3.5 |
| | 1.0 or more | 10 | 48 | 4.5 |

[a] Forms of stable K.

It follows from Table 11.2 that a ground-water level fall of more than 0.5 m leads to a three- to ten-fold reduction in the $^{137}$Cs content of vegetation. Interestingly, in 1972, when the summer was extremely dry and hot, the ground-water level naturally fell, and the $^{137}$Cs content of vegetation decreased by a factor of from 2 to 3.

Not only melioration but also tillage and cultivation of marshy and wooded virgin soils reduce $^{137}$Cs migration to vegetation (7,8). This results from a number of factors, including mechanical dilution due to mixing the surface layer with less contaminated soil from a depth of 20 to 30 cm, reduced water content of soil brought about by loosening, and soil enrichment with mobile K compounds applied with fertilizers. Therefore, the $^{137}$Cs content of crops from arable lands (potatoes, vegetables, grain) and of sown grass turns out to be considerably lower by a factor of 10 or more and from 4 to 5, respectively, as compared with that of wild grass from virgin lands, provided the soil type is similar.

In the region under discussion, the $^{137}$Cs content of pasture grass and, consequently, of local milk and meat is reported to exceed the USSR average values 10 to 100 fold. The $^{137}$Cs concentration in crops from arable lands is also observed to increase, but to a lesser degree. In recent years, a complex of agrotechnical measures taken and extensive use of cultivated pastures for cattle grazing have markedly reduced $^{137}$Cs transfer to milk.

However, for reasons given, in the BUP population the $^{137}$Cs content of the body from global fallout proved to be ten times as high as in other parts of the USSR.

## References

1. Lukashev KI (ed) (1966) Geochemical Characteristics of Lithogenesis and Landscapes of the Byelorussian Polessye (in Russian). Nauka i tekhnika, Minsk, 1966:319
2. Tyuryukanova EB (1957) Bog soils of the meshchyora lowland (in Russian). Vestn Mosk Univ No 4, 1957:115–123
3. Yudintseva EV, Gulyakin IB (1968) Agrochemistry of radioactive isotopes of strontium and caesium (in Russian). Atomizdat, Moscow, 1968:471
4. Russell RS (ed) (1966) Radioactivity and Human Diet. Pergamon Press, Oxford, Atomizdat, Moscow, 1971:374
5. Prokhorov VI (1973) Ion diffusion in soils and its role in radionuclide migration (in Russian). In Current Problems of Radiobiology, vol 2. Atomizdat, Moscow, 1973:118–145
6. Ryzhova LV (1965) Exchange patterns for univalent K, NH$_4$, Rb, Cs, Li cations and divalent Ca and Sr in clay minerals and soils (in Russian). Agrochimiya, No 3, 1965:106–115
7. Marei AN, Barkhudarov RM, et al (1972) Effect of natural factors on $^{137}$Cs accumulation in the bodies of residents in some geographical regions. Health Phys 22:9–15
8. Marei AN, Barkhudarov RM, Novikova NY (1974) Global Fallout of $^{137}$Cs and man (in Russian). Atomizdat, Moscow, 1974:168

CHAPTER 12

# Long-Lived Man-Made Radionuclides in the Soil–Plant System

F. A. Tikhomirov[1]

## Introduction

The soil is the most important link in the migration pathways, via trophic chains, of many long-lived radionuclides. An essential feature of soil is its ability to accumulate and to retain for a long time (tens and hundreds of years) radioactive isotopes coming from the outside. Therefore, for years, the contaminated soil becomes a source of radionuclides entering agricultural products and surface and subsurface waters.

In this respect, to work out reliable radiation forecasts concerning agricultural production and radionuclide biogeochemical distribution, it is especially important to evaluate the information on radionuclide behavior in soils and in the soil–plant system. This information includes the biological availability indices for radionuclides in the soil and changes of this availability with time and the parameters of radionuclide transfer into plants and of soil self-decontamination. All of these parameters vary over a wide scale, based on a number of factors. In the following, four groups of factors determining radionuclide mobility in the soil and entrance into plants are outlined. The significance of each group was studied in single-factor laboratory and field experiments (1).

## Biogeochemical Properties of Radionuclides

There are two main aspects determining the amount of radionuclides entering plants and migrating further by the trophic chain. One is the extent of radionuclide participation in plant and animal metabolism. The other is the physicochemical character of radionuclide compounds entering the soil.

From this point of view, we are especially interested in long-lived radionuclides belonging to biogenic macro- and microelements (such as $^{14}C$, $^{54}Mn$, $^{55}Fe$, $^{60}Co$, $^{65}Zn$, and $^{129}I$ or to the corresponding nonisotopic analogs (such as $^{90}Sr$

[1] Facultet pochvovedenia, MGU, 119899, Moscow, USSR.

**Table 12.1.** Concentration factors (CFs) for radionuclides in cereal grain on chernozem; radionuclides being introduced into the soil in a soluble form[a]

| Nuclide | CF | Nuclide | CF | Nuclide | CF | Nuclide | CF | Nuclide | CF |
|---------|-----|---------|------|---------|------|---------|----------------|---------|----------------|
| $^{22}$Na | 60 | $^{45}$Ca | 0.10 | $^{60}$Co | 0.05 | $^{106}$Ru | $3 \cdot 10^{-3}$ | $^{91}$Y | $3 \cdot 10^{-5}$ |
| $^{36}$Cl | 25 | $^{90}$Sr | 0.10 | $^{95}$Zr | 0.03 | $^{125}$I | $2 \cdot 10^{-3}$ | $^{144}$Ce | $3 \cdot 10^{-4}$ |
| $^{65}$Zn | 15 | $^{54}$Mn | 0.10 | $^{137}$Cs | 0.02 | $^{226}$Ra | $2 \cdot 10^{-3}$ | $^{147}$Pm | $3 \cdot 10^{-5}$ |
|  |  | $^{55,\,59}$Fe | 0.10 | $^{140}$Ba | 0.05 | $^{232}$Th | $4 \cdot 10^{-3}$ | $^{239}$Pu | $10^{-5}$ |
|  |  | $^{185}$W | 0.15 |  |  | $^{238}$U | $1 \cdot 10^{-3}$ |  |  |

[a] From refs. 2–8.

and $^{137}$Cs). Concentration factors (CFs) for these radionuclides in plants differ under equal conditions by several orders of magnitude (Table 12.1).

In most cases, the man-made radionuclides enter soil in a form that is mobile and relatively available for plant consumption. They are contained in fine aerosols, being readily leached by water, or in the form of water solutions (in rain- and groundwater) and gases. But, in some cases, the radionuclides entering the soil are characterized by low migration mobility, for instance, when they are contained in nuclear fuel particles (following radiation accidents), particles of melted silica (formed in surface nuclear explosions), or oxides (as a result of corrosion of reactor construction materials). In such cases, the CFs for radionuclides in plants decrease by one or two orders of magnitude.

## Soil Properties

The transfer of radionuclides from the soil into plants depends highly on the rate of radionuclide dilution by biogenic elements in the soil solution. The higher the availability of nutritive elements, the higher is the rate of dilution and the more pronounced is the competition among the stable elements and radionuclides for the sorption sites at the root surface. Therefore, soil fertility can be accepted, at first approximation, to be a general criterion reflecting the amount of radionuclide transfer from the soil into the crop. The higher the supply of nutritive mineral elements in the soil and the closer their composition to the optimum, the lower is the radionuclide content in the overground phytomass. The data available on the agrochemistry of $^{60}$Co, $^{65}$Zn, $^{90}$Sr, $^{125}$I, $^{137}$Cs etc., in general, confirm this assumption (Table 12.2). These radionuclides enter the plants at the highest rates in low-productivity soils (sandy and podzolic soils, in particular), which are typical for northern and some other regions (6,13,14). Still, for a more precise forecast of a particular radionuclide transfer into plants, a more individual approach is needed. This is partially due to the fact that consumption of radionuclides from the soil solution and their further transport in plant tissues depends, to a large extent, on the content in the soil solution of biogenic elements, which are the closest chemical analogs of the radionuclides. A reciprocal ratio connects these two values. For example, field experiments with $^{90}$Sr introduced under different soil and climate conditions revealed that 50-fold variations in CFs for wheat depend to an extent of 20% to 70% on the difference in the exchangeable Ca content in different soils (13). Similarly, the transfer of $^{137}$Cs from different soils into plants depends on the exchangeable K content in the soil. But, in this case, along with K an important role is carried out by the exchangeable Ca content of the soil: the $^{137}$Cs CF in the overground phytomass closely correlates here with the ratio $1/([K]_s + 0.05[Ca]_s)$, that is, the reciprocal to the sum of soil contents of exchangeable K and Ca (14). It is the difference of 10% to 50% in the last value that implies more than a 100-fold variation in $^{137}$Cs CFs for cereal cultures on different soils. Similar dependence is observed for the CFs in plants for $^{60}$Co and $^{65}$Zn, on one hand, and the soil content of stable Co and Zn in exchangeable form, on the other hand (8).

**Table 12.2.** Relative characteristics of loamy agricultural soils in accumulation values for $^{60}$Co, $^{65}$Zn, $^{90}$Sr, $^{125}$I, and $^{137}$Cs in an overground phytomass[a]

| Soil | Accumulation values, relative units | | | | |
|------|----------|----------|----------|----------|----------|
| | $^{60}$Co | $^{65}$Zn | $^{90}$Sr | $^{125}$I | $^{137}$Cs |
| Podsolic soil | 100 | 100 | 100 | | 40 |
| Soddy podsolic soil | 10 | 25 | 15 | 100 | 30 |
| Chernozem | 1.5 | 50 | 5 | 15 | 10 |
| Dark chestnut soil | 1.0 | 50 | 15 | 25 | 5 |
| Sierozem | 3.0 | 20 | 10 | 100 | 3 |
| Red soil | 65 | 25 | 15 | 45 | 100 |
| Yellow soil | | | 15 | 100 | 40 |

[a] From refs. 6 and 9–12.

The literature presents numerous data on other soil factors (pH, humus content, mineralogical and granulometric compositions, and soil moisture) influencing the transfer of various radionuclides into plants. However, independent action of these factors becomes evident only in special cases. For instance, the soil mineralogical structure (presence of loamy minerals) markedly influences the $^{137}$Cs availability for plants, which is explained by the fixation of this radionuclide in the soil. The dynamics depend on the soil type because of the specificity of a particular radionuclide interaction with separate components of the absorbing soil complex (Table 12.3 and 12.4). For instance, the radionuclide $^{90}$Sr is shown to be present in soil in an exchangeable form, its availability, according to 15-year experimental results, changing insignificantly during this period (15). A similar picture is received, according to a 2-year experiment, for $^{65}$Zn (8).

The dynamics of $^{137}$Cs availability depend on the mineralogical soil composition. For soils that are rich in loamy minerals, the availability of Cs, as a rule, decreases four- to eight-fold during the first 2 to 4 years (Table 12.4)

**Table 12.3.** Dynamics of $^{137}$Cs CFs for oat grain in a four-year pot experiment with soil moisture at 60% of the maximum water capacity[a]

| Vegetation period (years) | Leached chernozem, medium loamy | | Soddy podzolic, medium loamy | | Alluvial, sandy soil | |
|---|---|---|---|---|---|---|
| | Grain | Straw | Grain | Straw | Grain | Straw |
| 1 | 0.039 | 0.091 | 0.414 | 0.831 | 0.356 | 0.996 |
| 2 | 0.019 | 0.052 | 0.096 | 0.214 | 0.193 | 0.466 |
| 3 | 0.019 | 0.048 | 0.108 | 0.274 | 0.258 | 0.604 |
| 4 | 0.021 | 0.045 | 0.055 | 0.129 | 0.095 | 0.375 |

[a] Reprinted with permission from ref. 5.

**Table 12.4.** Dynamics of $^{90}$Sr and $^{137}$Cs concentrations in cereals on leached chernozem[a,b]

| $^{90}$Sr | | $^{137}$Cs | |
|---|---|---|---|
| Time passed from moment of soil contamination (years) | Concentration in grain (relative units) | Time passed from moment of soil contamination (years) | Concentration in grain (relative units) |
| 0.25 | 100 | 0.25 | 100 |
| 1 | 92 | 1 | 59 |
| 2 | 90 | 2 | 60 |
| 3 | 129 | 3 | 97 |
| 4 | 131 | 4 | 106 |
| 5 | 102 | 5 | 72 |
| 6 | 92 | 6 | 89 |
| 7 | 82 | 10 | 94 |
| 8 | 94 | 11 | 48 |
| 9 | 96 | 12 | 82 |
| 10 | 96 | 13 | 77 |
| 11 | 74 | | |
| 12 | 72 | | |
| 13 | 72 | | |
| 14 | 81 | | |

[a] Reprinted with permission from refs. 15 and 16.
[b] Data averaged for three plant species, taken from field experiments.

(11). But, under field conditions, this process would not be irreversible. In years that are extreme by agrometeorological standards, the $^{137}$Cs CF for plants may even exceed its initial value (6). For sandy and peaty soils, the decrease in $^{137}$Cs availability with time is much less pronounced.

A decrease of availability with time is also observed for $^{60}$Co in acid soils. The radionuclide transport into plants in the second year of vegetation after introduction into the soil decreases by two to five times compared to the first year. The availability of iodine radionuclides in soils decreases at an even higher rate (Fig. 12.1) (17). For some soils, the equilibrium CF value, typical

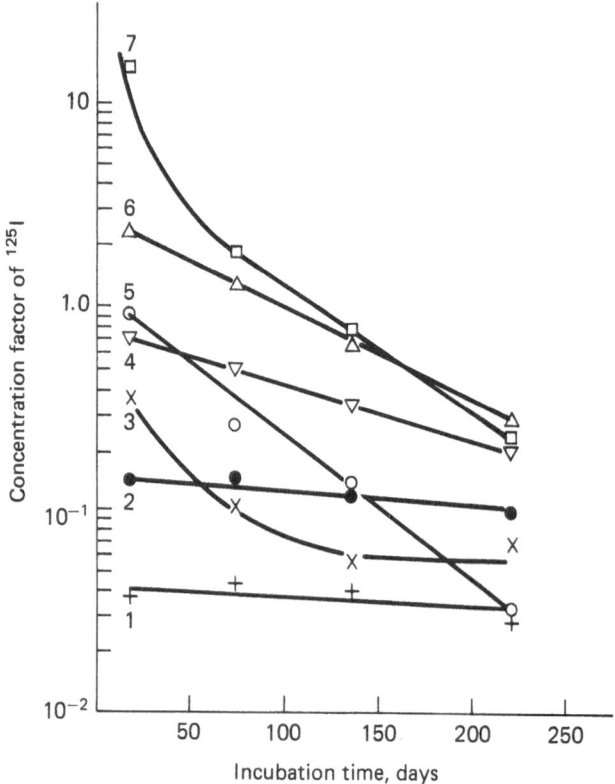

**Fig. 12.1.** Dynamics of $^{125}$I availability in soils for plants, depending on the time of radionuclide incubation in the soil: 1, typical chernozem, loamy; 2, red soil, light loamy; 3, soddy podsolic soil, light loamy; 4, yellow soil, heavy loamy; 5, calcareous chernozem, heavy loamy; 6, alluvial soil, sandy; 7, light sierozem, light loamy.

for native I, is achieved in two weeks. For other soils, it takes up to several months. These variations are caused by the different speeds of interaction of the I introduced into the soil with individual humus fractions. Therefore, we can outline that the process of achieving equilibrium between exchangeable forms of radionuclide compounds, after the introduction of radionuclides into the soil, as well as of radionuclide natural macroanalogs, may be variable in time. This is because of wide variations in compounds of a particular element, typical for the given soil; in the composition of the soil-absorbing complex; and in the rate of the isotopic exchange of elements contained in a crystalline lattice of loamy minerals (12). The pH value affects the entrance from the soil of $^{60}$Co, $^{239}$Pu and the rare earths (1,8). The humus content and its qualitative structure produce a special effect on the availability of I radionuclides for plants, because soil I is almost completely incorporated into I-organic compounds (21). In this case the availability of I depends on the humus fractional composition, because the I sorption varies for different humus fractions.

At present, there is a large amount of data available on the important role of the water-soluble metabolism products of soil organisms and the excretion products of plant roots for availability of polyvalent elements and the corresponding radionuclides for plants. The radionuclides, such as $^{35}$S, $^{45}$Ca, $^{59}$Fe, $^{65}$Zn, and $^{125}$I, are shown to be present in soil solution, not in an ionic form, but mainly in the form of water-soluble, nuclide-organic complexes. In this case, the radionuclide concentration factors for plants under cultivation on the corresponding soil (CFs) agree well with the following equation:

$$CFs = \alpha \sum_i n_i CF_i,$$

where $\alpha$ is the share of the total radionuclide content in the soil that corresponds to the soil solution; $n_i$ is the share of $\alpha$ that corresponds to radionuclide sorption by the $i$ fraction of organic compounds contained in the soil solution; and $CF_i$ is the radionuclide CF for plants under cultivation on the $i$ fraction of soil solution. This means that the variations in radionuclide availability for plants on different soils are caused by unequal radionuclide transfer into the soil solution and unequal distribution among individual fractions of water-soluble, radionuclide-organic compounds.

Another important problem in radionuclide agrochemistry is the dynamics of radionuclide availability after the introduction into these compounds.

The soil properties, especially its granulometric composition, affect the elution of radionuclides from the root-containing soil layer by the surface and ground waters. According to the data received, from 1% to 5% of $^{90}$Sr and up to 1% of $^{137}$Cs contained in the plough layer of the soil, typical for the European part of the Soviet Union, are removed by migration processes (18). On light (sandy) soils the rate of radionuclide removal from the upper soil layer is considerably higher compared to heavy (loamy) soils. The soil capacity for self-decontamination goes up with higher soil acidity.

## Interspecial and Intercultivar Variations of Plants in Radionuclide Accumulation

Among the most essential features of the plant organism affecting the radionuclide transfer into the phytomass is the physiological need for chemical elements that belong to isotopic and nonisotopic analogs of the radionuclides. This, to a certain extent, explains the wide scale of interspecial and intercultivar variations in radionuclide accumulation in the phytomass under equal conditions of root nutrition (Table 12.5).

Another important factor is the ability of plants to partially distinguish between radionuclides and their nonisotopic analogs during their transfer from the soil into plants and during transport inside the plant organism, the degree of distinguishing capability being a genetically dependent feature. This capability determines variations in radionuclide accumulation up to 1.5 to 2 times for cultivars of the same plant species (19).

**Table 12.5.** Relative accumulation of $^{90}$Sr, $^{125}$I, and $^{137}$Cs in grains of various agricultural plants on chernozem[a]

| Culture | Accumulation in grain, relative units | | |
|---------|-------|-------|--------|
| | $^{90}$Sr | $^{125}$I | $^{137}$Cs |
| Barley | 1 | | 1 |
| Wheat | 1 | 2 | 1 |
| Oat | 1.5 | 1 | 2 |
| Millet | 0.3 | | 2 |
| Buckwheat | 2.0 | | 3 |
| Chumiza | 1.0 | | 4 |
| Haricot | 2 | | 4 |
| Pea | 2 | 1 | 5 |
| Potato | 0.4 | 10 | 8 |
| Bean | 3 | | 8 |

[a] From refs. 6, 10, 16, 20, and 21.

The radionuclide transfer into the phytomass also depends on the root system distribution over the most contaminated soil layer. The plant species and their cultivars that have deeply penetrating root systems extract a considerable part of nutritive elements from uncontaminated soil horizons and accumulate radionuclides in lesser amounts compared to plants with surface-oriented root systems (19).

## Agrometeorological Conditions

Their role is expressed in large positive or negative variations of the radionuclide content in plants in different years that differ in the amount of rainfall and water distribution in the soil profile or that differ in temperature regime and water supply during the plant vegetation period. The values of $^{90}$Sr and $^{137}$Cs concentration in grain of the same cultivars on the same soil differ for some years by two to four times (Table 12.4) (15,16). In particular, the highest radionuclide entries into crops are recorded to take place in hot, dry years (20,22). The effect of agrometeorological conditions is most likely to be an indirect one. For instance, they cause changes in the quantity and composition of root excretion and in the metabolic products, of soil microorganisms which fix most of the soluble radionuclides, under temperature and water regimen changes.

The data presented on the significance of different factors in the transfer of long-lived radionuclides from the soil into agricultral plants can be used as a scientific basis for agrotechnical measures aimed at decreasing the radionuclide content in crops on contaminated soils. These measures include the dilution of radionuclides in the soil by their chemical macroanalogs; the increase of soil fertility by optimizing the composition of the nutritive elements contained in the soil in available form; the impregnation of the soil by compounds that selectively transform the radionuclides into less available forms (various minerals, organic substances, etc.); the choice of species and cultivars with minimal radionuclide accumulation in crops; and the transfer of the soil layer contaminated by radionuclides so that it becomes unavailable to plant root systems. Under large-scale agricultural industry conditions, the level of crop contamination can be considerably lowered by rational placement of different crops in areas differing in contamination levels. The implementation of these measures, taken as a whole, can provide a 10- to 100-fold decrease in the content of long-lived radionuclides such as $^{90}Sr$ and $^{137}Cs$, in agricultural production.

# References

1. Tikhomirov FA (1980) Theoretical and applied aspects of soil and plant environment protection from radioactive contamination (Russian). Biol Nauki (Moscow) 4:18–28

2. Arkhipov NP, Fedorov EA, et al. (1975) Soil chemistry and accumulation in crops via root systems of technogenic radionuclides (Russian). Agrokhimiya 12:40—52

3. Bufatin OI, Paraschukov NP et al. (1986) On the biological effect of $^{238}U$ and on its accumulation in rye plants (Russian).

4. Drichko VF, Lisachenko EP, et al. (1976) The transfer of some natural radionuclides from the soil into plants (Russian). Preprint, GKAE NKRZ

5. Moiseev FA, Tikhomirov FA et al. (1982) The soil properties and $^{137}Cs$ incubation time effects on dynamics of radionuclide forms and on its availability for plants (Russian). Agrokhimiya 8:109–113

6. Moiseev IT, Tikhomirov FA et al. (1984) The role of soil properties, interspecial plant differences, and other factors in radioactive iodine accumulation in agricultural plants (Russian) Agrokhimiya 8:77–82

7. Sansharova NI, Alexakhin RM (1982) $^{22}Na$, $^{32}P$, $^{65}Zn$, $^{90}Sr$, $^{95}Zr$, and $^{106}Ru$ entrance into agricultural plants (Russian), Pochvovedenie 9:59–65

8. Tikhomirov FA, Rerikh VI, Zirin NG (1979) Accumulation of natural and introduced cobalt and zinc by plants (Russian), Agrokhimiya 6:96–102

9. Arkhipov NP, Bondar PF (1978) Radioactive strontium accumulation by agricultural plants from soil under different soil and climate conditions (Russian). Preprint, GKAE NKRZ 7:24

10. Prister BS, Grigorjeva TA, Perevesentzev VM (1979) Iodine behavior in the plant and soil system (Russian). Agrokhimiya 3:93–99

11. Tikhomirov FA, Prokhorov BM et al. A search for the correlation between $^{137}Cs$ entrance into plants and soil properties.

12. Yudintzeva EV, Gulakin IV (1968) Agrochemistry of strontium and cesium radioactive isotopes (Russian). Atomizdat

13. Bondar PF (1983) The effect of soil and climatic conditions on $^{89}Sr$ accumulation in plants from soil and forecast of crop contamination levels (Russian). Agrokhimiya 7:69–80

14. Bondar PF, Yudintzeva EV (1984) Evaluation of different soil properties' effects on $^{137}Cs$ entry into plants and forecasting its accumulation in oat crops (Russian). Agrokhimiya 9:85–94

15. Marakushin AV, Fedorov EA (1977) Amounts of $^{90}$Sr accumulation by agricultural plants under continuous rotation (Russian)
16. Moiseev IT, Tikhomirov FA, Rerikh LA (1986) Dynamics oif $^{137}$Cs accumulation in agricultural plants under field experiment (Russian). Agrokhimiya 8:92–96
17. Tikhomirov FA (1983) Radioecology of iodine (Russian). Energoatomizdat (Moscow)
18. Alexakhin RM, Tikhomirov FA (1976) Biogeochemical aspects of natural and artificial radionuclides' migration (Russian). In: Biogeochemical cycles in the biosphere. Nauka, Moscow, 285–291
19. Korneeva NV (1974) The influence of special and cultivar peculiarities of plants on $^{90}$Sr accumulation in wheat and pea crops (Russian). Abstract of dissertation for degree of cand.sci., Moscow
20. Rerikh LA (1984) Agrochemical aspects of $^{137}$Cs behavior in soil and in agricultural plants (Russian). Abstract of dissertation for degree of cand sci., Moscow
21. Ridkiy SG (1969) $^{90}$Sr accumulation by different agricultural plants under its transfer from soil in a field experiment (Russian) Preprint, GKAE Atomizdat
22. Gromov VA, Nikolaeva EM, Marakushin AV (1982) The forecast for $^{90}$Sr accumulation in barley grain depending on weather conditions (Russian). Agrokhimiya 9:118–125
23. Moiseev IT, Tikhomirov FA et al. (1976) Behavior of $^{137}$Cs in soils and its accumulation in agricultural plants (Russian). Pochvovedenie 7:45–57
24. Tikhomirov FA, Agapkina GI (1987) The effect of the structure of iodine-organic compounds contained in soil solution on iodine availability for plants (Russian). Vestn Mosk Univ Ser 17, Pochvovedenie 2:18–22

# Part IV
# Consequences of Radionuclide Release to Health, Safety, and the Environment

CHAPTER 13

# Experience at Windscale—1957

H. J. Dunster[1,2]

## Introduction

In 1957 the Windscale Works of the United Kingdom Atomic Energy Authority
at Sellafield, Cumberland, was primarily concerned with the production of Pu.
It consisted essentially of two air-cooled, graphite-moderated, natural U reactors
together with the chemical processing plant required for the treatment of irradiated
U metal and the associated analytical and research laboratories.

The two reactors employed were identical in design and were first operational
in 1950 to 1951. The cooling air for each reactor was drawn through filters
from the atmosphere, blown through the reactor core, and exhausted through
a filter bank fitted near the top of a stack 125 m high. These exhaust filters
had to deal with a considerable flow of air and were not designed to have a
high efficiency for all particle sizes. They were intended, in the event of a
severe accident, to arrest the larger particles that might otherwise have caused
heavy contamination in the vicinity of the Works.

Windscale Works, now known as Sellafield, is situated on a low-lying coastal
strip in northwest England. Inland from the Works, the ground rises to about
600 m in a distance of about 8 km.

The accident that occurred October 1957 was the result of a deliberate release
of Wigner energy from the graphite moderator. At the operating temperatures
of those reactors, the neutrons being thermalized (or slowed down) in the graphite
caused lattice deformations. Energy was thus stored in the graphite matrix,
which could spontaneously relax, leading to dangerous overheating. This so-
called Wigner energy was therefore released by periodic controlled heating of
the graphite to restore the normal graphite lattice.

Because of the inadequate instrumentation, the operators mistakenly thought
the core was cooling without releasing all the stored Wigner energy. They
applied a second period of "nuclear heating," which overheated several fuel

---

[1] National Radiological Protection Board, Oxfordshire, Ox11 ORQ, United Kingdom.
[2] For consistency, abbreviations used for metric units conform to AMA style specifications,
which may vary from SI conventions.

channels leading to U oxidation and a spreading graphite fire that eventually involved some 150 channels of fuel. After several unsuccessful attempts to put out the fire, it was finally extinguished by flooding the pile with large volumes of water.

## Initial Assessments

Three significant routes of exposure were identified at the outset—direct radiation from the released activity, inhalation wherever the plume reached ground level, and ingestion of contaminated food and water (1). Air sampling and radiation dose-rate measurements soon established that neither direct exposure nor inhalation was likely to cause individual doses high enough to call for emergency action. Early measurements of $^{131}$I in milk, however, suggested that the thyroid dose to infants drinking fresh cow milk might become unacceptably high.

In 1957 there was no guidance on levels of dose or of activity concentration in foodstuffs above which action should be taken to protect individuals. Decisions had to be made quickly. It was decided to control milk, the only foodstuff likely to cause significant intakes, so that the thyroid dose to the milk-drinking infant would not exceed 20 rad (200 mSv). Current international advice (30 years later) suggests a range from 50 to 500 mSv. The conversion from thyroid dose to a concentration in milk requires information about the intake of milk, the duration of the contamination, the transfer to and retention in the infant thyroid, and the mass of the thyroid. Prudent judgments on these parameters suggested a peak concentration of milk of 0.1 μCi/L (3,700 Bq/L) with a mean life in milk of 11.5 days. Current judgments adjusted to a thyroid dose of 200 mSv would give about twice this figure (8,000 Bq/L) (2).

The control procedures were then based on replacing the milk in any area where the concentration on about day 3 was likely to be above 0.1 μCi/L. By good fortune, the line of demarcation ran mainly over mountainous areas and a major estuary. There was then no problem of requiring different treatment for adjacent farms. The milk was replaced by relatively uncontaminated supplies from outside the controlled area. This process was greatly simplified by the milk distribution system under which most milk was collected from farms and passed to retail outlets by a single body, the Milk Marketing Board.

## The Monitoring Measurements and Subsequent Action

The release took place over the period from about midday on Thursday, October 10, 1957, to midday on Friday, October 11.

The position by the evening of Friday, October 11, was as follows:

1. Radiation surveys in the district had demonstrated that there was no external radiation hazard.

2. Air sampling indicated that there was no significant inhalation problem.
3. Interpretation of the gamma radiation survey, on the assumption that the deposited radioactivity was a normal distribution of mixed fission products, indicated a possible marginal level of milk contamination.

In order to resolve the problem of a possible ingestion hazard from milk, samples of milk had been collected during milkings at local farms on the evening of Thursday, October 10, and the morning and evening of Friday, October 11. Arrangements were made to analyze these for both radiostrontium and radioiodine. These analyses were done in duplicate at Windscale and at the Atomic Energy Research Establishment (AERE) at Harwell some 500 km away because of possible contamination of the local analytical laboratories. The first set of results for the morning milking of October 11 at a Sellafield farm had an activity of 0.005 μCi/L $^{131}$I, and the second, the afternoon milking from the same farm, gave 0.78 μCi/L $^{131}$I. These results were followed by the results of the analyses at AERE, which confirmed those made at Windscale. By the afternoon of Saturday, October 12, it was evident that milk produced on Thursday and Friday morning showed only traces of radioiodine, whereas the afternoon milking of Friday, October 11, gave activity levels between 0.4 μCi/L and 0.8 μCi/L. On the basis of this information, it was decided to stop the use of local milk for human consumption. This decision was taken in consultation with the Ministry of Agriculture, Fisheries and Food, and the necessary arrangements were made by the local police force. It was also clearly shown that the problem was basically contamination of milk by $^{131}$I and that the release was not of a normal distribution of mixed fission products but had involved a preferential radioiodine content. This was further confirmed by the analyses for radiostrontium, which were now available and which showed that Sr was considerably less important than I in this accident.

When the decision to restrict the consumption of milk was made, it was also decided to extend the restriction to all areas where the radioiodine activity in milk exceeded 0.1 μCi/L. Consequently, on Sunday, October 12, a more widespread sampling program was initiated. The information collected by an extensive gamma dose-rate survey and milk monitoring program showed that the direction of travel of the released radioactivity had been predominantly down the coast in a south–southeasterly direction. On Monday, October 14, the area for milk restriction was finally extended to cover an area of approximately 500 km$^2$.

The deposition was such that considerable use could be made of the gamma survey as a guide to the areas from which milk samples were required. The gamma survey teams were used to guide the milk collection and, since the radiation measurements could be made more easily and more quickly than milk monitoring, these teams were well in the van of the monitoring surveys from Windscale. Surveys were made covering southern Scotland, Yorkshire, Lancashire, Westmorland, and North Wales. Samples of milk had also been collected from the Isle of Man, Northern Ireland, and down to the south of England.

In addition to milk, other foodstuffs were also monitored, and during the course of the survey more than one tonne of vegetables, about 700 eggs, and more than 50 kg of meat were monitored. Water samples from drinking water supplies in Cumberland, Lancashire, and North Wales were also analyzed for activity. In all these cases, it was found that the level of contamination was well below that which would constitute a hazard to individuals.

During the week beginning Monday, October 21, the survey and monitoring program was designed to forecast when derestriction might begin. On the advice of the Medical Research Council, the conditions for derestricting an area involved demonstrating that

1. all milk samples had an activity of less than 0.1 $\mu$Ci/L $^{131}$I, and
2. the decrease of the activity in successive samples showed a half-life of not more than 8 days.

## Some Comparative Measurements

The widespread measurements of activity in milk and of the gamma dose rate over pastureland gave useful information on the relationship between these quantities and on the way in which they varied with time. This is well illustrated by diagrams directly reproduced from material provided for a major review of the accident presented to the second United Nations Conference on the Peaceful Uses of Atomic Energy in September 1958 (1).

Figures 13.1 and 13.2 show the gamma dose rate and milk concentration for Sunday, October 13. The correlation between the two maps is striking. The quantitative relationship depends critically on the mixture of radionuclides deposited. The results for the gamma dose rate were obtained with scintillation gamma counters originally intended for geological survey work. They had been calibrated for Ra gamma rays and substantially overestimated the dose rate from $^{131}$I. The quantitative relationship between the measured gamma dose rate and the concentration in milk was on this occasion established empirically.

Successive measurements of the gamma dose rate and milk activity were made. The gamma dose rate showed a hint of short-lived nuclides but generally followed the decay of $^{131}$I (half-life of 8 days). The milk samples decreased more rapidly (half-life of about 5 days) as the cows' diet contained increasing proportions of newly growing grass. A similar initial reduction was found in the activity of $^{137}$Cs in milk and grass, but the longer term reduction became slower with a half-life of some tens of days.

## Long-Range Measurements

In September 1958, Stewart and Crooks of AERE Harwell published the results of longer range measurements in England to the south of Sellafield and in Europe (3). Many of the results were obtained by analyzing air filters originally

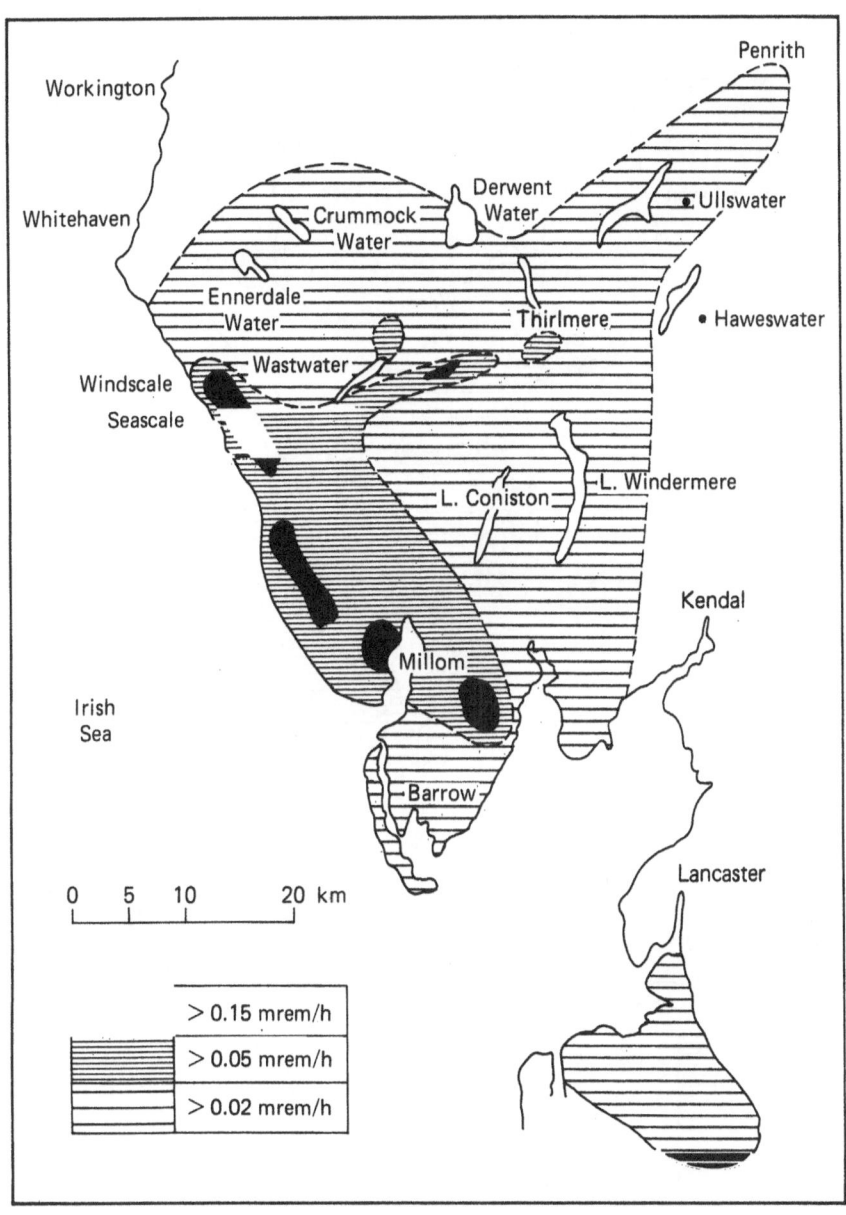

**Fig. 13.1.** Gamma background, October 13, 1957.

used for assessing industrial air pollution. In the units of the day, the time integral of air concentration ranged from around 1,000 pCi days/m³ in northwest England (some 50 to 100 km from Sellafield) through values of a few hundred pCi days/m³ in southern England, 30 to 50 pCi days/m³ in the Netherlands, 10 pCi days/m³ in Germany, and 0.2 pCi days/m³ in Switzerland. As an indication of the significance of these integrated concentrations, it may be noted that the

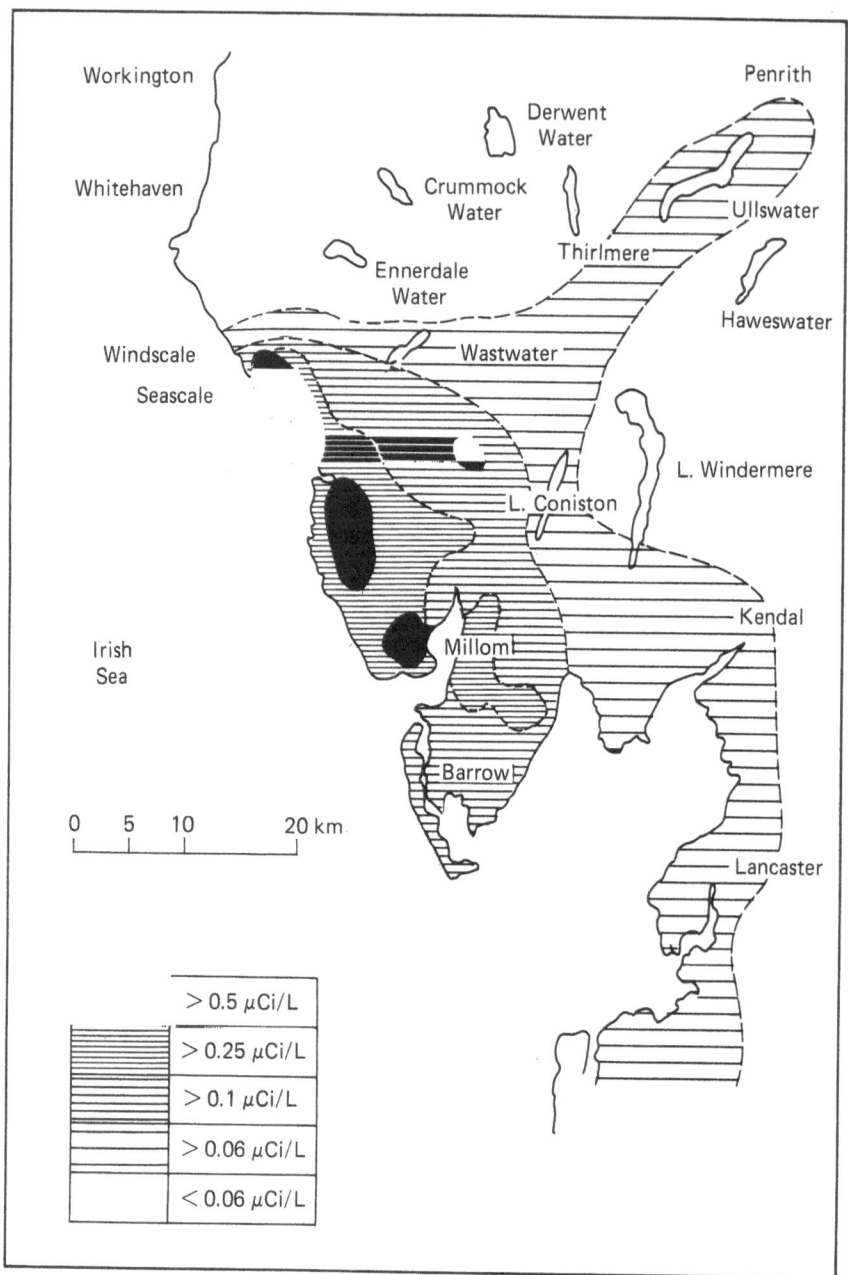

**Fig. 13.2.** Iodine-131 in milk, October 13, 1957.

inhalation by a child of an integrated concentration of 1,000 pCi days/m$^3$ would give a thyroid dose of about 30 mrem (0.3 mSv).

## Later Evaluations

By about 1980 the National Radiological Protection Board had the modeling capacity and the resources to undertake a review of the wider implications of the 1957 accident. The initial assessments had considered only individual dose, but the concept of a collective dose had developed since that time and the Board decided to evaluate collective doses to the thyroid and collective effective dose equivalents in an attempt to estimate the possible total of long-term health effects. The report (4), published in February 1983, and an addendum (5), published in September 1983, assessed the likely releases of some 45 nuclides. Using the geographical distribution of $^{131}$I in milk, together with the ratios of the released activities and environmental models to take account of the different behaviors of the nuclides, the authors, Crick and Linsley, assessed the dose distribution from the accident over the whole United Kingdom and, with less certainty, over the rest of Europe. They were then able to estimate the collective doses and their geographical distribution.

Table 13.1 shows the estimated releases of the principal nuclides. The values for $^{131}$I and $^{137}$Cs are the same as those estimated at the time (20,000 Ci and 600 Ci). The value for $^{90}$Sr was initially overestimated at 9 Ci (now 2 Ci).

Table 13.2 shows the estimated thyroid doses and those directly determined from measurements of $^{131}$I in the thyroids of a few people who volunteered for measurement soon after the accident. The agreement is unexpectedly good and confirms the belief that the main source of intake within the area of milk control was inhalation during the passage of the cloud.

Table 13.3 shows the collective thyroid doses and the collective effective dose equivalents reported by Crick and Linsley. The values take account of

**Table 13.1.** Releases from the Windscale Fire 1957[a]

| Nuclide | Quantity released (TBq) |
|---------|-------------------------|
| $^{90}$Sr | 0.07 |
| $^{106}$Ru | 3 |
| $^{131}$I | 740 |
| $^{132}$Te | 440 |
| $^{133}$Xe | 12,000 |
| $^{137}$Cs | 22 |
| $^{210}$Po | 8.8 |
| $^{239}$Pu | 0.0016 |

[a] Reprinted with permission from refs. 4 and 5.

**Table 13.2.** Comparison between measured and calculated thyroid doses to individuals living in the Windscale area[a]

| Distance from Windscale (km) | Calculated thyroid dose from inhalation (mSv) | Distance from Windscale (km) | Measured thyroid dose (mSv) |
|---|---|---|---|
| | | 3 | 5 |
| 0–10 | 18 | 6 | 14 |
| | | 7 | 14 |
| | | 9 | 18 |
| 10–20 | 13 | 18 | 14 |
| 20–30 | 5.5 | | |
| | | 31 | 4 |
| 30–40 | 3.1 | 37 | 5 |
| | | 38 | 3 |

[a] Reprinted with permission from refs. 1 and 4.

the milk restriction. Table 13.4 shows the distribution of the collective effective dose equivalent by geographical areas and by the exposure pathway. The geographical distribution of the dose from foods is uncertain because it has been assumed that the food is consumed in the area in which it is produced. This assumption does not affect the estimate of the collective dose in the whole country or for Europe as a whole.

If it can be assumed that radiation doses close to those received in a year from exposure to natural sources give rise to a proportionately increased risk of fatal cancer, it is possible to assess the long-term health effects in the United Kingdom. The number of effects in the rest of Europe would be small. In round figures, there might be about 30 additional fatal cancers over a period of 40 or 50 years while perhaps 300 nonfatal cancers, mainly of the thyroid. The number of cancer deaths in the United Kingdom over 50 years, if the present rates prevail, will be about 7 million, with about 30,000 thyroid cancers (mainly nonfatal).

The risk to the most highly exposed individuals is more difficult to assess because the doses depend on individual behavior, and the risks of natural and radiation-induced thyroid cancer are age dependent. It seems likely that the small natural risk of thyroid cancer in children might be increased by a factor of between 3 and 10 in the most highly exposed individuals. The number of children so exposed is such that it is extremely unlikely that even a single case will occur in this group.

**Table 13.3.** Collective dose in Europe from the Windscale fire[a]

| | |
|---|---|
| Thyroid dose equivalent | 26,000 person-Sv |
| Effective dose equivalent | 2,000 person-Sv |

[a] Adaptation from refs. 4 and 5.

**Table 13.4.** Collective effective dose equivalent from the Windscale fire[a]

| Pathway | Cumbria (person-Sv) | United Kingdom (person-Sv) | Europe (person-Sv) |
|---|---|---|---|
| Inhalation | 35 | 900 | 980 |
| Ingestion | | | |
| Milk | 88 | 570 | 590 |
| Other foods | 12 | 170 | 190 |
| External | | | |
| Ground deposit | 12 | 190 | 210 |
| Plume | 5 | 54 | 57 |
| Total (rounded) | 150 | 1900 | 2000 |

[a] Adaptation from ref. 5.

## Organizational Lessons

The Windscale accident was the first major nuclear accident to be reported. The lessons to be learned were spelled out at the time and covered the following principal points, all of which are taken from the review to the UN conference in 1958 (1):

1. There is an immediate need for predetermined action levels of radiation and radioactive contamination specified for a short-term hazard.
2. All the relevant factors must be considered and action levels for external radiation, inhalation, and ingestion must be determined, as far as possible, before the event, so that the appropriate action may be initiated immediately without diverting technical effort to the derivation of appropriate levels.
3. In an incident of this type, there is a need for a rapid increase in both monitoring and analytical facilities. Provided that the local survey and analytical resources can rapidly be augmented, it is unnecessary to have always available *on every site* sufficient personnel and equipment to carry out the comprehensive survey that is needed following the accident. However, the team available on the spot should be of sufficient strength to provide a quick estimate of all possible hazards within a period of a few hours. Close prior coordination between establishments can then provide for additional trained personnel and suitable instruments.
4. It may be desirable to establish fixed monitoring posts in the district surrounding a station and to coordinate these with mobile monitoring teams and survey aircraft. An aerial gamma survey, provided weather conditions permit flying at about 500 ft, is most useful in the early stages of an accident for broadly delineating the affected area. Besides providing information on which the controller can direct mobile ground survey teams, it will enable early definition of areas subject to restrictions. It is of importance to good public relations to overestimate rather than underestimate the area specified.

5. In many incidents of this nature, there is likely to be some milk contamination, since, as a food, milk is particularly sensitive to ground deposition; and it is also a primary food for children. Consequently, it is essential that any restriction applied to milk distribution must be applied quickly, certainly within a few days.
6. Every country should have a national plan that utilizes all possibile facilities and gives special consideration to the provision of analytical services; which is of paramount importance since local contamination may render inoperative laboratories in the vicinity of the accident.
7. Local meteorological information is an essential requirement, and there should be expert advice available to interpret this information so that an estimate can be made of the wind patterns and the likely path of the released activity.
8. Finally, there is the control organization. This has two aspects which might be termed the tactical and the strategic. The first of these is direct control of the mobile teams working in the field. In the control room, there are maps on which the current position is displayed and from which future trends may be predicted, enabling the controller to deploy monitoring survey teams to the best advantage. The second need is for strategic direction, which relies for its information on the tactical control room. Here, all inquiries from outside are answered and attention is directed to special problems. Longer term trends can be followed and the imposition or removal of restrictions forecast. Good communications with government departments and other establishments are essential for this second control room.

Most of these lessons have been learned, although experience following the Chernobyl accident in 1986 showed that the scale of preparedness was not always sufficient and that there were significant policy differences between countries. Perhaps the only new lesson concerns the importance of communications to the public. The fear of radiation has increased dramatically over the past 30 years and with it the need for an enormous increase in published information and reassurance. In this area there is still much to do.

## References

1. Dunster HJ, Howells H, and Templeton WL (1958) District surveys following the Windscale incident. Proceedings of the 2nd International Conference on the Peaceful Uses of Atomic Energy, United Nations, Geneva 18:296
2. National Radiological Protection Board (1986) Derived emergency reference levels for the introduction of countermeasures in the early to intermediate phases of emergencies involving the release of radioactive materials to atmosphere. NRPB-DL10, Chilton, UK
3. Stewart NG and Crooks RN (1958) Long-range travel of the radioactive cloud from the accident at Windscale. Nature 182:627–630
4. Crick MJ and Linsley GS (1982) An assessment of the radiological impact of the Windscale reactor fire, October 1957. National Radiological Protection Board, NRPB-R135, Chilton, UK
5. Crick MJ and Linsley GS (1983) Addendum to Report NRPB-R135. Chilton, UK

CHAPTER 14

# The Accident at Three Mile Island—1979

T. M. Gerusky[1]

## Introduction

At 7:04 A.M. on March 28, 1979, the Bureau of Radiation Protection's emergency duty officer, a nuclear engineer, was notified by the Pennsylvania Emergency Management Agency of an accident at the Three Mile Island Unit 2 reactor (TMI-2). He immediately contacted the control room to verify the notification and to receive detailed information. A site emergency had been declared because of high radiation levels detected on remote area radiation monitors in the auxiliary building.

That began what has been described, until Chernobyl, as the most serious accident in commercial nuclear power history. It was indeed a major malfunction of the control hardware and a major failure of plant personnel to recognize the problem. Unbelievably, the off-site consequences were minor. The reactor safety systems worked as designed. The only real effect of the accident was and continues to be the emotional trauma to the residents of the area as a result of misinformation, an uninformed press, and the scare headlines and stories in both the print and electronic media.

There have been many reports on the actual sequence of events that occurred at TMI-2 on the morning of March 28, but there are still some unanswered questions concerning some of the actions by the operators during those hours. A summary of the accident is presented in this chapter.

## The Accident

At 4:00 A.M. on March 28, 1979, a series of pumps supplying water to the steam generators tripped or shut down (1). When the flow of water stopped, the plant's safety system automatically shut down the steam turbine and steam generator. The pressure in the reactor increased because there was no way to

---

[1] Bureau of Radiation Protection, Pennsylvania Department of Environmental Resources, P.O. Box 2063, Harrisburg, PA 17120, USA.

get rid of excess heat, and a valve, the pilot operated relief valve (PORV), above the pressurizer opened as designed to relieve the pressure. The reactor automatically shut down, all within eight seconds. A schematic drawing of TMI-2 is shown in Fig. 14.1.

Heating was still occurring from the decay of the fission products in the reactor core. It had to be removed. Three emergency feedwater pumps automatically started to put water into the steam generators to take away some of the heat, but two closed valves prevented this water from reaching the system. The operators did not notice the warning lights indicating the problem.

The pressure was decreasing in the reactor because of the open PORV valve. Normally, the valve would close automatically when the pressure decreased, but in this case the valve was stuck open for a period of 2 hours and 22 min until an operator realized the valve was open and closed a backup valve. A "loss of coolant" accident was occurring.

The high-pressure injection system automatically came on to pour about 1000 gal per minute of water into the system to cool it down. The operators turned off the high-pressure pumps because they did not realize that water was being released through the valve that was stuck open.

Temperatures built up in the reactor core as water turned to steam and the cladding, and then the fuel itself, melted, releasing fission products to the reactor vessel and coolant system.

Coolant water was released through the PORV valve to a drain tank in the basement of the reactor building. It overflowed to a sump, which pumped the water to a tank in the auxiliary building. About a half an hour after the accident started, the operators turned off the sump pumps because of increasing radiation levels in the auxiliary building. The reactor coolant pumps were vibrating, and the operators also turned them off.

The operators finally realized that the PORV was still open and at approximately 6:25 A.M. closed the backup valve. The loss of coolant accident was over, but other methods were still needed to continue to cool the core. The high-pressure injection pumps were finally turned on an hour later.

## Misinformation

There was much misinformation and lack of knowledge at the utility and at the US Nuclear Regulatory Commission (NRC) during the first few days of the accident. No one really knew what had happened or what was happening inside the reactor. It is only now, during the billion-dollar clean-up of the reactor, that the core is being studied. Approximatley 50% of the fuel had melted. Figure 14.2 shows what the reactor core looked like when it was finally studied.

During the first week of the accident, a series of events took place that caused confusion among the "experts," much sensationalism in the press, and some panic among the population in the vicinity of the plant.

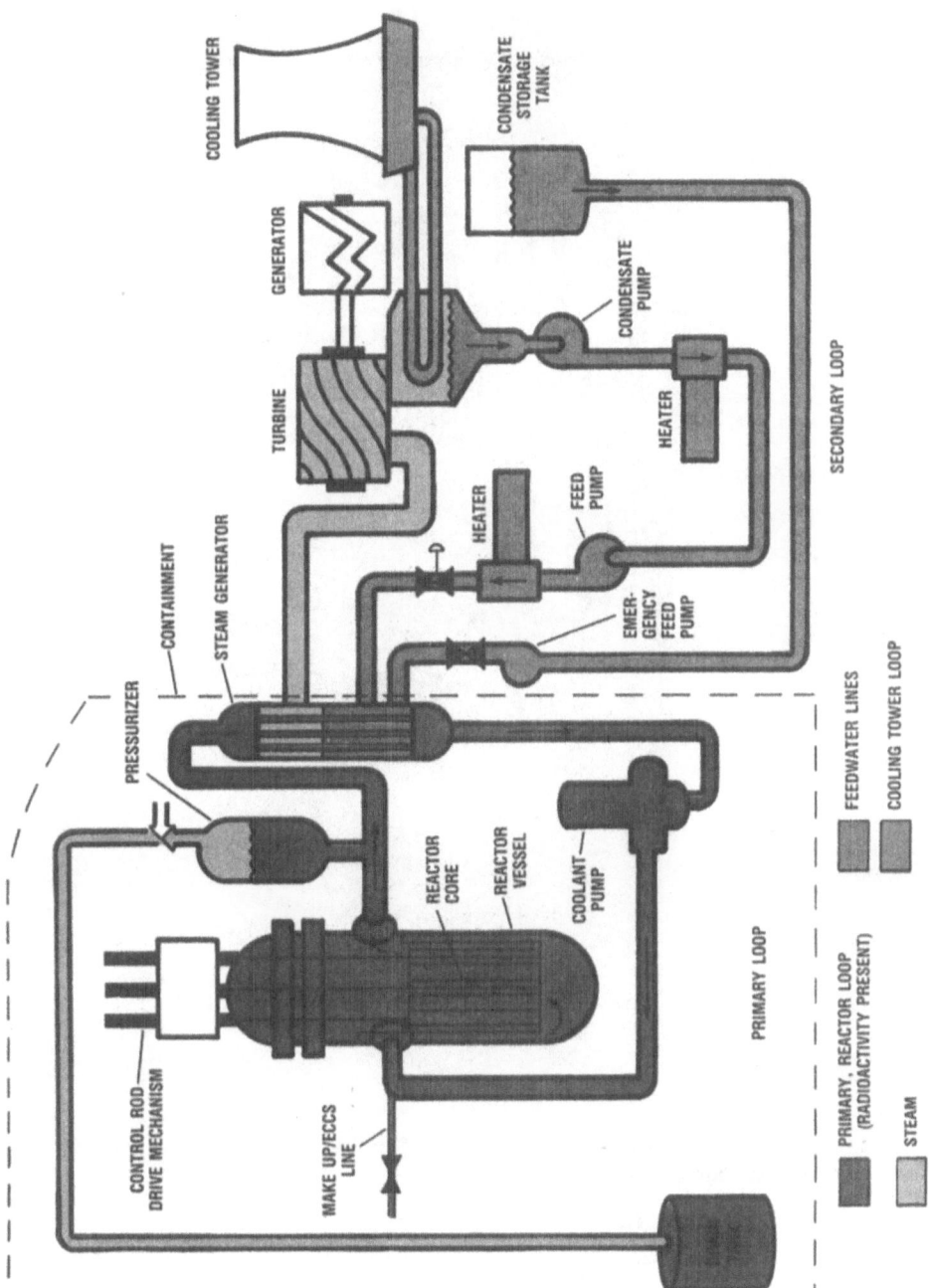

**Fig. 14.1.** Schematic of the TMI-2 facility. (Reprinted with permission from ref. 2.)

159

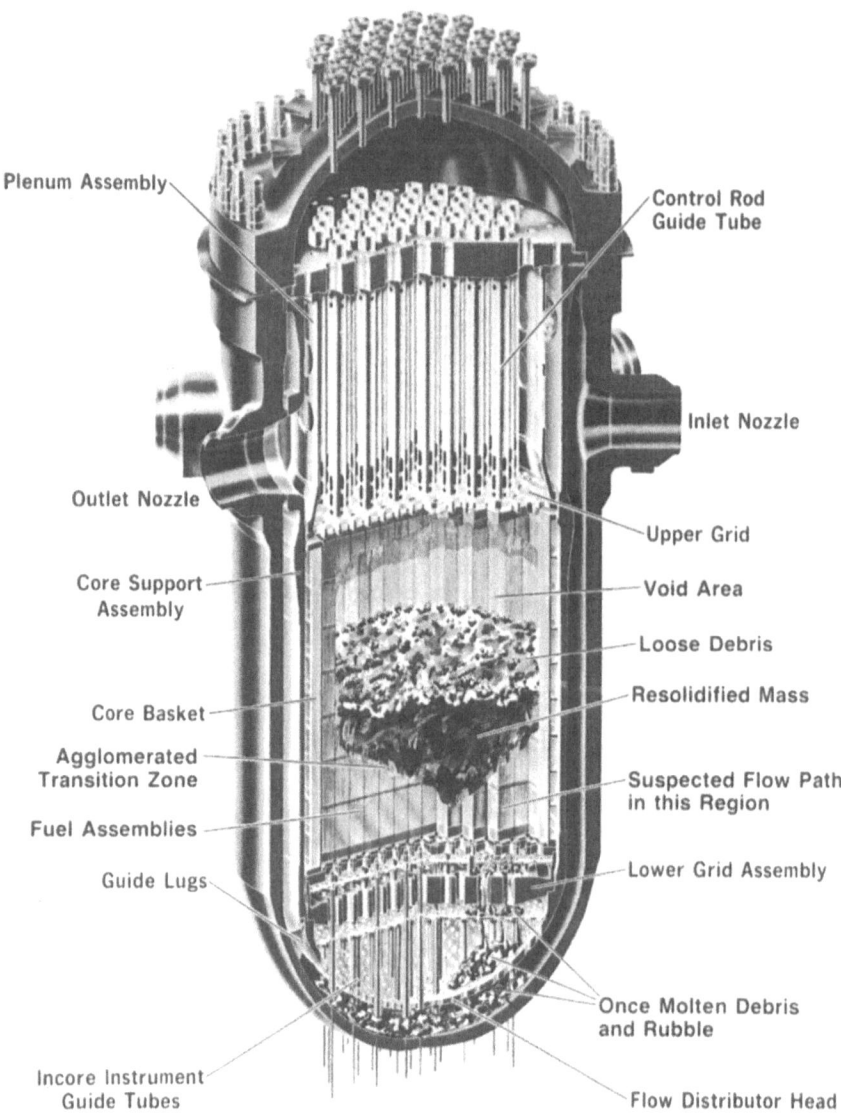

**Fig. 14.2.** TMI-2 Postaccident core conditions. Note that some components are not accurate representations of TMI-2 reactor dimensions.

Radioactivity was released from the plant beginning early on the first day of the accident. The radioactivity was being monitored and consisted mainly of the noble gases $^{133}$Xe, $^{133m}$Xe, and $^{135}$Xe. Some radioiodine was also detected, but well below levels that were seen during earlier nuclear weapons fallout periods. On Friday, March 30, the third day of the accident, an event occurred that caused the governor to recommend the evacuation of pregnant women and small children from an area within a five-mile radius of the reactor, to close all schools within the zone, and to set up temporary shelters outside the area to handle the evacuees. The TMI-2 reactor had a means of removing water from the primary reactor coolant system, called the let-down system, and a means to add water to the system, called the make-up system. Both systems had water storage tanks outside the reactor vessel in the auxiliary building, and the piping from these systems leaked. Dissolved gases, which are under pressure in the reactor coolant system, are released from the water upon depressurization and stored in the waste gas decay tanks for radioactive decay. Because of pressure buildup in these tanks during the accident, a controlled transfer of radioactive gases was made from the make-up tank to the waste gas decay tank. Leaks in the system allowed radioactive gases to escape into the auxiliary building and then into the atmosphere. To check on the releases, a helicopter was monitoring directly above the release point. At approximately 8:00 A.M., the helicopter was detecting 1,200 millirem (mrem) per hour at 130 ft above the vent stack or 600 ft above grade.

At NRC headquarters, there was concern that when the pressure in the waste gas storage tanks became too high, an uncontrolled and continuous release of radioactive gases would occur. NRC staff had calculated that the off-site ground level dose rate would be 1,200 mrem/h if the gas decay tanks were lost. On Friday morning, they had not been informed of the controlled transfer and when word came in that a level of 1,200 mrem/h had been detected, without confirming the reading, the source of the radiation, or the location of the reading, NRC officials recommended to the Pennsylvania Emergency Management Agency by telephone that the governor order an evacuation. They had assumed that the waste gas decay tanks had indeed been lost. They had *not*.

Information was passed on to local emergency management agencies that an evacuation may be imminent. When the Bureau office was contacted with that information, however, the staff verified that nothing unusual was happening at the reactor other than the controlled shift of gases to the waste gas tank. Levels off-site were in the same range as the previous days, little radioiodine was being detected, and the noble gases were still the main contributors to the off-site dose. The Bureau recommended against evacuation. Unfortunately, local emergency management agency officials had announced on a local radio station that an evacuation might be called.

Communications between the NRC officials on Three Mile Island and NRC headquarters were wholly inadequate, but when the NRC on-site staff learned of the evacuation notice, contact was made with headquarters to call off the recommendation.

In the Governor's office, a decision had been made not to evacuate immediately. The Governor had contacted the Chairman of the NRC, who had confirmed that they had erred on the evacuation request, but the NRC Chairman recommended that the Governor issue a warning to the public within 5 miles of the plant to stay inside. The Governor announced that all persons within 10 miles of the accident site stay inside until further notice.

In the meantime, the Governor and his top advisors continued to evaluate the situation. The Pennsylvania Department of Health had recommended an evacuation of pregnant women and children under the age of 2. President Carter had called the Governor and promised to send one expert from the NRC to direct federal activities and to set up a special communications link between the White House, the Governor's office, Three Mile Island, and the NRC. NRC Chairman Hendrie telephoned again. When asked by the Governor for his comments on the health department's recommendation, Chairman Hendrie responded "If my wife were pregnant and I had small children in the area, I would get them out, because we really don't know what is going on in the reactor and what will happen in the near future." Governor Thornburgh announced the evacuation recommendation for those parts of the population within 5 miles.

Another problem, again made in error by the NRC, became evident on Friday, March 30. Harold Denton, the NRC's director of reactor regulation, arrived as the President's expert and briefed the Governor. A hydrogen explosion had occurred during the afternoon of March 28 inside the reactor building. The NRC was aware that a gas bubble had been formed inside the reactor and was concerned that this bubble would grow from the additional generation of hydrogen and oxygen from radiolytic disassociation of water and possibly explode. The bubble problem was discussed in the press. Plans were made over the weekend for a massive evacuation of the population if oxygen levels grew to such an extent that an explosion was possible. The NRC and others were estimating that oxygen levels were increasing at a rate between 0.1% and 5% per day and that an oxygen–hydrogen explosion could occur within 4 to 12 days. Even without the problem resolved, the President and the Governor visited the plant on Sunday, April 1, to express the President's confidence that the federal government had the situation under control.

Most of the calculations were made assuming that no recombination of hydrogen and oxygen would take place. However, in pressurized water reactors, hydrogen is placed into the system to prevent radiolysis from occurring. When the concentrations and amount of hydrogen were taken into consideration with the pressures in existence in the reactor vessel at the time, it was obvious that no oxygen was being generated. That conclusion was finally reached by the NRC staff on Monday, April 2. Although Harold Denton's press briefing later that day included a statement that the bubble size was apparently getting smaller and that earlier estimates of oxygen generation rates were conservative, no statement was ever made that there was never a chance of an explosion inside the reactor vessel.

## Environmental Response

Early on the first day of the accident, the Bureau of Radiation Protection had requested assistance from the US Department of Energy's (DOE) emergency response team. The team arrived from Brookhaven National Laboratory in the early afternoon and assisted the state in monitoring the off-site radiation and radioactivity levels. DOE's presence increased as the situation evolved, and the team stayed until no longer needed. Other federal agencies also responded as the Federal Interagency Radiological Assistance Plan dictated and assisted their counterpart state agencies as required.

Bureau staff had been sent into the field to do monitoring on the morning of March 28. After the DOE arrived, the bureau staff was used only as a liaison between the Pennsylvania Department of Environmental Resources in Harrisburg and the DOE team, which set up a local airport. The DOE team brought with them one of their main contractors, E.G.& G., with aerial surveillance capabilities and other technical resources.

Following immediate notification from the plant around 7:00 A.M., contact was established by telephone with the health physics staff of the plant, and an open line was maintained for about a month. Over the next 30 days, the bureau office was staffed 24 hours a day with experienced health physicists. Information from the plant indicated that the dome monitor read 800 R/h. Based upon that reading, a calculated dose rate of 10 R/h was determined for Goldsboro, a town just west of the plant. This calculation was based upon a containment leak rate of 0.02% per day at design pressure. Since the containment pressure was not excessive, no driving force was available to produce a source term from the containment. There were no on-site readings that would indicate such a major release. But verification was requested and a survey team was dispatched across the river. No abnormal readings were found. Later in the morning, the operator indicated that the off-site monitoring team had detected radioiodine levels in air samples taken across the river west of the facility. Because of high background levels in the TMI laboratory, samples were taken to the bureau's environmental radiation laboratory in Harrisburg. No radioiodine was found.

During the period immediately after the accident to the present time, increased environmental radioactivity surveillance has been carried out in the vicinity of the reactor facility. The major radionuclides released from the reactor were the radioactive noble gases, $^{133}$Xe, $^{133m}$Xe, and $^{135}$Xe. Some small amounts of radioiodine were also released. The calculated release of radioactivity during the course of the accident was 10 million curies (Ci) of noble gases and less than 30 Ci of $^{131}$I (2).

Thousands of environmental samples were collected, analyzed, and reported. Samples of milk, air, water, produce, soil, vegetation, fish, and river sediment and silt in the vicinity of TMI were evaluated. The results showed that little or no radioactivity found its way into the environment with the exception of the noble gases, which resulted mainly in external beta and gamma exposures.

Table 14.1 provides a summary of the environmental data as reported by the Public Health and Safety Task Force of the President's Commission on the Accident at Three Mile Island (3).

External radiation from the noble gas cloud was the most significant source of exposure. The field survey data that the Bureau, DOE, the utility, and the NRC obtained were just that—field survey data, which indicated external beta–gamma levels at the survey location. The aerial monitoring done by the

**Table 14.1.** Summary of significant radionuclide concentrations in environmental samples off-site[a]

| Sample type | Isotope | Total samples analyzed | Number of samples with significant results | Mean | Maximum |
|---|---|---|---|---|---|
| Precipitation | $^{131}I$ | 7 | 0 | | |
| | $^{137}Cs$ | 5 | 0 | | |
| Air | $^{131}I$ | 1,476 | 176 | 5.8 pCi/m$^3$ | 119 pCi/m$^3$ |
| | $^{85}Kr$ | 34 | 34 | 70 pCi/m$^3$ | 1500 pCi/m$^3$ |
| | $^{133}Xe$ | 34 | 32 | 4900 pCi/m$^3$ | 140,000 pCi/m$^3$ |
| | $^3H$ | 9 | 4 | 1.5 pCi/m$^3$ | 3.3 pCi/L$^3$ |
| River sediment and silt | $^{131}I$ | 28 | 0 | | |
| | $^{137}Cs$ | 28 | 19 | 0.3 pCi/g (dry) | 0.52 pCi/g (dry) |
| Soil | $^{131}I$ | 59 | 0 | | |
| | $^{137}Cs$ | 59 | 17 | 0.6 pCi/g (dry) | 1.39 pCi/g (dry) |
| Grass | $^{131}I$ | 6 | 2 | 0.05 pCi/g (dry) | 0.063 pCi/g (dry) |
| | $^{137}Cs$ | 2 | 2 | 0.3 pCi/g (dry) | 0.32 pCi/g (dry) |
| Nondrinking water | $^{131}I$ | 943 | 0 | | |
| | $^{137}Cs$ | 148 | 0 | | |
| | $^3H$ | 249 | 136 | 190 pCi/L | 810 pCi/L |
| | $^{90}Sr$ | 35 | 0 | | |
| Drinking water | $^{131}I$ | 324 | 0 | | |
| | $^{137}Cs$ | 147 | 0 | | |
| | $^{90}Sr$ | 43 | 0 | | |
| | $^3H$ | 127 | 104 | 180 pCi/L | 810 pCi/L |
| Fish | $^{131}I$ | 16 | 0 | | |
| | $^{137}Cs$ | 16 | 7 | 0.35 pCi/g (wet) | 0.778 pCi/g (wet) |
| | $^{90}Sr$ | 5 | 0 | | |
| Goats' milk | $^{131}I$ | 52 | 40 | 30 pCi/L | 110 pCi/L |
| | $^{137}Cs$ | 5 | 0 | | |
| | $^{90}Sr$ | 5 | 0 | | |
| Cows' milk | $^{131}I$ | 1,724 | 134 | 9.4 pCi/L | 21 pCi/L |
| | $^{137}Cs$ | 1,483 | 10 | 4.7 pCi/L | 37 pCi/L |
| | $^{90}Sr$ | 375 | 0 | | |
| Food | $^{131}I$ | 552 | 0 | | |
| | $^{137}Cs$ | 541 | 0 | | |
| | $^{90}Sr$ | 225 | 0 | | |

[a] Reprinted with permission from ref. 3.

E.G.& G. helicopter was able to locate the plume and determine dose rates in and around the plume, but calculation of ground-level dose rates was an exercise in scientific guesswork.

The most important monitoring tools were thermoluminescent dosimeters (TLDs) that the Bureau and the utility had placed around the reactor as part of the routine environmental radiation surveillance efforts. Additional dosimeters were placed in the environment after the third day of the accident by the US Environmental Protection Agency (EPA), the NRC, and the US Food and Drug Administration (FDA).

The Public Health and Safety Task Force's analysis of the data reads as follows:

The procedures for calibration, processing and reading these dosimeters were reviewed. Adjustments were made for estimated background values and energy dependence. Data from TLD's placed by the NRC on the third day of the accident were rejected because their handling procedures were inappropriate for this evaluation. Because of their late deployment and distance from the source, the dosimeters placed by the other two Federal agencies did not provide useful data.

The population distribution used to calculate the collective dose is based on projections of 1970 census data to the year 1980, as given in the final Safety Analysis Report for TMI-2. Adjustments were made to account for the fact that only one person is known to have been at the many summer cottage sites on the islands near the plant at the time of the accident.

The doses measured by the TLD's would be applicable to people who were outdoors all during the first few days of the accident. Because most people spent most of that time indoors, some protection can be assumed due to absorption of gamma radiation in the structural materials for houses and offices. It is estimated that the average dose received indoors is about three-quarters that of outdoors.

Persons within a 2-mile radius of the plant probably received the highest doses. The dose to one person known to have been on the nearby islands for about 9½ hours during the first few days of the accident, is estimated to be about 50 millirem (mrem). In addition about 260 people living mostly on the east bank of the river may each have received between 20 and 70 millirem. All other people probably received less than 20 millirem.

In estimating the health effects of low doses to a population, it is important to know the collective dose—the sum of the doses received by every person in the collective area. This is usually given in person-rems. The collective dose was calculated by multiplying the average dose at each of 160 areas surrounding the TMI plant by the population in that area and summing the products. The average dose in each area was estimated by interpolating between the locations at which TLD measurements were available. The collective outdoors dose to people within a 50-mile radius of TMI was calculated to be about 2,800 person-rems. Assuming that doses indoors were three-quarters of those received outdoors, the actual collective dose to the population is estimated to be 2,000 person-rems (3).

A DOE estimate using aerial survey data was also 2,000 person-rems (4).

The Ad Hoc Population Dose Assessment Group, comprised of representatives from the NRC, the EPA, and the Department of Health, Education, and Welfare,

**Table 14.2.** Estimated collective dose to population 0 to 50 miles from Three Mile Island from March 28 through April 15, 1979[a]

| Radius (miles) | Population | Collective dose (person-rem) | Average dose (mrem) |
|---|---|---|---|
| 0.4–1.0 | 324 | 19 | 58.6 |
| 1–2 | 1,816 | 36 | 19.8 |
| 2–3 | 7,579 | 120 | 15.8 |
| 3–5 | 18,567 | 180 | 9.7 |
| 5–10 | 137,474 | 720 | 5.2 |
| 10–20 | 577,288 | 1,173 | 2.0 |
| 20–50 | 1,420,071 | 537 | 0.38 |
| Total | 2,163,579 | 2,785 | 1.3 |

[a] Reprinted with permission from ref. 4.

evaluated the population doses from TMI in a report entitled "Population Exposure and Health Impact of the Accident at the Three Mile Island Nuclear Station" (4). They used four different methods of obtaining population exposure, resulting in doses of 1,600, 2,800, 3,300, and 5,300 person-rems. The average of the four, or 3,300 person-rems, was considered acceptable.

It is also interesting to note that the beta component of the noble gases would have given a skin dose of as much as four times the gamma dose if any person was submersed in the plume. Because of clothing considerations and other variables, no real beta component can be calculated. The estimated external dose to the population within 50 miles of the reactor is shown in Table 14.2.

For internal exposures, $^{131}$I, $^{137}$Cs, tritium, $^{85}$Kr, $^{133}$Xe, and $^{90}$Sr were considered. The types of environmental samples considered in estimating the internal dose included cow's and goat's milk, drinking water, air, fish, river sediment and silt, grass, precipitation, nondrinking water, and food. The food included unprepared food products such as eggs, poultry, pork, beef, fruits, and vegetables. Prepared foods included baked goods, cheese, and candy.

The only increases in environmental radionuclide concentrations following the accident at TMI were $^{131}$I in cow's milk, $^{131}$I in goat's milk, $^{131}$I in nondrinking water on-site, $^{131}$I in air on- and off-site, and $^{137}$Cs in fish.

The following tables (Tables 14.3 and 14.4) summarize the doses calculated by the President's Commission Task Force (3).

## Other Postaccident Studies

During the weeks and months following the accident, the Bureau and other agencies attempted to determine if the estimates made during the accident were indeed correct. Whole-body counts were performed on 760 residents within 5 miles of the reactor. No reactor-produced radionuclides were found. It is interesting that radon daughter products were detected in a few of the residents. At

**Table 14.3.** Internal dose due to $^{131}$I[a]

| Intake mode | Organ | Dose (mrem) |
|---|---|---|
| Inhalation | Newborn thyroid | 2.0 |
| | 1-year-old thyroid | 6.5 |
| | Adult thyroid | 5.4 |
| | Ovaries | 0.0002 |
| | Testes | 0.0001 |
| | Red marrow | 0.0007 |
| | Total body | 0.003 |
| Cow's milk ingestion | Newborn thyroid | 6.5 |
| | 1-year-old thyroid | 4.7 |
| | Adult thyroid | 0.6 |
| | Ovaries | 0.00002 |
| | Testes | 0.00002 |
| | Red marrow | 0.00009 |
| | Total body | 0.0003 |

[a] Reprinted with permission from ref. 3.

that time it was assumed that well water may have been the cause, but recent studies in Pennsylvania on radon concentrations in homes indicate that the source may have been airborne radon (5).

Film or TLD badges worn by radiation workers in the area were evaluated to see if any anomalies could be uncovered. The data either verified the estimates or suggested that the estimates were too high. The Federal Bureau of Radiological Health evaluated rolls of camera film that were sitting on shelves in the area to determine if any of the film was fogged. The Eastman-Kodak Company analyzed the film and reported no fogging in excess of 10 mrem (6).

## Pennsylvania Department of Health Studies

Immediately following the accident, the Pennsylvania Department of Health assembled an expert advisory panel to plan for follow-up epidemiological studies on the population (2,7,8,9). The following is a summary of those studies.

**Table 14.4.** Internal doses from other environmental radionuclides[a]

| Source of exposure | Organ | Dose (mrem) |
|---|---|---|
| $^{137}$Cs in fish | Total body | 0.02 |
| Noble gases | Total body | Small compared to external doses |
| Natural background | Thyroid and gonads | 27/y |
| | Bone surfaces | 60/y |
| | Red bone marrow | 24/y |
| | Lungs | 124/y |

[a] Reprinted with permission from ref. 3.

## Pregnancy Outcome

A total of 3,582 pregnant women who delivered within one year following the accident and who had lived within 10 miles of TMI were compared with a control group of 4,000 pregnant women who were not exposed. There were no measurable differences for prematurity, congenital abnormalities, neonatal deaths, or any other factors studied. The TMI Mother–Child Registry will continue to follow the children and report at five-year intervals.

## Infant Hypothyroidism

One case was reported within the 10-mile radius, but seven cases were reported in Lancaster County. The health department concluded that these cases were not associated with TMI.

## Infant Mortality

No significant differences were found.

## TMI Cancer Study

The department concluded that, after studying cancer incidence among residents within 20 miles of the reactor, ". . . results of our epidemiological study including both mortality and morbidity data as well as cohort follow-up analysis do not provide evidence of increased cancer risks to residents near the TMI plant" (9). The study is continuing.

## Continuing Problems

To many people in the area around Three Mile Island, March 28, 1979, was just the beginning of a series of events. These included follow-up studies on the population by the Pennsylvania Department of Health, the massive 1-billion-dollar effort to clean up the facility, the continuing articles in the press claiming that there was a cover-up of the real exposures to the population and the environment and that serious environmental and health effects were occurring in the vicinity of the plant, and the more than 2,500 personal injury suits that have been filed in the Pennsylvania courts for compensation for diseases from AIDS to cancer.

## Venting of the Krypton-85

The first real crisis occurred when the utility proposed to vent the $^{85}$Kr from the reactor building during the summer of 1980 in order to be able to enter the building to begin the process of recovery. Many in the public did not want any more radiation exposure and did not trust the utility, the NRC, or the state. Many public meetings were held before the actual venting occurred; a community-based monitoring program was instituted to allow the residents in the immediate area to monitor the Kr levels themselves; and a major effort of environmental monitoring was undertaken by the utility, the Environmental Protection Agency, which had established an environmental monitoring facility in Middletown, close to the reactor, the state, and others who were interested in their own monitoring. The public meetings were extremely heated, but, in the end, reason prevailed and the venting was accomplished with little radiological impact. Population doses were well below the estimates, in the fractions of a mrem range. It is interesting to note that one of the television stations in the area presented the monitoring data to the public as part of their nightly weather forecast, and the newspapers carried daily accounts of the event.

## Concern Over Environmental and Biological Effects

Some members of the public were extremely upset about the accident and noticed things in their environments that they had never noticed before. Many of these strange occurrences were blamed on the accident. The concerns included hair falling out, redness of the skin when going outside, farm animals with deformities, extremely large weeds and other plants, cancer patterns that apparently were never checked by the health department in their studies, and a resultant cry of cover-up when the health department discounted the studies. Some magazines and television programs were eager to point out these issues and continue to do so.

### Personal Injury Lawsuits

Some personal injury lawsuits (10) were filed immediately following the accident. The utility's insurer felt that settlement of the first few would be better than a long, drawn-out court battle. The settlements included $855,000 for an infant who suffered cerebral palsy, $1,095,000 for a five-year-old, born just after the accident, who has Down's syndrome, and smaller amounts for ailments ranging from heart attacks and cancer to emotional trauma and genetic damage. Over 2,500 additional suits have been filed and are expected to go to court in 1988.

### Claims of More Radiation Released Than Reported

Some scientists have claimed that more radiation was released from TMI during the accident than reported in the official documents. Doctors Jan Beyea and Bernd Franke have claimed that the TLD monitors were so few and widespread that previously unaccounted for releases could have occurred and slipped through "windows" in the TLD coverage (11). Dr. Seo Takeshi of Kyoto University has estimated that radioiodine releases were 300 times the claim made by the government. Dr. Karl Morgan estimated that 3,000 Ci of radioiodines could have been released, instead of the 14 Ci publicly announced (10).

## Conclusion

The accident at Three Mile Island in March 1979 did not contribute significantly to increased levels of radioactivity in the food supply in the area. Only slight, temporary increases in radioiodine levels were found in milk supplies, less than was detected in the local area during earlier Chinese atmospheric nuclear testing. The external exposure of the population was small when compared to the normal background, with a collective population dose of 3,300 person-rems. The potential health effect on the population of 2 million residents is one or less total cancers and less than one neonatal or genetic effect.

To the residents of central Pennsylvania, the accident is continuing through the billion dollar clean-up. Over 50% of the reactor core has been removed, and it is anticipated that the plant will be mothballed in 1989. However, the people who resided in the area through the first 30 days of the accident will never forget it.

## References

1. Kemeny JG, et al. (1979) Report of the president's commission on the accident at Three Mile Island. US Government Printing Office, Washington, DC
2. GPU Nuclear Corporation (1986) Radiation and health effects—a report on the TMI-2 accident and related health studies. Middletown, Pennsylvania
3. President's Commission on the Accident at Three Mile Island (1979) Report of the task force on public health and safety. US Government Printing Office, Washington, DC
4. NUREG-0558 (1979) Population exposure and health impact of the accident at the Three Mile Island nuclear station. US Nuclear Regulatory Commission, Washington, DC
5. NUREG-0636 (1980) The public whole body count program following the Three Mile Island nuclear accident. US Nuclear Regulatory Commission, Washington, DC
6. Federal Bureau of Radiological Health (1979) Unpublished data. Department of Health, Education, and Welfare, Washington, DC.
7. Pennsylvania Department of Health (1981) The Three Mile Island population registry report 1, a general description. Harrisburg, Pennsylvania
8. Pennsylvania Department of Health (1982) Impact of the TMI nuclear accident on pregnancy outcome, congenital hypothyroidism, and mortality. Harrisburg, Pennsylvania

9. Pennsylvania Department of Health (1986) Cancer mortality and morbidity (incidence) around TMI. Harrisburg, Pennsylvania
10. Wasserman H (1987) Three Mile Island did it. Harrowsmith Magazine 2, 9, pp 41–55
11. Three Mile Island Public Health Fund (1985) Proceedings of the workshop on Three Mile Island dosimetry. Philadelphia, Pennsylvania

CHAPTER 15

# Food-Chain Contamination From Testing Nuclear Devices

M. W. Carter[1] and L. Hanley[2]

## Introduction

Our more than 42 years of experience in dealing with radioactive contamination of our environment produced by fallout from the testing of nuclear devices began in 1945 in the desert near Alamagordo, NM. This first nuclear test was detonated on a 100-ft steel tower and resulted in radioactive fallout that was tracked and measured some few thousand miles across the United States toward the northeast.

Two important radiological health consequences of this atmospheric nuclear test were the exposure of people to ionizing radiation from fallout at large distances from the initiating event and the very protracted periods of time of these exposures. The radioactive contamination of the world's population and environment is a legacy from this first test and the hundreds of subsequent nuclear tests that have resulted in atmospheric fallout.

Human radiation exposure to fallout is the sum of external exposure and that of an internal nature resulting from inhalation and ingestion of foods and water that are contaminated by fallout. Our emphasis will be on the ingestion of contaminated foods and liquids, and the other types of exposure will be used only as a basis for comparison. Emphasis will also be given to nuclear tests primarily conducted in the United States.

Several radionuclides of special interest will be considered from the viewpoint of their dosimetric importance to humans. These will include certain fission products, several radionuclides whose radioactivity is caused by neutron activation, and radioactive materials that result from radioactive decay of fission products. We shall also take a very brief look at food-chain modeling because of its importance and use in predicting future information of relevance in understanding contamination levels and their effects. Modeling plays a major role

[1] International Radiation Protection Association and Georgia Institute of Technology Atlanta, Georgia 30332, USA.

[2] Nuclear Engineering and Health Physics, Georgia Institute of Technology, Atlanta GA 30332, USA.

in projecting radiation doses and health effects from contaminating events that have occurred or may occur in the future. Finally, it should be worthwhile to examine our fallout experience over some four decades in light of the current situation with regard to nuclear weapons tests, the fallout they may produce, contamination of our food chain from this source, the present trends, and reasonable expectations for the future.

## Nuclear Device Tests

The world's first nuclear device test occurred on July 16, 1945, and was code named Trinity. It vividly demonstrated that humans could design, fabricate, and test a device that was based on the detonation of fissile materials. The test had a yield of some 19 kilotons (kt) due to the fissioning of nuclear fuel (1).

Glasstone (1) discusses the general principles of nuclear explosions, their scientific basis, and detailed characteristics, as well as their various effects, including those of a biological nature. In fact, he describes the biological effects in terms of "effects of early fallout" and "long-term hazard from delayed fallout." Appendix B in Glasstone's book (1) also lists all the announced nuclear detonations prior to September 1961. Other early and good discussions of fission and fusion processes and especially the production of radiation and radioactivity by nuclear weapons detonations were presented by Mills (2) and Graves (3). The article of the Hearings of the Joint Committee on Atomic Energy (JCAE), "The Nature of Radioactive Fallout and Its Effects on Man," is also a most useful reference in understanding fallout and its effects (4).

On October 31, 1952, the United States detonated an experimental thermonuclear device, code named Mike, at the Eniwetok Atoll, which was part of its Pacific Proving Ground. The device's yield was a product of fission as well as fusion. The yield was not announced but it was relatively high based on the height of the mushroom cloud it produced (1). This event signaled the ability of man to produce nuclear devices with truly enormous explosive yields.

The heights of the radioactive clouds produced from nuclear tests are very important in determining the types, distribution, and effects of fallout that will result. For example, the early fallout comes from radioactivity that is confined to the troposphere, whereas the delayed fallout is produced from radiation debris that has entered the stratosphere.

There is also a technical distinction between a nuclear weapon and a nuclear device. The former represents a practical configuration that could be used for military purposes, whereas the term device is much broader and would include experimental configurations that may require considerable engineering to render them of value as a military weapon. Of course, nuclear devices have been tested for several civilian purposes, such as earth moving, mineral recovery, and seismic research. In our discussion, we shall use these terms interchangeably.

The testing of nuclear devices has continued until the present. Five additional countries, the Soviet Union, United Kingdom, France, Peoples Republic of China, and India, have tested one or more nuclear devices. The number of announced tests had exceeded 800 by the end of June 1975 (5) and by the end of June 1978 had surpassed 900 (6). Information on nuclear tests is available periodically from the Nevada Operations Office of the US Department of Energy. For example, the report, NVO-209 (7), lists 787 nuclear tests conducted by the United States through December 1986. In reviewing the record of nuclear testing by all countries, it is obvious that peaks of atmospheric testing occurred between 1952 to 1958 and 1962 to 1963.

Nuclear tests are usually conducted at established testing sites but occasionally at other locations. They have varying yields, may be entirely fission or fission and fusion as their energy mechanism, and are emplaced in various ways for testing. For example, tests have been conducted underwater, underground, in the atmosphere, and at very high altitudes. The type of emplacement, its environment, or location and the yield of nuclear device tests are greatly influenced by pertinent international agreements and treaties. These aspects of nuclear weapons testing were reviewed by Carter and Moghissi (5) and are summarized in Table 15.1.

## Fallout From Nuclear Tests

Two references mentioned earlier (1,4) are a good source of fundamental information on fallout from nuclear weapons tests, its environmental transport, general distribution, and behavior. Our primary interest will be in selected fission products, such as $^{131}I$, $^{90}Sr$, $^{89}Sr$, and $^{137}Cs$, and neutron activation products, such as $^{14}C$ and $^{3}H$. These are produced or available in some abundance, easily transported through the environment, found in our food chain, and responsible for causing radiation doses and thus health risks to people.

Kellogg (8) described close-in fallout of radioactive debris from nuclear detonations. This type of fallout was identified previously as early fallout. It has also been described as local fallout, but it may occur over a relatively large regional area. In essence, it is derived from radioactive materials, released to the environment in a nuclear weapon's detonation, which do not rise above the troposphere. Early fallout tends to occur fairly soon (up to several months), remains in the band of latitude of the nuclear test, contains radionuclides with short half-lives, has relatively high intensity levels, and is deposited under the influence of precipitation. This type of fallout is primarily a problem because of external exposures and those resulting from inhalation. Categories, or classes of fallout, are arbitrarily defined; this paper uses two categories, tropospheric and stratospheric.

Contrasted to tropospheric fallout is that classified as stratospheric, late or delayed, or worldwide, global fallout. Machta (9) discussed this type of fallout, which is from relatively high-yield nuclear devices that have sufficient energy at detonation to rapidly inject radioactive debris into the stratosphere.

**Table 15.1.** International agreements regarding nuclear weapons testing[a]

| International agreement | Time | Principal features |
|---|---|---|
| Moratorium on nuclear testing | 11/58–9/61 | The United States and the Soviet Union did not test during this period. |
| Limited test ban treaty | Began 10/63 | Signatory nations would not test in the atmosphere, space, or underwater and would not cause radioactive debris to exit their borders. |
| Nonproliferation treaty | Began 3/70 | Nuclear nations would not supply nuclear weapons and nuclear weapons' materials and nonnuclear nations would not accept them. Also, international inspections were agreed upon, and the great nuclear powers agreed to enter sincere negotiations to limit the nuclear arms race. |
| Threshold test ban treaty | Began 3/76 | Prohibited all underground nuclear tests having yields greater than 150 kt. Exception made for tests for peaceful purposes. |
| Total test ban treaty | Talks began in 12/87 | Would prohibit any nuclear testing by signatory nations. |

[a] Reprinted with permission from ref. 5, copyright 1977, Pergamon Press.

For worldwide fallout, the residence time for radioactive materials is measured in many months or years. This fallout is usually distributed more uniformly over the earth's surface than local fallout, depleted in short-lived fission products because of the long-term stratospheric residence times, and primarily a biological hazard because of contamination of the food chain. It is, of course, also subject to the prevailing meteorological conditions as it slowly descends into the troposphere.

An article in *U.S. News and World Report* (10), captioned "H-Bomb Tests—They're Safe," indicated that weapons' testing fallout would have negligible effects on food, health, and heredity in the United States. This information, taken from a report of the US Atomic Energy Commission, also reviewed the four effects of nuclear detonations, namely, blast, heat, immediate nuclear radiation, and residual radioactivity; the influence of emplacement and yield on fallout levels; protection against fallout effects; and the admonition that radiostrontium and radioiodine constituted the principal internal hazards from the fallout of both fission and thermonuclear weapons.

Andrews (11) reviewed specific generation of fallout, control procedures used in the U.S. nuclear testing program, the potential hazards of fallout, and the enormous differences between the controlled testing of nuclear weapons versus their use in an all-out war, where they would be used in the most effective and destructive way possible. He stressed that radionuclides could enter the body via use of contaminated food or water and that $^{90}$Sr and its daughter $^{90}$Y were probably the most hazardous fission-product fraction because $^{90}$Sr is produced in high yields, has a long half-life, and tends to be retained by bone.

The weapons' fallout monitoring network of the US Atomic Energy Commission and analytical results through September 1955 were described and discussed by Eisenbud and Harley (12). They confirmed and documented the worldwide deposition of mixed fission products and $^{90}$Sr from nuclear device tests and reported results gathered from 26 U.S. and 62 foreign stations. Strontium-90 was selected for continuing attention because of its potential hazard to people.

Libby (13) described categories of fallout, the phenomenology of radionuclide formation and fallout, the specific radionuclides of concern, that is, $^{14}$C, $^{90}$Sr, and $^{3}$H, and the deposition and intakes of radionuclides, especially $^{90}$Sr. He considered $^{14}$C essentially safe because of its long half-life and the very large amount of diluting C available in the atmosphere. It was indicated that $^{3}$H was not hazardous because of its relatively short half-life and its low rate of production. He concluded that the primary radionuclides of concern from nuclear weapons tests are those produced in the weapons and not those produced from activation of components of the environment.

By the mid 1950s, it was obvious that the atmospheric testing of nuclear devices was continuing to contaminate our environment and humans through the inhalation and, more importantly, the ingestion of long-lived radionuclides produced in the nuclear detonations and distributed worldwide through fallout. The judgments were that the levels were not hazardous but should be watched carefully for any upward trends and the resulting increased human exposures.

## Food-Chain Contamination

### Role and Activities of the US Federal Radiation Council

Partly because of the concern about environmental contamination resulting from fallout from the explosion of nuclear devices in the atmosphere, the US Federal Radiation Council (FRC) was established in 1959 (14) to provide a federal policy on human radiation exposure. The Council (14) provided general philosophy on radiation protection and had broad responsibilities for providing guidance for federal agencies in activities designed to limit radiation exposure of members of the public (singly and collectively) from radioactive materials deposited in the body as a result of their occurrence in the environment (15).

Specific concepts, terminology, and limits were an inherent part of the FRC guidance. In Report 1 (14), the FRC gave the following definition: "Radiation Protection Guide (RPG) is the radiation dose which should not be exceeded without careful consideration of the reasons for doing so; every effort should be made to encourage the maintenance of radiation doses as far below this guide as practicable." Furthermore, the whole-body RPG for the general population was established as 0.5 rem per year applied to individual members of the general population, whereas the guide for the average exposure of a "suitable sample" of the exposed population should be one third the RPG for individual members of the group (or 0.17 rem). This latter concept was introduced as an operational technique where individual whole-body doses were not known and proved to be very useful.

The Council provided RPGs for certain organs; general principles of control applicable to all radionuclides occurring in the environment; specific guidance for $^{226}$Ra, $^{131}$I, $^{90}$Sr, and $^{89}$Sr in terms of the particular ranges of intake per day, and general guidance for the federal establishment of appropriate concentration values (15). Each of the ranges of daily intake for a specific radionuclide called for certain actions. Intakes in Range III (Table 15.2) would probably result in exposures exceeding the RPGs if continued for a sufficient period of time and therefore had to be evaluated for the need for appropriate positive control measures. Table 15.2 contains the RPGs established for several important fission products (15) and for several radionuclides that were not considered by the FRC (16).

For $^{131}$I, the level of 100 pCi per day is at the upper end of Range II and would be expected to produce the thyroid RPG of 0.5 rem per year to the average of suitable samples of an exposed group (i.e., children). Of course, of the specific radionuclides considered in FRC Report 2 (15), $^{226}$Ra is not relevant to this discussion as it is a naturally occurring radionuclide.

FRC Report 3 (17) contained a summary of exposures that had occurred through 1961 from fallout in the United States and from material from past tests yet to be deposited. The estimates were based on measurements of specific radionuclides in air, rain, soil, water supplies, food, and people. The data and information used in their summary were for delayed fallout (stratospheric), as most potential effects from local fallout (tropospheric) were controlled or mitigated as part of the weapons testing programs.

The FRC (17) compared the exposures from fallout with exposures from the natural background and with its RPGs and noted that the doses from fallout had generally been a small fraction of the RPGs for population groups. The exposures from fallout were also considerably lower than those from the natural background during the same time periods.

FRC (17) also estimated certain malignant diseases and genetic effects from weapon's test fallout. Carbon-14 effects were estimated in terms of somatic and genetic effects. The conclusions from this report are quoted here:

We cannot say with certainty what health hazards are caused by fallout from nuclear testing. We expect there will be some genetic effects; other effects such as leukemia and cancer are more speculative and may not occur at all. We can observe that, compared to the number of these same adverse biological effects occurring wholly apart from testing, the additional cases that might be caused by testing are a very small quantity. We conclude that nuclear testing through 1961 has increased by small amounts the normal risks of adverse health effects.

Federal Radiation Council Report 4 (18) updated the evaluation of fallout in the United States from nuclear weapons testing conducted through 1962, a

**Table 15.2.** Radiation protection guides for selected radionuclides

| Radionuclide | Target organ | RPG Range II upper limit (rem/y) | Rate of intake (pCi/d) | | |
|---|---|---|---|---|---|
| | | | Range I[a] | Range II[b] | Range III[c] |
| $^{131}$I[d] | Thyroid | 0.5 | 0– 10 | 10–100 | 100–1,000 |
| $^{90}$Sr[d] | Bone | 0.5 | 0– 20 | 20–200 | 200–2000 |
| $^{89}$Sr[d] | Bone | 0.5 | 0– 200 | 200–2,000 | 2000–20,000 |
| $^{137}$Cs[e] | Whole body | 0.17 | 0–1450 | 1450–14,500 | 14,500–145,000 |
| $^{140}$Ba[e] | Bone | 0.5 | $0$–$1.4 \times 10^4$ | $1.4 \times 10^4$–$1.4 \times 10^5$ | $1.4 \times 10^5$–$1.4 \times 10^6$ |
| $^{3+}$H[e] | Whole body | 0.17 | $0$–$2 \times 10^5$ | $2 \times 10^5$–$2 \times 10^6$ | $2 \times 10^6$–$2 \times 10^7$ |

[a] Range I requires periodic surveillance.
[b] Range II calls for active surveillance and routine control.
[c] Range III requires evaluation and, if needed, positive control measures.
[d] From ref. 15.
[e] From ref. 16.

Reprinted with permission from ref. 16, copyright 1981, Pergamon Press.

year during which both the Soviet Union and the United States conducted extensive atmospheric nuclear tests. This report also predicted the probable levels of fallout that might be expected in 1963 and in future years in the food supplies of the United States and drew conclusions about the suitability of contaminated food products for human consumption. It specifically noted the interest in the production and distribution of $^{90}Sr$, $^{137}Cs$, $^{131}I$, and $^{14}C$ and indicated that the dose from the latter radionuclide (produced in 1962) during the next 30 years would equal that occurring in fallout from all previous nuclear weapons tests.

The FRC (18) indicated that precipitation was the most important mechanism in depositing fallout on the earth's surface. It divided the United States into "wet" and "dry" areas based on the accumulated levels of $^{90}S$ deposited per unit surface area. Somewhat earlier, Anderson (19) had theorized as to the influence of weather patterns on annual average levels of $^{137}Cs$ in milk.

In December 1964, Straub et al. (20) substantiated the theory of Anderson (19) as well as the soundness of the wet and dry characterization of the FRC (18). This was done by examination of the concentrations of $^{90}Sr$ and $^{137}Cs$ from the pasteurized milk network of the US Public Health Service divided by the relevant precipitation in inches per year. The results from such a comparison demonstrated the influence of both precipitation and weather patterns as major mechanisms in the behavior, deposition, and uptake of certain fallout radionuclides in an important foodstuff. Other studies in the United Kingdom (21) and in the United States (22) showed that precipitation accounts for 90% or more of the total fallout deposited on the earth.

The FRC (18) identified $^{90}Sr$, $^{137}Cs$, $^{131}I$, $^{89}Sr$, and $^{14}C$ as the most significant contributions to internal doses in humans from fallout and pointed out the importance of the several diet-oriented programs established to collect and monitor data on radionuclide levels from fallout. These included the following:

Health and Safety Laboratory, quarterly survey in New York City, San Francisco, and Chicago based on consumption of 19 categories of food, began in late 1959

Consumers Union, complete two-week diets of teenagers in 24 cities, began in 1959

U.S. Public Health Service, monthly institutional diet sampling program for age groups 8 to 20 years in some 22 states, began in early 1961

Food and Drug Administration, total diet sampling program, began in May 1961 with continued regional sampling of major food items

These diet sampling programs furnished radionuclide intake data on a comprehensive basis and were thus more useful than the sampling and analysis of contaminated single foodstuffs or other materials taken into the body.

Strontium-90 was described in terms of its physical, chemical, and biological characteristics and the fact that it readily enters the body via the total diet with milk, wheat products, fruits, and vegetables as the major contributors. In contrast, water, meat, fish, poultry, eggs, sugars, and fats contributed negligible amounts of $^{90}Sr$ to the diet. Strontium, of course, behaved very similarly to

Ca and was deposited in bone with a long residence time. For these reasons, monitoring concentrated on those foods of most importance as well as the total diet.

Strontium-89 behaved like $^{90}$Sr except its effective half-life was much shorter. Thus, milk was its primary route to humans since the production of other foods containing $^{89}$Sr usually allowed more time for radioactive decay.

Cesium-137, similar to K chemically, has a relatively short residence time in the body and is found primarily distributed throughout soft tissue. Its main entrance into the body was via milk, meat, and vegetables. Gamma radiation from $^{137}$Cs allowed direct measurement in the body with whole-body counters.

Iodine-131 was important because of its high abundance in early fallout, its ready entrance to people primarily by the milk pathway, and its concentration in the thyroid gland. The effective half-life was fairly short so foods other than milk were of less importance. Like $^{137}$Cs, $^{131}$I gamma radiation permits its direct measurement in the body.

Carbon-14 has an extremely long half-life and is produced artificially by nuclear weapons because of the interaction of neutrons with nitrogen in the atmosphere. Whereas the fission products mentioned are not normal constituents of the environment, $^{14}$C is produced continuously by interactions of cosmic radiation with elements of the atmosphere.

The FRC (18) indicated that levels of radionuclides occurring in fallout and in humans through 1962 were not sufficiently large to justify measures to limit their intake. However, attention was directed at investigting protective actions should they be needed in the future. The Council noted the very large fission and total yields of weapons tested in 1961 and 1962 and the radionuclides that were associated with tropospheric fallout. Table 15.3 is a tabulation of radionuclides that were considered important in fallout as to diet contamination and lists several of their major radiological characteristics.

Report 5 of the FRC (23) provided background information regarding planning protective actions to reduce potential doses from radionuclides in foods and

**Table 15.3.** Radionuclides in nuclear test fallout of importance in the food chain[a]

| Radionuclide | Production method | Half-life | | Major environmental media | Principal organs |
|---|---|---|---|---|---|
| | | Physical | Biological | | |
| $^{90}$Sr | Fission | 28.8y | 49.3y | Milk, grains, fruits, and vegetables | Bone |
| $^{89}$Sr | Fission | 50.5d | 49.3y | Milk, grains, fruits, and vegetables | Bone |
| $^{131}$I | Fission | 8.0d | 138d | Milk, vegetables, and fruits | Thyroid |
| $^{137}$Cs | Fission | 30.2y | 70d | Milk, meats, vegetables, and fruit | Whole body |
| $^{14}$C | Activation | 5,730y | 10d | Water and milk | Whole body |
| $^{3}$H | Fusion | 12.3y | 12d | Water and milk | Whole body |

[a] Reprinted with permission from ref. 18.

doses at which implementation of protective actions might be appropriate. It defined the Protection Action Guide (PAG) as "the projected absorbed dose to individuals in the general population which warrants protective action following a contaminating event."

In its discussion of the transmission of radionuclides in the atmosphere (fallout) to humans through the food chain, the Council used Fig. 15.1 (23). It represented a simple transmission chain and by implication indicated the places where protective actions might be applied.

This report discussed the nature of protective actions, types of protective actions, their impact, as well as guidance applicable to [131]I, including recommended PAGs. These PAGs were 30 rads to the thyroid of individuals in the general population and 10 rads as the average projected dose to the thyroid of a suitable sample of the population. These later numbers corresponded to projected intakes of [131]I of 1,750 and 580 nCi to a 2-g thyroid (that of an infant of about 1 year's age).

Federal Radiation Council Report 6 (24) dealt mainly with verification of 1963 predictions (18) and estimates of fallout for 1964 to 1965. The FRC increased the predicted annual depositions in 1964 to 1965 by 50%; noted the generally good agreement between predicted and measured levels of several specific radionuclides; commented on the disappearance from the environment of [131]I by May 1963 and [89]Sr by June 1984, which had resulted from atmospheric testing of nuclear weapons; and reiterated its conclusion from FRC Report 4

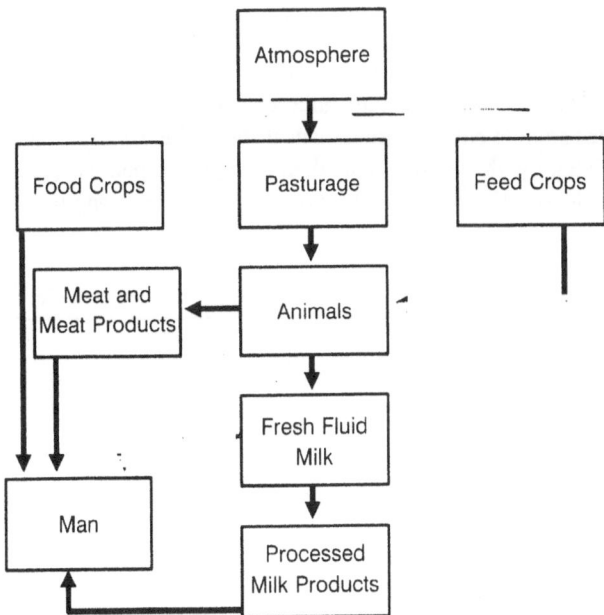

**Fig. 15.1.** Important steps in the transmission of radioactive material through the food chain to humans. (Reprinted with permission from ref. 23.)

(18) that the predicted dietary contamination levels (and the consequent health risks) were too small to justify protective actions.

This report predicted the amount of $^{90}$Sr occurring as fallout in 1967 would equal the amount lost in that year from radioactive decay and modified slightly the wet and dry areas discussed in Report 4 (18). It indicated the average diet data collected on fallout for New York City and San Francisco were comparable to the averages for the wet and dry regions of the United States, respectively.

The predictions of radionuclides in the diet and in people had been made using the methodology described in the article of the Hearings of the Joint Committee on Atomic Energy, "Fallout, Radiation Standards, and Countermeasures," (25) and various technical reports of the Health and Safety Laboratory. The Council (24) predicted estimates of only the long-lived radionuclides $^{90}$Sr and $^{137}$Cs for 1964 and 1965. Predicted levels of $^{90}$Sr were given for milk, total diet, and bone (divided into wet and dry regions) and for wheat and flour from a number of producer states. Fewer predictions were made for $^{137}$Cs, although its levels were predicted in milk and its decrease in the environment was expected to be faster than that for $^{90}$Sr.

Of special importance was the prediction that the year of maximum contamination of the total diet from fallout radionuclides was expected to be 1964 rather than 1963. For $^{90}$Sr, this translated into the observed values for 1959 to 1963 and predicted levels for 1964 and 1965 presented in Table 15.4, which is taken from FRC Report 6 (24) and modified slightly for clarity.

Federal Radiation Council Report 7 (26) was the last report of the FRC related to fallout and contained background information for the development of radiation protection standards and PAGs for $^{90}$Sr, $^{89}$Sr, and $^{137}$Cs. Following an acute contaminating event, the problem of evaluating when protective actions may be indicated was divided into three categories. Category I was limited to radionuclides gaining entrance to humans through the pasture–cow–milk–human pathway; Category II was for other than Category I pathways during the first year; and Category III was primarily concerned with the long-term transmission

**Table 15.4.** Average $^{90}$Sr content of the United States total diet[a,b]

| Year | Observed | |
| | New York (wet) | San Francisco (dry) |
| --- | --- | --- |
| 1959 | 15 | 10 |
| 1960 | 10 | 5 |
| 1961 | 5 | 5 |
| 1962 | 10 | 5 |
| 1963 | 30 | 10 |
| | Predicted | |
| 1964 | 40 | 20 |
| 1965 | 30 | 15 |

[a] Reprinted with permission from ref. 24, with modifications for clarity.
[b] Predicted and observed levels rounded to the nearest 5 (pCi $^{90}$Sr/gCa).

of $^{90}$Sr through soil into plants in the years following a contaminating event. For years 1 and 2, residual contamination of $^{137}$Cs could be a consideration.

For Category I, the PAGs were a mean dose of 10 rads for the first year to the bone marow or whole body of individuals for $^{90}$Sr, $^{89}$Sr, and $^{137}$Cs. (For $^{131}$I, 30 rads to the thyroid were used for individuals and 10 rads for a suitable sample of the population). Category II specified a mean dose of 5 rads in the first year to the bone marrow or whole body of individuals. These guides were deemed to be effectively met if the average dose to a suitable sample of the population was one third the PAGs or 3 and 2 rads, respectively. The PAGs for Category III were 0.5 rad to individuals or 0.2 rad to a suitable sample of the population; such situations were to be appropriately evaluated. Values for the PAGs in Category III are numerically equal to the RPGs recommended in FRC Reports 1 and 2 (14,15). The PAGs from FRC Reports 5 and 7 (23,26) are presented in Table 15.5.

This report (26) briefly reviewed Reports 3, 4, and 6, summarized the various factors of importance in assessing the effects of fallout radionuclides in the food chain and in humans and concluded that events requiring protective actions were most likely to involve $^{131}$I.

The maximum doses to bone and bone marrow from $^{89}$Sr were about the same in each year, 1962 and 1963, and amounted to about 3% to 4% of the numerical values of the RPGs for bone and bone marrow. The estimated doses from $^{89}$Sr were negligible in 1964. The peak value of about 40 pCi of $^{90}$Sr per gram of Ca in the total diet in the wet areas of the United States was reached in 1964. Annual doses to bone and bone marrow for the period this concentration is maintained are about 6% of the respective values of the RPGs.

Report 7 of the FRC (26) included a special section on the contamination of the environment, especially lichens and sedges, caribou, and reindeer, and

**Table 15.5.** Protective action guides for selected radionuclides

| Radionuclide | Target organ | Category of acute contaminating event and organ dose (rads)[a] | | | | | |
| | | Category I | | Category II | | Category III | |
| | | Individual | Suitable sample | Individual | Suitable sample | Individual | Suitable sample |
|---|---|---|---|---|---|---|---|
| $^{131}$I[b] | Thyroid | 30 | 10 | | | | |
| $^{90}$Sr[a] | Bone marrow | 10 | 3 | 5 | 2 | 0.5 | 0.2 |
| $^{89}$Sr[a] | Bone marrow | 10 | 3 | 5 | 2 | 0.5 | 0.2 |
| $^{137}$Cs[a] | Bone marrow | 10 | 3 | 5 | 2 | 0.5 | 0.2 |

[a] FRC Report 7 defined the three categories of an acute contaminating event as Category I, limited to the transmission of radionuclides through pasture–cow–milk–human pathway; Category II, dietary pathways to humans other than that in Category I during the first year; and Category III, long-term transmission of $^{90}$Sr and $^{137}$Cs through soil–plant–human pathway in years subsequent to first year. The values for the PAGs in Category III are numerically equal to the RPGs recommended in FRC Reports 1 and 2. Each of the three categories includes PAGs for individuals and a suitable sample of the population. From ref. 26.
[b] From ref. 23.

Eskimos and Indians by $^{137}$Cs and $^{90}$Sr in arctic Alaska. These radionuclides, concentrated in lichens and sedges, are contained in the diet of caribou and reindeer, especially in the winter. Thus, the meat of these animals had high concentrations of $^{137}$Cs and $^{90}$Sr, which were transferred to the individuals eating the contaminated meat.

In 1964, the highest body burdens of $^{137}$Cs in Eskimos were about 1,500 nCi, which is about one half the RPG equivalent of 3,000 nCi for whole body exposure of individuals in large population groups. The 3,000 nCi correspond to an applicable RPG of 0.5 rad per year. The $^{90}$Sr burdens in bone were about 4 times as high as those found in the conterminous United States. Of course, these data were for a specialized food chain, namely the lichen–caribou (reindeer)–human pathway.

The Council concluded in its seventh report (26), as it had in previous reports, that nationwide protective action programs to reduce potential exposures of the population to radionuclides in fallout were not currently necessary nor would they be for future levels of fallout from past nuclear weapons tests.

## Contamination of the Food Chain

In the United States, contamination of the environment and the food chain by fallout from weapons testing occurred both by early fallout from tests conducted in Nevada beginning in 1951 and delayed fallout from high-yield tests that were initiated in the Pacific in 1952. Thus, fallout levels and resulting contamination usually represent a combination of these effects.

Observations and measurements of fallout were usually evaluated and interpreted in terms of the framework and guidance of the FRC. Also, the vast majority of radiological surveillance information and data was published in Radiological Health Data (RHD), which later became Radiological Health Data and Reports (RHDR). This publication was monthly and began with its first issue in April 1960. It included environmental radiation levels as well as interpretive statements and evaluations. Information and data were provided by relevant federal agencies, such as the Health and Safety Laboratory through its Fallout Program Quarterly Summary reports; state and local government organizations; and many other groups and individuals. The last issue of Radiological Health Data and Reports was published in December 1974. Since 1975, environmental radiation data from across the United States have been and are published in Environmental Radiation Data, which is produced quarterly by the US Environmental Protection Agency, Office of Radiation Programs.

In many cases, individual items from the food chain were analyzed for fallout radionuclides. Most of the data were for contamination of air, water, milk, and diet. These, of course, were most directly related to the intake of radionuclides in people that led to potential biological effects. Most of the early monitoring was for public health purposes and resulted in the establishment of surveillance networks that monitored and assessed environmental contamination vis-à-vis its public health implications.

This chapter looks at results that were periodically reported in the literature, especially RHD (RHDR), and focuses attention on total diet results and most particularly the relatively small number of radionuclides considered of most concern in nuclear test fallout. The networks and programs providing data in 1960 were outlined in the first issue of RHD (27), and this type information was frequently repeated and updated in subsequent issues. The Food and Drug Administration (28) reported the $^{90}$Sr content of human and animal food collected in 1958 and 1959, and $^{90}$Sr data for wheat and bread from the Health and Safety Laboratory were also reported. Information on the food consumption of households in the United States was reported by the US Department of Agriculture in the August issue of RHD (29) and was necessary to be able to interpret radionuclide consumption by groups of people including populations.

Radiological Health Data (30) reported results of the Institutional Diet Sampling Program (children under 18) for January to August 1961, a period during and toward the end of the moratorium on nuclear testing. The average daily intake of $^{90}$Sr was less than 7 pCi from the seven geographically dispersed stations that were sampled. This report also indicated only $^{90}$Sr and $^{137}$Cs were measurable in milk, whereas, because of their half-lives, $^{131}$I, $^{89}$Sr, and $^{140}$Ba were not detectable at the time the samples were analyzed. This was a pattern that would occur many times with the periodic increases and decreases in levels of the shorter-lived fission products measured in the food chain.

A summary of data for fallout radionuclides in milk, total diet, and human bone for 1963 to 1964 as compared to FRC estimates and guidelines was published in RHD by Grundy et al. (31). The authors concluded the maximum intake period was 1963 to 1964 and that levels measured for $^{89}$Sr, $^{90}$Sr, $^{131}$I, and $^{137}$Cs were in relatively good agreement with predictions of the FRC. They noted that $^{131}$I in milk had become undetectable by mid-1963 and that its peak annual average of 104 pCi/L of milk had occurred at Palmer, Alaska, in 1962. This value, assuming milk consumption of 1 L per day, would correspond to the lower part of FRC Range III for $^{131}$I.

Also recorded was an estimated total dose to bone marrow of about 10 mrad during 1964 to children in the wet area of the United States. Compared to the RPG for bone marrow of 170 mrad per year, this amounted to approximately 6%. This estimate, based on a dietary network annual average of 32 pCi per day of $^{90}$Sr intake in 1964, could be compared to similar annual averages of 13 and 25 pCi per day in 1962 and 1963, respectively. These diet data and those from other sources confirmed predictions of the FRC and supported its conclusion that fallout levels were sufficiently low as to not justify any protective actions.

Setter et al. (32) summarized the results of $^{90}$Sr analyses on nine staple foods collected July 1962 through October 1963. They concluded $^{90}$Sr levels depended primarily on the type food, the location in which it was grown, the time sampled, fallout levels, and the season of sampling and that milk can be a good index for estimating $^{90}$Sr total intake if the diet consists of a reasonable balanced food selection.

The same authors, Setter et al. (33) reported on the [137]Cs levels of the 9 staple foods whose [90]Sr concentrations were reported previously. Similar trends were noted for [137]Cs, and it was pointed out by these authors that [137]Cs concentrations (easy to measure) might be important in estimating [90]Sr concentrations (difficult to measure).

Grundy (34) reviewed the use of fractional intake of [90]Sr in milk to estimate [90]Sr dietary intake. He concluded use of Sr units (pCi of [90]Sr/gCa) produced the best estimates but cautioned that care be used in interpretation. He also indicated that the dietary intakes of [90]Sr were within Range II (20 to 200 pCi per day) of the FRC guide and that maximum [90]Sr dietary intakes had always remained below 100 pCi per day. The FRC had always stressed the importance of determining internal exposures to humans from environmental radiation sources, especially from radionuclides in the diet.

Roecklein et al. (35) published a comparison of [90]Sr and [137]Cs data in total diet samples collected and analyzed by six federal, state, and private organizations for the period 1961 through 1967. The authors presented a brief history and overview of each program, identified sampling sites, summarized analytical data, compared results, and indicated the general agreement of results with some exceptions. Primary conclusions were that all the values reported were equally reliable and that the values reported for the diets of the special age groups and those reported for the diet of the general population were little different.

Strontium-90 data for the Health and Safety Laboratory quarterly tri-city diet surveys (New York, Chicago, and San Francisco) were reported for 1969 in RHDR (36) along with a summary of results from 1960 to 1969. The daily intakes of [90]Sr in San Francisco and New York in 1969 were approximately 5 and 12 pCi, respectively. This article also contained graphs of daily intake of [90]Sr in the tri-city diets from March 1960 through December 1969, which indicated the peak intakes had occurred in 1963 to 1964 in each city. Referenced in this paper were the two principal surveys (conducted in 1955 and 1965) of household diets used in sampling the total diet in the United States (37,38). Another summary of the tri-city diet network of the Health and Safety Laboratory appeared in RHDR (39) and indicated the slow decline in [90]Sr dietary levels.

Simpson et al. (40) reviewed monitoring programs for radionuclides in foods and concluded that radionuclide levels measured in 1973 were all within Range I of the appropriate guidelines, indicating a need for periodic surveillance. They noted the decision in 1969 of the US Food and Drug Administration to discontinue monitoring radionuclides in foods and the cancellation of the Institutional Diet Program of the US Environmental Protection Agency in 1973. These decisions were based on the low levels of radionuclides in the environment and their documented decreasing trend. Thus, certain monitoring networks were discontinued whereas others were reduced in scope and reoriented to other sources of radionuclides, rather than fallout from nuclear testing.

In 1981, Simpson, et al. (16) presented a summary of [90]Sr and [137]Cs data for 1961 to 1977 in total diet samples of the US Food and Drug Administration's monitoring program. These data confirmed the peak levels of these two fallout radionuclides in total diets during 1963 to 1965 and the fact that values of $^{90}$Sr and $^{137}$Cs had never exceeded the upper levels of Range II of the FRC guidelines for $^{90}$Sr or those derived from this guidance for $^{137}$Cs. The intake levels of $^{137}$Cs had become nondetectable by 1973, whereas the $^{90}$Sr intake had plateaued at 10 pCi per day in 1975, suggesting to the authors that they had reached a baseline for the years 1975 to 1977.

Klusek (41), in a 1984 report, reviewed the 1972 $^{90}$Sr data in the US diet as monitored by the Environmental Measurments Laboratory (formerly the Health and Safety Laboratory) of the US Department of Energy. She reported an average daily intake of about 5 and 3 pCi per day of $^{90}$Sr in New York City and San Francisco, respectively, in 1982, and noted the decline in intake from the peak years of 1963 to 1964. The author relates changes in the average annual daily intake of $^{90}$Sr during the period 1960 to 1982 to the effects of the moratorium on nuclear tests (1959 to 1961) and the limited test ban treaty, which began in October 1963, and discusses the relative importance of the major food categories to the daily intake of $^{90}$Sr. The primary change is that vegetables and, to a lesser extent, fruit have become more important in later years as the source of $^{90}$Sr for food crops because of a shift from fallout deposition to $^{90}$Sr accumulation in the soil. Thus, the relative importance of dairy products and grains has decreased somewhat.

In a note in *Health Physics,* Broadway and Strong (42) presented $^{90}$Sr data in bone samples over the period 1961 to 1975. The data generally confirm the predictions of the model of the International Commission on Radiological Protection (43). Such models predicted a peak in the early 1960s (1962 to 1964) with a slow decrease subsequent to the peak period. These results are not in complete agreement with the peak $^{90}$Sr data in the food chain and in the total diet where these peaks occurred in 1963 to 1965.

Stroube et al. (44) reviewed the levels of several radionuclides in foods during 1978 to 1982. They noted decreasing levels of $^{3}$H, $^{90}$Sr, and $^{137}$Cs for the five years reported as well as that the measured levels are well within Range I of the pertinent FRC RPGs and those derived from FRC guidance for $^{3}$H.

Klusek (45) presents an overview of 25 years of measurements of $^{90}$Sr fallout, its levels in US diet, and its transfer to human bone. She discusses annual intakes of $^{90}$Sr for New York City and San Francisco, the relative importance of various food categories to $^{90}$Sr intake, the influence of direct deposition versus soil accumulation of $^{90}$Sr, and uptake by bone. The importance of $^{90}$Sr in the assessment of the effects of fallout from nuclear weapons tests, significant of the ingestion pathway, effects of nuclear testing frequency and size of tests on food-chain contamination and dietary intake, and the relative low bone levels of $^{90}$Sr, compared to the pertinent guidance are also reviewed.

Figure 15.2 is a plot of $^{90}$Sr intakes for a 25-year period for the US population

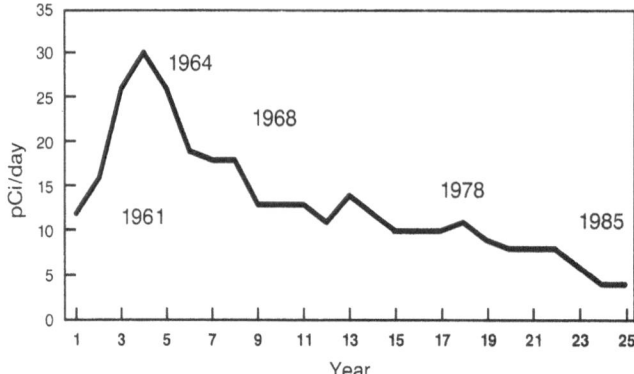

**Fig. 15.2.** US population [90]Sr intakes for a 25-year period. (Courtesy of Edward J. Baratta and William B. Stroube, 1987.)

based primarily on the diet analysis program of the US Food and Drug Administration. This information was compiled by Messrs. Edward J. Baratta and William B. Stroube and used with their permission.

A general review of food-chain contamination by radionuclides from the testing of nuclear devices would not be appropriate without mention of the large amounts of information and data that were published through the several congressional hearings of the Joint Committee on Atomic Energy (4,25,46,47). Also, modeling has played an important role in predicting future levels of radionuclides in the food chain from past or anticipated contaminating events (nuclear device tests) as well as potential levels and effects of radionuclides in humans. Figure 15.1, taken from FRC Report 5 (23), depicted the several steps in the transmission of radionuclides through the food chain to humans. This depiction and appropriate input data were used in the 1960s to model the transport of radionuclides through the food chain to humans.

It is of interest to contrast this rather simple picture of a model outline to one from a current state-of-the-art food-chain model as shown in Fig. 15.3. This figure is reproduced with permission of *Health Physics* from an article by Whicker and Kirchner (48), which describes Pathway, a dynamic food-chain model that predicts radionuclide ingestion after fallout deposition. A fairly large number of dynamic processes are included, and it covers some 20 radionuclides following a single deposition of fallout. The model was developed for nuclear detonations at the US Nevada Test Site but can be modified for other geographical areas and for chronic or other acute releases when the ground deposition can be estimated. Organ-specific dose estimates can be made if the deposition and a dose conversion factor are multipled by the output of Pathway, which is in Bq ingested per $Bq/m^2$ deposited.

Modeling has been used extensively by the United Nations Scientific Committee on the Effects of Atomic Radiation (UNSCEAR) in predicting effects of fallout from nuclear explosions. The most recent summary of its efforts in this

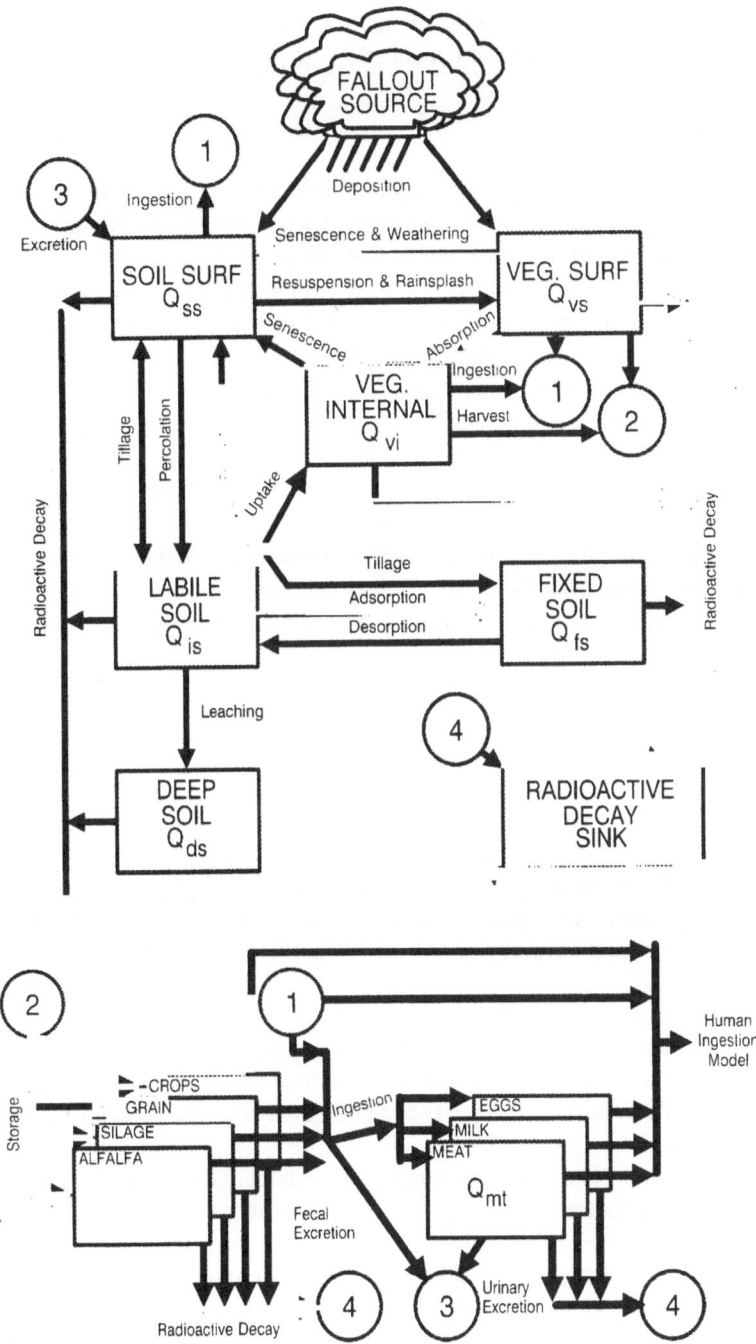

**Fig. 15.3.** Structural features of the pathway model. (Reprinted with permission from ref. 48, copyright 1987, Pergamon Press.)

area was presented in 1982 (49) along with exposures resulting from nuclear explosions. More details were contained in Annex C of an earlier report (50).

## Current Status

The rate of nuclear testing has greatly diminished from the 1961 to 1962 period and most of the tests are conducted underground. This greatly reduced rate of fallout has allowed levels of radionuclides in the environment to continuously decrease. In general, only relatively low levels of several long-lived radionuclides are presently being measured in the environment and in the food chain.

Several federal agencies, namely, the US Environmental Protection Agency, the US Food and Drug Administration, and the US Department of Energy, continue to monitor the food chain or its major components for weapons' fallout in addition to other sources. Many of the surveillance programs in the United States are source oriented and thus primarily concerned with regional monitoring. The various current monitoring programs are sophisticated and have available rapid systems for the collection and analysis of pertinent environmental samples. In addition, data collection, storage, and handling, as well as state-of-the-art modeling capabilities are computerized and automated for prompt response to current and future needs.

## Summary

There are still only six nations with a demonstrated nuclear device testing record. The last nation to test its first nuclear device was India in 1974. Furthermore, the frequency of nuclear testing has decreased appreciably, and the yields of nuclear tests have been greatly reduced. These facts coupled with the fact that most tests are conducted underground have led to much lower levels of fallout in the world.

Peak testing of nuclear devices occurred in 1961 to 1962 and the resulting peak worldwide fallout levels occurred in the period 1963 to 1965 in terms of radionuclide contamination of the world's food chain. Human exposures to this stratospheric fallout also peaked slightly later than the peak for food-chain contamination.

In the United States, the exposures from radionuclides in worldwide fallout and regional fallout never exceeded the relevant PAGs of the Federal Radiation Council for food-chain contamination. However, on several occasions levels of $^{131}$I in milk did exceed the FRC's RPG for this radionuclide. Such occasions were transient and limited to relatively small regional areas and their populations.

Physical decay, weathering, and other major factors have contributed to the large decline of radionuclides in the environment and the food chain and its major components. For example, rarely are short-lived radionuclides (such as $^{131}$I, $^{89}$Sr, and $^{140}$Ba) measured in the food chain from nuclear tests, and the levels of the several long-lived fission products (such as $^{90}$Sr and $^{137}$Cs) have been reduced to very low values. Carbon-14 levels have also decreased appreciably.

Emphasis has been placed on total diet effects because of the importance of the food chain in the assessment of exposures resulting from ingestion of radionuclides deposited by worldwide fallout. Thus, the focus has been public health evaluation rather than the many other important aspects of our long experience in dealing with fallout from nuclear weapons tests.

The vast amount of work, including research and development, done during the early and later years of monitoring the environment for weapon's fallout provided vast amounts of knowledge regarding our environment, meteorology, ecology, the food chain, and effects of radionuclides. This experience has led to the present state of understanding and furnished us with the ability and capacity to quickly measure and assess radionuclides in the food chain and to rapidly predict their behavior and effects, including those of a public health nature.

*Acknowledgments.* Appreciation is given to Charles R. Porter, Dr. Bernd Kahn, and Dr. Richard Blanchard for their generous assistance in terms of shared experiences. We also thank Barbara Williams for help with references and Betty Crumbley for her professional handling of the manuscript. Edward Baratta and William Stroube are thanked for allowing us to use Fig. 15.2 from work done over the years by the US Food and Drug Administration.

# References

1. Glasstone S (1962) The effects of nuclear weapons. United States Atomic Energy Commission, US Government Printing Office, Washington, DC
2. Mills M (1957) Statements on topic II radiation and radioactivity and topic III fission and fission reactors. In: The nature of radioactive fallout and its effects on man. Hearings of the Joint Committee on Atomic Energy, May 27, 28, 29 and June 3, US Government Printing Office, Washington, DC, pp 24–53
3. Graves AC (1957) The production of radiation and radioactivity by detonating nuclear weapons. In: The nature of radioactive fallout and its effects on man. Hearings of the Joint Committee on Atomic Energy, May 27, 28, 29 and June 3, US Government Printing Office, Washington, DC pp 53–88
4. Joint Committee on Atomic Energy (1957) The nature of radioactive fallout and its effects on man. Hearings of the Joint Committee on Atomic Energy, May 27, 28, 29 and June 3, US Government Printing Office, Washington, DC
5. Carter MW, Moghissi AA (1977) Three decades of nuclear testing. Health Physics 33:55–71
6. Carter MW (1979) Nuclear testing 1975–1978. Health Physics 36:432–437
7. NVO-209 (Rev. 7)(1987) Announced United States nuclear tests, July 1945 through December 1986. Office of Public Affairs, US Department of Energy, NTIS, Springfield, Virginia
8. Kellogg WW (1957) Atmopsheric transport and close-in fallout of radioactive debris from atomic explosions. In: The nature of radioactive fallout and its effects on

man. Heaings of the Joint Committee on Atomic Energy, May 27, 28, 29 and June 3, US Government Printing Office, Washington, DC, pp 104–134

9. Machta L (1957) Worldwide travel of atomic debris. In: The nature of radioactive fallout and its effects on man. Hearings of the Joint Committee on Atomic Energy, May 27, 28, 29 and June 3, US Government Printing Office, Washington, DC, pp 141–161

10. U.S News and World Report (1955) H-bomb tests—they're safe. US News and World Report 38:128–134

11. Andrews HL (1955) Radioactive fallout from bomb clouds. Science 122:453–456

12. Eisenbud M, Harley JH (1956) Radioactive fallout through September 1955. Science 124:251–255

13. Libby WF (1956) Radioactive fallout and radioactive strontium. Science 123:657–660

14. Federal Radiation Council (1960) Background material for the development of radiation protection standards, Report No. 1. Federal Radiation Council, US Government Printing Office, Washington, DC

15. Federal Radiation Council (1961) Background material for the development of radiation protection standards, Report No. 2. Federal Radiation Council, US Government Printing Office, Washington, DC

16. Simpson RE, Shuman FGD, Baratta EJ, Tanner JT (1981) Survey of radionuclides in foods, 1961–77. Health Physics 40:529–534

17. Federal Radiation Council (1962) Health implications of fallout from nuclear weapons testing through 1961, Report No. 3. Federal Radiation Council, US Government Printing Office, Washington, DC

18. Federal Radiation Council (1963) Estimates and evaluation of fallout in the United States from nuclear weapons testing conducted through 1962, Report No. 4. Federal Radiation Council, US Government Printing Office, Washington, DC

19. Anderson EC (1958) Radioactivity of people and milk. Science 128:882–886

20. Straub CP, Carter MW, Moeller DW (1964) Environmental behavior of nuclear debris. Proceedings of the American Society of Civil Engineers, Paper 4159, 90:25–40

21. Mercer ER, Burton JD, Bartlett BO (1963) Relationships between the deposition of strontium-90 and the contamination of milk in the United Kingdom. Nature: 198:662–665

22. Machta L (1963) Meteorological processes in the transport of weapon radioiodine. Health Physics 9:1123–1132

23. Federal Radiation Council (1964) Background material for the development of radiation protection standards, Report No. 5. Federal Radiation Council, US Government Printing Office, Washington, DC

24. Federal Radiation Council (1964) Revised fallout estimates for 1964–1965 and verification of the 1963 predictions, Report No. 6. Federal Radiation Council, US Government Printing Office, Washington, DC

25. Joint Committee on Atomic Energy (1963) Fallout, radiation standards, and countermeasures. Hearings of the Joint Committee on Atomic Energy, February 20 and 21 and April 2, 3, 4, and 5, US Government Printing Office, Washington, DC

26. Federal Radiation Council (1965) Background material for the development of radiation protection standards—protective action guides for strontium-89, strontium-90 and cesium-137, Report No. 7. Federal Radiation Council, US Government Printing Office, Washington, DC

27. Radiological Health Data (1960) Section II–Section V. Radiological Health Data 1:7–48
28. Radiological Health Data (1960) Section IV—Other data. Radiological Health Data 1:36–44
29. Radiological Health Data (1960) Food consumption of households in the United States. Radiological Health Data 1:26–34
30. Radiological Health Data (1962) Section II—Food other than milk. Radiological Health Data 3:42–45
31. Grundy RD, Chandler RP, Coleman JR, Terrill JG (1965) Fallout radionuclides in milk, total diet, and human bone compared to federal radiation council estimates, 1963–1964. Radiological Health Data 6:652–655
32. Setter LR, Smith D, Spector M (1966) Strontium-90 in food—a summary of results on selected foods in the United States, July 1962–October 1963. Radiological Health Data and Reports 7:64–78
33. Setter LR, Smith D, Spector M (1966) Cesium-137 in food—a summary of results on selected foods in the United States, July 1962 to October 1963. Radiological Health Data and Reports 7:145–156
34. Grundy RD (1976) Strontium-90 dietary intake estimates based on fractional intakes due to milk. Radiological Health Data and Reports 8:73–77
35. Roecklein RD, Smedley CE, Simpson RE (1970) Strontium-90 and cesium-137 in total diet samples—a comparative study of data. Radiological Health Data and Reports 11:47–64
36. Radiological Health Data and Reports (1970) Strontium-90 in tri-city diets January–December 1969. Radiological Health Data and Reports 11:297–299
37. United States Department of Agriculture (1956) Food consumption of households in the United States, household food consumption survey. US Department of Agriculture, Report 1, US Government Printing Office, Washington, DC
38. United States Department of Agriculture (1967) Food consumption of households in the United States, spring 1965, a preliminary report. US Department of Agriculture, ARS 62-16, US Government Printing Office, Washington, DC
39. Radiological Health Data and Reports (1974b) Strontium-90 in tri-city diets, January–December 1973. Radiological Health Data and Reports 15:780–782
40. Simpson RE, Baratta EJ, Jelinek CF (1974a) Radionuclides in foods: monitoring program. Radiological Health Data and Reports 15:647–656
41. Klusek CS (1984) Strontiumn-90 in the US diet, 1982. US Department of Energy, EML-429, New York
42. Broadway JA, Strong AB (1983) Radionuclides in human bone samples. Health Physics 45:765–768
43. International Commission on Radiological Protection (1973) Alkaline earth metabolism in adult man. International Commission on Radiological Protection, ICRP Publication 20, Pergamon, New York
44. Stroube WB, Jelinek CF, Baratta EJ (1985) Survey of radionuclides in foods, 1978–1982. Health Physics 49:731–735
45. Klusek CS (1987) Strontium-90 in food and bone from fallout. Journal of Environmental Quality 16:195–199
46. Joint Committee on Atomic Energy (1959) Fallout from nuclear weapons tests. Hearings of the Joint Commitee on Atomic Energy, May 5, 6, 7 and 87, US Government Printing Office, Washington, DC
47. Joint Committee on Atomic Energy (1962) Radiation standards, including fallout.

Hearings of the Joint Committee on Atomic Energy, June 4, 5, 6, and 7, US Government Printing Office, Washington, DC
48. Whicker FW, Kirchner TB (1987) Pathway: a dynamic food-chain model to predict radionuclide ingestion after fallout deposition. Health Physics 52:717–737
49. United Nations Scientific Committee on the Effects of Atomic Radiation (1982) Ionizing radiation: sources and biological effects. United Nations Scientific Committee on the Effects of Atomic Radiation, 1982 Report to the General Assembly, with annexes. United Nations Publication (Sales No. E.82.IX.8) New York
50. United Nations Scientific Committee on the Effects of Atomic Radiation (1977) Sources and effects of ionizing radiation. United Nations Scientific Committee on the Effects of Atomic Radiation, 1977 Report to the General Assembly, with annexes. United Nations Publication (Sales No. E.77.IX.I) New York

CHAPTER 16

# The Removal and/or Reduction of Radionuclides in the Food Chain

## M. J. Arnaud[1]

## Introduction

The general population is exposed to radioactive substances that originate from
a variety of sources, including naturally occurring terrestrial radionuclides, with
K, Ra, U, Th, and Rn, the most important among about 70 radioactive elements.
Other radionuclides are produced by the interaction of cosmic rays on atmospheric
nuclei. The dose delivered by these nuclides, where the most important are
[14]C and tritium, is negligible compared to terrestrial sources of natural radioactivity.

Man-made radionuclides have been developed by the atomic energy industries.
A great variety of radionuclides is produced for many industrial uses as well
as for medical and research applications. In addition, the generation of high-
level radioactive wastes by the nuclear industry and their transport for fuel
reprocessing constitute a risk of environmental contamination from many different
radionuclides. It was observed up to 1986, that accidental environmental contamination from civilian reactors was generally limited to small and controlled areas
and that the doses to people living nearby were far less than those observed
throughout the world as a result of fallout from nuclear weapons.

The most significant radionuclides present in nuclear weapon fallout were
[131]I, [90]Sr, [137]Cs, [14]C, and tritium. Exposure to the population was not from
inhalation, except in a few cases, but from the ingestion of fresh foods, mainly
dairy products.

To prevent human contamination or to remove radionuclides already incorporated in the body, studies were initiated as early as 1955. Some methods developed
for human radioprotection can be applied in vivo in the animal eating contaminated
grass in order to reduce or suppress contamination of milk and meat. Finally,
when food products are shown afterward to be contaminated, the removal of
radionuclides is also considered as the last issue. Decontamination processes
were studied and proposed as soon as the level of radionuclides was readily

[1] Nestlé Research Centre, Nestec Ltd., Vers-chez-les-Blanc, CH-1000 Lausanne 26,
Switzerland.

detectable in the environment after nuclear weapons tests. In the following sections, we will review the different processes of decontamination of food and of animals eating contaminated fodder, which differ according to the radionuclides present and also to the kind of food treated. The last sections will describe new research begun after the Chernobyl accident.

## Decontamination Processes of Food

### Food Treatment for Cesium Removal

Cesium-137 is one of the radionuclides of public health importance because of its long half-life (30 years) and its nearly complete gastrointestinal absorption in the human, followed by its distribution throughout soft tissues, especially muscle. The biological half-life in adult subjects varies from 70 to 100 days (1,2), while it appears to be shorter, 10 to 50 days, for children (2,3). The primary mode of entry of [137]Cs into the food chain is through its adsorption on plant surfaces and by foliar absorption. In most soil types, [137]Cs, which is strongly bound by clay particles, is not available for plant root absorption, and thus its concentration in plant material depends upon the amount deposited during the growing period. It has been shown that milk, grain products, meat, fruit, and vegetables provide more than 95% of the total [137]Cs intake of the average population from fallout from nuclear testing in the atmosphere (4). Moreover, the primordial importance of cow's milk on [137]Cs total intake, not only in infants but also in adults, was also demonstrated.

Cesium-137 levels in milk are directly proportional to the intake, and from 8% to 13% of the intake is transferred to the milk (5,6,7). Variations have been related to the lower milk yield (6) and to the crude fiber content of the diet (7). The distribution of radiocesium present in sheep's milk collected in Greece after the fallout following the Chernobyl accident has been studied. The results showed that only 7.5% and 0.3% of radiocesium from milk was found in cheese and cream, respectively (8). On a weight basis, the radioactivity in cheese, expressed in Bq/kg was reduced only by a factor of 2 when compared with the radioactivity present in milk, expressed in Bq/L. Cesium radioactivity will even increase because of the loss of weight of cheese during the ripening period. The importance of dairy products in the contamination of the food chain explains why most of the research for the removal of radionuclides has been performed on milk.

### Milk

Most of the patents (9,10,11) and publications on the decontamination of milk and water used for food preparation have employed ion-exchange resins and electrodialysis. Deionization of a solution through a bed of ion-exchange resin is a discontinuous process since the resin quickly exhausts itself and thus loses its exchange or sorption capacity. Different ion-exchange resins were compared for their capacity and the optimal working temperature, and more than 99% of $^{134}$Cs in milk was removed (12). Phenol-formaldehyde sulfonic acid resins gave the maximum $^{134}$Cs removal. After a contact of only 8 min, about 85% of the Cs from raw milk labeled in vivo was bound to this type of resin (13). This process also necessitates lowering the milk pH from about 6.6 to 6.8 to about 5.2 to 5.4. Acidification has been effected with citric acid, and, after neutralization, the preparation of cheese from this treated milk may need modifications in the manufacturing procedure to preserve the flavor and physical quality of the final product (14). Using a pilot plant, the effect of citric acid was compared with hydrochloric acid for pH reduction. Although the product was judged satisfactory and heavy metal remained well below maximum values, important losses of thiamine were recorded as well as a change of the flavor of the milk. Fooks et al. confirmed previous results (15), showing a more efficient removal of $^{137}$Cs from neutral than from acidified milk. Their study emphasized the important effect of the volume of treated milk so that when 30 resin bed volumes (rbv) and 15 rbv were treated, the concentration of residual Cs was reduced from 30% to 1% (16). This process was later improved (17) by the use of a double bed: the first, of a carboxylic acid type, retained all the metallic cations and the second, of a sulphonic acid type, restored the ionic composition of the milk. Some losses of vitamins were observed, with an increase in Pb, Cu, and Fe. The milk was treated at 40°C with only a pH reduction from 6.8 to 6.2 and 97% of the $^{137}$Cs was removed. Nutritional evaluation of the milks obtained by acidification to pH 5.2 reveals an unexplained lower nutritional value as compared to the untreated milk, while the double-bed process was found nutritionally satisfactory (18). Moreover, regeneration of the resin is time consuming, costly, and laborious; the alternative of electrodialysis through ion-exchange membranes, which is a continuous process, has been proposed (19). As other minerals are lost during the decontamination process, they must be reintroduced to reconstitute the mineral content of the milk. Such a treatment was shown to remove 90% of radiocesium from the milk, but in some cases, bitter flavor and the appearance of sediment were observed (20).

The mechanisms and capacity of sorption and fixation of Cs by clay minerals, such as mica, vermiculite, montmorillonite, and kaolinite, were studied. While Cs sorption by mica and vermiculite was governed by particle size and hydration, montmorillonite and kaolinite appeared to act through an electrostatic mechanism (21). Attempts to decontaminate milk with bentonite have recently been performed (Allgäuer Alpenmilch, Germany) but showed a Cs removal less efficient than with methods reviewed earlier in this paper.

## Vegetables

There are few reports on the decontamination of vegetables. Endres and Fischer (22) investigated the possibility of decontaminating green and bulb vegetables sprayed during growth with solutions containing different radionuclides. The

**Table 16.1.** Decontamination of vegetables by washings with water and citric acid at different temperatures and different lengths of treatment and rinsing[a]

| | % Decontamination | | | | | |
|---|---|---|---|---|---|---|
| Washings | Water | | | 0.2% Citric acid solution | | |
| Radionuclides | $^{144}Ce$ | $^{137}Cs$ | $^{85}Sr$ | $^{144}Ce$ | $^{137}Cs$ | $^{85}Sr$ |
| *Products and Treatments* | | | | | | |
| *Broad-leaved chicory* | | | | | | |
| 10 min at 60° | 20.3 | 51.6 | 20.7 | 29.8 | 57.8 | 31.6 |
| 15 min rinsing | | | | | | |
| 60 min at 60° | 21.9 | 62.3 | 59.7 | 30.6 | 57.8 | 49.6 |
| 15 min rinsing | | | | | | |
| *Cabbage lettuce* | | | | | | |
| 5 min at 90° | | 86.1 | 62.7 | | 86.2 | 68.0 |
| 15 min rinsing | | | | | | |
| 15 min at 90° | | 93.3 | 84.5 | | 89.6 | 70.3 |
| 15 min rinsing | | | | | | |
| *Swiss chard* | | | | | | |
| 30 min at 70° | 45.5 | 77.7 | 41.0 | 61.2 | 76.3 | 43.2 |
| 15 min rinsing | | | | | | |
| 60 min at 70° | 40.0 | 76.3 | 47.2 | 57.6 | 78.2 | 25.1 |
| 15 min rinsing | | | | | | |
| *Potatoes* | | | | | | |
| 60 min 70° | 55.6 | 73.9 | 49.1 | 63.5 | 72.5 | 50.9 |
| 180 min 70° | 58.2 | 87.3 | 45.8 | 75.3 | 91.0 | 62.5 |
| 350 min 70° | 75.5 | 96.1 | 60.3 | 80.2 | 98.1 | 84.3 |

[a] Reprinted with permission from ref. 22.

use of water or 0.2% citric acid solution gave similar results for Cs removal. Decontamination of curled lettuce, broad-leaved chicory, swiss chard, cabbage lettuce, turnip-rooted cabbage, and potatoes was controlled at different temperatures from 20°C to 90°C and for different lengths of treatment. Some results are presented in Table 16.1, showing that among the radionuclides studied, $^{137}Cs$ was easiest to wash off. However, with 30-min water treatments, Cs removal from lettuce was increased from 4% with water at 20°C to 72% with water at 80°C (22). The uptake of radionuclides by kale has been studied and decontamination with cold water for 1 hour or water at 37°C for 3 hours removed only 9% and 38% of $^{137}Cs$, respectively (23). Most of the studies published concerned radioactive material, absorbed primarily on the surface, so that washing and removal of the outer skin readily decreased the level of radionuclides in the vegetable studied. In the case of contaminated potatoes, skins must be removed because thermal treatment, such as blanching, results in a transfer of radioactivity from peelings to core. When peeled and sliced potatoes were stirred for 16 hours in a low-concentration KCl and NaCl solution with a cation-exchange resin, more than 95% of the $^{137}Cs$ was removed (24). When radionuclides have been absorbed through the roots, the level of contamination depends on the factor of transfer of each nuclide and decontamination is less efficient.

In the case of $^{137}$Cs, a 50% reduction was achieved in peas by broth flotation and water blanching while washing, and blanching spinach and broccoli reduced $^{137}$Cs by 60% and 90%, respectively. The combination of pickling and canning cucumbers reduced $^{137}$Cs by 94% and canning alone reduced the contamination by 63% for beans and 77% for kale. Freezing was shown to reduce Cs by 25% in kale (25).

## Fruits

Tomato juice enriched with $^{134}$Cs chloride and tomato puree obtained from tomatoes grown in $^{134}$Cs-enriched soils has been decontaminated by different ion-exchange resins. Some resins reduced radioactivity to the background level even at the 2% resin weight proportion, but in these studies, the modifications of the constituent concentrations and the nutritional quality of the treated tomato products were not controlled (26). The quality of apple juice does not seem to be affected by electrodialysis, which removed about 98% of the radioactive Cs added (27).

## Food Treatment for Strontium Removal

The release of long-lived radionuclides, such as $^{90}$Sr, into the biosphere from any source is potentially a serious health problem. A large fraction of the $^{90}$Sr deposited on the ground is retained by many soil types in the top of the soil. When ingested from contaminated food, the absorbed strontium accumulates in the bone structure in a manner similar to Ca. The half-life of $^{90}$Sr is 28 years, and its biological half-life depends on bone formation and turnover. The estimation of the biological half-life of the deepest compartment varied from two to five years in adult subjects but was only 60 days in 4 to 12-year-old children. Strontium is trapped in new crystals of the hydroxyapatite deposited in a collagen matrix, producing a localized bone irradiation with the beta particles emitted. Children represent the most critical segment of the population with regard to radiostrontium deposition because of their dynamic skeletal growth. An additional hazard related to infancy is that immature cell populations in the young can be more sensitive to radiation. Controlled balance studies in children have been reported, showing that 20% of $^{90}$Sr was absorbed through the intestine and that approximately half of that dose was retained in the body (28). The mode of contamination of the food chain by Sr is similar to Cs and depends on plant surface contamination. As shown in Table 16.2, dairy foods collected near nuclear power plants contributed to 40% of the total Sr intake (29). Strontium absorption from the intestine and its secretion in cows' milk is less efficient than for Cs, and the concentration of Sr in milk was 10 to 20 times lower than Cs. After 32 days, the total secretion of Sr in milk amounted to 0.88% of the dose administered. The existence of three elimination half-lives for Sr secreted in milk was also shown. The third half-life, of about 350 days, characterizes the mobilization of Sr from bones. Ilin and Moskalet (30)

**Table 16.2.** Adult total diet intake levels of $^{90}$Sr near US nuclear power plants[a]

| Composite | Consumption (kg /d) | $^{90}$Sr intake for 1980–1982 | |
|---|---|---|---|
| | | Bq /d | % |
| Dairy foods | 0.756 | 0.115 | 39.0 |
| Meat, fish, and poultry | 0.290 | 0.013 | 4.5 |
| Cereal foods | 0.369 | 0.041 | 13.8 |
| Potatoes | 0.204 | 0.027 | 9.1 |
| Leafy vegetables | 0.059 | 0.010 | 3.4 |
| Legumes | 0.074 | 0.023 | 7.9 |
| Root vegetables | 0.034 | 0.004 | 1.5 |
| Garden fruit | 0.088 | 0.007 | 2.3 |
| Fruits | 0.217 | 0.026 | 8.8 |
| Oil and fats | 0.052 | 0.006 | 2.2 |
| Sugar and adjuncts | 0.082 | 0.007 | 2.3 |
| Beverages | 0.697 | 0.014 | 4.9 |

[a] Reprinted with permission from ref. 29, copyright 1985, Pergamon Press.

showed also that 54% of the Sr ingested by cows was excreted in feces. In man, it has been estimated that 50% of dietary $^{90}$Sr is provided by milk.

## Milk

Synthetic cation-exchange resins in the $Na^+$, $K^+$, or $Ca^{2+}$ forms have been used with success for the removal of radiostrontium from milk labeled in vitro and in vivo. However, the removal of up to 90% of the Sr was always accompanied by important changes in the cationic composition of the milk and other physical properties. The first attempt to maintain the composition of the milk unaltered, by adding to the resin a synthetic mixed salt solution, led to a removal of only 60% of the Sr. These studies were reviewed by Murthy et al. (31), and they proposed a new method for the removal of radiostrontium in which milk was acidified to a pH of 5.4 to 5.3 with 0.1 or 0.5 $M$ citric acid, before passing through the ion-exchange resin. Most of the patents describing the use of the ion-exchange resin column or the electrodialysis process have been applied not exclusively to Cs but also to Sr (9,10,11,19). About 95% to 96% of the Sr was removed after milk acidification and using continuous regeneration of the ion-exchange resin (11).

Since 1965 several pilot plants capable of removing more than 90% of the environmental level of $^{90}$Sr have been built (32,33). Microbial population and chemical composition were not significantly changed, but a slight decrease in flavor was reported (34). Another plant was in operation over a period of 10 months, processing a total of 413,000 L of milk at a flow rate of 5,700 L/h during an 8-hour run, and $^{90}$Sr removal averaged 92% (35). Table 16.3 shows different commercial scale treatments of milk and the removal of $^{90}$Sr obtained (36).

The analysis of raw and processed milk constituents also showed no effect on the microbial population and other constituents except minor changes in

**Table 16.3.** Removal of $^{90}$Sr from milk using a fully automated plant[a]

| Type of processed milk | Total weight of milk (kg) | Averaged flow rate (kg/h) | Removal efficiency (%) |
|---|---|---|---|
| Skim milk | 45,060 | 5,625 | 93.1 |
| | 46,965 | 5,225 | 94.5 |
| | 47,290 | 5,740 | 94.5 |
| Whole milk | 47,120 | 4,020 | 96.9 |
| | 45,290 | 6,470 | 93.2 |
| | 46,720 | 5,445 | 89.5 |
| | 47,540 | 5,300 | 86.1 |
| | 91,700 | 6,060 | 88.7 |

[a] Reprinted with permission from ref. 36, copyright 1967, Pergamon Press.

the freezing point, curd tension, titratable acidity, and ash and a significant increase in K citrate concentration (37). The two-resin bed system described previously (16) for Cs removal reduced $^{85}$Sr introduced in vitro or in vivo in milk to 2% to 4% of the original value. A pulsed-bed ion exchange contactor was designed for continuous operation, and the improvements concerned the regenerating solution for simultaneous regeneration and sanitization (38). Electrodialysis has also been applied with success for Sr removal (20).

Other methods have been reported for the removal of Sr from milk. By the addition of Ca carbonate and Ca phosphate, 50% decontamination was obtained, but the final product had an unpalatable taste. Calcium and Sr pyrophosphates exhibit a good extraction capacity for radioactive $^{90}$Sr. When milk was placed in contact with silica–alumina pellets containing absorbed pyrophosphate, dried at 110°C and placed on a glass column, a continuous decrease in the activity of the milk was observed. The remaining $^{90}$Sr activity was 69% after 1 hour, 50% after 2.5 hours, 37% after 5 hours, 26% after 10.5 hours, and 17% after 24 hours (39). Total Ca and Sr precipitation from milk has been performed by the addition of oxalic acid and K oxalate at pH 6.4. Oxalate salts were separated by centrifugation. The Ca content was then restored to the normal milk concentration either by neutralization with Ca hydroxide and the addition of Ca lactate or through the contact with a cation-exchange resin saturated with Ca. In the two cases, total $^{90}$Sr removal varied from 85% to 90% (40). Phytates, especially Na phytate, have been used to remove $^{90}$Sr from cows' milk. By the addition and stirring of 80 g of Na phytate per liter of raw milk, 77% of the Sr was removed after 5 minutes, and the decontamination process improved continuously to attain 88% after 1 hour. Other phytates, such as Ca phytate, were less efficient (70% after 1 hour) (41).

### Other Food

In spite of the presence of phytate in cereals, the relative bioavailability of $^{90}$Sr from wholemeal bread was similar to $^{85}$Sr present in cows' milk (42). Treatment of vegetables by washing with water or 0.2% citric acid solution, at different temperatures (Table 16.1), showed that the efficacy of $^{85}$Sr removal was always lower than $^{137}$Cs decontamination (22). This observation is valid for most of the other methods of decontamination. Pickling and canning shown to reduce $^{137}$Cs by 94% removes only 65% of $^{90}$Sr. Weaver and Harris (25) extended these observations to canned beans where 63% of the Cs and 46% of the Sr was removed, canned kale 77% and 58%, frozen kale 25% and 7%, canned cucumbers 95% and 64%, and peeled, washed, and blanched sweet potatoes 26% and 15% removed, respectively. Cation-exchange resins have been used to remove Sr from tomatoes and strawberries, but a greater reduction of $^{90}$Sr was achieved on strawberry puree by precipitation with Ca carbonate. Although 97% of the radioactivity was removed, the authors pointed out that the recovery of liquid strawberry material was only 43% and that the final product had a salty taste (26). Electrodialysis has been shown to remove $^{90}$Sr from apple juice, and the analysis all along the process demonstrated that $^{90}$Sr removal was delayed compared with $^{137}$Cs (27).

## Other Radioisotopes

### Iodine

Fission produces a high $^{131}$I yield and in the case of the Windscale nuclear reactor accident, in October 1957, its concentration in milk, from some areas, exceeded 3700 Bq/L. Iodine-131 was detected in mozzarella cheese shipped from Italy after the Chernobyl accident. After the administration of labeled I as a single dose to cows, from 4% to 15% of the dose was recovered in milk in 7 days (43). Vacuum treatment did not remove $^{131}$I from the milk. Because of the short half-life of $^{131}$I, the contaminated milk may be transformed into dairy products suitable for storing sufficiently long to allow radioactive decay. In the case of $^{131}$I contamination, it thus seems important to prevent the population from consuming contaminated fresh food. However, the use of anion-exchange resins showed that 98% of $^{131}$I was removed from treated fresh raw milk and that only protein-bound $^{131}$I, which may vary from 0% to 10%, cannot be retained on the resin (44). Fixed-bed, pilot plant ion-exchange columns (32) as well as a fixed-bed contactor (33) were tested and from 93% to 97% of the $^{131}$I was removed.

### Barium

Low levels of $^{140}$Ba were detected in cheese after Chernobyl. Experiments on the removal of $^{133}$Ba and $^{140}$Ba present in fresh raw milk have been reported using ion exchange resins, showing that 88% and 85% to 95% of radioactive Ba was removed (13,31). Strong-acid resin (45) and weak-acid resin (46) have been applied to remove radionuclides of Ba and Ra from water and more than 95% of each radionuclide was removed in each case.

## Cerium, Zirconium, Niobium, and Ruthenium

Since the work of Hamilton (47), showing a negligible intestinal absorption of [144]Ce, [95]Zr, [95]Nb, and [106]Ru in the rat, Schroeder and Balassa (48) suggested in humans a gastrointestinal absorption of stable Nb greater than 60%. This result was not confirmed in a study on the metabolism of [95]Zr and [95]Nb in the rat, where gastrointestinal absorption was shown to depend on the chemical forms administered but did not exceed 0.2% (49). Bone appeared to be the critical organ for [95]Zr and testes for [95]Nb. Only one study, already mentioned (22), reports the decontamination of Zr present in potatoes. A mean value of 55% Zr was removed by washing with hot water (70°C) or 0.2% citric acid solution. The same treatment was more efficient for [144]Ce, and 80% was removed. The results obtained on other vegetables (Table 16.1) demonstrated that the efficiency of Ce and Sr decontamination was similar but lower than for Cs. Some studies have been performed on the removal of elements, such as Co and U, but for their concentration in radioactivity-bearing wastes and not in liquid food.

# In Vivo Decontamination

The study of decontamination processes that can be applied in the case of human exposure, preventing the absorption or increasing the release of the radionuclides already present in the body, has been a very important field of research. Active compounds have been tested in animals showing that the food-chain contamination can be interrupted.

## Elimination of Iodine

In vivo treatment to prevent I absorption or to increase the turnover of I already incorporated in the thyroid has been used in humans. Prophylactic administration of 100 to 200-mg of K iodide prior to radioiodine contamination reduced the dose delivered to the thyroid by more than 98%. The effect was less effective when iodide was given after the contamination but was not modified by the simultaneous administration of other antidotal agents, such as Ca alginate, ferric hexacyanoferrate, and Zn-diethylenetriaminepentaacetic acid (Zn-DTPA) (50).

## Elimination of Cesium

The amount of [137]Cs excreted in feces was positively correlated with the mass of fecal material (51). This effect, as well as that of alfalfa (52) and beet pulp, can be explained by a decreased intestinal absorption. A more important 50% reduction of radioactivity was observed in milk of cows fed a high hay ration as compared with a high grain ration (53). Richmond (54) published an extensive review on all antidotes tested for internal radioactive Cs decontamination in animals and humans. He demonstrated that the decrease of environmental temperature from 34°C to 5°C in mice increased the excretion of $^{137}$Cs, probably as a result of the enhancement of the metabolic rate. Cesium urinary and fecal excretion was enhanced when dietary K and Na intake was increased in rats, swine, rabbits, and sheep. In sheep, total $^{134}$Cs excretion increased from 12% in a diet deficient in Na and K to 52% when the dietary K content was increased to 2.5% (55). This observation was later confirmed in the rat by other laboratories. The drugs acetazolamide and chlorothiazide have been shown to modify Cs secretion through the renal tubule (56). Acetazolamide was the most potent and also increased the excretion of $^{86}$Rb, but as K excretion was affected, such treatment cannot be recommended in the case of radionuclide contamination.

Prevention of radiocesium intestinal absorption by the use of clay minerals or hexacyanoferrate derivatives presently constitutes the most efficient and safest treatment for the prevention of Cs contamination. Bentonite, given to sheep in the daily concentrate ration, at a level of 10%, was eaten without difficulty, and the average absorption of Cs was reduced from 64%–79% to 5% or less. On a short-term basis, the feeding of bentonite to cattle can be recommended, but long-term treatment is still questionable, because bentonite considerably decreased Ca absorption (57). Vermiculite (3% of the total ration) was also shown to reduce Cs secretion in goats' milk and urine excretion from 2.3% to 0.3% of the daily ingested dose and from 11.2% to 1.7%, respectively. In contrast, fecal Cs excretion increased from 26.6% to 64.6% of the dose (58). Feeding cows for 21 days with vermiculite (verxite flakes) in the morning grain ration, at a level of 0.82, 0.54, and 0.27 kg per day, reduced $^{134}$Cs secreted in the milk by 88%, 84%, and 68%, respectively, when compared with the control group. Milk mineral content was not affected by this treatment (59). The main limitation to the use of bentonite and vermiculite in the reduction of Cs transfer from fodder to the milk and meat is the high dosage that has to be fed to the animal. In this respect, the potency and selectivity of hexacyanoferrate derivatives in binding radiocesium make them the most promising material tested to date. Many reports since 1963 have shown the enhancement of $^{137}$Cs excretion when simultaneously fed with ferric hexacyanoferrate (60) and other derivatives (61) in the rat. These results also demonstrated that Cs appearing in the gut by secretory processes was prevented from reentering the body for recirculation. Since no toxic side effects were observed, Madshus et al. (62) extended these observations to dogs and human beings. In humans, a reduction of more than 50% in the biological half-life of Cs was reported with the administration of a dose of 3 g/day. Several laboratories have looked for more potent complexes, and ferric hexacyanoferrate in the colloidal form was shown to be more efficient than the insoluble form in removing $^{134}$Cs from the body (63). Several metal hexacyanoferrates were evaluated by Havlicek et al. (64), and their effectiveness, in increasing order, was K < Ca = Co(II) = Zn < Fe(III). Bismuth, Ni, Li, and Rb hexacyanoferrates were also tested, but Giese and Hantzsch (65) demonstrated that colloidal ammonium ferric hexacyanoferrate

(AFCF) was the most efficient. In laying hens, Nezel (66) showed that AFCF almost completely inhibits the uptake of $^{137}$Cs when given together with the contaminated food. This effect was confirmed in cows fed milk containing added $^{134}$Cs (67). The effectiveness and metabolism of AFCF will be discussed later.

## Elimination of Strontium

In a review published by Smith (68), all the published physiological parameters that can prevent Sr accumulation in bones, or increase its mobilization, were discussed. The inhibition of calcification with high doses of phosphate or fluoride and the stimulation of decalcification through ammonium-chloride-induced metabolic acidosis led to disappointing results. High citrate concentrations must be reached to induce excretion of $^{85}$Sr, whereas tricarballylate and salicylate enhanced Sr excretion through modification of Ca concentration in extracellular fluids. Oral and intravenous administration of stable Sr, as well as an increase of P, Mg, and Ca intake, will increase urinary excretion of $^{90}$Sr.

On a low Ca diet, Al phosphate gel decreased $^{85}$Sr absorption in humans by 43% when given 1 hour after Sr administration (69). A combined therapy with alginate and Ca phosphate supplementation gave the greatest reduction in Sr absorption and retention (70). Sodium and Ca montmorillonite and kaolin (71), as well as vermiculite (58), have no effect on Sr absorption. Combined treatment with ferric hexacyanoferrate, K iodide, and Ca alginate showed a reduction of whole-body $^{137}$Cs from 41% of the dose, 6 days after the oral administration, to 0.7%, $^{131}$I from 17.1% to 1.8%, and $^{85}$Sr from 14.3% to 2.7% (72).

## Other Radioisotopes

In the previous study (72), trisodium Ca salt of DTPA [Na$_3$(CaDTPA)] administration reduced liver contamination of orally fed $^{141}$Ce from 43% to 4.3% of the dose. DTPA also reduced the accumulation of $^{91}$Y, $^{65}$Zn, and $^{60}$Co in the liver, kidney, and skeleton of the rat (73) and lowered the $^{239}$Pu content of the liver and skeleton when intraperitoneally injected into the rat (74). After the injection of polymeric and monomeric solutions of Pu, DTPA therapy reduced by about half the level in bone in each case. However, while the monomeric form of Pu was nearly completely removed from the liver, the polymeric form was only reduced by one third (75). DTPA has also been proved superior to EDTA for the removal of $^{234}$Th. Because Ca-DTPA exhibits a higher toxicity than Zn-DTPA, it's product is recommended for long-term treatments necessary for the removal of $^{241}$Am and other radioisotopes from the skeleton (76). The comparative efficiency of Zn-DTPA and Ca-DTPA was reported for $^{241}$Am and $^{239}$Pu removal from the rat (77).

# Effectiveness of Ammonium Ferric Hexacyanoferrate in Complexing Cesium in Cows

A daily dose of 3 g of ammonium ferric hexacyanoferrate (AFCF) premixed with maize silage and concentrated feed was given to cows fed trefoil pasture (70 Bq/kg Cs) or fodder (3050 Bq/kg) collected from July to September 1986 in south Bavaria. Cesium radioactivity in milk increased from 20 Bq/L to a plateau value of 220 to 260 Bq/L 10 days after feeding contaminated fodder (Fig. 16.1). The administration of AFCF produced a significant decrease of milk radioactivity within 3 days. The pattern of the decline was similar to the multiexponential elimination of a single dose of $^{137}$Cs in cows' milk (6). When AFCF was given 2 days before the administration of contaminated fodder, radiocesium transfer from fodder to milk was completely prevented (78). This study demonstrated that radiocesium present in fodder contaminated from the Chernobyl fallout was not absorbed and did not appear in a significant amount in cows' milk when a single dose of 3 g/d AFCF was added through their concentrated feed.

# Metabolic Study of $^{14}$C AFCF in Cows

In spite of these in vivo experiments, in which the effectiveness of AFCF can only be explained by its fecal excretion and the absence of apparent toxic effects on animals fed various hexacyanoferrates, this additive could not be applied in cows until its stability in the rumen and its metabolic fate were known. It was particularly important to assess the possible release of cyanide and to demonstrate the safety of the milk and meat produced. For that purpose, AFCF labeled with $^{14}$C was synthesized and purified for administration to cows. The products of disintegration in the rumen, the metabolites excreted in feces and urine, secreted in milk and still present in the organs, 10 days after the administration were analyzed quantitatively and when possible identified (78). In the rumen juice, 0.13% of AF$^{14}$CF was recovered as labeled $CO_2$, and it was also shown to disintigrate partially into hexacyanoferrate. Most of the radioactivity (91% to 95% of the dose) was recovered in the feces during the first 3 days after oral administration. Along with the labeled AFCF found in feces, hexacyanoferrate and thiocyanate were also identified in urine at very low levels (0.2% to 0.5%). In milk, the radioactivity recovered was so low (0.07%) that the metabolites were not identified. They must be similar to those found in urine and in such quantities represent no risk for the milk consumer. These results demonstrated the efficiency and safety of AFCF as a cattle fodder additive. Food-chain contamination from fodder to cattle meat can be stopped, and its implications in milk and further processed dairy products avoided. Fecal excretion of AFCF was confirmed using $^{59}$Fe labeled AFCF, as was some intestinal absorption of $^{59}$Fe, which was recovered in urine and milk (Giese, personal communication).

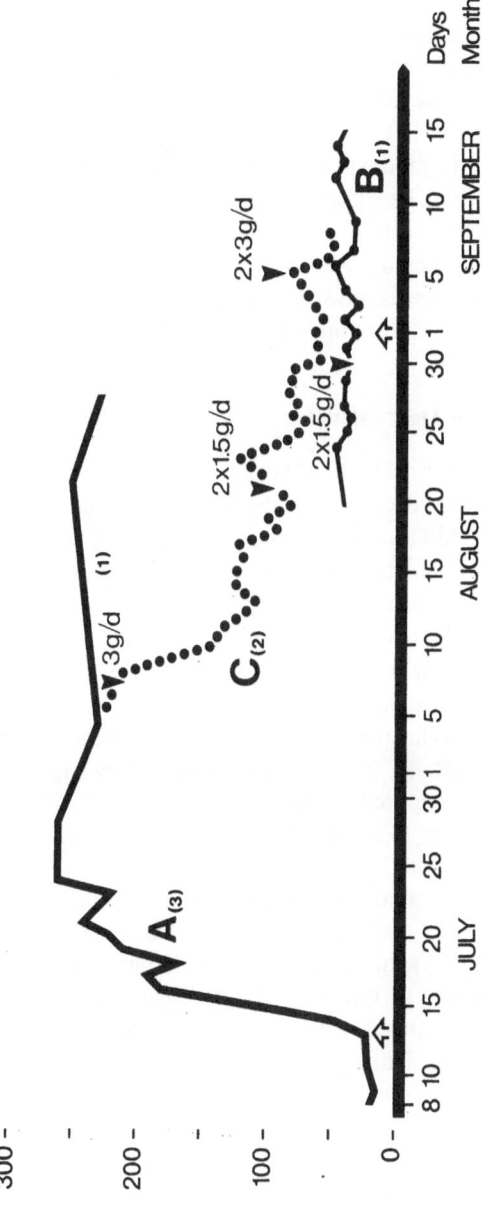

**Fig. 16.1.** Effect of AFCF administered (↓) to cows fed contaminated fodder (⇦) on total Cs radioactivity ($^{137}$Cs plus $^{134}$Cs) in milk, expressed in Bq/L. (Reprinted with permission from ref. 78.)

# New Research on Milk Treatment for Cesium Removal After Chernobyl

## Bentonite

It has been demonstrated that bentonite fed to cows and goats can bind the radiocesium that is present or added to the diet. The use of bentonite for milk decontamination has recently been shown to be effective in removing approximately 70% $^{134}$Cs and $^{137}$Cs. This process, however, requires a relatively high bentonite/milk weight ratio and several time-consuming steps to ensure an efficient decontamination. In addition, the final product always contains residual bentonite. Bentonite is a safe and accepted additive in animal feed, and thus its presence in milk would have no adverse consequences when given to pigs or calves. Such a process is, however, limited in application because it produces substantial quantities of bentonite that must then be treated as contaminated wastes.

## Hexacyanoferrate Derivatives

The high capacity and selectivity of hexacyanoferrate to bind Cs have stimulated the development of new decontamination processes for milk treatment. In contrast with the colloidal AFCF, which cannot be removed completely from milk, $(Fe^{3+})_4[Fe^{2+}(CN)_6]_3$, called Radiogardase-Cs(Heyl,Berlin), can be supplied as very resistant granular particles and was shown to be highly insoluble in water. The efficiency of this material, packed in columns, depends on the type of milk products, and it was shown to remove approximately 90% of the radiocesium (79). An important advantage of this treatment was that the mineral composition of milk was not significantly changed. Because of its very high capacity to bind Cs (up to 25 MBq/kg for skim milk in laboratory trials), the total quantity of Radiogardase-Cs required for the decontamination will be small and thus radioactive wastes kept to a minimum. We do not know yet if these columns can be regenerated.

Another product, recently developed and consisting of a polystyrol resin bed coated with AFCF, seems to offer the advantage of being regenerated with nitric acid or with a solution of ammonium nitrate (Giese, personal communication). If the stability of this resin during contact with milk and during the regeneration is demonstrated, then this decontamination process presents additional advantages for isolating radiocesium in a concentrated liquid form. However, the claim of a high Cs binding capacity has to be tested in practical applications.

## Conclusions

The removal and reduction of radionuclides in the food chain can be achieved either by directly treating the food or by preventing, in vivo, the transfer of radionuclides from contaminated fodder to dairy products, which have been shown to play the major role in human exposure. The efficiency of the methods used and the additives proposed will depend on the radionuclides present in the food. The decontamination processes must be studied and adapted according to the food considered (milk, vegetables, fruits) and whether or not the contamination is due to directly deposited radionuclides or to absorption from roots (vegetables, fruits). Washing, blanching, and canning have been shown to reduce contamination of vegetables for several radionuclides. Ion-exchange resins and electrodialysis need additional treatments to normalize the mineral composition of milk and are not easily applied on a large scale. Hexacyanoferrate derivatives can remove Cs with no effect on the mineral composition or nutritional quality of the milk. The development of such specific complexing products is the most promising field of research for the direct treatment of food.

The prevention in cattle of intestinal absorption of radionuclides present in fallout from nuclear power accidents by the use of complexing compounds (hexacyanoferrate, bentonite, alginate, etc.) has been shown to be safe and efficient and can be rapidly applied in case of an emergency. These combined treatments offer the easiest way to reduce radionuclide contamination in dairy products and meat.

*Acknowledgments.* The author wishes to thank Dr. I. Horman and Dr. M. W. Carter for reviewing the manuscript, J. Lindstrand for technical assistance, and C. Isom for secretarial assistance.

# References

1. Rundo J and Taylor BT (1964) In: The assessment of radioactive caesium in man, the assessment of radioactivity in man. Proc Symp, Heidelberg, May 1964, International Atomic Energy Agency, Vienna, vol 2, p3
2. Barth J and Bruckner BH (1969) Evaluation of clays as binding agents for reduction of radionuclides in milk. J Agr Food Chem 17:1340–1343
3. Wilson AR and Spiers FW (1967) Fallout caesium-137 and potassium in newborn infants. Nature 215:470–474
4. Gustafson PF and Miller JE (1969) The significance of cesium-137 in man and his diet. Health Phys 16:167–183
5. Kahn B, Jones IR, Carter MW, Robbins PJ, and Straub CP (1965) Relation between amount of cesium-137 in cows' feed and milk. J Dairy Sci 48:556–562
6. Sansom BF (1966) The metabolism of caesium-137 in dairy cows. J Agric Sci 66:389–393
7. Stewart HF, Ward GM, and Johnson JE (1965) Availability of fallout cesium-137 to dairy cattle from different types of feed. J Dairy Sci 48:709–713
8. Assimakopoulos PA, Ioannides KG, Pakou AA, and Paradopoulou CV (1987) Transport of the radioisotopes iodine-131, cesium-134 and cesium-137 from the fallout following the accident at the Chernobyl nuclear reactor into cheese and other cheese-making products. J Dairy Sci 70:1338–1343
9. Murthy GK, Campbell JE, Mazurovsky EB, and Edmondson LF (1962) US Patent No 3,020,161

10. Van der Stricht (1967) Dispositif et procédé de décontamination radioactive de produits alimentaires notamment liquides. Euratom Patent No 1.517.279, pp 3

11. Higgins IR (1968) Apparatus for the treatment of milk. US Patent No 3,415,377, pp 15

12. Poznanski S, Kornacki K, Chojnowski W, and Jedrychowski L (1972) Tests suitability of polish ion exchange resins for removal from milk of radioactive contamination. Medycyna Weterynaryjna 28:737–740

13. Walker JP and Edmondson FL (1969) Studies on ion exchange resins for the removal of radionuclides from milk. Health Phys 16:85–91

14. Murthy GK (1969) Preparation of product from milk treated with cationic resin for removing radionuclides from milk. J Dairy Sci 52:629–632

15. Edmondson LF (1964) Ion exchange process for removing radioactive contamination from milk. J Dairy Sci 47:1201–1207

16. Glascock RF and Bryant DWT (1968) A pilot plant for the removal of cationic fission products from milk. II. Efficiency of the process and composition of the product. J Dairy Res 35:269–286

17. Glascock RF and Bryant DWT (1971) A double-bed process for the removal of cationic fission products from milk. J Dairy Res 38:217–236

18. Cohen B, Ashworth M, Glascock RF, and Bryant DWT (1973) The removal of fission products from milk. The use of the baby monkey (macaca irus) for the nutritional evaluation of milk treated by two processes. J Dairy Res 40:53–61

19. Ionics Incorporated (1965) Process and apparatus for the decontamination of milk products by electrodialysis. Great Britain Patent No 1,005,125

20. Thiele D (1969) Die dekontaminierung von milch durch elektrodialyse unter praxisähnlichen bedingungen. Kieler-Milchwirtschaftliche-Forschungsberichte 21:447–459

21. Sawhney BL (1964) Sorption and fixation of microquantities of cesium by clay minerals: effect of saturating cations. Soil Sci Soc Am Proceedings 28:183–186

22. Endres O and Fischer E (1969) Untersuchungen zur Dekontamination von gemüse. Deut Lebensm Rundschau 65:1–5

23. Rohleder K (1972) Untersuchungen über die Aufnahme radioaktiver Stoffe durch Grünkohl aus dem Boden und aus der Atmosphäre und Versuche zur Dekontamination. Zeitsch für Leben Forsch 149:223–227

24. Perkins HJ and Strachan G (1964) Decontamination of potato tubers containing cesium-137. Science 144:59–60

25. Weaver CM and Harris ND (1979) Removal of radioactive strontium and cesium from vegetables during laboratory scale processing. J Food Sci 44:1491–1493

26. Ralls JW, Maagdenberg HJ, Guckeen TR, and Mercer WA (1971) Removal of radioactive strontium and cesium from vegetables and fruits during preparation for preservation. J Food Sci 36:653–656

27. Frindik O (1970) Elektrodialytische dekontamination von apfelsaft. Zeitsch für Leben Forsch 144:179–187

28. Wellman HN, Kahn B, Salem AJ, and Robbins PJ (1968) Assessment of deposited radiostrontium during childhood. In: Kornberg HA and Norwood WD (eds) Diagnosis and treatment of deposited radionuclides. Excerpta Medica Foundation, pp 386–394

29. Stroube WB, Jelinek CF, and Baratta EJ (1985) Survey of radionuclides in foods, 1978–1982. Health Phys 49–731:735

30. Ilin DI and Moskalev YI (1957) On the metabolism of cesium, strontium and a mixture of beta-emitters in cows. J Nuclear Energy II 5:413–420

31. Murthy GK, Masurovsky EB, Campbell JE, and Edmondson LF (1961) Method for removing cationic radionuclides from milk. J Dairy Sci 44:2158–2170

32. Walter HE, Sadler AM, Easterly DG, and Edmondson LF (1967) Pilot plant fixed-bed ion-exchange resin system for removing iodine-131 and radiostrontium from milk. J Dairy Sci 50:1221–1225

33. Stroup WH, Reyes AL, Murthy GK, Read RB, and Dickerson RW (1968) Combined process for removing radioactive iodine and strontium from milk by ion exchange. J Dairy Sci 51:1500–1502

34. Bales RE and Hickey JLS (1966) Commercial processing of milk for concurrent removal of cationic and anionic radionuclides. In: Radioisot radiat, Dairy Sci Technol, Proc Semin, pp 121–136

35. Sparling EM, Baldi EJ, Marshall RO, Heinemann B, Walter HE, and Fooks JH (1967) Large scale fixed bed ion-exchange system for removing strontium-90 from fluid milk, I Processing results. J Dairy Sci 50:423–425

36. Fooks JH, Terrill JG, Heinemann BH, Baldi EJ, and Walter HE (1967) Evaluation of full scale strontium removal system for fluid milk. Health Phys 13:279–286

37. Heinemann B, Baldi EJ, Marshall RO, Sparling EM, Walter HE, and Fooks JH (1967) Large scale fixed bed ion-exchange system for removing strontium-90 from fluid milk, II Compositional studies. J Dairy Sci 50:426–430

38. Dickerson RW, Reyes AL, Murthy GK, and Read RB (1968) Development of a pulse-bed ion exchange contactor for removing cationic radionuclides from milk. J Dairy Sci 51:1317–1323

39. Van't Riet B (1967) removal of strontium-90 from milk with calcium pyrophosphate, strontium pyrophosphate or a mixture thereof. US Patent No 3,359,117

40. Wenner V (1974) Contamination du lait par des éléments radioactifs et possibilité pratiques pour la décontamination du lait contenant des radio-éléments. Technicien du Lait, pp 5–18

41. Kudo T (1969) Procédés pour enlever des produits de contamination radioactifs de produits alimentaires ou de l'eau. Brevet Fr 1.582.677, pp 14

42. Carr TEF, Harrison GE, Loutit JF, and Sutton A (1962) Relative availability of strontium in cereals and milk. Nature 194:200–201

43. Demott BJ and Easterly DG (1960) Removal of iodine-131 from milk. J Dairy Sci 43:1148–1150

44. Murthy GK, Gilchrist JE, and Campbell JE (1962) Method for removing iodine-131 from milk. J Dairy Sci 45:1066–1074

45. Myers AG, Snoeyink VL, and Snyder DW (1985) Removing barium and radium through calcium cation exchange. J Am Water Assoc 77:60–66

46. Snyder DW, Snoeyink VL, and Pfeffer JL (1986) Weak-acid ion exchange for removing barium, radium and hardness. J Am Water Assoc 78:98–104

47. Hamilton JG (1947) The metabolism of the fission products and the heaviest elements. Radiology 49:325–343

48. Schroeder HA and Balassa JJ (1965) Abnormal trace metals in man: niobium. J Chron Dis 18:229–241

49. Fletcher CR (1969) The radiological hazards of zirconium-95 and niobium-95. Health Phys 16:209–220

50. Simonovic I, Kargacin B, and Kostial K (1986) The effect of composite oral treatment for internal contamination with several radionuclides on 131-I thyroid uptake in humans. J Appl Toxicol 6:109–111

51. Snipes MB and Riedesel ML (1969) Studies of diet as a factor in cesium-137 metabolism by rats. J Nutr 97:212–218

52. Mraz FR and Patrick H (1957) Organic factors controlling the excretory pattern of potassium-42 and cesium-134 in rats. J Nutr 61:535–546

53. Johnson JE, Ward GM, Firestone E, and Knox KL (1968) Metabolism of radioactive cesium-134 and 137, and potassium by dairy cattle as influenced by high and low forage diet. J Nutr. 94:282–288

54. Richmond CR (1968) Acceleration, the turnover of internally deposited radiocesium. In: Kornberg HA and Norwood WD (eds) Diagnosis and treatment of deposited radionuclides. Excerpta Medica Foundation, pp. 315–328

55. Mraz FR (1959) Influence of dietary potassium and sodium on cesium-134 and potassium-42 excretion in sheep. J Nutr 68:655–662

56. Sastry BVR and Bush MT (1964) Enhancement of the renal excretion of cesium-137 in rats treated with acetazolamide and related compounds. J Pharmacol Exp Ther 143:30–41

57. Van den Hoek J (1976) Cesium metabolism in sheep and the influence of orally ingested bentonite on cesium absorption and metabolism. Z Tierphysiol Tierernährg u Futtermittelkde 37:315–321

58. Hazzard DG (1969) Percent cesium-134 and strontium-85 in milk, urine, and feces of goats on normal and verxite-containing diets. J Dairy Sci 52:990–994

59. Hazzard DG, Withrow TJ, and Bruckner BH (1969) Verxite Flakes for in vivo binding of cesium-134 in cows. J Dairy Sci 52:995–997

60. Nigrovic V (1963) Enhancement of the excretion of radiocesium in rats by ferric cyanoferrate (II). J Rad Biol 7:307–309

61. Nigrovic V (1965) Retention of radiocesium by the rats as influenced by Prussian blue and other compounds. Phys Med Biol 10:81–91

62. Madshus K, Strömme A, Bohne F, and Nigrovic V (1966) Diminution of radiocesium body-burden in dogs and human beings by Prussian blue. Int J Rad Biol 10:519–520

63. Bozorgzadeh A and Catsch A (1972) Evaluation of the effectiveness of colloidal and unsoluble ferrihexacyanoferrates (II) in removing internally deposited radiocesium. Arch Int Pharmacodynamie Therapie 197:175–188

64. Havlicek F, Kleisner I, Dvorak P, and Pospisil J (1967) Die Wirkung von Zyanoferraten auf die Ausscheidung von Radiozäsium bei Ratten und Ziegen. Strahlentherapie 134:123–129

65. Giese W and Hantzsch D (1970) Vergleichende Untersuchungen über die Cs-137 Eliminierung durch verschiedene Eisenhexacyano-ferratkomplexe bei Ratten. Zentralblatt für Veterinärmedizin 11:185–190

66. Nezel K (1970) Ueber die Verhinderung der Cesium-137-Aufnahme bein Legehennen. Z Lebensm Unters Forsch 144:25–31

67. Giese W (1971) Das Verhalten von Radiocesium bei Laboratoriums und Haustieren sowie Möglichkeiten zur Verminderung der Radioaktiven Strahlenbelastung, Tierärztliche Hochschule Hannover, Habilitationsschrift, pp 99

68. Smith H (1968) Experiences in the removal of radioactive strontium from animals by modifying physiological parameters. In: Kornberg HA and Norwood WD (eds) Diagnosis and treatment of deposited radionuclides. Excerpta Medica Foundation, pp 372–385

69. Spencer H, Lewin I, Samachston J, and Belcher MJ (1969) Effect of aluminum phosphate gel on radiostrontium absorption in man. Radiat Res 38:307–320

70. Kostial K. Maljkovic T, Kadic M, and Manitasevic R (1967) Reduction of the absorption and retention of strontium in rats. Nature 215:182

71. Boni AL (1969) Variations in the retention and excretion of Cs-137 with age and sex. Nature 222:1188–1189
72. Kostial K, Kargacin B, and Simonovic I (1983) Efficiency of a composite treatment for mixed fission products in rats. J Appl Toxicol 3:291–296
73. Nigrovic V and Catsch A (1965) Dekorporation von Radionukliden. Strahlentherapie 128:283–287
74. Seidel A, Volf V, and Catsch A (1971) Effectiveness of Zn-DTPA in removal of plutonium from rats. 19:399–400
75. Schubert J, Fried JF, Rosenthal MW, and Lindenbaum A (1961) Tissue distribution of monomeric and polymeric plutonium as modified by a chelating agent. Radiat Res 15:220–226
76. Lloyd RD, Mays CW, McFarland SS, Taylor GN, and Atherton DR (1976) A comparison of Ca-DTPA and Zn-DTPA for chelating [241]Am in beagles. Health Phys 31:281–284
77. Taylor DM and Volf V (1980) Oral chelation treatment of injected Am-241 or Pu-239 in rats. Health Phys 38:147–158
78. Arnaud MJ, Clement C, Getaz F, Tannhauser F, Schoenegge R, Blum J, and Giese W (1988) Synthesis, effectiveness and metabolic fate in cows of the caesium complexing compound, ammonium ferric hexacyanoferrate labeled with [14]C, in cows. J. Dairy Res, 55:1–13
79. Allgaüer Alpenmich (1987) Patent 370446.4, Verfahren zum entfernen von radioaktiven Metallen aus Flüssigkeiten, Lebens. und Futtermitteln

# Part V
# Effects of Radionuclides in Food and Water Supplies

# CHAPTER 17

# Long-Term Health Effects of Radionuclides in Food and Water Supplies

## A. C. Upton[1] and P. Linsalata[1]

## Introduction

More than 90 years have elapsed since the discovery of ionizing radiation and its capacity to cause various types of injury. In the meantime, studies of such injuries have received continuing impetus from the growing uses of radiation in medicine, science, industry, and nuclear energy. As a result, the biomedical effects of ionizing radiation have received greater study and are better known than those of any other environmental agent.

On the basis of the extensive knowledge of such effects that now exists, the health impacts of low-level irradiation from various sources have been assessed in detail (1,2). The long-term effects that are attributable to irradiation via the food chain—other than the heritable effects, which are considered elsewhere (3)—are reviewed briefly in the following. The reader is referred to ref. 4 for a comparative treatment of the various sources of ionizing radiation in the environment.

## Sources of Radioactivity in Foods

### Natural Radioactivity

Weathering of the earth's crust is the ultimate mechanism for the release into soils of primordial radionuclides and their decay products, which constitute the principal sources of internal natural background radiation (Table 17.1). Terrestrial plants acquire these nuclides via their roots or leaves, and farm animals acquire them through consumption of plants, phosphate-based mineral feed supplements (important for U series), and soils.

Radionuclides ingested in food and, to a lesser extent, water account for a substantial part of the average radiation dose to the various organs of the human

---

[1] Institute of Environmental Medicine, New York University School of Medicine, 550 First Avenue, New York, NY 10016, USA.

**Table 17.1.** Average concentrations of major primordial radionuclides in rocks and soils[a]

| | Concentration | | | | |
|---|---|---|---|---|---|
| | $^{40}K$ | $^{232}Th$ | | $^{238}U$ | |
| Type of rock, soil | Bq/kg | Bq/kg[b] | μg/g | Bq/kg[c] | μg/g |
| Basalt (crustal average) | 259 | 11–15 | 3–4 | 7–11 | 0.5–1.0 |
| Granite (crustal average) | >1100 | 70 | 17 | 37 | 3.0 |
| Shale | 814 | 48 | 12 | 46 | 3.7 |
| Carbonate rock | 74 | 8 | 2 | 25 | 2.0 |
| Soils[d] | 370 | 26 | 6 | 26 | 2.1 |

[a] Reprinted with permission from ref. 5.
[b] Series equilibrium alpha and beta activity are calculated by multiplying the values by 6 and 4, respectively.
[c] Series equilibrium alpha and beta activity are calculated by multiplying the values by 8 and 6, respectively.
[d] World average (6).

body, especially the skeleton. The identities and sources of the relevant radionuclides and their environmental pathways to humans via food and water are discussed in this chapter, as well as elsewhere in this volume [Harley, 1987 (7); Pentreath, 1987 (8)].

Human intake through food consumption can be estimated indirectly by the use of steady-state concentration factors (CFs) (9,10), which relate, by simple ratio, the concentrations in edible plant or animal tissues to the levels in the soil or animal feed. These factors vary by several orders of magnitude among the chemically different long-lived members comprising the U and Th series. In terms of food-chain processes leading to human intake, the nuclides decrease in order of importance as $^{210}Pb$-$^{210}Po \geq {}^{226}Ra = {}^{228}Ra > {}^{234}U = {}^{238}U \geq {}^{228}Th = {}^{230}Th = {}^{232}Th$. A more direct, and ultimately less variable, method of determining human intake from foods is provided by the sampling and analysis of representative diets. With the exception of Th isotopes, which are highly discriminated against in the food chain and at the level of the gastrointestinal tract, estimates of the dietary intake of these radionuclides are available from several countries (6,11).

Fresh weight concentrations of U-series radionuclides among various foodstuffs of the New York City diet (12,13) have been shown to vary by about two orders of magnitude for $^{238}U$ and $^{226}Ra$ (e.g., 1 to 100 mBq/kg) and by about a factor of 10 for $^{210}Pb$ (e.g., 10 to 100 mBq/kg) (Table 17.2). Activity concentrations of U, Ra, and Pb are not correlated among the foodstuffs analyzed because of differences in environmental chemistry, donor compartment concentrations, and metabolic behavior.

Based on the rates of food consumption and the radionuclide concentrations shown in Table 17.2, the annual intakes (Bq per year) of $^{238}U$, $^{226}Ra$, $^{210}Pb$, $^{228}Ra$ (14), and $^{40}K$ in the New York City diet are calculated to be 5.8, 21.8, 17.0, 15.8, and $4.2 \times 10^4$, respectively (Table 17.3). Intakes of $^{234}U$, although

**Table 17.2.** Typical concentrations of certain stable elements and naturally occurring radionuclides in the New York City diet[a]

| Diet item | Item (kg/y) | Intake Ca (g/y) | K (g/y) | Natural radionuclide concentrations (mBq /kg) $^{238}$U | $^{226}$Ra | $^{210}$Pb[b] |
|---|---|---|---|---|---|---|
| Dairy products | 200 | 216.0 | 280 | 0.9 | 9.3 | 10.7 |
| Fresh vegetables | 48 | 18.7 | 110 | 7.4 | 18.5 | 40.7 |
| Canned vegetables | 22 | 4.4 | 29 | 2.5 | 24.1 | 16.3 |
| Root vegetables | 10 | 3.8 | 29 | 8.6 | 51.8 | 7.8 |
| Potatoes | 38 | 3.8 | 171 | 30.8 | 103.6 | 55.5 |
| Dried beans | 3 | 2.1 | 42 | 18.4 | 40.7 | 28.1 |
| Fresh fruit | 59 | 9.4 | 112 | 14.8 | 15.9 | 14.8 |
| Canned fruit | 11 | 0.6 | 13 | 2.5 | 6.3 | 88.8 |
| Canned fruit juice | 28 | 2.5 | 53 | 1.2 | 15.5 | 8.5 |
| Bakery products | 44 | 53.7 | 53 | 22.1 | 103.6 | 66.6 |
| Flour | 34 | 6.5 | 34 | 6.2 | 70.3 | 48.1 |
| Whole grain products | 11 | 10.3 | 38 | 18.4 | 81.4 | 81.4 |
| Macaroni | 3 | 0.6 | 5 | 4.9 | 77.7 | 34.0 |
| Rice | 3 | 1.1 | | 23.4 | 28.1 | 32.6 |
| Meat | 79 | 12.6 | 261 | 13.5 | 0.4 | 18.1 |
| Poultry | 20 | 6.0 | 54 | 2.5 | 28.1 | 16.7 |
| Eggs | 15 | 8.7 | 22 | 2.5 | 225.7 | 9.6 |
| Fresh fish | 8 | 7.6 | 27 | 4.9 | 24.8 | 14.4 |
| Shellfish | 1 | 1.6 | | 116.9 | 29.6 | 125.8 |
| Water | 400 | 2–10 | < 1 | 0.4 | 0.2 | 1.5 |
| Total (food) | 637 | 375 | 1,333 | | | |

[a] Reprinted with permission from ref. 12 and from ref. 13, copyright 1971, Pergamon Press, plc.
[b] From ref. 13.

not specifically measured, should be approximately equal on an activity basis to those of $^{238}$U, as indicated by the ratios measured in human bone (18). Most of the $^{210}$Po ($t_{1/2}$ = 138 days) in human bone results from the intake and decay of $^{210}$Pb ($t_{1/2}$ = 22.3 years), which is a bone seeker. Polonium-210, which is an alpha-emitting, soft-tissue seeker with a short biological half-life in the body [50 days (19)], has an associated dietary intake range similar to that of $^{210}$Pb (18 to 62 Bq per year[1], mean = 42 Bq per year) (6).

Based on the patterns of intake in the New York City diet, two categories of plant-derived foods (vegetables, potatoes, and beans) and bakery products (cereals, rice, and grains) account for 56% to 66% of the total intake of U-series radionuclides and $^{228}$Ra (Table 17.3). For the remainder of the intakes, the contributions of the categories decrease in the following order: meats and eggs ≥ fruits and juices ≥ dairy products > fish > water. The data from several countries indicate variations in the annual intake of U-series radionuclides over a range of about 4 to 5. Clearly, the relative contributions to total intake

**Table 17.3.** Average annual intakes of certain naturally occurring radionuclides in various foods in the New York City diet

| Type of food | $^{238}$U[a] Bq/y | % | $^{226}$Ra[b] Bq/y | % | $^{228}$Ra[c] Bq/y | % | $^{210}$Pb[d] Bq/y | % | $^{40}$K Bq/y × 10$^3$ | % |
|---|---|---|---|---|---|---|---|---|---|---|
| Dairy products | 0.17 | 3 | 1.85 | 8 | 0.68 | 4 | 2.15 | 13 | 8.84 | 21 |
| Vegetables, potatoes, and beans | 1.72 | 30 | 6.03 | 28 | 7.56 | 49 | 4.55 | 27 | 12.02 | 29 |
| Fruits and juices | 0.93 | 16 | 1.44 | 7 | NA[e] | | 1.94 | 11 | 5.62 | 13 |
| Bakery products, cereals, rice, and grains | 1.48 | 26 | 8.18 | 38 | 5.67 | 36 | 5.60 | 33 | 4.10 | 10 |
| Meat and eggs | 1.16 | 20 | 4.00 | 18 } | 1.89 | 12 | 1.92 | 11 | 10.64 | 25 |
| Fin and shellfish | 0.16 | 3 | 0.22 | 1 } | NA[e] | | 0.24 | 1 | 0.85 | 2 |
| Water | 0.15 | 2 | 0.07 | <1 | NA[e] | | 0.59 | 3 | <0.03 | |
| Total intake (NY) | 5.8 | | 21.8 | | 15.8 | | 17.0 | | 4.21 × 10$^4$ | |
| Range for various countries[f] | 2.7–12.2 | | 5.4–23.0 | | | | 17–84[g] | | | |
| Arithmetic mean for various countries[f] | 5.4 | | 12.2 | | | | 52 | | | |

[a] From ref. 15 as summarized by ref. 12. Intake estimates are appropriate for $^{234}$U as well. The percentages are of intake derived from different foods.

[b] From ref. 16 as summarized by ref. 12.

[c] From ref. 17 as summarized by ref. 14.

[d] From ref. 13.

[e] NA refers to not available.

[f] From ref. 6.

[g] For $^{210}$Po, ref. 6 provides a range of 18–62 Bq/y$^1$ and an arithmetic mean of 42 Bq/y$^1$.

made by each foodstuff also vary among geographic regions as a result of differences in dietary composition and soil radionuclide concentration.

Potassium, an essential element for plants and animals, is under homeostatic control. The $^{40}K/^{39}K$ atom ratio in nature (which is constant, equaling $1.18 \times 10^{-4}$, and which gives a specific activity of $3.2 \times 10^4$ Bq/kg K) determines the intake of $^{40}K$ from food consumption and its levels in humans. Thus, an average dietary intake of 3.3 g K per day (20) corresponds to a $^{40}K$ intake of 104 Bq per day ($3.8 \times 10^4$ Bq per year) and results in an equilibrium body burden of about $4.4 \times 10^3$ Bq.

The intake of $^{14}C$ in food is likewise determined by the $^{14}C/^{12}C$ ratio in nature. Carbon-14, which exists in the atmosphere primarily as $CO_2$, is produced naturally by cosmic radiation [$^{14}N$ $(n,p)$ $^{14}C$] at a rate of approximately $1.4 \times 10^{15}$ Bq per year, resulting in a natural global $^{14}C$ inventory of about $9 \times 10^{18}$ Bq, 92% of which resides in the deep ocean (6,21). The specific acitivy of atmospheric $^{14}C$ prior to dilution as a result of the Suess effect is 227 Bq/ kg C. Thermonuclear weapons tests in the atmosphere have added a total of $3.6 \times 10^{17}$ Bq of $^{14}C$, and reactors are estimated to contribute about $6 \times 10^{14}$ Bq annually, as of the mid-1980s (21). After reaching a peak in the early 1960s to about 400 Bq/kg C reflecting the global fallout maximum, the $^{14}C$ atmospheric specific activity has declined rapidly over the last two decades (in spite of rapidly increasing emissions during this period from nuclear power production) to near the equilibrium value resulting primarily from $^{14}CO_2$ exchange with the oceans (21).

Like $^{14}C$, tritium (99% HTO) is also produced cosmogenically ($7.4 \times 10^{16}$ Bq per year), as well as released to the environment from weapons tests ($2.4 \times 10^{20}$ Bq and reactor operations (about $8 \times 10^{16}$ Bq per year projected for the mid-1980s) (4,22). As with $^{40}K$ and $^{14}C$, the ratio of $^3H/H$ determines human intake of tritium. Since $^3H$ is a very weak beta emitter, the equilibrium dose equivalent rate from natural sources of $^3H$ is less than $10^{-3}$ that of natural $^{14}C$ and thus will not be considered further in this chapter.

## Anthropogenic Radioactivity

There have been a number of pre-Chernobyl episodes of widespread contamination from radionuclides intentionally or unintentionally released into the environment, but none has affected the world's intake of radionuclides as much as the global fallout from atmospheric nuclear and thermonuclear weapons tests. In terms of effective dose equivalent commitment (EDEC) from ingestion of fallout-derived radionuclides, $^{14}C$ (by virtue of its 5,730-year half-life) and, in order, by $^{137}Cs$, $^{90}Sr$, $^{131}I$, and $^3H$ have been estimated to contribute 98% of the ingestion EDEC and 70% of the total (ingestion + inhalation + external irradiation) EDEC (4). Accordingly, attention herein is focused on the ingestion of these radionuclides.

The deposition history of fallout-produced radionuclides is best exemplified by that of $^{90}Sr$, which has been monitored worldwide. Annual deposition patterns

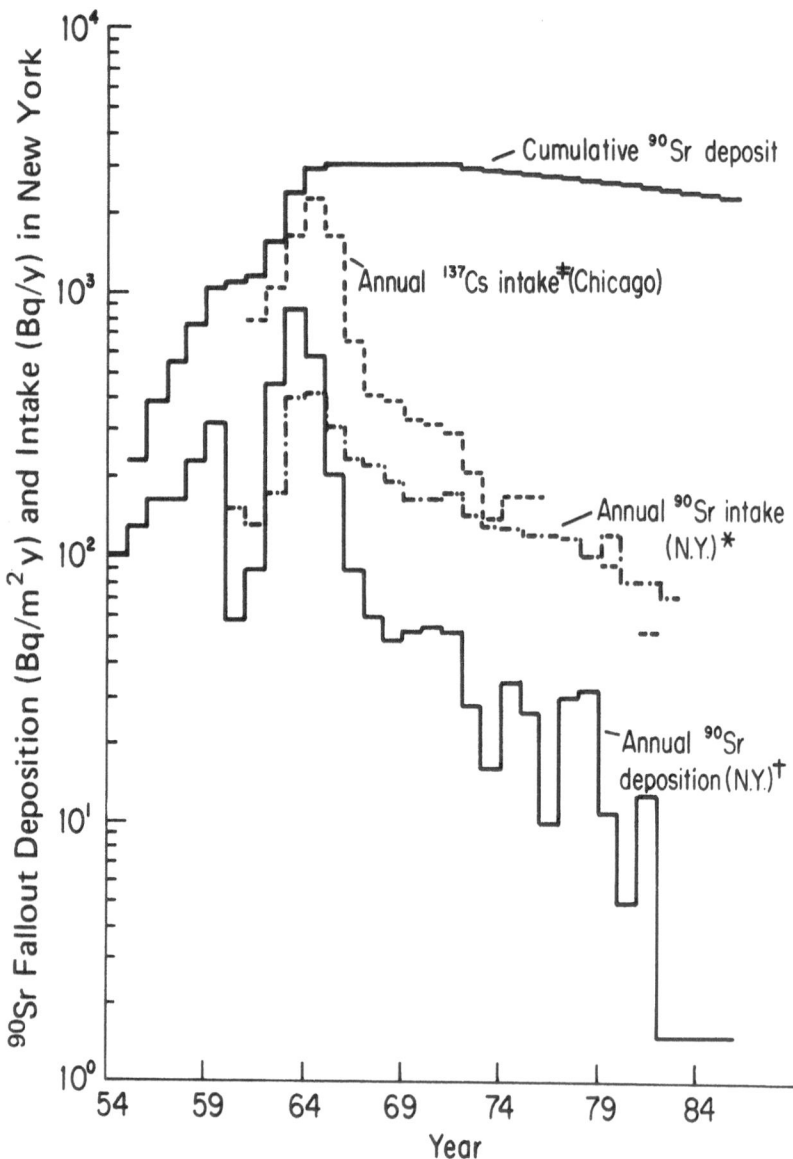

**Fig. 17.1.** Deposition of $^{90}$Sr (NY) and intake of $^{90}$Sr (NY) and $^{137}$Cs (Chicago) from global fallout over time. (*From ref. 23; †from refs. 24 and 25; ‡adapted from ref. 26.)

for all fallout-produced particulate radionuclides resemble the pattern shown for $^{90}$Sr (Fig. 17.1), with maxima in 1959 and 1963, reflecting the periods of heaviest atmospheric nuclear weapons testing.

While dietary intakes and annual dose rates derived from the primordial radionuclides can be considered to be approximately constant over time, ranging by factors of 4 to 5 among diets from different countries in normal background areas, those derived from fallout radionuclides show a marked dependence in time and location, well correlated with the temporal and spatial variations in the fallout deposition exemplified in Fig. 17.1 for $^{137}$Cs and $^{90}$Sr. The strong correlations between annual deposition and intake are reflective of the rapidity of processes inherent in the transfer of atmospherically deposited pollutants via the food chain (e.g., grass–cow–milk). If differences in dietary composition and points of origin for the New York City and Chicago diets are neglected, the annual intakes of $^{90}$Sr and $^{137}$Cs in the two cities (Fig. 17.1) are more or less directly comparable because the deposition of $^{90}$Sr in New York City (shown) is about equal to that of $^{137}$Cs in Chicago. It is evident from the slopes of the intake curves that, with time, $^{137}$Cs has become increasingly less available (relative to $^{90}$Sr) for incorporation into foodstuffs, probably as a result of the relatively greater binding capacity of $^{137}$Cs to soil clay minerals. Changes in the relative dietary contributions of $^{137}$Cs and $^{90}$Sr intakes over time indicate a shift from dairy products (grass–cow–milk) as the principal source for both nuclides during active deposition periods to meat and fish products ($^{137}$Cs) and vegetables and dairy products ($^{90}$Sr) at later times when food-chain transfer processes from the cumulative soil burden predominate (Table 17.4).

With the use of the population-weighted integral depositions and transfer factors given by the United Nations Scientific Committee on the Effects of Atomic Radiation (UNSCEAR (4)], the integrated per-capita intakes (northern hemisphere) for fallout radionuclides are estimated to be on the order of about $5 \times 10^6$, $3 \times 10^6$, $1.5 \times 10^4$, and $4.6 \times 10^3$ Bq for $^{14}$C, $^3$H, $^{137}$Cs, and $^{90}$Sr. Intakes of fallout $^{131}$I have traditionally been difficult to assess since its short radiological half-life (8 days) has generally precluded detailed environmental measurements in foodstuffs other than milk from relatively few locations.

Estimates of the average doses delivered to different organs by the major radionuclides in food and water are listed in Table 17.5. In comparison with the doses from natural background radiation, the additional doses received through ingestion of radionuclides released by atomic weapons fallout or nuclear accidents are small except in those persons residing in the near vicinity of such releases (e.g., Table 17.6).

**Table 17.4.** Relative contributions of different food stuffs to annual intake of $^{90}$Sr and $^{137}$Cs from fallout

| Food stuff | $^{90}$Sr intake (%)[a] | | $^{137}$Cs intake (%) | |
|---|---|---|---|---|
| | 1963 | 1982 | 1963[b] | 1981[c] |
| Dairy products | 56 | 32 | 39 | 20 |
| Grains | 22 | 18 | 21 | 20 |
| Vegetables | 13 | 36 | 6 | 12 |
| Fruits and juices | 7 | 11 | 8 | 4 |
| Meats, eggs, fish | 2 | 3 | 22 | 44 |

[a] From ref. 23; pertains to the New York City diet.
[b] From ref. 27; pertains to the Chicago diet.
[c] From ref. 28; pertains to the Chicago diet.

**Table 17.5.** Average annual dose equivalents received by members of US population from internally deposited radionuclides[a]

| Radionuclide | Average annual dose equivalent ($\mu Sv/y$) | | |
|---|---|---|---|
| | Soft tissue | Bone marrow | Bone surfaces |
| Naturally occurring | | | |
| $^{40}K$ | 180 | 270 | 140 |
| $^{226}Ra$ | 3 | 15 | 90 |
| $^{228}Ra$, $^{224}Ra$ | 2 | 22 | 120 |
| $^{210}Pb$, $^{210}Po$ | 140 | 140 | 700 |
| $^{14}C$ | 10 | 30 | 8 |
| Other | 25 | 23 | 42 |
| Subtotal (rounded) | 360 | 500 | 1,100 |
| Fission products[b] | | | |
| $^{137}Cs$ | 3 | 3 | 3 |
| $^{90}Sr$ | 1 | 9 | 13 |

[a] Reprinted with permission from ref. 29.
[b] Rounded yearly averages are for the period 1950–2000, based on the estimated dose commitments to the year 2000, most of which have already been received.

**Table 17.6.** Estimated average dose equivalent received by inhabitants of different countries as a result of ingestion of radionuclides released by the Chernobyl accident[a]

| Country | Dose to thyroid ($\mu Sv/y$)[b] | Dose to whole body[c] ($\mu Sv/y$) |
|---|---|---|
| Albania | 54 | 15 |
| Austria | 56 | 14 |
| Belgium | 30 | 4 |
| Canada | 0.1 | 0.03 |
| China | 0.14 | 0.05 |
| Czechoslovakia | 88 | 24 |
| Denmark | 2.2 | 0.6 |
| France | 0.17 | 0.05 |
| German Democratic Republic | 94 | 22 |
| Greece | 16 | 4.2 |
| Hungary | 34 | 7.6 |
| Italy | 32 | 8.0 |
| Sweden | 48 | 10.4 |
| United Kingdom | 38 | 10 |
| United States | 0.08 | 0.03 |

[a] Reprinted with permission from ref. 30.
[b] Values (rounded) based on estimated annual average fraction of 50-year dose-equivalent commitment.
[c] From $^{137}Cs$.

# Nature and Mechanisms of Biological Effects of Irradiation

The molecular mechanisms of radiation effects remain to be fully elucidated (31); however, damage to DNA is thought to play an important role in most effects, since alteration of a single gene may kill or profoundly injure the affected cell. A dose of x-radiation that is large enough to kill the average dividing cell (e.g., 1 to 2 Gy) suffices to produce dozens of lesions in the DNA molecules of the cell (32). Although most of the DNA lesions are potentially reparable, depending on the effectiveness of the cell's DNA repair processes, some are likely to remain unrepaired or to be misrepaired, and these may give rise to mutations. The frequency of mutations typically increases as a one-hit function of the dose of radiation in the low-to-intermediate dose range. For example, mutations to thioguanine resistance in human lymphocytes exposed in vitro to 50 to 220 mGy of X rays increase at a rate of about 6 mutations per $10^6$ cells per Gy, whether the dose is delivered in several exposures or in a single brief exposure (33). Likewise, the rate of glycophorin mutations in the bone marrow cells of persons surviving A-bomb irradiation increases similarly with the dose (34). These dose–effect relationships indicate that mutation can result from traversal of the genetic material by a single ionizing particle. It is noteworthy, however, that the frequency of mutations per unit dose may vary as a function of the dose rate and the linear energy transfer (LET) of the radiation (2).

In addition to causing point mutations, radiation-induced damage to DNA can alter the number and structure of chromosomes in the cell. Over the dose range from 0.05 to 1.0 Gy, the frequency of chromosome aberrations increases roughly in proportion to the dose, approximating 0.1 aberration per cell per Gy in human blood lymphocytes x-irradiated in culture (35). The frequency of such aberrations has also been observed to be increased in the lymphocytes of radiation workers (36). At the same time, however, only a small percentage (less than 2%) of all chromosome aberrations is attributable to natural background irradiation, most resulting instead from the action of certain viruses, drugs, chemicals, or other factors. Thus, the health implications of an increase in the frequency of such aberrations cannot be predicted without further information.

Radiation-induced damage to the genes, chromosomes, or other vital structures of a cell may lead to the death of the cell. In general, the susceptibility of a cell to radiation-induced killing increases with its rate of proliferation and decreases with its degree of differentiation. As measured by the ability to proliferate, the survival of dividing cells tends to decrease exponentially with an increasing dose, the rate of decrease being characteristically steeper with high LET radiation than with low LET radiation. Moreover, with low LET radiation, a given dose tends to cause less killing of cells if spread out over a period of hours or days than if delivered in a few seconds or minutes, owing to the repair of some of the damage with the passage of time (4).

Because of the generally high radiosensitivity of dividing cells, such cells may be depleted rapidly by irradiation in tissues that are characterized by a high rate of cell turnover. The resulting interference with cell renewal in such tissues, if severe enough, may lead to aplasia, atrophy, and impairment of function. If, for example, a large enough percentage of hemopoietic cells is killed by irradiation of the bone marrow, as can occur with rapid exposure of the whole body to a dose as large as 5 Gy, fatal granulocytopenia and thrombocytopenia may ensue within 4 to 6 weeks. Other tissues in which interference with cell renewal can lead to early abnormalities include the skin, gastrointestinal mucosa, and testis. Lymphoid organs may also undergo rapid injury, as a result of the prompt killing of nondividing lymphocytes. In general, however, the doses of radiation that are required to cause clinically significant depletion of cells in normal tissues are too large to be accumulated under the conditions of low-level irradiation that are encountered by radiation workers or the general public, except during radiation accidents or emergencies (37). Hence, there is little likelihood of their occurrence as a result of radionuclide contamination of the food supply.

In contrast to the types of radiation-induced lesions that do not occur unless a sufficiently large number of cells are killed to compromise normal function in affected organs, cancer (and perhaps teratogenic effects) may conceivably be induced by appropriate damage to a single cell in a suitably susceptible individual. It is such effects, therefore—for which no threshold is presumed to exist—that are of chief concern in populations exposed to low-level radiation.

## Dose–Incidence Relationship for Radiation-Induced Cancer

Although the incidence of various forms of cancer has been observed to increase systematically with the dose of radiation, the available dose-effect data do not suffice to establish with confidence the shape of the dose-incidence curve in the low-dose domain. Hence, the possible risks of low-level irradiation can be estimated only by extrapolation from observations at higher dose levels, based on assumptions about the relevant dose–incidence relationships (1,2,4).

Epidemiological studies of A-bomb survivors, patients exposed to radiation for diagnostic or therapeutic purposes, certain radiation workers, and various other irradiated populations have indicated that many types of cancer can be increased in frequency by ionizing radiation, depending on the conditions of exposure. The cancers do not appear until years or decades later, however, and they have no distinguishing features by which they can be recognized to have resulted from radiation per se. Hence, the occurrence of cancer in an irradiated individual can never be attributed with certainty to previous radiation exposure (38). Furthermore, the epidemiological data came predominantly from observations at relatively high doses (0.5 to 2.0 Gy) and do not define the shape of the dose-incidence curve in the low-dose domain.

Although the existing data do not exclude the possibility of a threshold for radiation carcinogenesis, several lines of evidence are consistent with a non-threshold dose–incidence relationship. First, the incidence of various types of human cancer has been observed to be increased at doses in the range of 0.01 to 0.25 Gy; for some such cancers, moreover, the available dose–incidence curves appear consistent with linear nonthreshold functions (1,2). Second, the additivity between successive small increments of dose appears relatively complete in the induction of some types of cancer, especially with high LET radiation, implying strict proportionality between the incidence and the dose in these instances (39). Third, a 30% to 50% increase in the risk of childhood leukemia has been observed to be associated with prenatal exposure to a dose of the order of 0.01 Gy (40,41). Finally, on theoretical and radiobiological grounds, there is no reason to assume the existence of a threshold dose for the carcinogenic effects of radiation (42).

In view of the foregoing, experts generally advocate the use of a nonthreshold dose–incidence extrapolation model in estimating the carcinogenic risks of low-level irradiation for public health purposes. With the use of such models— which generally assume that the incidence of a given type of cancer is proportional to the dose of radiation at low doses—various numerical estimates of the carcinogenic risks of low-level radiation have been formulated (e.g., Table 17.7). In view of the many uncertainties inherent in such estimates, however, they must be considered crude approximations at best. Nevertheless, they provide a framework that enables the potential health impacts of various types of radiation exposure, including irradiation from radionuclides in the food chain, to be assessed in perspective with other risks.

## Long-Term Carcinogenic Effects Attributable to Radionuclides in Food and Water

As noted previously, internal irradiation by radionuclides ingested in food and water accounts for only a small part of the average total dose received by most tissues of the body from natural and man-made sources of radiation. The carcinogenic effects that may be attributed to such internal irradiation are thus correspondingly small (Table 17.8). At the same time, however, it is clear that the doses received from internally deposited radionuclides may be appreciably higher than average in persons residing in areas of high natural background radioactivity—such as certain parts of China (45), Brazil (46), and India (47)— or in areas where contamination of the food supply by man-made radioactivity may be unusually high. Examples include certain arctic regions (48), areas downwind from atomic weapons test sites (2), and areas contaminated by radionuclides released in nuclear reactor accidents (30). Nevertheless, except for an excess of thyroid cancer in Marshall Islanders resulting from an extraordinarily high intake of radioactive iodines (1), the carcinogenic risks to these populations have not been demonstrably increased (1,2). In Alaskan natives subsisting on a high intake of caribou, for example, the greatly elevated body burdens of $^{137}$Cs and $^{90}$Sr resulting from fallout released in atmospheric weapons tests during the 1950s and 1960s were estimated to pose an undetectably small carcino-

**Table 17.7.** Estimated risks of radiation-induced cancers of different organs in relation to age at exposure[a]

| Cancer type or site | Age at exposure (years) | | |
|---|---|---|---|
| | 20–34 | 35–49 | 50+ |
| | (Excess cancers per million per year per Sv)[b] | | |
| Leukemia[c] | | | |
| (males) | 80 | 110 | 160 |
| (females) | 50 | 70 | 100 |
| Esophagus | | | |
| (both sexes) | 5 | 8 | 22 |
| Stomach | | | |
| (both sexes) | 300 | 50 | 130 |
| Colon | | | |
| (both sexes) | 20 | 30 | 90 |
| Liver | | | |
| (both sexes) | 30 | 30 | 30 |
| Pancreas | | | |
| (both sexes) | 20 | 30 | 80 |
| Lung | | | |
| (both sexes) | 60 | 90 | 120 |
| Breast | | | |
| (females) | 490 | 310[d] | 80 |
| Urinary | | | |
| (both sexes) | 20 | 40 | 60 |
| Thyroid[e] | | | |
| (males) | 50 | 50 | 50 |
| (females) | 150 | 150 | 150 |

[a] The estimates may need to be increased, pending revision of the dosimetry for the A-bomb survivors (43, 44).

[b] Risk (values rounded) expressed from 10th to 30th year after exposure unless otherwise specified.

[c] Risk expressed from 5th to 26th year after exposure (chronic lymphocytic leukemia excluded).

[d] Value interpolated from figures in original report.

[e] Risk expressed from 10th to 34th year after exposure.

genic risk (Table 17.9). By the same token, the additional cancers in European USSR that have been projected to result from the ingestion of radionuclides released in the accident at Chernobyl constitute an undetectably small increase in the natural incidence (Table 17.10).

# Dose–Incidence Relationships for Radiation-Induced Teratogenic Effects

The tissues of the embryo, fetus, and growing child are highly radiosensitive. Exposure to 0.25 Sv during a critical stage in organogenesis has been observed to cause malformations of many types in laboratory animals (6), and comparable effects have been observed after larger doses in prenatally irradiated children. In children who were exposed to atomic-bomb radiation at Hiroshima and Naga-

**Table 17.8.** Estimated contribution of irradiation by ingested radionuclides, as compared with irradiation from other sources, to the lifetime risk of mortality from cancer in the US population[a]

| Source of radiation | Annual collective dose (person-Sv) | Lifetime cancer mortality commitment[b] | |
|---|---|---|---|
| | | Number of fatal cancers[c] | % of natural number of fatal cancers |
| Ingested radionuclides | 52,000 | 1,200 | 0.3 |
| Cosmic rays | 56,000 | 1,700 | 0.4 |
| Terrestrial | 52,000 | 1,600 | 0.3 |
| Medical and dental | 212,000 | 6,400 | 1.4 |
| Nuclear weapons fallout | 10,000 | 300 | 0.06 |
| Nuclear power | 400 | 12 | 0.002 |

[a] Reprinted with permission from refs. 1 and 2.
[b] Per year of exposure.
[c] Values rounded; lower limit estimate in each case is zero.

saki between the 8th and 15th week of prenatal development, for example, the incidence of mental retardation was observed to increase as an apparently linear nonthreshold function of the dose, at a rate approximately 40% per Gy (2,50).

Although, in principle, damage to a single progenitor cell may suffice to cause gross malformation or impairment of function in a developing organ, the available dose–response data imply that damage to more than one such cell is usually required and that a threshold probably exists for most such effects (1,6). Furthermore, since the period of greatest susceptibility to a particular effect is sharply circumscribed in time, the frequency with which the effect is induced by a given dose decreases markedly as the dose is protracted over a period longer than the few hours or days of maximal radiosensitivity (1,6). Hence, the extent to which teratogenic effects may result from irradiation at the low doses and dose rates normally attributable to radionuclides in the food chain is highly uncertain.

## Estimated Teratogenic Impact of Radionuclides Ingested in Food and Water

Although, as noted previously, it is questionable whether irradiation at the low dose rates characteristic of natural background radiation can cause teratogenic effects, the possibility that mental retardation may increase in frequency as a linear nonthreshold function of the dose has prompted estimation of the increase in risk of mental retardation attributable to the radioactivity released by the accident at Chernobyl. On the basis of the assumption that the frequency of mental retardation is increased by 40% per Gy in children irradiated prenatally between the 8th and the 15th week of gestation, the extent to which the risk

**Table 17.9.** Maximal expected annual numbers of cancers in northern Alaskan villages attributable to uptake of $^{137}Cs$ and $^{90}Sr$ derived from nuclear weapons fallout[a]

| | Anaktuvuk Pass | Kotzebue | Barrow | Point Hope | River Villages | Total |
|---|---|---|---|---|---|---|
| 1980 native population | 191 | 1,574 | 1,720 | 434 | 1,796 | 5,715 |
| Cancer or tumor type | | | | | | |
| Leukemia | 0.002 | 0.01 | 0.01 | 0.004 | 0.002 | 0.049 |
| Breast cancer | 0.004 | 0.04 | 0.04 | 0.01 | 0.04 | 0.13 |
| Bone sarcoma | 0.00004 | 0.0003 | 0.0003 | 0.0001 | 0.0003 | 0.001 |

[a] Reproduced with permission from ref. 49.

**Table 17.10.** Estimated numbers of fatal cancers induced in inhabitants of European USSR by radionuclides released in the Chernobyl accident[a]

| Population at risk | Lifetime cancer mortality commitment | |
|---|---|---|
| | Number of fatal cancers[b] | % of natural number of fatal cancers |
| Chernobyl evacuees (135,000) | 410 | 1.8 |
| European USSR (75 million) | | |
| External irradiation | 6,400 | 0.05 |
| Internal irradiation | | |
| (radiocesium) | 4,600 | 0.04 |
| Thyroid irradiation ($^{131}$I) | 200 | 0.1 |
| Total | 12,000 | |

[a] Reproduced with permission from ref. 30.
[b] Values rounded; lower limit estimate is zero.

of this disorder in European countries may be increased by irradiation from the accident has been estimated to range from 0.1% in children exposed in the United Kingdom to 140% in children exposed within 30 km of Chernobyl (Table 17.11). Of the total estimated potential increase, however, less than one-half is attributable to the dose received from internally deposited radionuclides, the remainder being attributable to irradiation from external radioactivity. If the same risk coefficient for induction of mental retardation is assumed to be applicable to natural background radiation, fewer than 0.002% of all cases of severe mental retardation in the United States may be attributed to internal irradiation of the embryo by radionuclides in the food chain.

## Public Health Implications

From the foregoing analysis, it may be inferred that for most of the world's population the long-term carcinogenic and teratogenic effects attributable to radionuclides in food and water are not large. Depending on the dose–effect

**Table 17.11.** Estimates of the potential increase in the risk of severe mental retardation attributable to prenatal irradiation by radionuclides released at Chernobyl[a]

| Region | Population dose (person-Gy)[b] | Estimated radiation-induced excess of mental retardation | |
|---|---|---|---|
| | | Number of cases | % Increase[b] |
| Chernobyl (30-km radius) | 16 | 18 | 140 |
| European USSR | 220 | 230 | 3 |
| Europe (USSR excluded) | 330 | 350 | 0.9 |
| United Kingdom | 5 | 5 | 0.1 |

[a] Reproduced with permission from ref. 30.
[b] Thousands of person-Gy; all values rounded.

model that is used to estimate the risks of such effects, their incidence in most parts of the world may be inferred to range from zero to a fraction of 1% of the natural incidence. In a few populations, however, the risks may be appreciably higher because of increased levels of natural and/or anthropogenic radioactivity in food and water.

Although such effects may be relatively rare, they are likely to have serious, if not fatal, consequences for the individuals who are affected. Hence, any unnecessary increase in their frequency is obviously to be avoided. Although it is beyond the scope of this report to review methods for minimizing the intake of radionuclides via food and water, the most effective strategies involve control of the release of radioactivity at its source. Barring the possibility of such control, alternative strategies exist for minimizing the consumption of potentially contaminated food and water through systematic monitoring of the environment and food chain coupled with appropriate interventions. Since radioactivity released into the environment respects no territorial boundaries, optimal countermeasures call for close cooperation at the international level as well as at the local and national levels (36,51).

*Acknowledgment.* The authors are grateful to Dr. Naomi H. Harley for helpful discussions and to Lynda Witte and Fran Lupino for assistance in the preparation of the manuscript. Preparation of this report was supported in part by Grants ES 002650 and CA 13343 from the US Public Health Service and Grant SIG 09 from the American Cancer Society.

# References

1. National Academy of Sciences, Advisory Committee on the Biological Effects of Ionizing Radiation [NAS/BEIR] (1980) The effects of populations of exposure to low levels of ionizing radiation. National Academy of Sciences/National Research Council, Washington, DC
2. United Nations Scientific Committee on the Effects of Atomic Radiation [UNSCEAR] (1986) Genetic and somatic effects of ionizing radiation. Report to the General Assembly, with annexes: Forty-first Session, Supplement No. 16 (A/41/16). United Nations, New York
3. Sankaranarayanan K (1988) Radionuclides and genetic risks. In: Carter MW (ed) Radionuclides in the food chain. Springer-Verlag, New York, pp 236–263 (see Chap 18 of this monograph)
4. United Nations Scientific Committee on the Effects of Atomic Radiation [UNSCEAR] (1982) Ionizing radiation: sources and biological effects. Report to the General Assembly, with annexes: Thirty-seventh Session, Supplement No. 45 (A/37/45). United Nations, New York
5. National Council on Radiation Protection and Measurement [NCRP] (1975) Natural background radiation in the United States. NCRP Report 45, NCRP, Washington, DC
6. United Nations Scientific Committee on the Effects of Atomic Radiation [UNSCEAR] (1977) Sources and effects of ionizing radiation. Report to the General Assembly,

with annexes: Thirty-second Session, Supplement No. 40 (A/32/40). United Nations, New York

7. Harley JH (1988) Naturally occurring sources of radioactive contamination. In: Carter MW (ed) Radionuclides in the food chain. Springer-Verlag, New York, pp 58–71 (see Chap 6 of this monograph)

8. Pentreath RJ (1988) Radionuclides in the aquatic environment. In: Carter MW (ed) Radionuclides in the food chain. Springer-Verlag, New York, pp 99–119 (see Chap 9 of this monograph)

9. Ng YC (1982) A review of transfer factors for assessing the dose from radionuclides in agricultural products. Nucl Saf 23:57–71

10. Ng YC, Colsler CS, Thompson SE (1982) Soil-to-plant concentration factors for radiological assessments. Lawrence Livermore National Laboratory, NUREG/CR-2975, UCID-19463

11. National Council on Radiation Protection and Measurement [NCRP] (1984) Exposures from the uranium series with emphasis on radon and its daughters. NCRP Report 77, NCRP, Washington, DC

12. Harley JH (1969) Radionuclides in food. In: Biological implications of the nuclear age. Symposium held at Lawrence Radiation Laboratory, March 5–7, 1969, pp 189–200

13. Morse BS, Welford GA (1971) Dietary intake of $^{210}$Pb. Health Phys 21:53–55

14. Holtzman RB (1980) Normal dietary levels of $^{226}$Ra, $^{228}$Ra, $^{210}$Pb and $^{210}$Po for man. In: Natural radiation environment III. Available from NTIS, Springfield; CONF-780422 (Vol. 1)

15. Welford GA, Baird R (1967) Uranium levels in human diet and biological materials. Health Phys 13:1321–1324

16. Hallden HA, Fisenne IM, Harley JH (1963) Radium-226 in human diet and bone. Sci 140:74–75

17. Petrow HG, Schiesle WJ, Cover A (1965) Dietary intake of Radium-228. In: Radioactivity studies, 1965. US Atomic Energy Commission Report NYO-3086–1. New York Operation Office, NTIS, Springfield, Virginia, pp 1–10

18. Fisenne IM, Perry PM, Chu NY (1983) Measured $^{234,238}$U and fallout $^{239,240}$Pu in human bone ash from Nepal and Australia: skeletal alpha dose. Health Phys 44:457–467

19. International Commission on Radiological Protection [ICRP] (1979) Limits on intake of radionuclides by workers. ICRP Publication 30, Part 1, Pergamon, New York

20. International Commission on Radiological Protection [ICRP] (1975) Report of the task group on reference man. ICRP Publication 23, Pergamon, New York

21. National Council on Radiation Protection and Measurement [NCRP] (1985) Carbon-14 in the environment. NCRP Report 81, NCRP, Bethesda, Maryland

22. National Council on Radiation Protection and Measurement [NCRP] (1979) Tritium in the environment. NCRP Report 62, NCRP, Washington, DC

23. Klusek CS (1984) Strontium-90 in the US diet. US Department of Energy Report. Available from NTIS, Springfield: EML-429

24. United States Department of Energy Health and Safety Laboratory Environmental Quarterly (1977) Final tabulation of monthly $^{90}$Sr fallout data: 1954–1976. Available from NTIS, Springfield: HASL-329

25. United States Department of Energy Environmental Report (1982). Available from NTIS, Springfield: EML-405, UC-11

26. Evans C, Bennett BG (1976) Transfer of $^{137}$Cs through the food chain to man. US Department of Energy Report. Available from NTIS, Springfield: HASL-310

27. Gustafson PF Miller JE (1969) The significance of $^{137}$Cs in man and his diet. Health Phys 16:167–183

28. Karttunen JO (1982) Cesium-137 in various Chicago foods. US Department of Energy. Available from NTIS, Springfield: EML-405

29. National Council on Radiation Protection and Measurement [NCRP] (1987) Ionizing radiation exposure of the population of the United States. NCRP Report 93, NCRP, Bethesda, Maryland

30. Committee on the Assessment of Health Consequences in Exposed Populations (1987) Health and environmental consequences of the Chernobyl nuclear power plant accident. Report to the US Department of Energy, Office of Health and Environmental Research from the Interlaboratory Task Group on Health and Environmental Aspects of the Soviet Nuclear Accident. DOE/ER-0332, UC-41 and 48, Washington, DC

31. Silini G (1988) Biological effects of ionizing radiation. In: Carter MW (ed) Radionuclides in the food chain. Springer-Verlag, New York, pp 35–44 (see Chap 4 of this monograph)

32. Cole A, Meyn RE, Chen R, Corry PM, Hittelman W (1980) Mechanisms of cell injury. In: Meyn R, Withers HR (eds) Radiation biology in cancer research. Raven, New York, pp 33–58

33. Grosovsky AJ, Little JB (1985) Evidence for linear response for the induction of mutations in human cells by x-ray exposure below 10 rads. Proc Natl Acad Sci USA 82:2092–2095

34. Langolis RG, Bigbee WL, Kyoizumi S, Nakamura N, Bean MA, Akiyama M, Jensen RH (1987) Evidence for increased somatic cell mutations at the glycophorin A locus in atomic bomb survivors. Sci 236:445–448

35. Lloyd DC, Purrott RJ (1981) Chromosome aberration analysis in radiological protection dosimetry. Rad Protect Dosi 1:19–28

36. Lloyd DC, Purrott RJ, Reeder EF (1980) The incidence of unstable chromosome aberrations in peripheral blood lymphocytes from unirradiated and occupationally exposed people. Mutat Res 72:523–532

37. International Commission on Radiological Protection [ICRP] (1984) Protection of the public in major accidents. Annals of the ICRP, Vol. 14, No. 2, Pergamon, Oxford

38. Rall JE, Beebe GW, Hoel DG, Jablon S, Land CE, Nygaard OF, Upton AC, Yavlow RS, Zeve VH (1985) Report of the National Institues of Health Working Group to Develop Radioepidemiological Tables. NIH Publication No 85–2748, US Government Printing Office, Washington, DC

39. Upton AC (1986) Cancer induction and non-stochastic effects. Br J Radio 60:1–6

40. Monson RP, MacMahon B (1984) Pre-natal x-ray exposure and cancer in children. In: Boice JD Jr, Fraumeni JF Jr (eds) Radiation carcinogenesis: epidemiology and biological significance. Raven, New York, pp 97–105

41. Harvey EB, Boice JD Jr, Honeyman M, Flannery JT (1985) Prenatal x-ray exposure and childhood cancer in twins. New Eng J Med 312:541–545

42. Upton AC (1987) Evolving perspectives on the biology and mechanisms of carcinogenesis. Leukemia Res 10:727–734

43. Preston DL, Kato H, Kopecky KJ, Fujita S, (1987) Studies of the mortality of A-bomb survivors. 8. Cancer mortality, 1950–1982. Rad Res 111:151–178

44. Preston DL, Pierce DA (1987) The effect of changes in dosimetry on cancer mortality risk estimates in the atomic bomb survivors. RERF Technical Report TR9–87, RERF, Hiroshima, Japan

45. Wei L et al. (1985) Report of third stage (1982–1984) health survey in high background area in Yangjiang, China. Chin J Radiat Med Prot 5:144–153

46. Brazilian Academy of Sciences (Academia Brasileira de Ciencias) (1977) Proceedings of International Symposium on Areas of High Natural Radioactivity, Rio de Janeiro, Brazil

47. Sunta CM, David M, Abani MC, Basu AS, Nambi KSV (1982) Analysis of dosimetry data of high natural radioactivity areas of southwest coast of India. In: Vohra KG, Mishra US, Pillai KC, Sadasivan S (eds) The natural radiation environment IV. Wiley Eastern, Bombay/New Delhi

48. Hanson WC (1986) Ecological processes in the cycling of radionuclides with Arctic ecosystems. In: Stonehouse B (ed) Arctic air pollution. University Press, Cambridge, pp 221–228

49. Stutzman CD, Nelson DM (1986) Cancer incidence and risk in Alaskan natives exposed to radioactive fallout. In: Stonehouse B (ed) Arctic air pollution. University Press, Cambridge, pp 229–238

50. Otake M, Schull W (1984) In utero exposure to A-bomb radiation and mental retardation: a reassessment. Brit J Radio 57:409–414

51. International Atomic Energy Agency [IAEA] (1985) Principles for establishing intervention levels for the protection of the public in the event of a nuclear accident or radiological emergency. Safety Series No. 72. International Atomic Energy Agency, Vienna

CHAPTER 18

# Radionuclides and Genetic Risks

K. Sankaranarayanan[1]

## Introduction

Ionizing radiation from natural sources has always been part of the human environment, and that from artifical (man-made) sources, since the beginning of the twentieth century. The fact that exposure to ionizing radiation—irrespective of whether it is from external sources or from internally deposited radionuclides—increases the risk of incurring adverse health effects was dramatically brought to the attention of the public by the atomic bombing of Hiroshima and Nagasaki in 1945. Since then, considerable attention has been paid over the years to the levels and effects of radiation.

The aim of this chapter is to present a broad overview of the genetic risks of ionizing radiation in humans, with emphasis on internally deposited radionuclides. The term "genetic risks" as used in this paper denotes those risks associated with mutations and chromosomal aberrations induced in germ cells and transmitted to the progeny, both close and remote. To provide a background and to place this theme in perspective, the radionuclides present in the human environment and their transfer to humans and the magnitude of radiation exposures sustained from these and other sources are to be kept in mind. These subjects are treated in the contributions by Jacobi, Harley, and Upton and Linsalata [see Chaps 5, 6, and 17, respectively, of this monograph].

## Considerations Relevant in the Context of Genetic Risk Estimation

The radiation dose received by the human populations from natural sources and from various uses of nuclear energy is small and is greatly protracted. In

[1] Department of Radiation Genetics and Chemical Mutagenesis, Sylvius Laboratories, and the J. A. Cohen Institute, Interuniversity Research Institute for Radiopathology and Radiation Protection, University of Leiden, Wassenaarseweg 72, 2333 AL Leiden, The Netherlands.

estimating genetic risks to the population as a whole, therefore, the relevant radiation conditions are low doses and chronic (or low dose-rate) exposures. The germ cell stages at risk are the stem cell spermatogonia in males (these constitute a permanent germ cell population in the testes and continue to multiply throughout the reproductive life span of the individual) and immature oocytes in females (the predominant germ cell population in females).

In spite of extensive efforts so far, attempts to detect genetic damage in the progeny of irradiated individuals have not been successful (1,2). However, the wealth of data that has accumulated from various mammalian radiation genetic studies and similar ones in a variety of biological systems leave no room to doubt that ionizing radiation is capable of causing damage to the genetic material. Thus, there has been—and the situation is still true—no alternative to the use of experimental data for genetic risk estimation. Such extrapolations necessitate a number of assumptions, among which the following are important: (1) the rates of induction of specific kinds of genetic damage are the same in humans and in the test species, for specific germ cell stages and under similar conditions of radiation exposure; (2) at low doses and low dose rates, there is a linear relationship between the absorbed dose and the frequency of genetic effects studied; and (3) the known relationship between spontaneously occurring gene mutations and chromosomal abnormalities to genetic disease in humans provides a basis for translating genetic effects actually measured in experimental studies into cases of genetic disease in humans.

## Radionuclides of Genetic Importance and Their Properties

The radionuclides that have been used in genetic studies with mammals and some of their properties are summarized in Table 18.1. Radionuclides such as $^3H$ and $^{14}C$ are distributed uniformly in all the body tissues, whereas those such as $^{90}Sr$, $^{131}I$, and $^{239}Pu$ concentrate in specific tissues or organs. These distribution patterns and particularly the extent to which the radionuclides reach the gonads and are retained in them are some of the important factors that determine the magnitude of dose received by the germ cells and, thus, of genetic effects. Furthermore, beta emitters such as $^3H$, $^{14}C$, and $^{32}P$ can become incorporated into DNA; when such incorporated atoms decay, the resulting change in atomic number, recoil or excitation, collectively referred to as "transmutation," may give rise to genetic effects beyond those induced by the attendant beta radiation. These aspects are briefly considered in this section.

### Retention and Turnover Rates

Exposure to an atmosphere containing HTO vapor (tritiated water; the common form in which environmental $^3H$ is found) results in total absorption of the inhaled radioactivity through the lungs and absorption of about 50% of the amount through the skin. Ingested HTO is completely absorbed from the gastroin-

**Table 18.1.** Some properties of internal emitters used in mammalian genetic studies[a]

| Radionuclide | Half-life | Mode of decay | Mean track length (μm) in soft tissue[b] | Location/distribution in body |
|---|---|---|---|---|
| $^3$H | 12.3 y | beta | 1 | In all (including gonadal) tissues; possible transmutation effects when in DNA |
| $^{14}$C | 5,730 y | beta | 40 | As for $^3$H |
| $^{22}$Na | 2.6 y | beta | 300 | General |
| $^{32}$P | 14 d | positron | 2,000 | Retention in bone, but may enter DNA, with possible transmutation effect |
| $^{35}$S | 88 d | beta | 30 | General |
| $^{89}$Sr | 52 d | beta | 2,000 | Concentrates in bone |
| $^{90}$Sr[c] | 28 y | beta | 400 | Concentrates in bone |
| $^{131}$I | 8 d | beta | 400 | Concentrates in thyroid |
| $^{239}$Pu | 24,390 y | alpha | 40 | Concentrates mainly in bone and liver |
| $^{241}$Am | 458 y | alpha | 40 | As for Pu |
| $^{222}$Rn[d] | 3.8 d | alpha | 40 | Concentrates in lung and in upper respiratory tract |

[a] Reproduced with permission from refs. 3 and 4.
[b] Smoothed values.
[c] Decays to $^{90}$Y with a mean track length of 4,000 μm.
[d] The short-lived decay products of $^{222}$Rn (Rn daughters) are $^{218}$Po (3.05 min), $^{214}$Pb (26.8 min), $^{214}$Bi (19.7 min), and $^{214}$Po (164 μs); decay by alpha-particle emissions.

testinal tract and rapidly distributed throughout the body via blood. The biological half-life of $^3H$ in the human body following HTO intake has been found to range from 2.4 to 18 days among 300 individuals (5,6).

The absorption of $^{90}Sr$ by the body is relatively high. Its mean residence time in the human bone has been estimated to be in the range from 3.4 to 6.7 years corresponding to bone turnover rates of 12% to 23% per year (7). For radioiodine, the absorption by the blood from the gastrointestinal tract is complete and rapid. The International Commission on Radiological Protection (ICRP) (8) assumes that 30% of the I entering the blood is translocated to the thyroid while the remainder is excreted. Iodine in the thyroid is assumed to be retained with a biological half-life of 120 days and to be lost from the gland in the form of organic I.

The metabolism of Pu has been recently reviewed by the ICRP (9). As is well known, liver and bone are the primary sites of deposition of Pu, accounting for about 80% of that which reaches the bloodstream. The amount transferred to the gonads is very small (testes: 0.035%; and ovaries: 0.011%). Once deposited systematically, Pu is tenaciously retained (biological half-life: 50 years in the skeleton and 20 years in the liver). In the gonads, it is assumed to be retained without loss.

In the mouse, $^3H$ administered as HTO has a half-life of 1.1 days in the body-water pool (10) with less than 0.5% of the total activity incorporated into macromolecules and less than 0.1% into DNA (11). Intravenously injected Na $^{32}P$ phosphate was found to be taken up by the mouse testis to 20% of the concentration in the blood and retained in that organ with an approximate biological half-life of 20 days (12). In inhalation studies with Pu nitrate and $PuO_2$ in nonhuman primates, Brooks et al. (13) noted that the $PuO_2$ was retained in the lung with an effective half-life of greater than 1000 days, while the nitrate form was cleared from the lungs with an effective half-life of 150 days; by 1000 days, $PuO_2$ particles were about equally distributed between the lung and lung-associated lymph nodes, and little activity was detected in the liver and bone. The nitrate form was translocated from the lungs to the blood and through it to the liver and bone, and the ratio of activity in the liver to bone was rather constant from 45 to 365 days after exposure at about 1 to 1.5. The level in the testes was very small following the inhalation of the nitrate form and undetectable after $PuO_2$ inhalation.

## Distribution in Gonads

The distribution of Pu within the testis has been extensively studied in a number of rodents (14,15,16,17,18,19,20,21) and to a limited extent in beagle dogs (22) and in some nonhuman primates (13). The general finding is that Pu initially tends to concentrate in the interstitial tissue in the testis. As a result, the absorbed dose to the spermatogonial stem cells (which lie at the base of the seminiferous epithelium adjacent to interstitial tissue) will be higher (by factors of between

2 and 4 in the mouse) than that to the testis as a whole (16,19). Subsequently, the early diffuse deposit in the interstitial tissue gradually becomes transferred to macrophages in which it aggregates (14).

There are marked differences between the rodent testis on the one hand and that of the dog on the other with respect to volume, composition, and organization of interstitial tissues and the relative number and distribution of interstitial tissue macrophages. However, in many respects, the morphological and histological characteristics of the canine testis are similar to those of human testis and unlike those of the rodent testis (17,22,23,24,25,26,27,28).

In a more recent study, Miller et al. (22) compared the testicular distribution of intravenously injected $^{241}$Pu in the rat and the dog (beagle). In the testis preparations studied one week after the injection, the cell distribution factor for the spermatogonial stem cells (defined as the number of cells within the alpha-particle range observed, divided by the number of cells within the alpha-particle range expected if Pu deposits are uniform) was about 2.2 for the rat and 1.6 for the dog; furthermore, the rat and dog differed in the percentage of spermatogonial stem cells within the alpha-particle range from Pu in the interstitial tissue: 37% in the rat and about 11% and 18%, respectively, in the two dogs studied. As the authors point out, extrapolations from rodents to humans may overestimate the potential for radiation exposure to spermatogonial stem cells as well as the fraction of the spermatogonial stem cell population at risk to exposure from internally deposited plutonium.

Information on the distribution of Pu in the mammalian ovary is limited. In the mouse (29,30,31), 1 to 2 days after intravenous injection of $^{239}$Pu, alpha tracks were found to be uniformly distributed over all the ovarian tissue, with some concentrations in atretic follicles. Autoradiographs of ovaries taken from animals killed after the birth of three litters (i.e., about 2 months after injection) showed a complete redistribution: the uniform deposition and concentration in atretic follicles had disappeared and the main features were aggregates associated with macrophages located mainly in the stroma of the organ. As Searle et al. (30) point out, (1) initially, the concentration in atretic follicles would not lead to any exposure of the oocytes in the primordial and developing follicles, and (2) the subsequent concentrations in macrophages might result in some oocytes receiving a supralethal dose and others, none at all. The autoradiographs of ovaries from animals killed when sterile show that the Pu distribution seen after three litters is maintained until sterility.

## Transmutation Effects

The biological effects of transmutation and decay of radioactive isotopes have been the subject of some earlier (32,33) and recent (34) reviews. The current view is that, in mammals, the contribution of transmutations involving $^{3}$H, $^{14}$C, or $^{32}$P to the genetic effects is probably quite small and can be neglected. In view of this, this aspect is not further discussed in this chapter.

# Genetic Studies on Radionuclides in Mammals

## End Points

Three principal end points, namely, dominant lethals, chromosomal aberrations (primarily reciprocal translocations), and recessive mutations at specific gene loci, have been used in assessing the magnitude of genetic effects of radionuclides in experimental mammals. The mouse has been about the only test species, and even here, the focus has been on the responses of male germ cells.

For the measurement of dominant lethality, more often, the prenatal method has been used (35). Induced dominant lethality occurring after implantation is generally the result of chromosomal breakage in the germ line, which leads to the production of inviable zygotes. The method of choice to study the induction of reciprocal translocations (exchange of chromosomal segments between nonhomologous chromosomes) has been the cytogenetic one (36). In some studies, the genetic method was used, namely, the ascertainment of the incidence of semisterility in the offspring of those exposed, with cytological confirmation of translocation heterozygosity. To measure the induction of recessive visible mutations, the specific locus method is employed in which irradiated normal mice are mated to those of a special tester stock that is homozygous for a set of seven recessive genes (37,38). The $F_1$ progeny are screened for the appearance of mutants.

## Experimental Results

Summaries of results obtained with different beta and alpha emitters are given in Tables 18.2 and 18.3. In most of these studies (with the exception of some $^3H$ experiments), the radionuclides were directly injected into the test animals in order to optimize the chances of their reaching the gonads. The subject has recently been reviewed by Searle (3,77).

### Beta Emitters

The data on beta emitters (Table 18.2) permit three important conclusions. First, despite the large number of beta emitters studied in the mouse, most of the investigations have been rather limited in scope, and data on the genetic responses of the spermatogonial and oocyte stages are scarce; most of the available information pertains to irradiated postmeiotic and meiotic male germ cell stages only.

Second, with $^3H$ for which there are relatively more data, the impression one gains is that its RBE (relative to chronic gamma rays) is likely to be dose dependent and may be higher than 1 at low absorbed doses. On the basis of their specific locus results, Russell et al. (39) pointed out that, although it is difficult to rigorously establish that the RBE of $^3H$ (relative to chronic gamma rays) is higher than 1, "for risk estimation purposes, it seems prudent to use

**Table 18.2.** Some genetic effects of beta emitters in rodents[a]

| Nuclide | Study number | Activity/route of administration | Principal findings | Reference |
|---|---|---|---|---|
| | | | *A. Specific locus mutations* | |
| ³H | 1 | 18.5–27.8 MBq /g; IP (interperitoneal) injection as HTO to male mice | Significant induction in pre- and postmeiotic stages; relative to chronic gamma rays, RBEs (relative biological effectiveness) of about 1 for postmeiotic and about 2 for spermatogonial stages | Russell et al. (39) |
| | | | *B. Dominant lethals* | |
| ³H | 2 | 8.1 MBq of ³HTdR/male mouse; IP injection | Percent lethals correlated with degree of sperm labeling; highest in week 5 (spermatocytes) | Bateman and Chandley (40) |
| ³H | 3 | 81 kBq/mL HTO in drinking water to male and female mice from weaning | Tests conducted in second generation animals; significant reductions in numbers of live embryos when either or both parents exposed, but no reduction in breeding efficiency | Carsten and Commerford (41); Carsten et al. (42) |
| ³H | 4 | HTO or ³HTdR to male mice; IP injection | Main effect on spermtids (weeks 2–4) after HTO and in earlier stages (weeks 5–7) after ³H-thymidine; yield of dominant lethals higher after HTO, relative to gamma irradiation | Kudritskaya and Balanov (43) |
| ³H | 5 | 1.85–13.3 MBq/g; IP injection as HTO in male mice | Significant induction in postmeiotic and meiotic stages; spermatids most sensitive; RBE relative to chronic gamma irradiation increases with decreasing doses reaching a value of about 2.5 at doses less than 0.1 Gy | Balanov and Kudritskaya (44) |
| ³H | 6 | 3.3 MBq/g of ³H-glucose to males (oral administration) | Results generally similar to those with HTO | Balanov et al. (45) |

**Table 18.2.** (*Continued*)

| Nuclide | Study number | Activity/route of administration | Principal findings | Reference |
|---|---|---|---|---|
| ³H | 7 | HTO administration to mice (details not given) | RBE relative to chronic gamma rays decreasd with increasing dose | Wu De Chang (46) |
| ³H | 8 | 37 kBq/g of ³HTdR single IP injection to males every generation or 0.37 MBq/mL of HTO in drinking water for 35 days every generation; animals bred through sib mating | Average dose from ³HTdR was 0.039 Gy and from HTO 0.037 Gy over the 5-week period of spermatogenic differentiation and maturation; progressive impairment of reproductive capacity in both the ³H series and ³HTdR line became extinct by generations 19–23; in the HTO line overall premating mortality increased; reduction in body size and susceptibility to infections; more congenital anomalies | Mewissen and Ugarte (47) <br> Mewissen et al. (48) <br> Ugarte et al. (49) |
| ¹⁴C | 9 | 37 kBq ¹⁴C-glucose; IP injection to male mice | Significant induction in meiotic and postmeiotic stages | Goud et al. (50) |
| ¹⁴C | 10 | 54 kBq/g to 2.5 MBq/g of ¹⁴C-glucose; oral administration to male mice | Significant induction in postmeiotic stages; RBE relative to X or gamma rays, not different from 1 | Shevchenko et al. (51) |
| ³²P | 11 | 1.9–5 MBq as orthophosphate; IP injection to male mice | Significant induction; spermatids most sensitive | Reddi and Vasudevan (52) |
| ³²P | 12 | 85 kBq–0.9 MBq as orthophosphate; IP injection to female mice | Significant induction in oocytes | Krishna and Reddi (53) |
| ⁸⁹Sr | 13 | 37 kBq/g IP injection to male rats | Significant induction in spermatogonia | Baev et al. (54) |
| ⁹⁰Sr | 14 | 0.67 MBq; IP injection to male mice | Induction in postmeiotic stages with peaks in matings 3–4 weeks after injection | Lüning et al. (55, 56) |
| ⁹⁰Sr | 15 | 0.67 MBq; IP injection to male and female mice | Evidence for induction in spermatogonia (matings 11–14 weeks after injection) and in oocytes (weeks 1–4 after injection) | Lüning et al. (57) |
| ⁹⁰Sr | 16 | 0.93 MBq; IP injection to male mice | Increased rates of intrauterine deaths in mating 10–40 weeks after injection | Reddi (58) |

243

**Table 18.2.** (*Continued*)

| Nuclide | Study number | Activity/route of administration | Principal findings | Reference |
|---|---|---|---|---|
| 137Cs | 17 | 0.67 MBq; IP injection to male mice | No clear evidence for induction in postmeiotic stages | Lüning et al. (55) |
| 131I | 18 | 7.4–259 kBq; IP injection to male and female mice | Significant induction in both cases, especially in spermatids | Reddi (59); Reddi et al. (60) |
| 3H | 19 | 3.3–12.6 MBq/g HTO IP injection to male mice | Induction in spermatogonia; RBE relative to acute gamma rays increases from 1 at a dose of 1 Gy to 2 at a dose of 2 Gy | Pomerantseva et al. (61) |
| 14C | 20 | 15–70 MBq/g of 14C-glucose oral single administrations and daily (for 33 days) 1.8–3.6 kBq/g and long-term (6 or 12 months): 0.18 kBq–0.9 kBq/g | Only in the single and long-term exposure groups were the frequencies significantly higher than in the controls; RBE relative to chronic gamma rays is 1 | Pomerantseva et al. (62) |
| 14C | 21 | Same as in experiment 10 | Translocation yields at absorbed doses of 0.22, 0.5, and 1.01 Gy consistent with an RBE of 1, relative to chronic gamma rays | Shevchenko et al. (51) |
| 90Sr | 22 | 0.9 MBq; IP injection to male mice | Induction in postmeiotic stages | Reddi (63) |
| 131I | 23 | 0.37–0.9 MBq; IP injection to male mice | Induction in spermatogonia after 0.5 and 0.9 MBq | Ebenezer et al. (64) |
| 129I | 24 | 0.185, 0.37, and 0.555 MBq; IP injection | Animals killed after 60 days; significant increases in translocations in all treated groups (frequencies of 1.2%, 1.8%, and 2.3%) | Lavu et al. (65) |

a Reprinted with permission from ref. 3, with some additions.

**Table 18.3.** Some genetic effects of alpha emitters in mice[a]

| Nuclide | Study number | Activity /route of administration | Principal findings | Reference |
|---|---|---|---|---|
| | | *A. Specific locus mutations* | | |
| $^{239}$Pu | 1 | 0.37 MBq/kg of Pu-citrate, i.v. injection into tail vein of males | Treated mice mated after 13 weeks; 11 mutations in 54,679 $F_1$ progeny (significantly higher than in controls); induction rate of 7.1 $10^{-7}$ per 0.01 Gy (on the assumption of linearity and a testicular dose of at least 0.3 Gy); RBE of about 10 relative to chronic gamma rays | Russell et al. (66) |
| | | *B. Dominant skeletal mutations* | | |
| $^{239}$Pu | 2 | 0.37 MBq/kg of Pu-citrate, i.v. injection into tail vein of males; estimated mean gametic dose 0.58 Gy | No difference between the frequencies of presumed dominant skeletal mutations in the Pu series (0.18%; 6/3322) and in controls (0.20%; 4/1971) | Selby et al. (67) |
| | | *C. Translocations and fragments* | | |
| $^{239}$Pu | 3 | 0.37 MBq/kg of Pu-citrate, i.v. injection into tail vein of males | Mice killed 6, 12, or 18 weeks after injection; frequencies of spermatocytes with fragments higher than in controls in all experimental groups, but not significantly so; while translocations (spermatogonial) were induced, the yield decreased at the longest exposure period; for 6- and 12-week exposure, rates of induction of 1.43 $10^{-3}$ per 0.01 Gy (based on average testicular dose) or 0.58 $10^{-3}$ per 0.01 Gy (dose inhomogeneity factor of 2.5 taken into account); the latter rate similar to that after chronic fission neutron exposures | Beechey et al. (68) |

**Table 18.3.** (*Continued*)

| Nuclide | Study number | Activity/route of administration | Principal findings | Reference |
|---|---|---|---|---|
| ²³⁹Pu | 4 | 0.148 MBq/kg of Pu-citrate, i.v. injection into tail vein of males (comparison with chronic gamma rays) | In the Pu-series, mice killed 21, 28, or 34 weeks after injection; in the gamma series, after 28 weeks (and an accumulated dose of 10.5 Gy); for fragments, Pu-induction rate of about twice that obtained by Beechey et al. (68) and an *RBE* (relative to chronic gamma rays) of *about 24*; for Pu-induced translocations, no significant increases between weeks 21 to 34; mean rate of $3.4 \, 10^{-4}$ per 0.01 Gy (average testicular dose) and an RBE of 24, relative to chronic gamma rays | Searle et al. (69) |
| ²³⁹Pu | 5 | 0.19 and 0.37 MBq/kg Pu-Citrate, i.v. injection into tail vein of males (comparison with single or weekly gamma and neutron exposures) | Mice killed 8 to 60 weeks after injection; Pu: frequencies of translocations plotted against total accumulated testes dose showed *no dose-dependent increase*; RBE of about 38 after a few weeks exposure, relative to chronic gamma irradiation | Grahn et al. (70) |
| ²³⁹Pu | 6 | 0.37 MBq/kg Pu-citrate, i.v. injection into tail vein of males (comparison with single or weekly gamma and neutron exposures) | Mice killed 15–62.5 weeks after injection; Pu: nonlinear (concave downward) dose response for translocations; frequencies of fragments increased over the first 20 weeks then falling off and at 60 weeks not different from controls; RBE estimates variable: for translocations, the neutron/gamma ratio between 10 and 24, depending on weekly dose interval and the ratio is 1 or less for Pu relative to neutrons; for fragments, the neutron/gamma ratio is 18–22, depending on age factors and Pu/neutron is 1.5 | Grahn et al. (71) |

246

**Table 18.3.** (*Continued*)

| Nuclide | Study number | Activity/route of administration | Principal findings | Reference |
|---------|--------------|----------------------------------|--------------------|-----------|
| $^{239}$Pu | 7 | 0.37 MBq/kg Pu-citrate, i.v. injection into tail vein of males | Semisterility tests; intrvals of 23.5 to 52.5 weeks (series 1) and of 29 to 56.5 weeks (series 2) between injection and mating; frequencies of semisterile $F_1$ males: series 1, 0.20% and 0.39% (1st and 2nd intervals); series 2, 0.23% and 0.56% (1st and 2nd intervals); thus evidence for an increase with dose; rate of induction per 0.01 Gy in the range of $1.45$–$2.91/10^5$ gametes (dose inhomogeneity factors of 2 to 4 taken into account); efficiency not markedly different from acute X rays (rates per 0.01 Gy of $3.42$–$3.89/10^5$ gametes) | Generoso et al. (72) |
| $^{239}$Pu | 8 | 7.4 kBq/kg Pu-citrate, i.v. injection into tail vein of male mice; long-term retention in the testes, 0.04% of injected activity; accumulated average testicular dose of 0.042 Gy in a mean duration of 724 days | No significant difference in translocation frequencies between the Pu (0.5%) and the age-matched control (1.1%) series | Pacchierotti et al. (73) |

### D. Dominant lethals

| Nuclide | Study number | Activity/route of administration | Principal findings | Reference |
|---------|--------------|----------------------------------|--------------------|-----------|
| $^{239}$Pu | 9 | 1.85–18.5 kBq Pu-nitrate of Pu-citrate; i.v. injection into tail vein of males | Injected males mated to females for 12–24 weeks; no measurable effects with Pu-nitrate; with Pu-citrate, (9) significant excess of intrauterine deaths in females mated to exposed males, especially in later matings; (b) marked increase in late intrauterine deaths, as compared to controls; (c) excess intrauterine death also in matings of $F_1$ males with, again, a high proportion of late deaths; and (d) effects unrelated to the amount of Pu injected | Lüning et al. (74, 75) |

247

**Table 18.3.** (*Continued*)

| Nuclide | Study number | Activity/route of administration | Principal findings | Reference |
|---|---|---|---|---|
| $^{239}$Pu | 10 | Same as study 4 | Irradiated males mated to females during the last 4 weeks; Pu: amount of embryonic lethality (induced in postmeiotic stages) unrelated to the duration of exposure; significant increase in the proportion of late deaths as compared with gamma rays; RBE of about 22 relative to chronic gamma rays (with Pu, no correction for dose inhomogeneity) | Searle et al. (69) |
| $^{239}$Pu | 11 | Same as study 5 | Findings generally similar to those of Searle et al. (69) mentioned above; rate of dominant lethal induction (postimplantation mortality following postmeiotic male germ cell irradiation) rate 0.0064 per gamete per 0.01 Gy with a similar effectiveness to that of fission neutrons and with an alpha/chronic gamma ratio of 13 (RBE) | Grahn et al. (70) |
| $^{239}$Pu | 12 | 0.37 and 0.74 MBq/kg of Pu-citrate; i.v. injection into tail vein of females | Mating after 6 days (0.37 MBq) or after 3, 6, or 12 weeks; no evidence for postimplantation mortality in the 0.37 MBq series, but with 0.74 MBq, significant increases in preimplantation mortality (3, 6, 12 weeks exposure); only after 12 weeks exposure, postimplantation mortality increased significantly (estimated dose 0.59 Gy) | Searle et al. (31) |
| $^{241}$Am | 13 | 0.37 MBq/kg; i.v. injection into tail vein of males | In 10-week dominant lethal tests conducted between 125 and 195 days after injection, no significant induction of postimplantation mortality | Grahn et al. (76) |

$^a$ Reproduced with permission from ref. 3, with some additions.

248

the RBE value of 2.'' Further, as Searle (77) suggested ''it seems very probable that the RBE of $^3$H for the induction of deleterious dominant mutations will be very similar to that for specific locus mutations, . . . while that for the induction of reciprocal translocations is likely to be similar to figures for dominant lethal mutations . . .''

Third, the limited data on $^{14}$C that are currently available would permit the conclusion that the RBE for this radionuclide, relative to chronic gamma irradiation is probably close to unity. For all the other radionuclides, the data are insufficient to estimate RBEs.

## Alpha Emitters

The data on these are summarized in Table 18.3. It is obvious that the focus has been on Pu, there are clear genetic effects attributable to testicular burdens of this radionuclide, and by and large, the Pu alpha rays appear to have approximately the same effectiveness as fission neutrons. For translocations induced in spermatogonia, for which the data are relatively more extensive, the findings are such that their interpretation poses considerable problems: Their frequencies rise irregularly over the first approximately 120 days (following injection) and then decline to low levels [see Fig. 2 in Pacchierotti et al. (73)]. This decline in frequency (observed up to a retention time of about a year) seems independent of both the initial activity and total accumulated dose. With an injected activity of 7.4 kBq/kg and a mean retention time of about two years, there was no evidence for induced translocations (73).

Grahn et al. (71), who made a detailed analysis of these data, have argued that these observations cannot simply be attributed to cell killing and a concurrent loss of cells bearing translocations; rather, the overall response needs to be interpreted taking into account microdosimetric heterogeneity, the nearly invariant pattern of deposition in the testis, the kinetics of germ cell differentiation in the male, the high sensitivity of differentiating spermatogonia to cell killing, and the capacity of stem cells in relatively radiation-free areas to progressively assume the major spermatogenic role. As the authors point out, ''these rationalizations do not permit the development of a quantitative model or a basis for accurate prediction.''

Equally baffling is the observation of excess intrauterine deaths among the offspring to $F_1$ sons sired by Pu-injected males (74,75). The $F_1$ involvement would suggest that some of these males were carrying inherited translocations and therefore should exhibit semisterility, but none showed any indications of this.

The very limited data on dominant lethals induced by $^{241}$Am (0.37 MBq/kg; injection of males and tests between 125 and 195 days after injection) show that the amount of intrauterine mortality was low and not significantly different from that in controls. Thus, if anything, these data suggest that $^{241}$Am is probably less effective than $^{239}$Pu (76).

The only data for female mice come from the limited dominant lethal studies (31) and show that $^{239}$Pu induces oocyte killing (measured as preimplantation losses) and some postimplantation mortality after longer periods of exposure. Thus, the general picture is one of apparently low sensitivity of female mice to the induction of genetic damage by protracted exposure to Pu alpha particles, as had been demonstrated earlier for protracted gamma or neutron irradiation [reviewed in Sankaranarayanan (78)].

Turning now to Table 18.4, most of these RBE estimates have been made on the assumption of linearity between the alpha-particle dose and the effects studied or by fitting the data to a linear quadratic equation [translocations; Grahn et al. (71); the estimated value of the quadratic component was negative]. In some studies, the dose inhomogeneity factors were taken into account in RBE estimates [e.g., Beechey et al. (68) in translocation studies involving 6 or 12 week exposures; Generoso et al. (72), in translocation studies involving 23 to 56 week exposures] while in others, the average testicular doses were used [e.g., Grahn et al. (71) and Searle et al. (69) in translocation studies involving 15 to 60 week exposures]. Thus, the overall impression that one gains from inspection of Table 18.4, namely, that Pu alpha particles may have approximately the same effectiveness as fission neutrons, needs to be qualified.

## Oocyte Killing

Although oocyte killing as such is not a genetic effect, in view of the fact that such killing affects reproductive performance and, more importantly, the interpretation of some genetic results, it is briefly considered here. Female mice chronically exposed to HTO in body water during early life (from conception to 14 days after birth = 33 days), show an exponential decrease in oocyte survival with an $LD_{50}$ of about 74 kBq/mL (82,83,84,85). When these results are compared with those obtained after similar chronic low-level exposure to gamma rays, the RBE of tritium increases from about 1.5 (at about 0.035 Gy/d; dose: 0.5 to 0.6 Gy) to nearly 3 at low dose rates and low doses (less than 0.005 Gy/d) and an effective dose of less than 0.05 Gy (82,85). Similar results have been obtained by Wu De Chang (46). The increase of RBE with decreasing dose is not due to any increased effectiveness of tritium but to a decreased effectiveness of gamma rays at lower doses. The exceptionally high sensitivity of the mouse immature oocytes to the killing effects of HTO and gamma rays, together with the finding that $^3$HTdR incorporated in the chromosomes is less effective in this regard (86), led the authors to suggest that the target for the killing effects in these cells is not in the nucleus but most probably in the plasma membrane and experiments to verify this suggestion (87).

In other studies, Dobson and colleagues (85) found that the reproductive capacity of female mice that had chronic 33-day HTO exposures (via the mother during pre- and postnatal development up to 14 days of age) is less affected than oocyte number, as has been reported for other radiation exposure (88). The germ-cell-deficient female shows peak litter production at a younger age, and she prematurely runs out of oocytes.

**Table 18.4.** A comparison of the estimated RBE values for different end points measured in genetic experiments with mice with respect to protracted exposures to Pu and to fission neutrons, relative to similar gamma-ray exposures[a]

| End point | Cell stages | Pu | | | Fission neutrons | | |
|---|---|---|---|---|---|---|---|
| | | Estimated values | Mean | References | Estimated values | Mean | References |
| Specific locus mutations | All male germ cell stages | 10 | | 66 | 17,23 | 20 | 79,80 |
| Dominant lethals (postimplantation mortality) | Postspermatogonial stages | 13, 22 | 18 | 69,70 | 11 | | 70 |
| Chromosomal fragments | Early meiotic stages | 10, 24, 28, 33 | 24 | 68, 69, 70 | 18–22 | 20 | 71 |
| Translocations (cytogenetic studies) | Spermatogonia | 24, 38, 63 | 42 | 68, 69, 71 | 10–24, 20–25, 38 | 23 | 70, 71, 81 |
| Translocations (genetic studies) | Spermatogonia | Probably around 10 or less[b] | 72 | | | | |
| Testis mass reduction | | 10–15 | 69 | | | | |
| Female reproductive performance[c] | Oocytes | 2.5 | 30 | | | | |

[a] From ref. 3, with some additions.
[b] No genetic studies on translocation have been done with chronic gamma rays; the statement "around 10 or less" is based on the assumption that chronic gamma rays may be at least 10 times less effective than acute X rays; the rates of $1.45–2.91/10^5$ gametes per 0.01 Gy for Pu divided by the estimated rates for chronic gamma rays [based on the acute X-ray results (see ref. 72) of $0.34–0.39/10^5$ gametes per 0.01 Gy] give values in the range 3.7 to 8.6.
[c] Measured as duration of fertility and offspring per litter in 4 successive weekly periods; attributed to germ cell killing.

With primates, the sensitivity of primordial oocytes depends on the species (84,89,90). In squirrel monkeys (*Saimiri sciureus*) exposed continuously in utero to $^3$H in drinking water from conception to birth of the young (153 days), there was a massive oocyte loss that increased exponentially with the dose; the $LD_{50}$ level was 18.5 kBq/mL (giving 1 mGy/d), which is about one fourth of that found in mice. With other primates (the rhesus monkey, *Macaca mulatta*, and the bonnet monkey *Macaca radiata*) with similar treatment schedules (from the time of proof of pregnancy, day 25 to parturition, day 164; HTO in maternal drinking water at 185 kBq/mL; $^3$H body-water concentration level by day 60 of 144 kBq/mL; dose rate of 8.6 mGy/d during the second and third trimesters), the ovaries of the newborn showed about a 20% loss of oocytes (and thus much less than in squirrel monkeys). By 6 months, however, more fully expressed losses were striking: 59% in the rhesus monkeys and 95% in bonnet monkeys.

Dobson (89) suggests that, judging from these data, fertility problems may not be expected in young women following low-level prenatal exposure, but early germ cell losses exceeding 50% may lead to premature menopause in women. Since effective oocyte $LD_{50}$ values for the prenatal rhesus monkey, bonnet monkey, and squirrel monkey were about 600, 200, and 100 mGy, respectively, and if such doses were accumulated over the second half of human gestation (160 days), doses of 3.8, 1.2, or even 0.6 mGy/d may already suffice to produce subsequent premature menopause.

In the studies of Searle et al. (30) with $^{239}$Pu, the breeding performance of female mice injected with the radionuclide (0.185 or 0.37 MBq/kg body weight) was compared with those kept in a $^{60}$Co gamma irradiation field. Ovarian dose rates from the injected Pu were initially 0.008 and 0.017 Gy/d, changing little thereafter; actual gamma-ray dose rates to breeding females averaged around 0.08 and 0.16 Gy/d, respectively. Most of the females were mated to males and were allowed to continue breeding until sterility ensued. The approximate RBE for effects on reproduction (measured in terms of duration of fertility and offspring per litter in successive 4-week periods) and attributed to germ cell killing was found to be 2.5 for the alpha particles relative to chronic gamma rays; this value is much lower than 10 to 15 found by Searle et al. (69) for testis mass reduction after chronic exposure to injected $^{239}$Pu, in terms of average testis dose. In subsequent work, Searle et al. (31) adduced further evidence for oocyte killing by alpha particles from injected $^{239}$Pu.

# Genetic Risk Estimation

## Methods and Data Base

Genetic risk estimation involves two steps, namely, (1) an assessment of the magnitude of transmissible genetic damage likely to result from the exposure of the human population to radiation (through extrapolation from animal data), and (2) translating these effects into increased load of genetic and partially genetic diseases in humans. The approaches that have been used thusfar can be broadly grouped under two headings: the "doubling-dose method" or the "relative mutation risk method" and the "direct methods."

The doubling-dose method is used to provide an estimate of risks to a population under continuous irradiation, relating them to the "natural" prevalence of genetic and partially genetic diseases. What one does is to (a) estimate the doubling dose (the amount of radiation necessary to produce as many mutations as those already occurring naturally in a generation) from experimental data on spontaneous and induced mutations and chromosomal aberrations; (b) assume that the doubling dose so estimated is applicable to humans in the context of estimating the risk of induction of genetic disease; (c) estimate the expected increase at equilibrium as a product of natural prevalence, relative mutation risk, and the dose sustained by the population; and (d) estimate the expected increase in the first generation as a fraction of that at equilibrium.

The doubling dose that has been used by UNSCEAR since 1977 is 1 Gy for the exposure conditions (i.e., low-dose or low dose-rate, low LET irradiation) applicable in this context. This is a crude average of the estimates obtained for different genetic end points in experiments, primarily with male mice (4, 91,92). The estimates of prevalence of Mendelian, chromosomal, and irregularly inherited diseases in humans derive from (a) the results published by Trimble and Doughty (93) for the Canadian province of British Columbia, (b) those from several ad hoc studies for specific dominant and recessive conditions, (c) several cytogenetic surveys of newborns for chromosomal anomalies, and (d) data of Czeizel and Sankaranarayanan (94) and of Czeizel et al. (99) for congenital anomalies and other diseases of complex etiology, respectively. These data are discussed in the 1977 (91), 1982 (4), and 1986 (92) reports of UNSCEAR and may be consulted for details.

In contrast to the doubling-dose method, the direct methods aim at providing estimates of risk to the first generation progeny of those exposed using the rates of induction of mutations or of chromosomal aberrations estimated from animal data (mouse, nonhuman primate species). More specifically, the data on the induction of dominant skeletal and dominant cataract mutations in male mice and those on the induction of reciprocal translocations in some nonhuman primate species have been used (4,91,92). It bears reiterating here that (a) most, if not all, of the basic experimental data on rates of induction were collected in studies involving X- or gamma-ray exposures (i.e., external sources of low LET irradiation) delivered as high doses and at high dose rates, which were then "corrected" using a number of correction factors to make them applicable to radiation conditions applicable in the risk estimation context; and (b) the gap between the estimated induction rates for mutations and translocations in experimental species and tangible genetic defects in humans (diseases attributable to induced dominant mutations or translocations) was bridged using, again, a number of correction factors and assumptions (4,91,92).

## Current Risk Estimates for Low LET Irradiation
## From External Sources

The current risk estimates are those given by UNSCEAR in its 1986 report (92) and reproduced in Tables 18.5 and 18.6, respectively [see also Searle (95)]. Inspection of Table 18.5 will reveal that, if a population is continuously exposed to low-dose-rate, low LET irradiation at a rate of 0.01 Gy per generation, the expected increase at equilibrium is about 100 cases of dominant genetic disease per million live births; of this, about one sixth will be manifest in the first generation (1 generation = 30 years). Note that no risk estimates are given for congenital anomalies and other diseases of complex etiology.

The estimates of risk arrived at using the direct methods (Table 18.6) show that if human males are exposed to low LET, low-dose or low-dose-rate irradiation, there will be about 10 to 20 cases of dominant genetic disease per million live births per 0.01 Gy; for irradiation of females, the comparable estimates lie in the range between 0 and 9 cases per million live births per 0.01 Gy. The risks associated with the induction of balanced reciprocal translocations are estimated to be about 1 to 15 and 0 to 5 cases of congenitally abnormal

**Table 18.5.** Estimated effect of 0.01 Gy per generations of low LET irradiation on a population of one million live births according to the doubling-dose method. The doubling dose assumed in these calculations is 1 Gy and is based on mouse studies involving external radiation sources[a]

| Disease classification[b] | Current incidence per million[c] | Effect of 0.01 Gy per generation | |
| --- | --- | --- | --- |
| | | First generation[d] | Equilibrium |
| Autosomal dominant and X-linked diseases[e] | 10,000 | 15 | 100 |
| Autosomal recessive diseases | 2,500[f] | Slight | Slow increase |
| Chromosomal diseases due to | | | |
|     Structural anomalies | 400[g] | 2.4 | 4 |
|     Numerical anomalies | 3,000[h] | Probably very small | Probably very small |
| Congenital anomalies and other irregularly inherited disorders | 660,000[i] | No estimates given | No estimates given |

[a] Reprinted with permission from ref. 92.

[b] Follows that given in the 1972 BEIR report (96), except that chromosomal diseases are divided into those with structural and those with numerical anomalies.

[c] Based on the results of British Columbia and other studies; for details, see ref. 91.

[d] The first generation increment is assumed to be 15% of the equilibrium value for autosomal dominant and X-linked diseases and about 3/5 of the equilibrium value for structural anomalies; see ref. 4 for details.

[e] Includes diseases with early and late onset, but excludes fragile X syndrome.

[f] Also includes diseases maintained by heterozygous advantage.

[g] Unbalanced structural rearrangements.

[h] Excludes mosaics.

[i] Includes congenital anomalies and other irregularly inherited diseases (see ref. 92 for details).

**Table 18.6.** Risk of induction of genetic damage in humans per 0.01 Gy at low dose rates of low LET irradiation according to the direct methods. The data used are from mouse or primate experiments involving external radiation sources[a]

| Genetic damage | Expected frequency (per $10^6$ live births) of genetically abnormal children in the first generation after irradiation of | |
|---|---|---|
| | Males | Females |
| Mutations having dominant effects[b,c] | ~10 to ~20[d] | 0 to ~9[e] |
| Recessive mutations[f] | 0 | 0 |
| Unbalanced products of reciprocal translocations | ~1 to ~15[g] | 0 to ~5[h] |

[a] Reprinted with permission from ref. 92.

[b] Includes risk from the induction of dominant mutations, deletions, and balance reciprocal translocations with dominant effects and from that of the most detrimental effects of recessives.

[c] Does not include the risk of induction of genetic changes with dominant sublethal effects that may kill between birth and early childhood (between 5 and 10 cases per $10^6$ live births per 0.01 Gy of paternal irradiation; estimated on the basis of data on litter-size reductions in mice); for maternal irradiation a comparable estimate is not available.

[d] The lower limit is derived from the data on dominant cataract mutations and the upper limit from those on dominant skeletal mutations (both in mice). See the 1977 and 1982 UNSCEAR reports (4, 91) for details.

[e] The lower limit is based on the assumption that the mutational sensitivity of the human immature oocytes is similar to that of mouse immature oocytes; the upper limit is based on the assumption that the sensitivity of the human immature oocytes is similar to that of the mouse mature and maturing oocytes and that the latter is 0.44 times that of spermatogonia. See the 1982 UNSCEAR report (4) for details.

[f] Although the risk of recessive disease from the induction of recessive mutations is zero in the first generation, about 1 extra case per million live births would be expected in the following 10 generations from partnership effects and, on certain assumptions, about 10 extra cases per $10^6$ would be expected by the tenth generation from effects due to identity by descent. See Searle and Edwards (97) for details.

[g] The lower limit is based on combined cytogenetic data from chronic low LET irradiation experiments involving the rhesus monkey and the crab-eating monkey and the upper limit on the combined human and marmoset (*Saguinus fuscicollis*) cytogenetic data. It has been assumed that 9% of unbalanced products of reciprocal translocations will result in congenitally malformed live births.

[h] The lower limit is based on the assumption that the sensitivity of the human immature oocytes to the induction of heritable reciprocal translocations will be similar to that of the mouse immature oocytes with respect to the induction of chromosomal aberrations. The upper limit is based on the assumptions that: (a) the sensitivity of the human immature oocytes will be one-half of that of human and marmoset spermatogonia (based on results with mice on heritable translocations); (b) the frequency of unbalanced products will be 6 times that of recoverable balanced translocations; and (c) about 9% of unbalanced products will result in congenitally malformed children.

children per million live births per 0.01 Gy for irradiation of males and females, respectively.

## Problems and Uncertainties

Although it has been possible to provide quantitative estimates of genetic risks associated with the exposure of human populations to ionizing radiation, they cannot be considered very precise. Considering first the estimates of risk obtained with the doubling-dose method, it may be recalled that (a) the doubling dose of 1 Gy used in the calculations is based almost exclusively on data obtained in experiments with male mice; the extent to which this doubling-dose estimate is valid for mouse females (and, by extrapolation, to human females) is not known; (b) data from genetic studies of the Hiroshima and Nagasaki populations (1,2) lend credence to the notion that the doubling dose for "genetic effects" may perhaps be of the order of 4 Gy (i.e., lower relative risks per unit dose); however, none of the indicators of "genetic damage" used by Schull and colleagues (1) (untoward pregnancy outcomes, death of live-born children and frequency of children with sex-chromosomal aneuploidy) in their analysis is really comparable with those used in the mouse; furthermore, the large standard deviations associated with the estimates preclude their meaningful use in the context of risk estimation; and (c) risk estimates for diseases of complex etiology (numerically the most frequent) remain elusive in view of the fact that the mechanisms of their maintenance in the population and the extent to which they will respond to mutation pressure are unknown.

The estimates made using the so-called "direct" methods are not without problems either. They rely entirely on mouse data on dominant skeletal and cataract mutations (collected at high dose rates) and those on the induction of reciprocal translocations in nonhuman primate species. There are no data on the induction of dominant skeletal or cataract mutations in mouse females, there are no dose–response curves, and there is no experimentation to verify whether the correction factors used to estimate effects at low doses and at low dose rates are in fact valid. Furthermore, the questions of the sensitivity of human oocytes to the induction of genetic effects by radiation, as well as whether the mouse is a proper model for human females, remain unresolved. Notwithstanding these problems and uncertainties, there is, at present, no evidence that we have underestimated the risks.

## Radionuclides and Genetic Risks: Some Personal Views

In principle, it should be possible to assess genetic risks associated with internally deposited radionuclides on the basis of experimental data provided (a) the gonadal doses can be reliably estimated, and (b) the effectiveness of the radiation from the radionuclide of interest relative to chronic gamma irradiation (i.e., the RBE) is known. These statements are valid for radionuclides such as $^3H$ or $^{14}C$ because their distributions in the body, including the gonads, are uniform. For those, such as $^{239}$Pu, known to be inhomogeneously distributed in the testis, it is additionally necessary to estimate the influence of these factors on dose to stem cell spermatogonia. If these conditions are satisfied and if there are no other complicating factors, one can use the "basic risk estimates" for low LET irradiation from external sources (Tables 18.5 and 18.6) as a framework and assess risks for any given radionuclide by multiplying the numerical values given in Tables 18.5 and 18.6 by the RBE values deemed to be applicable. Admittedly, this is a crude procedure, but there is no alternative at present.

For $^3$H, the weight of evidence (from germ cell studies summarized in Table 18.2 and from those using somatic cells, not considered in this paper) suggests that there is enough justification to use an RBE value of 2 (i.e., the risks per unit dose from tritium will be twice those for chronic gamma irradiation). For other beta emitters, at present, there are no compelling empirical data to contradict the assumption that their RBE is 1. This means that the risks will be the same as those given in Tables 18.5 and 18.6. The follow-up of female patients treated with $^{131}$I for thyrotoxicosis showed no significant genetic effects [see Hennemann et al. (98) for a listing of the relevant references], but the sample sizes were very small.

For $^{239}$Pu, the estimation of the dose to the spermatogonial stem cells and of the RBE for relevant genetic effects poses some problems. It is worth recalling that in the mouse, (a) the estimated RBEs for the induction of specific locus mutations, presumed dominant skeletal mutations, and heritable translocations are probably not more than 10; (b) with injected activities of 0.15 to 0.37 MBq/kg of Pu and retention times of up to about 120 days, the estimated RBEs for translocations (studied cytogenetically) are in the range of 24 to 63; (c) under conditions of chronic (retention time of up to 2 years), low-level (injected activity of 7.4 kBq/kg) exposure to testicular Pu, there is no evidence for the induction of translocations; and (d) the limited data from female mice are consistent with only a low effectiveness of Pu alpha rays in the induction of genetic effects. On the basis of these, it is suggested that, for genetic risk estimation, an effective RBE range of 10 to 20 can be used. This means that the basic risk estimates given in Tables 18.5 and 18.6 need to be multiplied by this range of values to obtain risk estimates for gonadally deposited Pu. In considering genetic risks from inhaled or ingested Pu, it is worth reiterating the fact that the amount reaching the gonads is very small, although once deposited, Pu is tenaciously retained.

## Summary

Ionizing radiation, irrespective of whether it is received from external sources or from internally deposited radionuclides, is capable of causing genetic damage that will be transmitted from one generation to the next. Since strictly relevant human data are limited, at present, there is no alternative to the use of data collected from experimental mammals for estimating genetic risks associated with the exposure of human populations to ionizing radiation.

Animal studies reveal that Pu concentrates in the testicular interstitial tissue (thus placing the stem cell spermatogonia at special risk) but is later aggregated by macrophages, apparently leading to a reduction in effectiveness. For protracted alpha irradiation from testicular Pu in the mouse, the RBE for genetic effects of concern may be in the range between 10 and 20, relative to chronic gamma irradiation. In female mice, Pu appears to be much less effective, but the data are very limited.

Studies with beta emitters provide reasonable grounds for believing that for chronic exposures to tritium, the RBE value may be about 2, whereas for others, such as $^{14}$C, it may not be significantly different from unity. No evidence for clear-cut transmutational effects has been found with the beta emitters studied.

These and other results support the view that for Pu alpha rays, the genetic risks following the exposure of human males may be 10 to 20 times that estimated by UNSCEAR (92) for chronic, low-level, low LET irradiation per unit absorbed testicular dose; for $^3$H exposures, the risks may be higher by a factor of 2, while for $^{14}$C the risk will be the same as for chronic gamma irradiation. For estimating genetic risks following exposure of human females to radionuclides, it would be prudent to use the same multiplication factors as those suggested for males to derive estimates from those of UNSCEAR for chronic, low-level, low LET irradiation.

*Acknowledgments.* I am grateful to my colleagues, Dr. A. D. Tates for placing his reprint collection on radionuclides at my disposal, Dr. P. P. W. van Buul for drawing my attention to some important publications, and Professor P. H. M. Lohman for his warm encouragement. The writing of this paper was supported by EURATOM Contract No. BI 6-0226-NL (GDF) with the University of Leiden.

# References

1. Schull WJ, Neel JV, Otake M, Awa AA, Satoh C, Hamilton HB (1982) Hiroshima and Nagasaki. Three and a half decades of genetic screening. In: Sugimura T, Kondo S, Takebe H (eds) Environmental mutagens and carcinogens. Tokyo University Press, Tokyo, and AR Liss, New York, pp 687–700

2. Awa AA, Honda T, Neriishi S, Sofuni T, Shimba H, Ohtari K, Nakano M, Kodama Y, Itoh M, Hamilton HB (1987) Cytogenetic study of the offspring of atomic bomb survivors, Hiroshima and Nagasaki. In: Obe G, Basler A (eds) Cytogenetics, basic and applied aspects. Springer-Verlag, Berlin, pp 166–183

3. Searle AG (1983) Cytogenetic effects of incorporated radionuclides on mammalian germ cells. In: Ishihara T, Sasaki MS (eds) Radiation-induced chromosome damage in man. AR Liss Inc, New York, pp 347–367

4. United Nations Scientific Committee on the Effects of Atomic Radiation (UNSCEAR)

(1982), Ionizing radiation: sources and biological effects, 1982 Report to the General Assembly, with annexes. United Nations, New York

5. Butler HL, LeRoy JH (1965) Observations of biological half-life of tritium. Health Phys 11:283–285
6. Wylie KF, Bigler WA, Grove GR (1963) Biological half-life of tritium. Health Phys 9:911–914
7. World Health Organization (WHO) (1983) Selected radionuclides, IPCS International Programme on Chemical Safety. Environmental Health Criteria 25, WHO Geneva
8. International Commission on Radiological Protection (ICRP) (1979) Limits for intakes of radionuclides by workers. ICRP Publication 30, Part 1, Ann ICRP 2, No 3/4, Pergamon, Oxford, UK
9. International Commission on Radiological Protection (ICRP) (1986) The metabolism of plutonium and related elements. ICRP Publication 48, Ann ICRP 16, No 2/3, Pergamon, Oxford
10. Van den Hoek J, Kirchmann R, Juan NB (1979) Transfer and incorporation of tritium in mammals. In: Behaviour of tritium in the environment. IAEA Vienna, pp 433–444
11. Cumming RB, Sega GA, Walton MF (1979) Radiation dosimetry in experimental animals exposed to tritiated water under different conditions. In: Behaviour of tritium in the environment, IAEA, Vienna, pp 463–468
12. Mian TA, Meistrich ML, Hayne TP, Glenn HJ (1981) Testicular radiation dose and cytotoxic effects of $^{32}$P. Radiat Res 87:445 (Abstract)
13. Brooks AL, Redman HC, Hahn FF, Mewhinney JA, Smith JM, McClellan RO (1983) The retention, distribution, dose and cytogenetic effects of inhaled $^{239}$PuO$_2$ or $^{239}$Pu(NO$_3$)$_4$ in non-human primates. In: Broerse JJ, Barendsen GW, Kal HB, Van der Kogel AJ (eds) Proc VII Int Cong Rad Res, session B (Biology) p B4-04 (Abstract)
14. Ash P, Parker T (1978) The ultrastructure of mouse testicular interstitial tissue containing plutonium-239 and its significance in explaining the observed distribution of plutonium in the testis. Int J Rad Biol 34:523–536
15. Brooks AL, Diel JH, McClellan RO (1979) The influence of testicular micro-anatomy on the potential genetic dose from internally-deposited $^{239}$Pu citrate in Chinese hamster, mouse and man. Radiat Res 77:292–302
16. Green D, Howells GR, Humphreys ER, Vennart J (1975) Localization of plutonium in mouse testes. Nature (London) 255:77
17. Miller SC (1982) Localization of plutonium-241 in the testis. An inter-species comparison using light and electron microscope autoradiography. Int J Rad Biol 41:633–643
18. Miller SC, Bowman BM (1983) Tissue, cellular and subcellular distribution of $^{241}$Pu in the rat testis. Radiat Res 94:416–426
19. Russell JJ, Lindenbaum A (1979) One-year study of non-uniformly distributed plutonium in mouse testis as related to spermatogonial irradiation. Health Phys 36:153–157
20. Priest ND, Jackson S (1978) The uptake and redistribution of $^{241}$Pu within the gonads. Int J Radiat Biol 34:49–65
21. Taylor DM (1977) The uptake, retention and distribution of plutonium-239 in rat gonads. Health Phys 32:29–31
22. Miller SC, Rowland HG, Bowman BM (1985) Distributions of cell populations

within alpha-particle range of plutonium deposits in the rat and beagle testis. Radiat Res 101:102–110

23. Clark RV (1976) Three-dimensional organization of testicular interstitial tissue and lymphatic space in the rat. Anat Rec 184:203–225

24. Connell CJ, Christensen AK (1975) The ultrastructure of the canine testicular interstitial tissue. Biol Reproc 12:368–382

25. Fawcett DW, Burgos MH (1960) Studies on the fine structure of the mammalian testis. II The human interstitial tissue. Am J Anat 107:245–269

26. Fawcett DW, Neaves WB, Flores MN (1973) Comparative observations on intertubular lymphatics and the organization of the interstitial tissue of the mammalian testis. Biol Reprod 9:500–532

27. Kerr JB, de Kretser DM (1981) The cytology of the human testis. In: Burger H, de Kretser DM (eds) The testis, Raven, New York, pp 141–169

28. Stover BJ, Atherton DR, Bruenger FW, Buster DS (1962) Further studies on the metabolism of $^{239}$Pu in adult beagles. Health Phys 8:589–597

29. Green D, Howells Gr, Vennart J, Watts R (1977) The distribution of plutonium in the mouse ovary. Int J Appl Rad Isotopes 28:497–501

30. Searle AG, Beechey CV, Green D, Howells GR (1980) Comparative effects of protracted exposures to $^{60}$Co-gamma radiation and $^{239}$Pu-alpha radiation on breeding performance in female mice. Int J Radiat Biol 37:189–200

31. Searle AG, Beechey CV, Green D, Howells GR (1982) Dominant lethal and ovarian effects of plutonium-239 in female mice. Int J Radiat Biol 42:235–244

32. International Atomic Energy Agency (IAEA) (1968) Biological effects of transmutation and decay of radioactive isotopes. IAEA, Vienna

33. Krische RE, Zelle MR (1969) Biological effects of radioactive decay: the role of the transmutation effect. Adv Radiat Biol 3:177–213

34. The National Council on Radiation Protection and Measurements (NCRP) (1979) Tritium and other radionuclide labeled organic compounds incorporated in genetic material. NCRP report 63, NCRP Publications, Bethesda, Maryland

35. Bateman AJ (1977) The dominant lethal assay in the male mouse. In: Kilbey BJ, Legator M, Nichols W, Ramel C (eds) Handbook of mutagenicity test procedures, Elsevier, Amsterdam, pp 325–334

36. Evans EP, Breckon G, Ford CE (1964) An air-drying method for meiotic preparations from mammalian testes. Cytogenetics 3:289–294

37. Russell WL (1951) X-ray-induced mutations in mice. Cold Spring Harb Symp Quant Biol 16:327–336

38. Searle AG (1975) The specific locus test in the mouse. Mutation Res 31:277–290

39. Russell WL, Cumming RB, Kelly EM, Phipps EL (1979) Induction of specific locus mutations in the mouse by tritiated water. In: Behaviour of tritium in the environment, IAEA Vienna, pp 489–497

40. Bateman AJ, Chandley AC (1962) Mutations induced in the mouse with tritiated thymidine. Nature (London) 193:705–706

41. Carston AL, Commerford SL (1976) Dominant lethal mutations in mice resulting from chronic tritiated water (HTO) ingestion. Radiat Res 66:609–614

42. Carston AL, Commerford SL, Cronkite EP (1977) The genetic and late somatic effects of chronic tritium ingestion in mice. Current Topics Radiat Res Quarterly 12:212–224

43. Kudritskaya OY, Balonov MI (1980) Dynamics of dominant lethal mutation yield in mice affected by tritium. Radiobiol 220:881–885

44. Balanov MI, Kudritskaya OV (1984) Mutagenic action of tritium upon the germ cells of male mice. I. Induction of dominant lethal mutations by tritium oxide and estimation of RBE. Genetika 20:224–232

45. Balanov MI, Pomerantsova MD, Ramaiya LK (1984) The mutagenic effects of tritium on germ cells of male mice. Consequences of $^3$H-glucose incorporation. Radiobiologia 24:753–757

46. Wu De Chang (1986) The experimental studies on the biological effects of low level radiation. In: Chinese Medical Association (ed) Proc Int Symp Biol Effects of Low Level Radiation, 24–26 Nov 1986, Nanjing (People's Republic of China), pp 41–59

47. Mewissen DJ, Ugarte AS (1979) Cumulative genetic effects from exposure of male mice to tritium for 10 generations. In: Biological implications of radionuclides released from nuclear industries, Vol I, IAEA Vienna, p 215

48. Mewissen DJ, Ugarte AS, Rust JH (1983) Radiation-induced heritable multiple intestinal adenocarcinoma exhibiting Mendelian inheritance with chromosomal abnormality. In: Broerse JJ, Barendsen GW, Kal HB, Van der Kogel AJ (eds) Proc VII Int Cong Rad Res, Book of Abstracts, Sessions C, p C6–12

49. Ugarte AS, Mewissen DJ, Rust JH (1983) Cumulative genetic effects from exposure of male mice to tritium for 23 generations. In: Broerse JJ, Barendsen GW, Kal HB, Van der Kogel AJ (eds) Proc VII Int Cong Rad Res, Book of Abstracts, Sessions C, pp C4–14

50. Goud SN, Reddi OS, Reddy PP (1981) Dominant lethal mutations induced by $^{14}$C in mice. Experientia 37:448–449

51. Shevchenko VA, Pomerantseva MD, Ramaiya LK, Vasilenko II, Liaginskaia AM (1981) Genetic effects of incorporated $^{14}$C in the male germ cells of mice. 1 The single administration of $^{14}$C-glucose. Radiobiologia 21:780

52. Reddi OS, Vasudevan B (1968) Increased dominant lethality in mice by phosphorus-32. Nature (London) 218:283

53. Krishna M, Reddi OS (1974) Genetic effects of phosphorus-32 in female mice. Radiat Res 59:266 (abstract)

54. Baev I, Bairakova A, Benova D, Vassillev G (1972) Genetic effects of radioactive isotopes. I. Dominant lethals from strontium-89 or iodine-131 in rat spermatogonia. Strahlentherapie 144:338–341

55. Lüning KG, Frölén H, Nelson A, Rönnbäck C (1963) Genetic effects of strontium-90 injected into male mice. Nature (London) 197–304–305

56. Lüning KG, Frölén H, Rönnbäck C, Nelson A (1965) Further studies of the genetic effects of $^{90}$Sr on various stages of spermatogenesis in mice. A preliminary report. FOA 1 Report C1164-F17

57. Lüning KH, Frölén H, Nelson A, Rönnbäck C (1963) Genetic effects of strontium-90 on immature germ cells in mice. Nature (London) 199:303–304.

58. Reddi OS (1971) Long term genetic effects of strontium-90 in mice. Ind J Med Res 59:1754–1757

59. Reddi OS (1970) Genetic effects of $^{131}$I in mice. Nature (London) 227:961–962

60. Reddi OS, Reddy PP, Krishna M (1974) Iodine-131 induced dominant lethal mutations in mice. Radiat Res 59:265 (Abstract)

61. Pomerantseva MD, Balanov MI, Ramaiya LK, Vilkina GA (1984) Mutagenic effect of tritium on the germ cells of male mice. II Genetic damages in stem spermatogonia induced by tritiated water and gamma irradiation. Genetika 20:782–787

62. Pomerantseva MD, Ramaiya LK, Vilkina GA, Shevchenko VA, Vasilenko IJ, Lyagin-

skaya AM, Istomina AG (1983) Genetic effects of radiocarbon in reproductive cells of male mice. Mutation Res 122:341–346

63. Reddi OS (1971) $^{90}$Sr-induced translocations in mice. Ind J Med Res 59:574–577

64. Ebenezer DN, Reddy SB, Reddy PP, Reddi OS (1980) Effects of acute and fractionated doses of $^{131}$I induced radiation damage to mouse spermatogonia. IRCS Medical Sci 8:912–914

65. Lavu S, Reddy PP, Reddi OS (1985) Chromosomal abnormalities induced by iodine-125 in mouse germ cells, Int J Rad Biol 48:603–607

66. Russell WL, Cumming RB, Kelly EM, Lindenbaum A (1978) Plutonium-induced specific locus mutations in mice. Genetics 88 s 85 (Abstract)

67. Selby PB, McKinley TW Jr, Raymer GF (1985) Dominant skeletal mutation study shows that the quality factor commonly used for alpha irradiation probably leads to a sizeable overestimation of genetic risk in males. Genetics 110 (suppl): s118

68. Beechey CV, Green D, Humphreys ER, Searle AG (1975) Cytogenetic effects of plutonium-239 in male mice. Nature (London) 256:577–578

69. Searle AG, Beechey CV, Green D, Humphreys ER (1976) Cytogenetic effects of protracted exposures to alpha particles from plutonium-239 and to gamma rays from cobalt-60 compared in male mice. Mutation Res 41:297–310

70. Grahn D, Frystak BH, Lee CH, Russell JJ, Lindenbaum A (1979) Dominant lethal mutations and chromosome aberrations induced in male mice by incorporated $^{239}$Pu and by external fission neutron and gamma irradiation. In: Biological effects of radionuclides released from nuclear industries, Vol I, IAEA Vienna, pp 163–184

71. Grahn D, Lee CH, Farrington BF (1983) Interpretation of cytogenetic damage induced in the germ line of male mice exposed for over 1 year to $^{239}$Pu alpha particles, fission neutrons, or $^{60}$Co-gamma rays. Radiat Res 95:566–583

72. Generoso WM, Cain KT, Cacheiro NLA, Cornett CV (1985) $^{239}$Plutonium-induced heritable translocations in male mice. Mutation Res 152:49–52

73. Pacchierotti F, Andreozzi U, Russo A, Metalli P (1983) Reciprocal translocations in ageing mice and in mice with long-term low level Pu-239 contamination. Int J Rad Biol 43:445–450

74. Lüning KG, Frölén H, Nilsson A (1976) Genetic effects of $^{239}$Pu salt injections in male mice. Mutation Res 34:539–542

75. Lüning KG, Frölén H, Nilsson A (1976) Dominant lethal tests of male mice given $^{239}$Pu salt injections. In: Biological and environmental effects of low level radiation, Vol I, IAEA Vienna, pp 39–49

76. Grahn D, Frystak BH, Russell JJ (1978) Genetic effects of americium-241. In: Argonne Nat Lab Rep ANL-79-90, p 29

77. Searle AG (1984) Genetic effects of tritium in mammals. In: Gerber G, Myttenaere C (eds) European Seminar on the risks from tritium exposure. Commission of European Communities, Brussels, pp 303–312

78. Sankaranarayanan K (1982) Genetic effects of ionizing radiation in multicellular eukaryotes and the assessment of genetic radiation hazards in man. Elsevier Biomedical Press, Amsterdam

79. Batchelor AL, Phillips RJS, Searle AG (1966) A comparison of the mutagenic effectiveness of chronic neutron and gamma irradiation of mouse spermatogonia. Mutation Res 3:218–229

80. Searle AG (1967) Progress in mammalian radiation genetics. In: Silini G (ed) Proc III Int Cong Rad Res North Holland, Amsterdam, pp 469–481

81. Searle AG, Evans EP, West BJ (1969) Studies on the induction of translocations

in mouse spermatogonia. II Effects of fast neutron irradiation. Mutation Res 7:235–240

82. Dobson RL (1979) The toxicity of tritium. In: Biological implications of radionuclides released from nuclear industries, Vol I, IAEA, Vienna, pp 203–211

83. Dobson RL, Cooper MF (1974) Tritium toxicity: effect of low level $^3$HOH exposure on developing female germ cells in the mouse. Radiat Res 58, 91–100

84. Dobson RL, Kwan TC (1977) The tritium RBE at low-level exposure-variation with dose, dose-rate and exposure duration. Curr Top Radiat Res Quarterly 12: 44–62

85. Dobson RL, Kwan TC, Straume T (1984) Tritium effects on germ cells and fertility. In: Gerber G, Myttenaere C (eds) European seminar on the risks from tritium exposure. EUR 9065 EN, Commission of the European Communities, Brussels, pp 285–298

86. Baker TG, McLaren A (1973) The effect of tritiated thymidine on the developing oocytes of mice. J Reprod Fertil 34:121–130

87. Straume T, Dobson RL, Kwan TC (1987) Neutron RBEs and the radiosensitive target for mouse immature oocyte killing. Radiat Res 111:47–57

88. Oakberg EF (1966) Effect of 25 R of X-rays at 10 days of age on oocyte number and fertility of female mice. In: Lindop PJ, Sachers GA (eds) Radiation and ageing, Taylor and Francis, London, pp 293–306

89. Dobson RL (1985) Delayed reproduction consequences of low level irradiation early in life. Paper presented at the Am Nucl Soc 1985 Winter Meeting, San Francisco, 10–14 Nov 1985, UCRL-92866 Summary

90. Dobson RL, Felton JS (1983) Female germ cell loss from radiation and chemical exposure. Am J Indust Med 4:175–190

91. United Nations Scientific Committee on the Effects of Atomic Radiation (UNSCEAR) (1977), Sources and effects of ionizing radiation, 1977 Report to the General Assembly, with annexes. United Nations, New York

92. United Nations Scientific Committee on the Effects of Atomic Radiation (UNSCEAR) (1986), Genetic and somatic effects of ionizing radiation, 1986 Report to the General Assembly, with annexes. United Nations, New York

93. Trimble BK, Doughty JH (1974) The amount of hereditary disease in human populations. Ann Hum Genet (London) 38:199–229

94. Czeizel A, Sankaranarayanan K (1984) The load of genetic and partially genetic disorders in man. I. Congenital anomalies: Estimates of detriment in terms of years of life lost and years of impaired life. Mutation Res 128:73–103

95. Searle AG (1987) Radiation—the genetic risk. Trends in Genetics 3:152–157

96. BEIR Report (1972) The effects on populations of exposure to low levels of ionizing radiation. Natl Acad Sci, Natl Res Council, Washington DC

97. Searle AG, Edwards JH (1986) The estimation of risks from the induction of recessive mutations after exposure to ionizing radiation. J Med Genet 23:220–226

98. Hennemann G, Krenning EP, Sankaranarayanan K (1986) Place of radioactive iodine in treatment of thyrotoxicosis. Lancet i:1369–1371

99. Czeizel A, Sankaranarayanan K, Losonci A, Rudas T, Keresztes M (1988) The load of genetic and partially genetic diseases in man. II. Some selected common multifactorial diseases: Estimates of population prevalence and of detriment in terms of years of lost and impaired life. Mutation Res, in press.

# CHAPTER 19

# Evaluation Procedures

## E. D. Rubery[1]

Many foods and drinks contain small quantities of chemicals that are not in themselves nutritive components. Some of these chemicals will have been intentionally added during production and therefore can be described as "additives" in food (e.g., colors, preservatives). Others are unintentional and can be described as "contaminants" [e.g., aflatoxins, Pb, and substances entering food from contact materials (Table 19.1)].

Most countries have a regulatory structure to control the levels of additives and contaminants in food. This structure necessarily involves a mechanism for making a judgment about the likely health effects of exposures to these adulterants. The general structure of the controlling organizations in the United Kingdom are shown in Fig. 19.1. The evaluation of the toxicity of additives and contaminants and assessment of the public risk associated with exposures is the particular responsibility of the Department of Health and Social Security's (DHSS) Committee on the Toxicity of Chemicals in Food, Consumer Products and the Environment (COT) with the assistance of two more specialized committees, the Committee on Carcinogenicity (COC) and the Committee on Mutagenicity (COM).

In addition, the Veterinary Products Committee (VPC) and the Advisory Committee on Pesticides (ACP) are responsible for advising the government with respect to individual veterinary products and pesticides. Part of that assessment consists of consideration of the levels of chemical residues in plant or animal tissues intended for human consumption. When a particular problem is identified, they may refer the product to the COT for more detailed consideration before finalizing their advice.

These committees are coordinated by the Steering Group on Food Surveillance, which consists of officials from each of the areas outlined above and from a number of other areas related to the safety of food. This group is responsible

[1] Department of Health and Social Security, Elephant and Castle, London SE1 6TE, UK.

The contents of this paper represent the author's views alone and in no way commit the Department of Health and Social Security.

**Table 19.1.** Chemicals in food

| Additives | Contaminants |
|-----------|--------------|
| Colors | Veterinary residues |
| Flavors | Pesticide residues |
| Preservatives | Mycotoxins |
| Stabilizers | Heavy metals |
| Sweeteners | Radionuclides |
|  | Food contact materials |

for ensuring that the advice from the different expert committees is carried forward with due consideration of the different priorities involved.

Finally, before dealing with the different chemical and radioactive agents that may be found in food in more detail, it may be worth stressing that this chapter does not deal with microbiological contamination of food. Therefore,

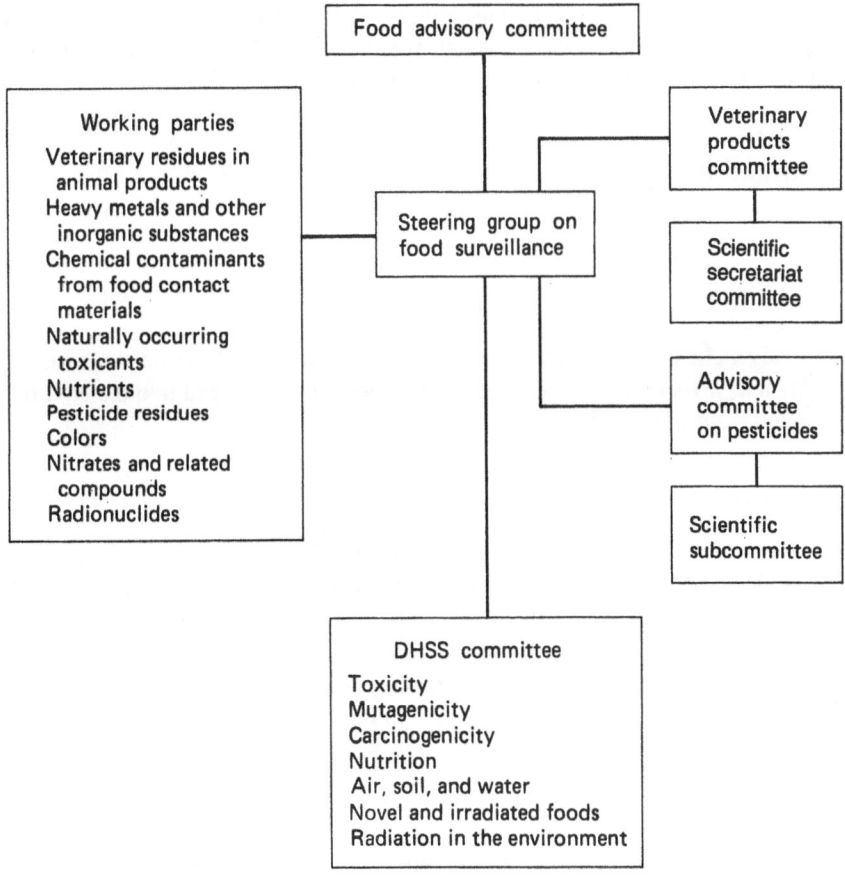

**Fig. 19.1.** United Kingdom organizational structure for assessing and managing food safety.

the magnitude of any risk from a chemical or radioactive agent will be related to the product of the level of exposure to that agent and the duration of the exposure.

## Additives

In the United Kingdom most classes of additives are permitted only after a case of need is established and a full toxicological assessment is made (an exception to this is flavors). Generally, therefore, the presence of an additive will confer some beneficial property on the food (e.g., prolong the shelf life). Since additives are intentionally added to food, the required toxicological assessment must be thorough and must demonstrate, as far as possible, that the chemical will not produce adverse health effects at the intended level of use. It follows that a carcinogen having a genotoxic effect on cells (i.e., damaging the DNA of the cell) is most unlikely to be acceptable as a food additive. Indeed, even the demonstration of possible nongenotoxic carcinogenic activity (as has been the case for saccharin) results in an intensive toxicological reassessment, including further experimental work and a review of the levels of use.

## Contaminants

Contaminants in food can be considered as two subclasses:

1. residues arising from the use of pesticides or veterinary drugs during food production; and
2. inadvertent environmental contaminants that enter the food unintentionally, such as mycotoxins, heavy metals, or chemicals discharged into the environment from industrial sites.

### Residues

Residues can be dealt with by exercising regulatory control on the type of pesticide or veterinary drug permitted for use and on the conditions under which it may be used in commodities or in animals that are intended to enter the food supply. Since these compounds will have been specifically designed to have physiologic effects at low concentrations, it is obviously important to have an adequate regulatory framework to control the conditions under which they might persist in food destined for human consumption and that those foods be regularly monitored.

### Inadvertent Environmental Contaminants

Inadvertent environmental contaminants such as mycotoxins and heavy metals, which are already in the environment, are more difficult to control, and action to reduce levels in food is likely to be more difficult. Either expensive environmental activities (which have rarely been found to be justified) will be required, or an upper "tolerable" level for the chemical in the food will have to be established.

## Radionuclides in Food

Radionuclides may be present in food. They may be present from natural sources, e.g., $^{210}$Po from Rn decay products or $^{40}$K, a constant fraction of natural K (Table 19.2); or they may be present as a result of discharges of radioactivity into the environment from industry, hospitals or research laboratories.

In the United Kingdom the discharges from a nuclear site are controlled by the authorizing departments (Department of the Environment and Ministry of Agriculture Fisheries and Food in England). Liquid or aerial site discharges may be transferred through the environment and result in exposures to the local populations from ingestion, inhalation, or by external irradiation. The extent to which this happens is carefully monitored by both the site operator and the authorizing departments. Site operators in the United Kingdom are required to carry out monitoring of the environment and local food produced near their site. The UK government also carries out check monitoring. In this way the impact of site discharges on population exposures is assessed. From the data, estimates of exposures of the local and distant populations to radionuclides from the discharges can be made (1). Estimates of possible maximal doses to the most exposed or critical groups are considered when discharge authorizations are decided. In addition to an overriding requirement to keep the doses below the annual permitted dose limits for members of the public, as recommended by the International Commission on Radiological Protection (ICRP), there is also a requirement to keep doses "as low as reasonably achievable" (the so-called ALARA principle), economic and social circumstances being taken into account.

Radionuclides are also in the environment from man-made sources (Table 19.3) and from nuclear weapons test fallout. These other sources of exposures to radioactivity should not be forgotten when considering exposures from food.

Levels of radioactivity in food from nuclear weapons testing was carefully

**Table 19.2.** Natural radionuclides found in food[a]

| $^{40}$K (22) |
| --- |
| Potassium is an essential component of the body and is under close homeostatic control. |
| The average concentration of K in the body is 2 g/kg body weight. |
| Potassium-40 emits beta and gamma rays. |
| The average concentration of $^{40}$K in the body is 60 Bq/kg. The highest concentration is in the bone marrow (130 Bq/kg). |
| The annual effective dose equivalent from $^{40}$K is about 180 μSv. The highest dose from $^{40}$K is to the bone marrow (annual effective dose equivalent about 270 μSv). |

[a] Reprinted with permission from ref. 22.

**Table 19.3.** Man-made alpha emitters in food[a]

| | μBq/g | | |
|---|---|---|---|
| | $^{210}$Po | $^{240}$Pu and $^{239}$Pu | $^{241}$Am |
| Plaice | | | |
| Muscle | 814–3,219 | 21.5[b] | 21.5[b] |
| Liver | 16,872–125,763 | 48.5[c] | 74.4[c] |

[a] Reprinted with permission from the Controller of Her Brittanic Majesty's Stationery Office from ref. 23, © British Crown copyright 1979.
[b] Whitehaven Cumbria, UK: representative figure in fillets (range 8.5–48.5).
[c] Windscale Cumbria, UK: representative figure in fillets (range 1.1–74.4).
The dose from $^{210}$Po is about 1% of the then current ICRP recommended dose limit for members of the public (5 mSv) and two orders of magnitude greater than that from $^{239,240}$Pu and $^{241}$Am.

monitored during the late 1950s and the 1960s in many countries. The levels in food were such that, in the United Kingdom at least, no action was felt to be necessary to control the food supply. The estimates of doses likely to be received by various members of the population were generally small in comparison to doses from natural background and were only a small proportion of the ICRP recommended annual dose limit for members of the public.

Before 1986 the only time that controls on food had been necessary in order to reduce exposures to radionuclides in the United Kingdom was following the Windscale Fire in 1957, when milk contaminated with radioiodine at levels above 3,700 Bq/kg was not permitted to be sold for human consumption (2).

The widespread contamination of western Europe by radionuclides following the Chernobyl nuclear accident has highlighted the desirability of a fuller consideration of the basis of controls on levels of radionuclides in food, especially in the context of controls on chemical contaminants in food. Most of the chapters that form the rest of this book outline the actions proposed or taken to control levels of radionuclides in food. This chapter attempts to put these actions into context by reviewing the criteria used to control the contamination of food with genotoxic carcinogenic chemicals and by comparing these with the principles that are beginning to emerge in the context of controlling levels of radionuclides in food.

## Stages in the Regulatory Decision-Making Process

When considering a particular exposure, the first step is to perform a toxicological evaluation or assessment of the agent to define the nature of the potential hazard as clearly as possible, and to obtain information on the likely shape of the dose-response curve. The second step is to assess the likely exposure by taking account of the levels of the agent likely to occur in food and the food consumption patterns of the exposed population. From these two sets of information some estimate of the magnitude of the risk and the size of the population likely to be exposed to that risk can be made. Finally, a decision whether or not any action is necessary and how that action should be implemented will have to be made. At that time the likely cost of any proposed or possible action will be a relevant factor in the decision-making process.

This chapter concentrates mainly on the toxicological and risk assessments of the regulatory process. Since it is generally accepted that exposure to ionizing radiation is associated with an increased risk of cancer and that this effect is mediated via a direct effect of radiation on cellular DNA, the toxic effect of greatest relevance when comparing chemical and radiation hazards is the risk of carcinogenicity. Therefore, this chapter concentrates on the management of genotoxic carcinogens. Possible effects on the germ cells are also an important aspect of the toxicological profile of such agents but will not be considered further in this chapter. There are two different conditions under which decisions need to be made; normal and accident situations. The normal or chronic situation is considered first, and then the possible additional factors relevant in an accident situation is briefly touched upon.

## The Toxicological Assessment of a Carcinogen

The first information needed is the possible health effects of the agent. Table 19.4 identifies the sources of data that may be available when assessing a genotoxic chemical carcinogen. Epidemiological data and lifetime animal feeding studies are generally the most relevant sources of information for the assessment of carcinogenicity, although mutagenicity tests, especially in vivo, are becoming an increasingly important part of the assessment package. The other tests are generally able to provide supporting evidence only, which may assist in the interpretation of the epidemiological or lifetime feeding studies.

Human epidemiological data are obviously the most appropriate and relevant data, but are rarely available and certainly not when a new chemical is being considered. There can also be problems associated with the interpretation of the results of any studies. The association between exposure and health effects can only be circumstantial and retrospective. Because of the long latency period

**Table 19.4.** Risk estimation database

Epidemiology
  Occupational
  Population

Animal long-term feeding studies

Mutagenicity tests

Animal short-term feeding studies

Other in vitro studies

Chemical structure and structure activity relationships

for cancers in humans, it is likely to be 10 to 30 or more years before the full effects of any exposures are detected. By that time so many other factors may have changed that assessment of the relevance to present-day exposures can be very difficult. Furthermore, in situations where there is no group of relatively highly exposed people, or no unexposed group, human epidemiology studies just may not be possible. In the chemical field, epidemiological studies have generally been most useful in the context of worker exposures, where exposures tend to be rather higher than for the general population. However, this is not always the case. For example, good quality studies on exposure to fluoride in the water supply have been possible, and studies of the wives of workers exposed to asbestos have revealed an unexpected health hazard. More recently, epidemiological studies around certain UK nuclear sites have revealed as yet unexplained excesses of childhood leukemias (3–7).

The situation is, perhaps more satisfactory for radiation exposures than for chemical exposures because there are a number of exposed groups that have been followed for a number of years, for example, Japanese atomic bomb survivors (8) patients with ankylosing spondylitis receiving radiotherapy (9) and patients receiving internal exposures to $^{224}$Ra as therapy for tuberculosis (10). Although most of the studies in the literature relate to exposures at higher levels than are likely to occur from environmental levels of radiation, at least the uncertainty due to species differences is not present.

It is important, however, not to minimize the other uncertainties still present in the risk quantification process. One should realize that there are different types of radiation (i.e., alpha, beta, gamma, neutron) with differing biological effects; that the radioactive element will frequently be present as part of a molecule with its own specific chemical characteristics and metabolic features within the body; that the penetration of the radiation emitted by different isotopes can vary widely; that routes of exposure are also important and that different radionuclides can have very variable patterns of behavior in the environment or the body. Therefore, extrapolating from one type of radiation or route of exposure, or from exposures to one type of radionuclide to predict effects from other radionuclides in food also involves many assumptions and uncertainties.

A further problem, common to chemical and radionuclide risk assessments, is the difficulty encountered in extrapolating from data obtained at high doses to possible effects at lower doses, since this involves making assumptions about the shape of the dose response curve below the lowest level where health effects were observed (11).

## Animal Feeding Studies

In the absence of human epidemiology data, lifetime animal feeding studies are frequently the only in vivo source of information on the carcinogenicity of a compound. For reasons of economics and convenience, such studies are usually carried out on small rodents such as rats and mice. To maximize the chances of observing an effect, it is usual to use higher doses than are likely to occur in human exposures. It is important to bear in mind the inevitable species differences. Ideally, when assessing a chemical, two lifetime studies are undertaken, each including males and females in equal numbers. Interpretation of such studies, even when well performed, is not always easy, and there is an extensive literature on this (12–14).

## Mutagenicity Data

These are useful in identifying a compound as a potentially genotoxic carcinogen, but are not usually helpful in quantifying the risk from human exposures.

## The Quantification of Risk

Expert assessment of well-conducted animal feeding studies is usually accepted as a sufficient basis for a decision as to whether or not a chemical is a human carcinogen. Extrapolating the effects of different carcinogens, as determined in animal feeding studies, to provide some assessment of the risks of exposures in man is much more difficult. The different size and life expectancy of the rodent compared with the human, together with the metabolic differences between different species, mean that direct extrapolation of risk as established in a rodent feeding study to risk in man is not possible. In addition, there is a need to guess the likely shape of the dose-response curve, since human exposures will usually be occurring at much lower doses than those used in animal feeding studies. Some experts (including our own committee members) take the view that for a genotoxic carcinogen, an assessment of potency is not possible on the basis of animal feeding studies. They, therefore, advise that unless there is clear evidence of metabolic differences between the test animal and man, the only sensible regulatory action is to keep exposures to genotoxic carcinogens as low as is feasible. This will include considering actions such as exploring the way that the contaminants enter the food and then developing technical procedures to reduce their levels whenever possible.

Others, however, believe that some attempt at quantifying risk from exposures to carcinogens is useful in identifying those that should be regulated first and in permitting the setting of levels at which certain contaminants in food (such as pesticides) can be tolerated. The US Environmental Protection Agency (USEPA) has tended to adopt this approach. Using available toxicological data, Travis et al. (15) recently reviewed the basis for 132 regulatory decisions made by USEPA. Their work is used in this chapter as the basis of an assessment of the levels of risk that have been found to be acceptable when controlling exposures to a number of chemicals by one major regulatory agency (i.e., USEPA) and for comparing the levels of risk considered acceptable within the chemical field with the levels of risk considered acceptable in the radiation protection field. To simplify the discussion, the validity of the criteria upon which the estimates of carcinogenic risk are based (which are somewhat controversial in some of the examples mentioned below) is not dealt with in any detail. Neither is any weight given to the likely mechanism of action of the carcinogens considered, since they have been chosen merely to illustrate the criteria for an acceptable level of risk.

When making risk assessments based on animal feeding studies, USEPA considered a number of models to develop the shape of the dose-response curve below the lowest dose at which effects were observed in the most sensitive test species. All of these models assume that there is no threshold for genotoxic carcinogens. (This is likely to overestimate any carcinogenic risk because, among other reasons, it is possible that some chemicals may be broken down at low exposures before they reach the target tissue.) The agency then used safety factors ranging from 100 to 2,000 to allow for interspecies differences and differing sensitivity in different members of the public. The calculations were conservatively worked from the 95% upper confidence limit of the dose-response line and back-extrapolated to zero. The USEPA then used the data to calculate lifetime risks of developing cancer from exposures to the test chemical (11).

One should not underestimate the uncertainties in such calculations. Different groups can come up with widely differing (by as much as several orders of magnitude) risk estimates using the same database but employing different assumptions. These models, however, are probably felt to overestimate the risks. The USEPA argues that it has to set priorities on some basis and that such an assessment is preferable to something more arbitrary. Table 19.5 lists some of the decisions in which USEPA has used quantitative risk assessment.

Travis and his colleagues (15) provide some interesting insights into the levels of risk that the USEPA has found "acceptable" in the past. Thus, of the 132 regulatory decisions reviewed by them:

1. An estimated lifetime risk of developing cancer of less than one in a million usually resulted in the compound being approved;
2. an estimated lifetime risk of more than one in 10,000 usually meant that the compound was not approved; and
3. at levels of risk between these two, USEPA usually took steps to reduce dietary exposure and to confirm that risks from other routes of exposure were also small.

In the context of this final group, USEPA has generally taken account of the benefits of the use of the compound and of the number of people likely to be exposed to the agent.

Based on the data (Table 19.5) and on certain other considerations, Travis et al. concluded that when the population was relatively small, USEPA felt that the tolerable risk was larger, especially if no alternative existed or if the cost of reducing exposures was very great (e.g., with the exposures from an elemental phosphorus factory where an individual lifetime risk of developing cancer of $1 \times 10^{-3}$ to a small local population was found to be acceptable).

The application of these general principles by USEPA can be illustrated by

**Table 19.5.** Compounds and estimated risks from their use[a]

| Chemical or radionuclide | Individual lifetime risk $1 \times 10^{-5}$ | Exposed population $1 \times 10^{6}$ | Annual[b] cancer cases |
|---|---|---|---|
| Aflatoxins | | | |
| Corn | 70 | 34 | 90 |
| Peanuts | 3 | 220 | 35 |
| Pesticides | | | |
| Amitraz—apples | 0.2 | 220 | 8 |
| —pears | 0.2 | 220 | 6 |
| Chlorobenzilate | 0.3 | 210 | 8 |
| Dioxane | 0.0000004 | — | — |
| Ethylene dibromide | | | |
| Food | 1 | 230 | 36 |
| Ethylene oxide | 200 | — | 60 |
| Formaldehyde foam-insulated | 5 | 1.8 | 1.3 |
| Decaffeinated coffee (methylene chloride) | 0.1 | 3.7 | 0.05 |
| Saccharin | 40 | 220 | 600–1200 |
| Radionuclides | | | |
| DOE facilities | 70 | 64 | 0.07 |
| Non-DOE facilities | 2 | — | 0.001 |
| Elemental phosphorus | 100 | 3 | 0.06 |
| $^{222}$Rn uranium | 1,000 | — | 3–6 |

[a] Reprinted with permission from ref. 15, copyright 1987, American Chemical Society.
[b] Estimated annual cancer cases according to model selected. DOE = US Department of Energy.

considering its actions in respect to the three pesticides chlorobenzilate, amitraz, and chlordane (16) (Table 19.6).

USEPA based its assessment of the carcinogenic potential of chlorobenzilate on effects in the liver of both male and female mice; the studies in rats being negative. The population exposure was felt to be tolerable here, but the exposures of the workers at a lifetime risk of $5 \times 10^{-5} - 2 \times 10^{-4}$ was less acceptable. As no alternative was available for citrus fruit, USEPA eventually permitted the use with appropriate labeling requirements.

Amitraz was requested for use on apples and pears. The USEPA considered that the single carcinogenicity study performed gave weak evidence of carcinogenicity; it therefore permitted a temporary 3-year registration (for pears for which no substitute compound was available) and required further studies to be submitted before permanent registration.

In the case of chlordane, USEPA considered that strong evidence for carcinogenicity was available based on liver carcinomas in both mouse and rat. The chemical was also known to accumulate in adipose tissue of humans. The quantitative risk assessment demonstrated a risk an order of magnitude greater

**Table 19.6.** Upper-bound risk estimates for population exposure to suspected carcinogenic pesticides[a]

| Pesticide | Population exposed | Upper-bound lifetime probability of cancer death due to exposure | Number of cancer deaths per year at upper bound |
|---|---|---|---|
| Chlorobenzilate | 220 Million | | |
| | Citrus consumption | $2 \times 10^{-6}$ | 7 |
| | Citrus applicatory | $4 \times 10^{-4} -$ $1 \times 10^{-3}$ | |
| Amitraz | 220 Million | | |
| | Apple consumption | $3 \times 10^{-6}$ | 8 |
| | Pear consumption | $2 \times 10^{-6}$ | 6 |
| Chlordane/Heptachlor | 220 Million | $2 \times 10^{-4b}$ | 500 |
| | | $5 \times 10^{-5c}$ | 150 |

[a] Reprinted with permission from ref. 16.
[b] Based on total tumors.
[c] Based on large carcinomas.

than that for the other two chemicals. In addition, the environmental impact was possibly greater. The agency therefore cancelled most uses except for the underground application of chlordane for termite control.

## Exposure Data

Implicit in the USEPA risk assessment procedures is some estimate of the likely exposures of the population to the chemical. Exposures are the product of the level of the chemical in food and the quantity of food affected and consumed. Estimates of exposure therefore depend on measurements or predictions of levels of contaminants in food and information on probable dietary intake.

Exposure estimates usually include some maximizing assumptions. It is frequently assumed, for example, that all of the food is contaminated at the maximum permitted level; however, in a situation where a regulatory limit is being complied with, it will generally be found that most of the food in circulation is, in fact, contaminated at a much lower level.

Exposure is usually equated with levels in food as sold. Preparation of food for consumption may involve washing, peeling, or removing inedible portions. Cooking can also result in losses due to either degradation of the chemical or leaching out into surrounding fluids. All of these may reduce levels of chemicals in food as consumed below the levels found "in the market." However, it is rare to take these factors into account in an initial assessment. They may be given some weight in a reassessment when it appears that exposures may be reaching a level of concern.

When estimating consumption, a further decision has to be made as to whether to use average or extreme consumption data. Within the United Kingdom we have generally used extreme consumption data when they are available.

Finally, one has to consider special subgroups of the population, such as neonates and the aged, who may exhibit special sensitivities to certain chemicals. In the United Kingdom we take account of the sensitivities of such population subgroups. However, we usually do not take account of subgroups of people who might experience idiosyncratic responses (such as hypersensitivity reactions). An exception to this is made, however, in the case of veterinary residues, where a more restrictive approach is generally felt to be necessary because it is not possible to identify or label affected food in the same way as is possible when a processed food contains a particular additive (such as tartrazine).

## Radionuclides in Food and Their Control

As in the chemical field, the important quantities when assessing the possible risk are the exposure data that can be used to calculate a risk.

In the radiation protection field a much clearer separation is made between what is acceptable in a normal or controlled situation and what is acceptable in an accidental or uncontrolled situation. Although such considerations must also apply in the chemical field, the criteria for decision-making in the context of a chemical accident are much less clearly defined. All the criteria mentioned in the previous sections on chemicals apply to "normal" or chronic exposure situations.

In the chemical field exposures from different chemicals are usually assessed separately, although occasionally tolerable levels for "total organochlorines" or for other groups of chemicals with similar modes of action are developed.

In the radiation field a rather different philosophy has developed in which exposures from different types of radiation—that is, by different radionuclides and via different routes—are converted into whole-body effective dose equivalents. These "effective dose equivalents" are then added to give a total "effective dose" that can then be equated with a "risk of fatal cancer and genetic effects," which has been developed from a consideration of all relevant available data on the effects of radiation.

This system, developed by the ICRP, enables risks from different routes of exposure, different radionuclides, and different sources to be considered on the same basis. However, it should be emphasized that in order to do this it is necessary to make many assumptions about exposures, sensitivities, and risks. The risk assessment procedure is really designed for planning purposes rather than as a means of assessing actual risks from specific exposures. When a particular population with clearly identified routes and types of exposures is being considered, it is frequently possible to use more accurate risk estimates from more specifically relevant data in the literature or from measurements of environmental materials. When this is possible, it should always be done.

The unit of biological effect from dose to which "risk" can be equated using available data on doses and health effects in the ICRP system is called the sievert (Sv). In a recent statement after its meeting at Como, Italy, ICRP has modified its advice with respect to the risk associated with exposures of members of the public (17). For the purpose of this chapter, we shall consider that exposure to 1 Sv carries with it a lifetime risk of fatal cancer of the order of $2 \times 10^{-2}$. A sievert is quite a large unit, and most environmental exposures (e.g., from background radiation) are around or below the millisievert (mSv) level, which therefore corresponds to a risk of the order of $2 \times 10^{-5}$.

The ICRP has recommended that exposures from artificial sources of radiation, excluding medical sources, should not exceed the *principal* dose limit of 1 mSv annual effective dose equivalent when averaged over a lifetime. The ICRP also recommends a *subsidiary* dose limit of 5 mSv annual effective dose equivalent for some years provided the average lifetime dose does not exceed 1 mSv per year.

Radionuclides (e.g., radioiodine) may be concentrated in individual organs (e.g., the thyroid), and in such cases there is also a need for a limit on doses to individual organs. Therefore, ICRP recommends an annual limit of 50 mSv for individual organs when necessary.

Recognizing that radiation is a genotoxic carcinogen and therefore assuming that there is no threshold below which no adverse effect will occur, ICRP also recommends that exposures should at all times be kept as low as reasonably achievable, taking economic and social factors into account, and that no process that results in exposures to radiation should be permitted unless the process can be justified. To control exposures it is necessary to translate these doses to the body into levels of radionuclides in the environment or in food.

Levels of radioactivity in a material are measured in becquerels (Bq)—1 Bq equaling one disintegration of a radioactive atom per second. A becquerel is a fairly small unit. Different radioisotopes emit different types of radiation of differing energies which will have different biological effects. The metabolic parameters (absorption, distribution, metabolism, and excretion factors) for the different radioisotopes will also differ. The number of becquerels from a given radionuclide in a particular physical or chemical form that will result in a millisievert of dose to the whole body will also vary according to whether the material is inhaled, ingested, or merely stays on a nearby surface irradiating a person externally. Absorbed radiation doses resulting from intakes of radionuclides by ingestion and inhalation or by external exposure can be calculated. These calculations, however, usually involve making a large number of assumptions about the metabolic fate of the radionuclides.

When controlling doses from an artificial source, ICRP recommends that population doses are calculated to average members of "critical groups," (i.e., extreme consumers if food is being considered) and that compliance with the annual recommended dose limits should be based on the doses calculated to these groups with extreme habits. This is similar to the practice in the chemical field described above.

## Acceptable Risks in the ICRP System

As mentioned above, ICRP recommends that exposures from artificial sources of radiation should be kept below a lifetime average of 1 mSv per year. This corresponds to a lifetime maximum possible exposure from artificial sources of radiation of approximately 70 mSv and corresponds to a lifetime risk of the order of $1.4 \times 10^{-3}$. However, it would be rare for a member of the public to achieve this level of risk because the other two principles in the ICRP system (justification of the activity and ALARA) mean that, generally, members of the public receive doses considerably less than the ICRP lifetime limit.

When comparing this level of risk with the levels mentioned in connection with the USEPA regulatory system, it is important to realize that USEPA is controlling in each case for a single chemical and taking no account of exposures from other chemicals; whereas the ICRP system is intended to keep total exposures from artificial radiation below the dose limit. Finally, the ICRP system relates to a *fatal* risk of cancer, whereas the USEPA risk, because it is based on cancer incidence in animal feeding studies, relates to a risk of *developing* cancer.

## Accident Situations

The management of an accident situation will involve additional considerations and may require the development of accident-specific criteria. ICRP has always clearly differentiated between the two situations in the field of nuclear safety and radiation protection (18,19) and suggests different criteria may be necessary when the source (i.e., the nuclear reactor) is out of control.

In the chemical field no generally accepted criteria for controlling exposures to chemicals in food after a chemical accident have been developed, although there are now European community regulations requiring chemical sites to develop accident plans. The lack of agreed guidelines for controlling contamination of food following a chemical accident probably in part results from the wide range of possible chemicals that could be involved. Decisions will, of necessity, have to be taken on an ad hoc basis. A further consideration is that although chemical accidents may have devastating local consequences, they rarely have the potential for the widespread contamination of the food chain that has been experienced in western Europe following Chernobyl. However, it is interesting to note that in the past, for example, following widespread contamination of corn in the United States with aflatoxins, the permissible level of this contaminant was raised temporarily and contaminated food was deluted with uncontaminated food to enable it to be used for human consumption (20).

Subsequent chapters deal with the bases of the management of radioactively contaminated food that have begun to be developed following Chernobyl, but they can be summarized briefly as follows: Most systems that have been discussed since Chernobyl have developed levels in food relating to a maximum whole-body dose of 5 mSv in the first year for populations outside the near-field.

Using the revised ICRP risk estimate, this corresponds to a risk of fatal cancer (which will be expressed over the subsequent lifetime of those exposed) of the order of $1 \times 10^{-4}$. When considering such estimates, one has to bear in mind that actual exposures and risks are likely to be considerably smaller because of the maximizing assumptions built into the process that estimates the risk. In the radiation field, following Chernobyl, one study of levels of radionuclides in consumed food as opposed to levels in food monitored (21) found that actual levels of exposure were between 10 and 100 times less than the levels at which the controls were designed to restrict the dose. This was probably because of the extreme assumptions used in predicting doses (e.g., assuming 100% of food consumed was contaminated at the maximum permissible level and not allowing for the fact that much of the food consumed in an area is not produced locally).

## Discussion

There are obvious similarities between the biological effects of exposure to radiation and exposure to genotoxic chemical carcinogens. The two disciplines have tended to follow separate paths in the past and to develop their own philosophies. The widespread contamination of food with radionuclides following the Chernobyl nuclear accident has meant that the two disciplines have been forced to try to reconcile their rather different approaches.

The problems that have had to be addressed in the two disciplines have been rather different and partly relate to the different emphases in the two fields. The time seems right for an attempt to establish a dialogue between the two fields so that each can learn from the other's special areas of expertise. This chapter is an attempt to get such a dialogue started and does not aim to do more than start discussion on the different problems and approaches being used.

It can be seen that the criteria for risk assessment in the two fields have developed along broadly similar pathways. However, there are some differences of emphasis largely due to the various types of data available for assessment in the two fields (Table 19.7). Thus, because of the general lack of human exposure data in the chemical field, there has been a reluctance to quantify risks. In the radiation field, where human epidemiological data have been available, there has been a greater emphasis on quantification of risk.

However, in both fields there is difficulty in dealing with predicting effects at low levels of exposure below those at which biological data are available. When dealing with genotoxic carcinogens, both fields have therefore tended to make the conservative assumption that there are no thresholds below which it can be assumed that there is no increased risk of carcinogenesis. Because of this, both fields have tended to try to keep exposures to direct-acting carcinogens as low as reasonably achievable. Although this concept has perhaps been more

**Table 19.7.** Summary of parameters considered in risk-assessment procedures

| Chemicals | Radionuclides |
|---|---|
| 1. Each chemical assessed separately (occasionally chemically related groups considered together). | Tendency to summate exposures from all routes and types of radiation. |
| 2. Risk estimated in terms of "lifetime risk of cancer." | Risk estimated in terms of "fatal risk of cancer per year." |
| 3. Human data usually sparse or absent. Risk estimates based on animal data. | Risk estimates for fatal cancer based on human exposure data at relatively high exposures. |
| 4. Exposure estimates usually make maximizing assumptions. | Exposure estimates usually make maximizing assumptions. |
| 5. High-dose chronic exposure data extrapolated to low-dose chronic situation. | High-dose acute or relatively acute exposure data used in many risk estimation procedures. |

clearly stressed in the radiation field, it is also a principle widely used in the chemical field.

On the other hand the radiation protection field has perhaps tended, when dealing with nuclear site discharges in particular, to push their extrapolations to the limit. It has not always stressed sufficiently the areas of uncertainties and the generalizations that are necessary in order to end up with the seemingly simple universal risk estimate for 1 Sv exposure to radiation. Against this one has to consider the greater uncertainties in the chemical field in trying to quantify human risk from rodent exposure studies.

In the context of controlling nuclear site discharges there has been a tendency to assume that all exposures from one radionuclide, or even from groups of radionuclides, can be treated as similar without always taking full account of the chemical form in which the radionuclide exists. More research has been undertaken into organic forms of radionuclides in recent years with some interesting and relevant results and has shown that this is an area that perhaps needs to be taken account of more in the future.

Because of the simplifying approach used in the radiation field, it has been possible to develop limits for exposures from all artificial sources and to develop an international approach to standard setting. The chemical field is more complex, mechanisms of action are more varied, and it is not possible to sum effects from different chemicals so easily. This is probably part of the reason for the more fragmented approach to regulation in the chemical field.

The levels of risk found acceptable in chronic or normal exposure situations differ somewhat in the two fields, however. The ICRP system recommends annual permitted dose limits from radiation (excluding background and medical exposures) of 1 mSv per year, equivalent to a risk of fatal cancer of the order of $2 \times 10^{-5}$ per year. (It should be mentioned that, in fact, exposure for most members of the public are likely to be only a small fraction of this level because of the concurrent need to keep exposure "as low as reasonably achievable.") In the chemical field, because of the use of lifetime animal feeding studies, risks are expressed in terms of lifetime risks of developing cancer, and risks found to be acceptable by USEPA were between $10^{-4}$ and $10^{-6}$ for members of the public.

However, the similarities in the underlying approaches, even if some of the assumptions used to reach the risk estimates differ, can be underlined by a consideration of the implicit value placed on a life in the two fields. The USEPA has calculated that it valued a life at about $2 million (15). The International Atomic Energy Agency has developed a suggested cost for reducing exposures by 1 Sv for use in planning nuclear sites and their levels of permitted discharges. They suggest a minimum value of $2,000, but also state that highly developed countries are likely to be willing to spend rather more than this value. If one assumes (as was done in one of the examples considered at the recent World Health Organization meeting on intervention levels) that the United States is prepared to spend ten times more than a developing country, then the cost of saving 1 Sv of exposure would be $20,000, and, using the ICRP risk estimate of $2 \times 10^{-2}$, this gives a value of around $1 million for a life saved. Given the many other uncertainties in such calculations, these two figures are broadly similar.

## References

1. Hunt GJ (1985) Radioactivity in surface and coastal waters of the British Isles. Aquatic Environmental Monitoring Report 14 MAFF HMSO
2. Dunster HJ, Howells H, Templeton WL (1958) District surveys following the Windscale incident October 1957. In: Proceedings of the 2nd UN International Conference on the Peaceful Use of Atomic Energy. Publication P/316 UK, pp 296–300
3. Black D (1984) Investigation of the possible increased incidence of cancer in West Cumbria. Report of the Independent Advisory Group. London, HMSO
4. Gardner MJ, Hall AJ, Downes S, Terrell JD (1987) Follow up study of children born elsewhere but attending schools in Seascale, West Cumbria (schools cohort). Br Med J 295:819–822.
5. Gardner MJ, Hall AJ, Downes S, Terrell JD (1987) Follow up study of children born to mothers resident in Seascale, West Cumbria (birth cohort). Br Med J 295:822–827
6. Forman D, Cook-Mozzafari P, Darby S, et al. (1987) Cancer near nuclear installations. Nature 329:499–505
7. Roman E, Beral V, Carpenter L, Watson A, Barton C, Ryder H, Aston DL (1987) Childhood leukaemia in the West Berkshire and Basingstoke and North Hampshire District Health Authorities in relation to nuclear establishments in the vicinity. Br Med J 294:597–602
8. Preston DL, Pierce DA (1987) The effect of changes in dosimetry on cancer mortality risk estimates in the atomic bomb survivors, RERF Technical Report 9.87
9. Smith PG, Doll R (1982) Mortality among patients with ankylosing spondylitis after a single treatment course with X-rays. Br Med J 284:449–460
10. Mays CW, Spiess H, Chmelevsky D, Kellerer A (1986) Bone sarcoma cumulative

tumour rates in patients injected with radium-224. In: Gossner W, Gerber GB, Hagen U, Luz A (eds) The Radiology of Radium and Thorotrast. Urban & Schwarzenberg, Munchen 1986, pp 27–31

11. Cothern CR, Coniglio WA, Marcus WL (1986) Estimating risk to human health: trichlorethylene in drinking water is used as an example. Environ Sci Tech 20(2):111–116

12. Peto R (1978) Carcinogenic Effects of Chronic Exposure to very low levels of Toxic Substances. Environ Health Perspect 22:155–159

13. Peto R, Roe FJC, et al. (1975) Cancer and aging in mice and men. Br J Cancer 32:411

14. Peto R, Gray R, Brantom P, Grasso P (1984) Nitrosamine concentration in 5120 rodents. In: O'Nieill IK, Von Borstal RC, Miller CT, Long J, Bartsch H (eds) N-Nitroso Compounds: Occurrence, Biological Effects and Relevance to Human Cancer: IARC Scientific Publication No 57. Lyon IARC

15. Travis CC, Richter SA, Crouch EAC, Wilson R, Klema ED (1987) Cancer risk management: A review of 132 federal regulatory decisions. Environ Sci Tech 21:415–420

16. Anderson EL, Carcinogen Assessment Group of USEPA (1983) Quantitative approaches in use to assess cancer risk. Risk Anal 3:277–295

17. International Commission on Radiological Protection (1987) Statement from the 1987 Como meeting of the International Commission on Radiological Protection: Supplement to Radiological Protection Bulletin 86

18. International Commission on Radiological Protection (1977) Recommendations of the International Commission on Radiological Protection. Vol 1, No. 3 Publication 26. Pergamon Press, Oxford

19. Annals of the International Commission on Radiological Protection (1984) Vol 14, No 2 Publication 40. Protection of the public in the event of a major radiation accident: Principles for planning. Pergamon Press, Oxford

20. Food Chemical News (1984) FDA discretion to set aflatoxin action levels upheld by court. Washington DC, March 5, 1984

21. Meeking G (1987) EEC Meeting on levels of radioactivity in food, Luxemburg, May 1987. European Commission Report

22. United Nations Scientific Committee on the Effects of Atomic Radiation (1982) Ionising radiation: Sources and biological effects. p 90

23. Pentreath RJ, Lovett MB, Harvey BR, Ibbett RD (1979) Alpha emitting nuclides in commercial fish species caught in the vicinity of Windscale, UK, and their radiological significance to man. IAEA-SM-237/1. In: The Biological Implications of Radionuclides released from Nuclear Industries, vol 2, IAEA

# Part VI
# Risk Management of Food and Water Supplies

# CHAPTER 20

# Procedures on Assessment and Measures of Safety for the Population After the Chernobyl Accident

## Y. A. Izrael[1] and V. N. Petrov[1]

The Chernobyl accident and consequent radioactive contamination of the environment, including soil surface, water, and food products raised a number of problems concerning measures and guides on assessment and provisions for safety of the population.

Detailed investigations into the radioactive contamination of the atmosphere, terrain, and water were conducted under the USSR State Committee for Hydrometeorology (Goskomhydromet) just after the Chernobyl accident.

The information obtained was immediately directed to the Governmental Commission and to all organizations concerned. Based on this information, conclusions were made and recommendations were adopted jointly with other ministries and agencies as to what measures should be taken to protect the population against radiation and to reduce radioactive material in water bodies and food products, as well as agricultural products.

The air survey carried out by the institutions under the Goskomhydromet and analysis of soil samples made it possible to clarify the dynamics of the radiation situation and to construct maps of soil contamination for significant isotopes over the USSR's European territories. These contaminants were $^{131}$I, $^{137}$Cs, $^{90}$Sr, and to some extent $^{239\,+\,240}$Pu (the deposition of plutonium was insignificant in the immediate vicinity of the accident).

Recent measurements indicate that the quantity of $^{137}$Cs deposited over the European USSR (close-in and remote fallout) was 1.0 MCi or about 12% of the $^{137}$Cs in the reactor at the time of the accident.

The Councils of Ministers of Republics (Russian Soviet Federated Socialist Republic, Ukranian, Belorussian), jointly with Goskomhydromet and the USSR Ministry of Health, recommended partial evacuation and provision of the population with uncontaminated food products to prevent internal and external radiation doses of the population from exceeding 5 rem in the first year.

Calculation of the potential radiation doses to the population from the radioactive fallout (taking into account the actual isotope composition) allowed the

[1] USSR State Committee for Hydrometeorology, 12 Pavlik Morozov, 123376, Moscow, USSR.

formulation of criteria for long-term residence of the population in the contaminated region. Consequently, criteria were developed for temporal and permanent evacuation of the population and restrictions on access to the affected territory.

Table 20.1 shows calculated external gamma-radiation doses from the terrain (ground) just outside the 30-km zone boundary for various sectors with the assumption that the gamma dose on May 10, 1986, at boundary was 1 mR/h.

With allowance for radionuclide migration in soils, the external gamma-dose from the terrain outside the 30-km zone boundary did not practically exceed 5 rad for the first year after the accident. In subsequent years the annual dose would be substantially lower.

The results of the air survey and soil analysis demonstrated that the effects of $^{90}$Sr and $^{239}$Pu are localized almost completely in the zone of evacuation. The area unfit for living and agriculture, which is to be restricted for a long time, is about 1,000 km$^2$.

Radioactive contamination of river water in the region of the accident and of the Kiev reservoir was first caused by radioactive jets of volatile and aerosol products and then (possibly) by radionuclides washed from the surface by precipitation (and in the river flood land by flood waters).

The calculated total amount of radioactive material in water was $2 \times 10^5$ Ci. The amount of $^{90}$Sr was less than $10^2$ Ci. Reduced precipitation in the nuclear power plant (NPP) zone (to preclude radioactive washout into rivers) was achieved through four Goskomhydromet aircraft used throughout May to modify weather conditions and to produce rain fallouts outside the zone.

Based on the experimental data and calculations, high radioactive concentrations in the Pripyat and Kiev water reservoir would be expected to result from runoff from precipitation in the area of the industrial site. This situation necessitated special measures to prevent such contaminated water supplies.

It was calculated that the case of reliable isolation of the industrial site (an area of about 10 km$^2$) and cooling pond, the radioactive runoff from the whole fallout area during a year would be substantially lower compared with the water radioactivity (by both total activity and individual isotopes) from primary atmosheric fallout. The radioactivity levels even during the flood would not exceed permissible values (subsequent measurement results confirmed these calculations).

**Table 20.1** External gamma-radiation dose from over the terrain for various sectors of the radioactive fallout (roentgens), assuming the Gamma-dose on May 10, 1986, at boundary was 1 mR/h

| Sector | Dose from the total radionuclides | | Dose from $^{137}$Cs | |
| --- | --- | --- | --- | --- |
| | From the second day to 1 year | From 1 year to 50 years | From the second day to 1 year | From 1 year to 50 years |
| North | 2.5 | 7–9 | 0.21–0.25 | 6–8 |
| South | 2.4 | 1.6–3.7 | 0.04–0.11 | 1.1.–3.1 |
| West | 2.3 | 2.0–6.0 | 0.05–0.18 | 1.6–5.2 |

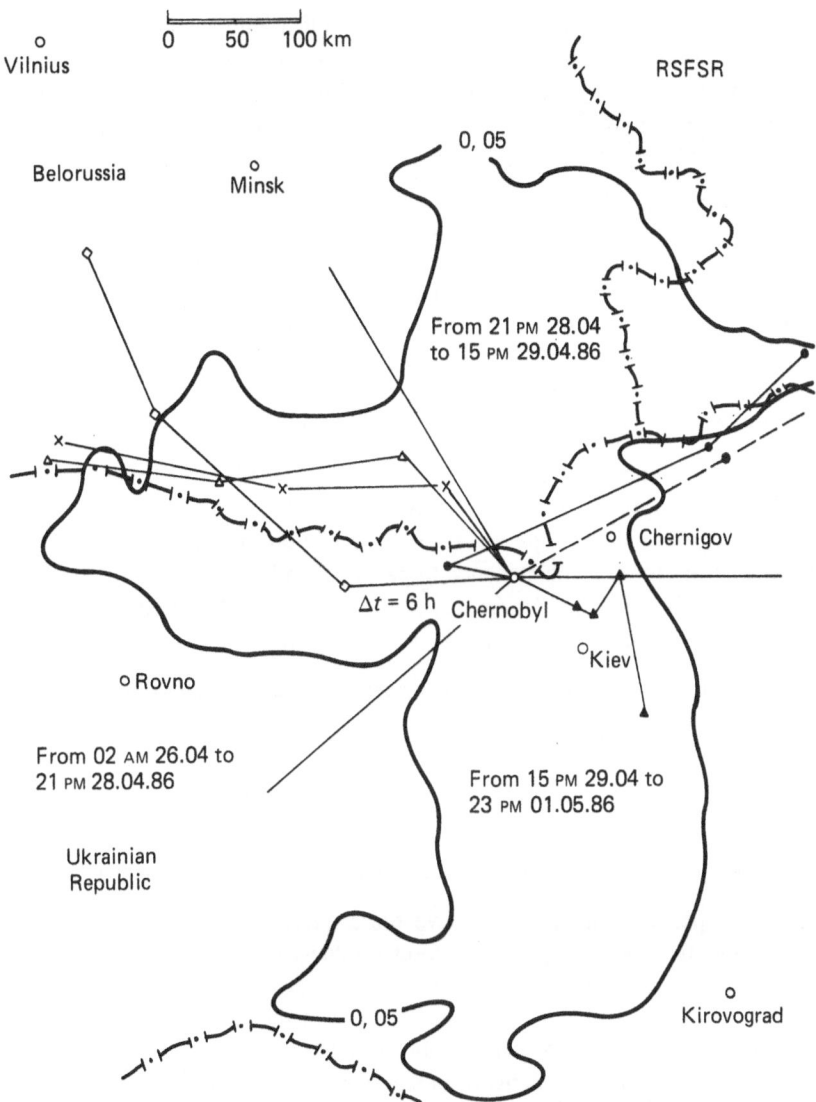

**Fig. 20.1.** Gamma-field distributing within the USSR territory according to the isolevel of dose rate 0.05 mR/h on June 10, 1986, and sectors of particle trajectory distributions at 925 mb level (×—×) from 3 AM, April 26, 1986; ◇—◇ from 15, PM, April 26, 1986; △—△ from 3 AM, April 27, 1986; ▶—◀ from 15 PM, April 27, 1986; ———— from 3 AM, April 29, 1986; ▲—▲ from 15 PM, April 29, 1986).

The meteorological peculiarities of the radioactive substances spread by air masses during the first few days resulted in separate patches of enhanced contamination of the terrain in the northeastern and western directions over the Ukranian SSR and Belorussian SSR (Fig. 20.1). The spotted nature of the terrain contamination is explained mostly by rain precipitation.

The area where the radiation level was 0.2 mR/h amounted to 200,000 km$^2$ (by May 5, 1986). Apparently the temperature regimen in the reactor promoted release of relatively volatile Cs which resulted in a unique (from the viewpoint of the isotopic composition) radioactive contamination of the terrain in zones where the $^{137}$Cs and $^{134}$Cs content was above 50% of the total activity.

Beginning with April 30, under conditions of steady northern winds (Fig. 20.1), increased radiation levels were observed in the southern direction. In accordance with calculations, increased doses were expected in some populated areas. Therefore, a thorough control of radioactivity in milk and other food products was introduced in different regions of the Ukranian SSR and Belorussian SSR.

Two peculiarities of the Chernobyl accident that significantly affected the contamination of agricultural products should be noted:

1. The radioactive release took place at the end of April through the beginning of May when main agricultural activities (sowing, planting) had been over in the region; winter crops had enough vegetation mass and cattle had been moved from in-door maintenance to pastures; and
2. The soil of the region predetermines a rather intensive radioactive incorporation in agricultural migration chains (first of all by root uptake).

Radioecologists were requested to answer the following questions:

1. What will be the levels of aerosol radioactive contamination of crops and will their use be possible?
2. Is it possible to obtain acceptable animal products in radioactively contaminated regions?
3. Is plant-growing acceptable on contaminated lands in subsequent years?

It was especially important to solve these questions since the contribution of radiation from locally produced food products to the total radiation dose can change at various stages of the postaccident period, and, on the whole, it is rather significant. That is why the decrease in the total population dose is significantly determined by the possible limitation of internal radiation dose from food consumption. A network of stations for controlling food and agricultural products was established.

Regular observations of radioactive concentrations in crops and animals were carried out mainly within the system of experimental farms where the monitoring of the most important agrometeorological parameters was provided, such as soil moisture and temperature, precipitation, vegetation phases, etc. (all the animals were moved from the evacuation zone).

A thorough analysis of the radiation situation with consideration of the isotope composition of the atmosphere and terrain contamination was carried out jointly with the USSR Ministry of Health, State Committee for Agriculture and Related Industries (Gosagroprom), and with participation of the USSR, Ukranian, and Belorussian Academies of Sciences. This made it possible to formulate recommendations on safe residence, food use, and farming in zones of increased radioactive contamination.

A system of monitoring of radioactivity in natural environment has been established and is in operation. A program on investigations into the radioactive contamination of the biosphere has been developed. Organizations of the USSR State Committee for Hydrometeorology and Control of Natural Environment (Goskomgidromet), the USSR Academy of Sciences, the USSR Gosagroprom, and other agencies take part in its realization.

The following procedures have been adopted (completely or partially) and developed for future use:

1. On the basis of operational aerogamma survey of the zones of contamination, a decision has been made to evacuate people from a 30-km zone as well as from some settlements outside this zone.
2. To prevent contamination of underground waters over the territory adjacent to the Chernobyl NPP site and the river Pripyat, the NPP site was surrounded by an antifiltration "wall" and screened by filtration wells for ground water sampling. The "wall" penetrated into the soil to 30 to 35 m and stretched up to 20 km in length.

Low dams have been constructed along contaminated river banks (first of all along the Pripyat River) and filtration weirs have been erected on some small rivers.

3. The investigation of the isotope composition in combination with total gamma-radiation over the contaminated terrain allowed formulation of criteria for population residence in these zones and for evacuation, as well as criteria for long-term control of entry into the zones.

Based on the total radiation doses of 5 rem for the first year and 50 rem for the whole life, which were authorized by the USSR Ministry of Health for the given accidental conditions, the following recommendations were developed:

1. A zone of complete alienation (unfit for residence and economy for a long time) is limited by a radiation isopleth of 20 mR/h on May 10, 1986.
2. Zones of temporal (but long-term) evacuation of population are limited by a radiation isopleth of 5 mR/h on May 10, 1986.
3. Zones of severe control (with temporal evacuation of children and pregnant women) are limited by a radiation isopleth of 3 mR/h on May 10, 1986.
4. Based on the isotope composition in different zones of radioactive contamination of terrain and the isotope behavior and its penetration into food products, proposals have been formulated jointly with the USSR Ministry of Health and Gosagroprom on the use of major agricultural products (including animal products): meat, milk, green vegetables, berries, fruit over the territory where the radiation level did not exceed 2.0 mR/h on May 10, 1986, and for potatoes and other root crops—5.0 mR/h on the same date.
5. The investigation of vast territory contaminated by $^{137}Cs$ and by other long-lived radionuclides along with the associated gamma-field allowed the prediction of doses of external and internal radiation to the population outside the zones of evaluation.

6. Continued activities for several years on localization and decontamination of soils (in a 1,000 m$^2$ zone of alienation) will be necessary in zones heavily contaminated by long-lived isotopes.
7. The monitoring of radioactive contamination of the environment will be carried out over a vast contaminated territory and related water systems.

CHAPTER 21

# Radioactivity in Food: Surveillance Procedures in the United Kingdom

## G. F. Meekings[1]

## Introduction

In the United Kingdom, the Ministry of Agriculture, Fisheries and Food (MAFF) is an authorizing department for routine disposals of low-level radioactive waste from the major civil nuclear establishments in England and Wales. In addition to requiring nuclear operators to undertake environmental monitoring around their establishments, including both marine and agricultural produce of local importance, validation monitoring in the terrestrial environment has recently been extended by MAFF into a fully independent and comprehensive surveillance program for agricultural food chains at each site. This Terrestrial Radioactivity Monitoring Program (TRAMP) and the procedure for calculating upper estimates of exposures are described and emergency monitoring requirements are reviewed in the light of experience following the Chernobyl accident.

Duplicate diet studies and their use as a complementary surveillance technique for validating assessments of intake based on conventional environmental monitoring results are discussed together with preliminary results of two recent studies.

## TRAMP Program

In the United Kingdom strict controls are exercised over authorized emissions of low-level radioactive waste, and discharges are carefully monitored at the source. Since active air streams are generally required to incorporate high-performance filtration or clean-up plant, the environmental and radiological impact of routine atmospheric discharges from most establishments is minimal.

Until recently, therefore, the Ministry undertook only limited check monitoring to satisfy itself of the reliability of the operators' measurements. A few years ago, however, the Ministry decided to set up its own independent agricultural surveillance program in the vicinity of the major nuclear sites to complement

[1] Ministry of Agriculture, Fisheries and Food, London SW1, UK.

**Fig. 21.1.** TRAMP sites in England and Wales.

the extensive aquatic monitoring program carried out by the Ministry's Fisheries Laboratories at Lowestoft. The locations of the sites covered are shown in Fig. 21.1, Table 21.1 lists the range of radionuclides covered by the program at the various types of establishment, and Table 21.2 lists the types of material currently analyzed.

The most valuable food commodity for monitoring purposes is milk. This is because the cow, grazing large areas of pasture each day, acts as a very efficient sampling machine. Milk therefore accounts for the largest number of samples in the program. The Milk Marketing Board undertakes centralized milk collections 365 days a year in the United Kingdom and is able to ensure prompt delivery

**Table 21.1.** Types of radionuclides covered[a]

| | |
|---|---|
| Nuclear power stations | Activation products and low-energy beta-emitters |
| Reprocessing (Sellafield) | Transuranic alpha-emitters, fission and activation products |
| Fuel preparation plants | Natural and enriched uranium |
| Amersham international sites | Wide range of specialist radiopharmaceutical products |

[a] TRAMP involves analysis of 17 radionuclides in total.

of samples from selected farms and dairies. However, not all radioactive species concentrate in or even reach milk, and the precise metabolic behavior in different animals determines which tissues need to be tested in order to provide adequate assurance in relation to various types of meat and meat products. Procurement of the correct tissues and organs clearly requires the assistance of veterinarians working directly with farmers to ensure that the precise origin of each animal is known.

Sampling and analysis of crops must also be properly targeted in the light of the relative importance of each radionuclide in the different crop types, including root crops, leafy vegetables, fruit, cereals, etc. Local crop and animal specialists, based in Ministry offices throughout the country, and in regular contact with the farming community, provide assistance with sample selection and procurement. Routine operation of the new program started on January 1, 1986, and the 1986 report will be the first in a regular annual series.

Most samples are subjected to high-resolution gamma-ray spectrometry, in which the principal nuclides routinely quantified are $^{134}$Cs, $^{137}$Cs, $^{103}$Ru, and $^{106}$Ru, although other gamma-emitting fission and activation products will also be detected if present. In addition, following radiochemical separation, $^{3}$H, $^{14}$C, and $^{35}$S are determined in appropriate samples by liquid scintillation counting; $^{90}$Sr, $^{99}$Tc, and $^{131}$I by gas-flow proportional counting; $^{125}$I and $^{129}$I by solid scintillation counting; and $^{239}$Pu, $^{240}$Pu, $^{241}$Am, $^{235}$U, $^{238}$U by alpha-spectrometry.

**Table 21.2.** Terrestrial radioactivity monitoring program (TRAMP)

| Materials routinely sampled | |
|---|---|
| Power station sites | Milk—weekly (2–12 farms)<br>Crops—in season (8–12 per year)[a] |
| British Nuclear Fuels, Ltd., Sellafield | Milk—weekly (13 farms, 1 dairy)<br>Crops—in season (12 per year)[a]<br>Animals—5 per year |
| Other BNFL sites | Milk—weekly (5 farms)<br>Grass/soil—28 per year<br>Animal feces—28 per year |
| Other sites | Milk—weekly (4–7 farms) |

[a] Crops will include fruit, vegetables, root crops, and cereals, depending on local availability, which may vary from year to year as well as from site to site.

## Dose Assessment

Regulatory control in the United Kingdom is based on the general dose limitation principles of the International Commission on Radiological Protection (ICRP) (1). So far as exposure of the general public is concerned, ICRP recommends the identification, for control purposes, of "critical groups" whose exposure may be considered to be representative of those individuals within the population most at risk.

In view of the fact that the major items of concern within this program are staple agricultural foodstuffs for which a substantial body of data about consumption patterns within the United Kingdom is available from the National Food Survey (2), a new approach to defining "critical group" consumption rates has been possible. This builds on work carried out over the years by the UK Committee on Toxicity who has taken the 97.5 percentile consumption rate within the consuming population as an appropriate benchmark for assessing the importance of dietary contaminants (in particular heavy metals). Work by Sherlock and Walters (3) has shown that where a sufficiently large national database exists, a generally well-characterized ratio will exist between the median consumption rate for consumers and various higher percentage rates (Fig. 21.2). The 97.5 percentile is close to three times the median. The 97.5 percentile rates for a range of dietary components in the United Kingdom are listed in Table 21.3.

The establishment of this new program turned out to be very fortunate, for being operational in January 1986 facilitated a switching of resources at very short notice in response to the Chernobyl accident to achieve a degree of national monitoring coverage almost instantaneously. A year earlier this would have been virtually impossible.

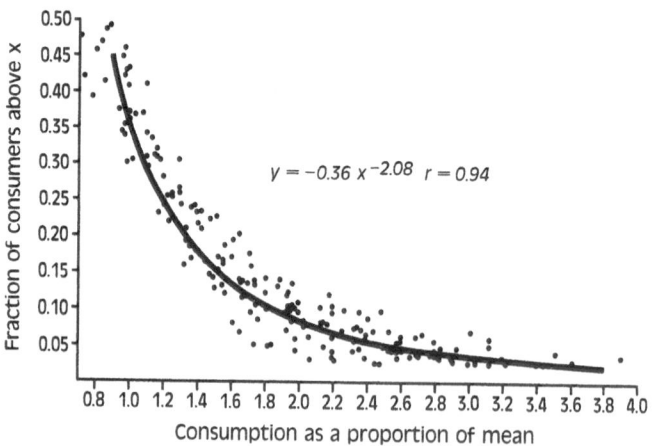

**Fig. 21.2.** Frequency distribution of food consumption. Data from 22 surveys covering 13 foods ($0.025 < y < 0.50$) (Reprinted with permission from ref. 3).

**Table 21.3.** Critical group
(97.5%) consumption rates for
selected foodstuffs[a]

| Foodstuff | kg/y |
|-----------|------|
| Milk | 365 |
| Beef | 38 |
| Pork | 43 |
| Lamb and mutton | 25 |
| Poultry | 28 |
| Leafy vegetables | 44 |
| Legumes | 29 |
| Potatoes | 140 |
| Soft fruit | 10 |
| Tree fruit | 39 |

[a] All rates are for adults, except
for milk, which is for 6-month-
old infants.

# Emergency Response Plans

The lessons learned after the Windscale accident in 1957 provide the basis for standing contingency plans within the Ministry that have been in existence, subject to periodic updating, ever since. The plans are exercised regularly in conjunction with nuclear site operators, the emergency services, and other local and central government agencies. The overriding objective is to ensure the safety of food produced in the area around the affected site by exercising any necessary controls so far as possible at the point of production. Should restrictions prove necessary, Ministers have powers under Part I of the Food and Environment Protection Act 1985 to designate areas within which activities such as the harvesting, distribution, or sale of contaminated produce may be prohibited for as long as necessary. This Act was used for the first time in June 1986 to introduce controls on sheep within certain areas in Cumbria and North Wales where relatively high deposition of radioactivity from Chernobyl had occurred.

In milk, following a single deposition event, both I and Cs levels rise quickly to a peak and then decline over the succeeding days or weeks. Figure 21.3 illustrates the temporal variation of $^{137}$Cs concentrations, which peak after approximately 5 to 7 days. In the case of I, this process is even more rapid. The peak level is reached as early as day 2, and by day 10, levels have fallen back to about 20% of the peak, and, more importantly, between 80% to 90% of the dose will already have been committed. If one is to intervene in any meaningful way in such a situation, therefore, one has to act very quickly. No significant delays can be tolerated in procuring samples, transporting them to the laboratory, undertaking analysis, reporting and interpreting results, nor, finally, can they be tolerated in making and implementing any appropriate decisions about the need for intervention.

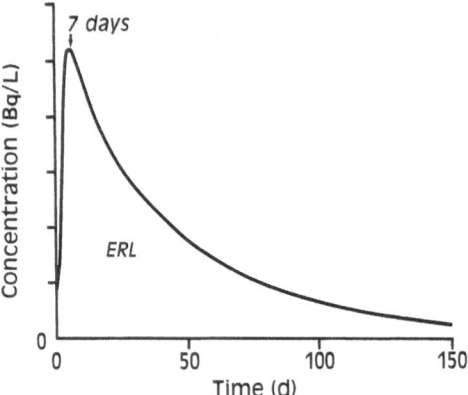

**Fig. 21.3.** The concentrations of [137]Cs in milk following a spike deposition to ground.

## Monitoring After Chernobyl

When news of the Russian accident reached the United Kingdom, daily testing of fresh milk was instituted from several bulk supplies and individual farms in southern and eastern counties of England. As soon as we started to detect radioactivity, we activated our monitoring network to obtain daily milk and vegetation samples from all parts of the country to determine as quickly as possible the scale and nature of the deposition and its implications for the national food supply. We had previously alerted the Meteorological Office to our intrerest in rainfall while the cloud was passing over the country and had arranged to receive regular rainfall summaries during this period. It soon became clear that deposition had been greatest in areas in the north and west where rainfall had been heaviest.

During the first few weeks after this fallout was detected, it proved possible, by hiring aircraft to fly samples to our laboratories from distant parts of the country, and by working staff round the clock (over what was a Bank Holiday weekend), to arrange for daily monitoring results to be available within 24 hours of taking the sample. In addition to milk, the other category of food for which it is important to have rapid information is fresh vegetables, which may be harvested and consumed very soon after the deposition has occurred. Fortunately, the range of field crops in the United Kingdom ready for harvest in early May last year was extremely limited—basically, spring greens, cauliflower, and a few others—and no restrictions for these crops were necessary. Such good fortune cannot be relied on, however, and it is essential that any emergency plan makes adequate provision for the rapid collection, preparation, and analysis of exposed fruit and vegetable samples.

Attention must also be given to a range of longer-term problems. Following Chernobyl, it was necessary to test all home-produced food crops throughout the year as they came into season. This included the whole range of field and

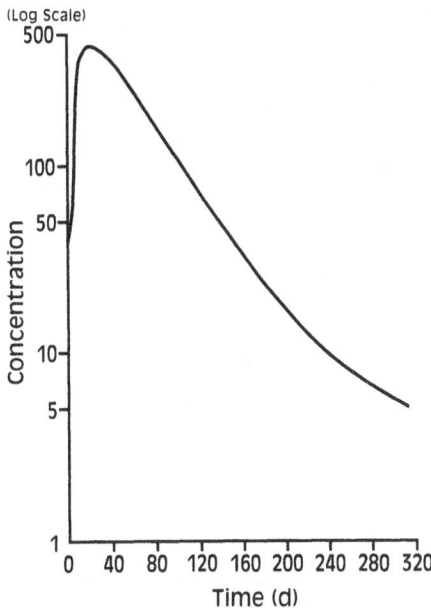

**Fig. 21.4.** Concentration of $^{137}$Cs in beef following a spike deposition to ground (Reprinted with permission from refs. 5 and 6).

glasshouse fruits and vegetables, wild fruits and berries, fungi, cereals, etc., as well as meat and animal products of all types. Comprehensive compilations of data from this program have already been published for 1986 (4) and data for 1987 is currently in press.

Beef and lamb are particularly important because the animals often graze freely outdoors. The MAFF operates a Soil, Plant, Animal Dynamic Evaluation (SPADE) model (5,6) which can predict the time profile of concentrations in plant and animal tissues following atmospheric deposition (Figs. 21.4 and 21.5). It is important to realize that in most circumstances, the time dependence of the radioactivity concentration in meat is "driven" by the dynamics of the activity in the pasture rather than in the animal, and Fig. 21.5 illustrates the direct dependence of results on grazing density following a single short-term deposition event. These models describe very closely what we observed in most parts of the United Kingdom following Chernobyl as they are based on typical lowland agricultural conditions. Unfortunately, however, the highest levels of deposition occurred in upland areas where ecological conditions are radically different. We, therefore, watched as levels in animals from these areas continued slowly rising well into August. In mid-June, anticipating a continuing upward trend, the government introduced precautionary controls on the movement and slaughter of sheep from these areas to ensure protection of consumers both at home and abroad. This experience serves to underline the fact that in an emergency, model predictions should never be used as an alternative to direct monitoring of the foods in question.

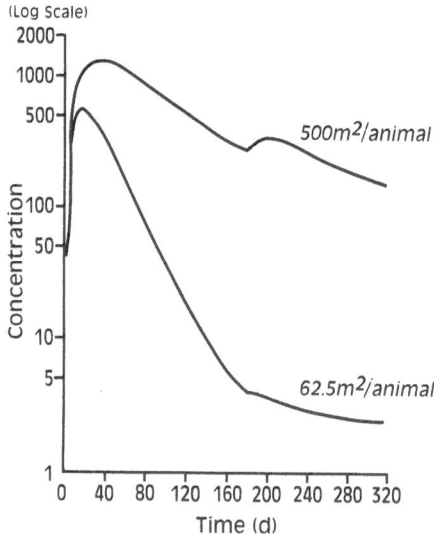

**Fig. 21.5.** Concentrations of $^{137}$Cs in lamb following a spike deposition to ground for animals at two grazing densities (Reprinted with permission from refs. 5 and 6).

## Live Monitoring

Our testing initially required animals to be slaughtered to obtain meat samples for laboratory analysis, and an expensive and laborious program was set up for this purpose. The most helpful single development in relation to food-chain monitoring following Chernobyl, however, was the bringing into service of a live-monitoring technique for upland sheep. Using large Na I detectors with portable single-channel scaler/rate meters, it has been possible to check live animals in field conditions at a rate of approximately one per minute. At this throughput, the limit of detection was in the region of 100 to 200 Bq/kg, total radiocesium, well below the interim action level of 1,000 Bq/kg.

In the case of an accident leading to such widespread geographical fallout as Chernobyl, no type of foodstuff can be left out of consideration. Among the other types of meat and animal products tested in the United Kingdom were venison, pigs, rabbits, hare, grouse, poultry, eggs, cheese, fish, goats (including dairy products), duck, pheasant, and many others, including migrating birds that might, in this case, have been bringing contamination into the country directly from the area close to Chernobyl itself.

## Duplicate Diet Studies

This technique is aimed at direct determination of radioactivity in the diet, and within the last 18 months the Ministry has conducted two such studies. The first was carried out in the vicinity of the Sellafield reprocessing plant in

**Table 21.4.** Measured intakes based on duplicate diet study

| Group | Total Cs (Bq/kg) | Weight of diet Average (kg/wk) | Intake (Bq/wk) Average | Intake (Bq/wk) Maximum |
|---|---|---|---|---|
| Adults | 8.5 | 6.9 | 58.7 | 135 |
| Children | 9.9 | 4.6 | 45.5 | 133 |

Cumbria to check that routine environmental monitoring programs were not overlooking any significant dietary exposure pathways. The second study was designed to quantify the contribution of Chernobyl-derived fallout to the diets of people in areas that were known to be high- and low-deposition areas from environmental monitoring results.

In both studies approximately 50 participants were selected by house-to-house visits and were interviewed to determine the amount of locally produced food consumed and to obtain three roughly equal groups of men and women and 3- to 5-year-old children. Each participant weighed and set aside a complete duplicate of every item of food eaten over a 7-day period and kept a diary itemizing the content of each meal, including a note of all foods known to be of local origin. The diets were then analyzed for a range of alpha-, beta-, and gamma-emitting radionuclides.

In view of the substantially enhanced levels of Chernobyl activity, it has proved possible to make a direct comparison between measured individual intakes and theoretical intakes calculated on the basis of conventional assumptions about dietary habits and local environmental monitoring data.

Actual intakes of $^{137}$Cs and $^{134}$Cs measured by the duplicate diet study in the high-deposition area are given in Table 21.4 for adults and children grouped separately. The average weekly intakes for adults and children were 59 Bq and 46 Bq, respectively, and the maximum values were 135 Bq and 133 Bq.

Table 21.5 records the national weekly consumption rates for major items of diet. The values given for a "typical" individual are those used by an expert group established within the European Community, while the "critical group" consumption rates correspond to those of extreme (97.5 percentile) consumers within the United Kingdom. These represent the range of values generally used for regulatory assessment. Table 21.6 lists "typical" total Cs levels recorded in the same range of foods produced within the local high-

**Table 21.5.** Theoretical adult consumption rates (kg/wk) used in regulatory assessments

| Foodstuff | Typical | Extreme |
|---|---|---|
| Dairy produce | 2.4 | 5.8 |
| Meat | 1.5 | 3.8 |
| Fruits and vegetables | 1.9 | 5.2 |
| Cereals | 1.9 | 4.0 |
| Total | 7.7 | 18.8 |

**Table 21.6.** Environmental monitoring data for high-deposition area and resultant estimated intakes

| Foodstuff | Bq/kg | Intakes (Bq/wk) | |
| --- | --- | --- | --- |
| | | Typical | Extreme |
| Dairy produce | 100 | 240 | 580 |
| Beef | 100 | 75 | 190 |
| Lamb | 1,000 | 750 | 1,900 |
| Fruits and vegetables | 50 | 95 | 260 |
| Cereals | 10 | 19 | 40 |
| Calculated total intake | | 1,179 | 2,970 |

deposition area during the period of the study (June 1986) and the corresponding weekly intakes that would be calculated for typical and extreme consumers, respectively.

The predicted intakes for adults were 1,179 Bq/wk and 2,970 Bq/wk for typical and extreme consumers, respectively. The ratios by which these values exceed the measured intakes from the duplicate-diet studies were therefore 20 : 1 and 22 : 1, respectively. When the results of direct measurement of radionuclide intake by duplicate diet studies are compared with estimates that would have been made by calculations based on known or assumed concentrations in environmental materials, the substantial and systematic overestimates of intake therefore illustrate the degree of conservatism built into the conventional assessment methodologies.

## Conclusion

Within the United Kingdom, the Ministry of Agriculture now undertakes a substantial routine monitoring program close to all the major civil nuclear establishments for foodstuffs of agricultural origin. Milk remains the single most important component of this program, but a whole range of other foods, including crops and animal products of all types, is also monitored to ensure that no significant pathways are overlooked. The precise composition of the program will, therefore, vary from site to site and from year to year. It is also necessary to target within any surveillance program those foodstuffs particularly relevant to the range of radionuclides likely to be encountered in the discharges from a given nuclear site. These same technical considerations are of equal importance in deciding the scope of any emergency food surveillance program. In addition, speed of response is of crucial importance in the procurement and analysis of fresh milk and of vegetable samples of harvestable quality. As a complement to both routine and emergency environmental food-chain surveillance, duplicate diet surveys are found to provide valuable direct confirmation of total actual dietary intakes by individual members of population groups of especial interest.

# References

1. International Commission on Radiological Protection (1977) ICRP Publication 26, Ann ICRP, 1, 3, Pergamon Press, Oxford
2. Ministry of Agriculture, Fisheries and Food (1987) Household food consumption and expenditure: 1985. HMSO, London
3. Sherlock J, Walters B (1983) Dietary intake of heavy metals and its estimation. Chem J Ind (London) 13:505–508
4. Ministry of Agriculture, Fisheries and Food (1987) Radionuclide levels in food, animals and agricultural products. HMSO, London
5. Jackson D, Coughtrey PJ, Crabtree DF (1985) Dynamic models for application to soil-plant-animal systems. Nucl Eur 4/1985, 29
6. Meekings GF, Walters B (1986) Dynamic models for radionuclide transport in agricultural ecosystems: Summary of results of a UK code comparison exercise. J Soc Radiol Protect 6,2:83–89

CHAPTER 22

# Methodology for Surveillance of the Food Chain as Conducted by the United States

C. R. Porter,[1] J. A. Broadway,[1] and B. Kahn[2]

## Introduction

The Environmental Radiation Ambient Monitoring System (ERAMS) was organized in 1973 by the US Environmental Protection Agency (EPA) to provide data on levels of radioactive pollutants for determining environmental trends and calculating the radiation dose and health risk on a national scale, to monitor pathways from major sources of population exposure, and to respond to accidental releases of radioactivity into the environment. This system was formed from the consolidation and redirection of separate monitoring networks that had been oriented primarily to measurements of fallout. Earlier networks were modifed by changing collection and analysis frequencies and sampling locations and by increasing the analyses for some specific radionuclides.

The ERAMS normally involves several thousand individual analyses per year on samples of airborne particles, precipitation, milk, and surface and drinking water. Samples are collected at about 268 locations in the United States and its territories, mainly by state and local health agencies (Table 22.1). These samples are forwarded to the Eastern Environmental Radiation Facility (EERF) in Montgomery, Alabama, for analyses (Table 22.2). The analyzed radionuclides are those that emit gamma rays, as well as tritium, $^{14}$C, $^{85}$Kr, $^{90}$Sr, and $^{89}$Sr, U, and Pu. The emphasis of the current system is on identifying trends in the accumulation of long-lived radionuclides in the environment. However, ERAMS, by design, is flexible and can provide short-term assessments of large-scale contaminating events such as industrial releases or fallout. Results are presented in *Environmental Radiation Data*, which is published quarterly by the EPA Office of Radiation Programs (ORP). This chapter describes the operation of and results from ERAMS with emphasis on monitoring radioactivity in food.

---

[1] US Environmental Protection Agency, Office of Radiation Programs, Eastern Environmental Radiation Facility, Montgomery, AL 36109 USA.
[2] Nuclear Engineering and Health Physics Programs, School of Mechanical Engineering, Georgia Institute of Technology, Atlanta, GA 30332 USA.

**Table 22.1.** ERAMS sampling stations

| ERAMS component | Number of stations | Type of sample | Routine sampling frequency | Alert sampling frequency |
|---|---|---|---|---|
| Airborne particles and precipitation | 67 | | | |
| Particles | | Filters from positive displacement air samplers | Filters are changed twice weekly | Changed once per day |
| Precipitation | | Precipitation | Collected as precipitation occurs and composited into single monthly sample | Composited into daily sample after precipitation |
| Pasteurized milk | 65 | Composite samples representing milk consumed in major population centers | One daily sample per month | Twice weekly |
| Drinking water | 78 | Grab samples from major population centers or selected nuclear facility environs | Quarterly | |
| Surface water | 58 | Grab samples downstream from nuclear facilities or from background sites | Quarterly | |

The ERAMS is only a part of the federal program for monitoring environmental radioactivity in the United States. The Food and Drug Administration (FDA) has a total-diet market-basket program for gamma-ray-emitting radionuclides, $^{90}$Sr, and $^3$H in foods collected near some nuclear power stations (1, 2). The Department of Energy (DOE), through its Environmental Measurements Laboratory, collects and analyzes worldwide deposition, airborne particles and gases in surface and high-altitude air, and food components in New York and San Francisco (3). In addition, each DOE nuclear facility and each commercial nuclear power plant operates an extensive environmental radioactivity monitoring program. Data from these programs are reported annually to the DOE or the Nuclear Regulatory Commission.

The ERAMS was developed from several programs undertaken by the US Public Health Service in the late 1950s and the 1960s to monitor radioactive fallout from atmospheric nuclear tests performed mostly by the United States and the USSR. The major programs that related to foods are listed in Table 22.3. The main conclusions from these early measures that guided the design of ERAMS were the following:

1. The most significant radionuclides from fallout, in regard to radiation dose, are short-lived [131]I (8-day half-life), intermediate-lived [89]Sr (50 days), and long-lived [90]Sr (29 years) and [137]Cs (30 years).
2. The most significant pathway for short-lived fallout radionuclides through ingestion is milk, and milk is also an important component of food for intake of intermediate- and long-lived radionuclides.
3. The radiation dose from fallout, to a certain extent, is predictable from measurements of radionuclide deposition, since this is the initial step in the movement of radionuclides into food, directly through exposed surfaces or indirectly through soil into roots, or through vegetation into food or dairy animals.

**Table 22.2.** ERAMS sample radiochemical analysis

| ERAMS component | Analysis | Analytical frequency |
|---|---|---|
| Airborne particles and precipitation | | |
| Particles | 5- and 29-hour GM field estimates | Each of twice weekly samples |
| | Gross beta- | Each of twice weekly samples |
| | gamma spectrometry | All samples showing $> 1.14$ Bq/m$^3$ gross beta |
| | $^{238}$PU, $^{239}$Pu, $^{234}$U, $^{235}$U, $^{238}$U | Semiannually on composite samples |
| Deposition | $^3$H | Monthly on composite sample |
| | Gross beta- | Monthly on composite sample |
| | gamma spectrometry | Monthly on composite samples showing $> 1.37$ Bq/L gross beta |
| | $^{238}$Pu, $^{239}$Pu, $^{234}$U, $^{235}$U, $^{238}$U | Annually on spring quarter composites |
| Air | $^{85}$Kr | Annually |
| Pasteurized milk | $^{131}$I, $^{140}$Ba, $^{137}$Cs, $^{40}$K | Monthly |
| | $^{89}$Sr, $^{90}$Sr, Ca | Annually on July samples |
| | $^{89}$Sr, $^{90}$Sr | January, April, and October intraregional composites of each of EPA's 11 regions |
| | $^3$H | Annually on April samples |
| | $^{14}$C | Intermittently |
| Drinking Water | $^3$H | Quarterly |
| | Gamma spectrometry | Annually on composite samples |
| | Gross alpha and beta | Annually on composite samples |
| | $^{226}$Ra | Annually on composite samples |
| | $^{228}$Ra | Annually on composite samples with $^{226}$Ra between 0.07–0.119 Bq/L |
| | $^{90}$Sr, $^{89}$Sr | Annually on composite samples |
| | $^{238}$Pu, $^{239}$Pu, $^{234}$U, $^{235}$U, $^{238}$U | Annually on composite samples with gross alpha $> 0.07$ Bq/L |
| | $^{131}$I | Annually on one individual sample |
| Surface water | $^3$H | Quarterly |
| | Gamma spectrometry | Annually on spring samples |

**Table 22.3.** Major programs prior to 1970 for measurement of radioactivity in foods within the United States

| | | |
|---|---|---|
| FDA | Radionuclides in diet for teenagers | May 1961–Nov 1965 |
| US Public Health Service | Radionuclides in institutional total diet samples | Jan 1961–Dec 1967 |
| Consumers Union | Selected results from total diet studies and radionuclide levels in teenage diets | May 1961–June 1964 |
| Atomic Energy Commission Health and Safety Laboratory | Tri-city study | Mar 1960–Dec 1967 |
| California State Department of Public Health | Estimated daily intake of radionuclides in California diets | Jan 1964–Dec 1967 |
| Connecticut State Department of Health | Estimated daily intake of radionuclides in Connecticut standard diet | Mar 1963–Dec 1967 |

Source: Rocklein et al. (14).

As numerous nuclear power plants began operating in the 1970s, it appeared that the sampling media and radionuclides considered for monitoring fallout would also be appropriate for monitoring these new potential sources of radionuclides.

Extensive research efforts have guided the monitoring programs to provide reliable sample collection, develop efficient methods to analyze numerous samples in a short period of time (4–8), indicate significant pathways and radionuclides, and develop calculational models for relating environmental measurements to radionuclide intakes and radiation doses. Comparisons have been made regarding filter retention efficiencies for airborne radionuclides (9–11), deposition collector retention (12), sampling of raw versus pasteurized milk (13), and sampling of food basket versus combined diet (14). The United Nations Scientific Committee on the Effects of Atomic Radiation (UNSCEAR), in periodic reports, has presented relatively simple models to predict intakes and doses for $^3$H, $^{14}$C, $^{89}$Sr, $^{90}$Sr, $^{131}$I, and $^{137}$Cs, and Pu on the basis of deposition in the current year and in previous years, that is, moving directly and via soil and roots into foods (10a,15,16). The UNSCEAR estimated that by 1980 the effective dose equivalent commitment worldwide due to fallout averaged 3.8 millisievert (mSv), of which 3.0 mSv was due to ingestion—mostly $^{14}$C, $^{90}$Sr, and $^{137}$Cs, with 0.7 mSv from external radiation and 0.1 mSv from inhalation.

Studies of the contributions by the various components of the diet to the total radionuclide ingestion indicate that in the seventies approximately one fourth of the total intake of $^{90}$Sr and $^{137}$Cs was from milk (16). The other major contributing categories during this period were grains and vegetables for both radionuclides, and, additionally, meat for $^{137}$Cs. During periods of fresh fallout in the early sixties, however, milk contributed as much as one half of total dietary $^{90}$Sr (17,18).

The combination of soil type and vegetation has been shown to be an important factor in the level of $^{137}$Cs in foodstuffs. As an example, unusually high levels of $^{137}$Cs in deer in the southeastern United States were shown to be related to a combination of soil type and vegetation endemic to the area (19). Studies of elevated or low radioactivity levels in milk relative to network averages consistently showed that correspondingly high or low levels in dairy herd feed were responsible for these levels (20–22).

The operation of ERAMS is described here to provide guidance for developing or expanding such a program. Obviously, extensive planning and operational efforts are required to arrange for collecting the many and varied samples, achieve prompt transportation to the central laboratory, control sample and data handling at the laboratory, perform consistent and reliable analyses, compute doses, and compile and distribute the information. Considerable effort must be devoted to maintaining continuity for all these aspects of a program that operates for many years. New nuclear tests or accidents that release radionuclides into the atmosphere require, in addition, organization for applying the available information concerning the source and environmental transport of radioactivity to expanding ERAMS activities for increased geographical coverage, sampling frequency, analytical effort, and information transfer. The results of these efforts are briefly described below for some selected examples, and the applicability of the component programs is discussed, together with possible improvements.

## Food-Chain Surveillance Program Under ERAMS

Emphasis of the ERAMS system is on identifying trends in the accumulation of long-lived radionuclides in the environment. However, ERAMS can also assess brief events.

Stations are widely dispersed throughout the United States and its territories, covering each geographical region, most individual states, and all major population centers. Many stations are located near major potential environmental release points.

During routine operation, samples are shipped by US regular mail; during alert or emergency conditions, samples are shipped by US Express Mail for overnight delivery. When the samples are received in the laboratory, they are entered into the ERAMS database. Detailed information on sample collection and the overall operation of ERAMS is available in the *ERAMS Manual* (23). Specific analytical procedures used for ERAMS analyses are published in the *Eastern Environmental Radiation Facility's Radiochemistry Procedures Manual* (24). All procedures currently in the manual have been developed and tested over time and cover all radionuclides and media addressed by ERAMS.

At least 10% of all samples are analyzed in blind duplicate as part of our

Intralaboratory Quality-Control Program. The laboratory also participates in quality-control programs operated by the EPA; the World Health Organization (WHO) through the International Reference Center, Paris, France; and the International Atomic Energy Agency.

Analytical results are stored automatically in the ERAMS database and are published quarterly in *Environmental Radiation Data,* which is distributed to all station operators, 50 State Radiation Control Program Directors, and other appropriate officials. When a radiological event occurs that results in an increase in sampling frequency, all data are verified and transmitted to ORP headquarters in Washington, DC, and to State Radiation Control Program Directors on the same day that results are obtained.

## Milk Program

The ERAMS milk program is a cooperative effort between ORP and the Lipid Products Branch of the FDA. It consists of 65 sampling locations (Fig. 22.1) that represent the population as indicated by the US Census Bureau. Each month each station collects a composite example that represents greater than 80% of the milk consumed within the respective population center. This composite sample is weighted by the daily fraction of total milk volume produced at each geographical location.

## Air and Precipitation Program

The ERAMS Air Sampling Program reflects the air particulate and precipitation exposure received by about 30% of the US population. It consists of 67 stations

**Fig. 22.1.** Pasteurized milk sampling sites.

**Fig. 22.2.** Air and precipitation sampling sites.

(Fig. 22.2) that twice weekly submit a sample of any precipitation that has occurred and an air particulate filter obtained from continuous sampling. Annually, dry compressed air (150 atm) samples (nominal 3-m volume at standard temperature and pressure) are purchased at 12 locations from commercial air suppliers and shipped to the EERF for [85]Kr analysis (Fig. 22.3). Table 22.2 details the various analyses that are performed on all ERAMS air and precipitation samples.

## Drinking Water Program

Four-liter grab samples are taken at 78 sites that represent the drinking water of major population centers (see (34) for details). The samples are finished water supplies and have no end-use treatment such as water softeners or filters. The station operator receives the collection supplies from the laboratory as notice to collect each quarterly sample.

## Surface Water Program

Surface water grab samples are collected quarterly at 58 locations. These samples are collected from sources where public water supplies might be affected by major nuclear facilties. Recently, continuous sequential samplers have been installed at five ERAMS locations that are downstream from major nuclear facilities.

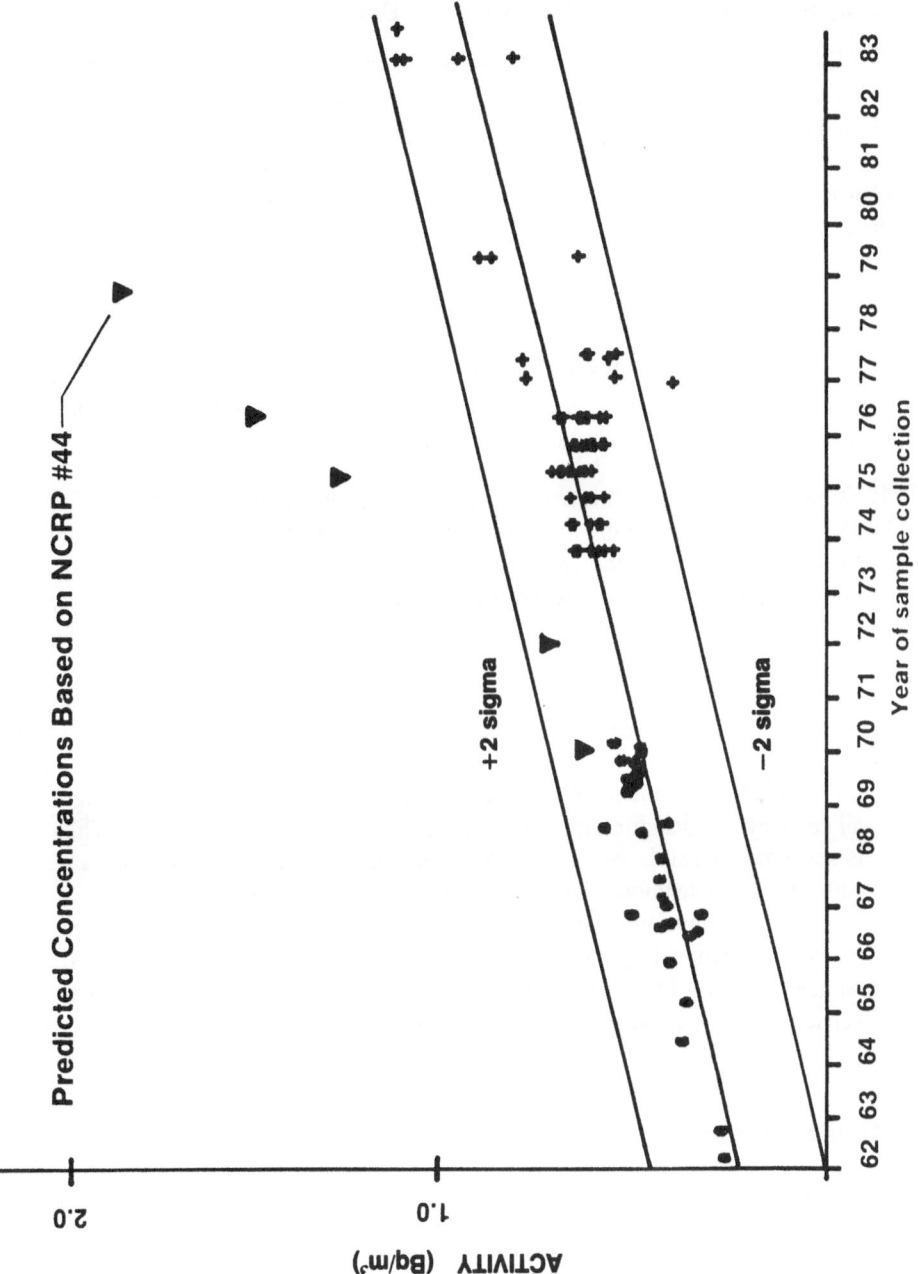

Fig. 22.3. Krypton-85 in air measurements in the United States from 1962 to 1983.

# Dosimetry and Risk Analysis From the ERAMS Database

## Structure and Function of the ERAMS Database

The ERAMS database provides EPA the opportunity to assess the hazards from technologically enhanced radiation levels (such as industrial operations that elevate environmental radiation levels) and short-term regional or global impact (such as waterborne unplanned release events and atmospheric fallout episodes). Concentrations are measured through human receptor pathways and, ultimately, dose and health impact are calculated.

### Steps Involved in Dose and Risk Assessment

The methodology for analyzing ERAMS data may be applied to assess short-term events or persistently elevated environmental concentrations of radionuclides (26). The steps in performing a dose assessment are as follows:

1. Define the assessment location(s), time interval, and sample media.
2. Determine concentrations for each sample type and location. This is done by removing background contributions from the measured value. Plots or color graphical displays may be produced to show the time dependency of the measured levels. These data are passed to the next step for the calculation of dose and risk values.
3. Calculate media concentrations integrated over time from the concentration profiles. These are used to estimate inhalation or ingestion of radionuclides by persons.
4. Calculate the movement of the radionuclides to human receptors through all pathways—inhalation, ingestion, submersion, and external exposure. This is achieved by using data on the time-integrated activity for each exposure pathway in conjunction with an environmental pathways model.
5. Apply dose-equivalent and risk factors. Dose-equivalent factors are based on the International Commission of Radiological Protection (ICRP) report number 30 (27). Dosimetry and risk factors are obtained from the current version of the RADRISK (28) computer code.

## Results Obtained

Since the inception of the nationwide monitoring program preceding ERAMS in the early 1960s, several radiological events of international significance have received special attention. These events originated with the early episodes of high fallout during the atmospheric weapons testing era of the 1950s and 1960s and continued through the monitoring of fallout from the Chernobyl reactor accident in April 1986 (Table 22.4).

**Table 22.4.** Summary of significant fallout episodes

| Occurrence | Collective dose equivalent (person-Sv) | Health risk estimates (committed fatalities) |
|---|---|---|
| Early 1960s fallout | NA | NA |
| Chinese fallout, September 1976 | 687[a] | 0.4[b] |
| Chinese fallout, September 1977 | 1,280[c] | |
| | 172[d] | 10[e] |
| Three Mile Island, Unit 2 | 50[f] | Negligible |
| Chernobyl | 3,300[g] | 2[h] |
| | 4,500[i] | 8[j] |
| | 8,500[k] | 20[l] |

NA = Dose analysis not available.

[a] Collective thyroid dose, milk ingestion only (34).

[b] Fatal thyroid cancers (34).

[c] Collective thyroid dose, milk ingestion, and inhalation (25).

[d] Collective effective dose equivalent (25).

[e] Fatalities based on all cancers (25).

[f] Estimated collective thyroid dose, milk ingestion and inhalation (24).

[g] Collective thyroid dose (24).

[h] Fatal thyroid cancers (24). *Note:* This value is proportionally less due to the reduced fatality risk of ingested $^{131}$I (see ref. 24).

[i] Collective dose equivalent (all organs) (May 1986–April 1987 exposure).

[j] Fatalities from all cancer, all organs (May 1986–April 1987 exposure).

[k] Collective dose equivalent due to $^{137}$Cs ingestion in milk for 45 years.

[l] Fatalities due to all cancers from $^{137}$Cs ingestion in milk.

## Fallout Following Atmospheric Weapons Tests in the Early 1960s

Atmospheric fallout of long-lived nuclides peaked sharply in the United States during the early 1960s. For example, $^{90}$Sr in total diet rose from 0.2 Bq/d in 1961 (29,30) to approximately 1.1 Bq/d in 1964 (1) and decreased to 0.3 Bq/d in 1982 (2). During the same interval, the US average milk concentration according to ERAMS decreased from a high of 1.1 Bq/L to 0.07 Bq/L (Fig. 22.4). Assuming an average daily milk consumption of 0.5 L/d, this corresponds to daily $^{90}$Sr consumption in milk decreasing from a high of 0.5 Bq/d to a low of 0.07 Bq/d. These data suggest that milk has contributed as much as 50% of the total dietary intake during periods of relatively high fallout to a low of 12% to 20% during periods when daily intake of $^{90}$Sr was low.

The UNSCEAR (16) reports estimates of average values of $^{90}$Sr in milk as a fraction of total diet during the 1960s and 1970s as measured in Denmark, Argentina, and New York. The above observations are generally consistent with that reported by Porter et al. (31) concerning dietary intake during a period of heavy fallout in 1963 and 1964. Based on a survey of dietary practices, their work proposed the use of an empirical factor to relate total dietary intake

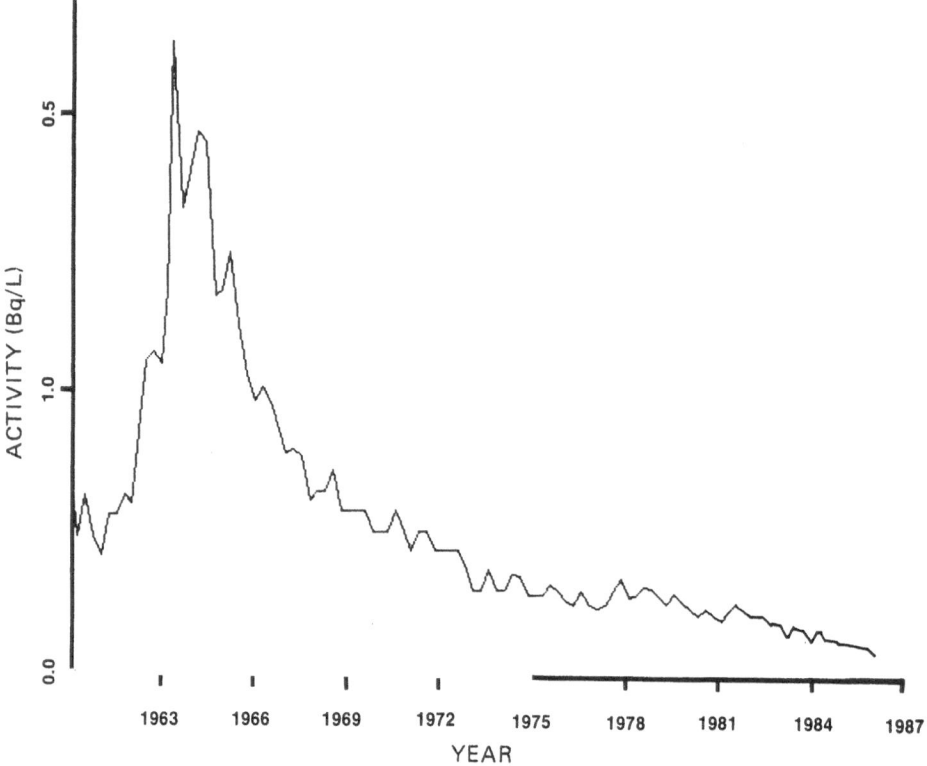

**Fig. 22.4** Strontium-90 in pasteurized milk: Network averages.

of $^{90}$Sr to intake via milk (Fig. 22.5). When converted to units of fractional intake using data on daily Ca intake in diet (32), Porter et al. (31) show a range of 20% to 60% of the dietary intake being contributed by milk. A summary of the sources of data referenced above indicates that between 0.2 and 0.4 of the current dietary $^{90}$Sr is received via milk ingestion.

Average values of $^{137}$Cs in milk within the United States have decreased from a peak of 6 Bq/L during 1964 to 0.1 Bq/L prior to the Chernobyl accident in April 1986 (Fig. 22.6). Following the accident, the average value increased to about 0.3 Bq/L and appears to be decreasing with an effective half-life of approximately 4 years. According to Cunningham (33), measurements during 1986 of food products of US origin showed no measurable increase in $^{137}$Cs following the Chernobyl accident.

## Contribution From Nuclear Power Plants

Although locations of ERAMS stations are not directed toward monitoring specific nuclear reactor sites, the stations are positioned to record ambient trends that

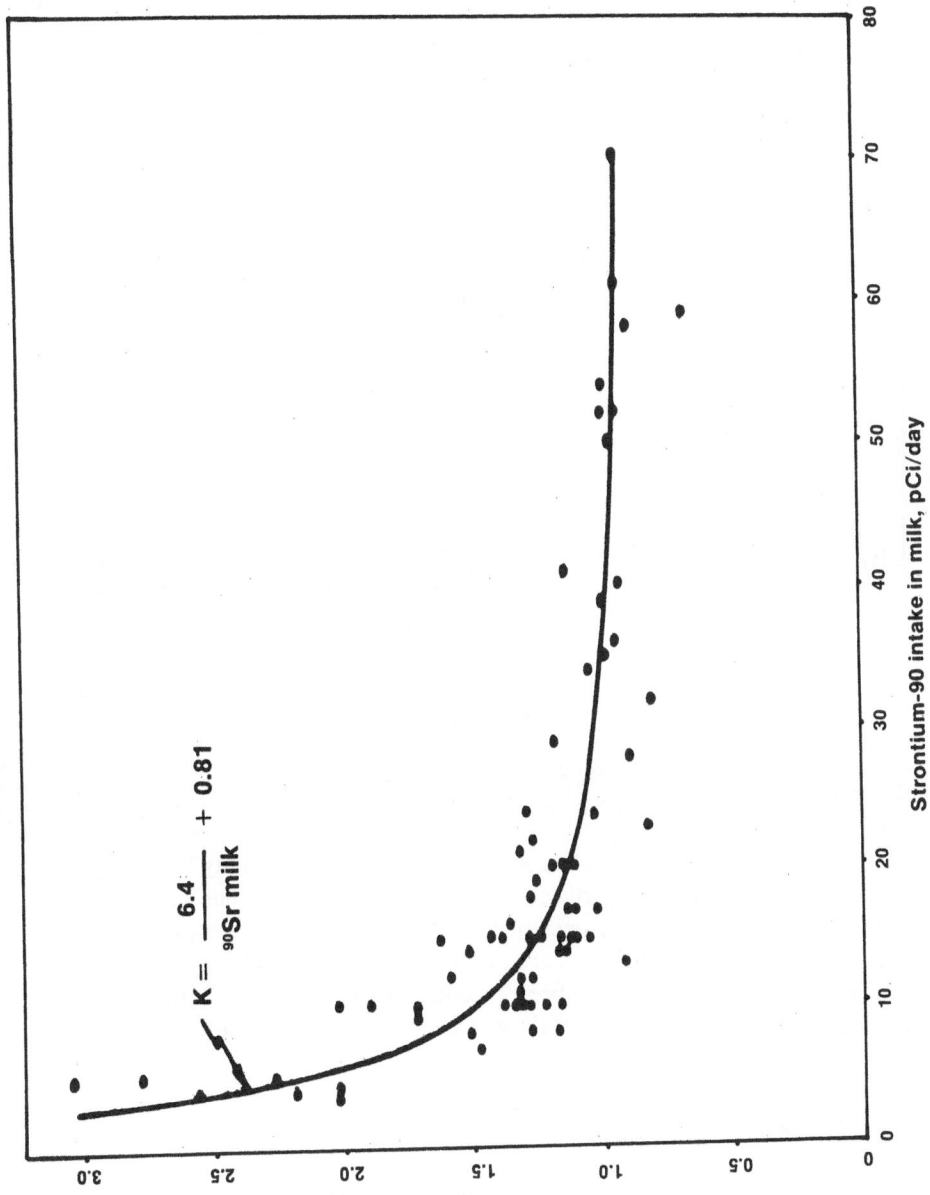

$$K = \frac{6.4}{^{90}\text{Sr milk}} + 0.81$$

**Coefficient K,** $^{90}$**Sr/Ca in diet** $\div$ $^{90}$**Sr/Ca in milk**

**Strontium-90 intake in milk, pCi/day**

**Fig. 22.5.** Ratio of $^{90}$Sr in total diet to $^{90}$Sr in milk.

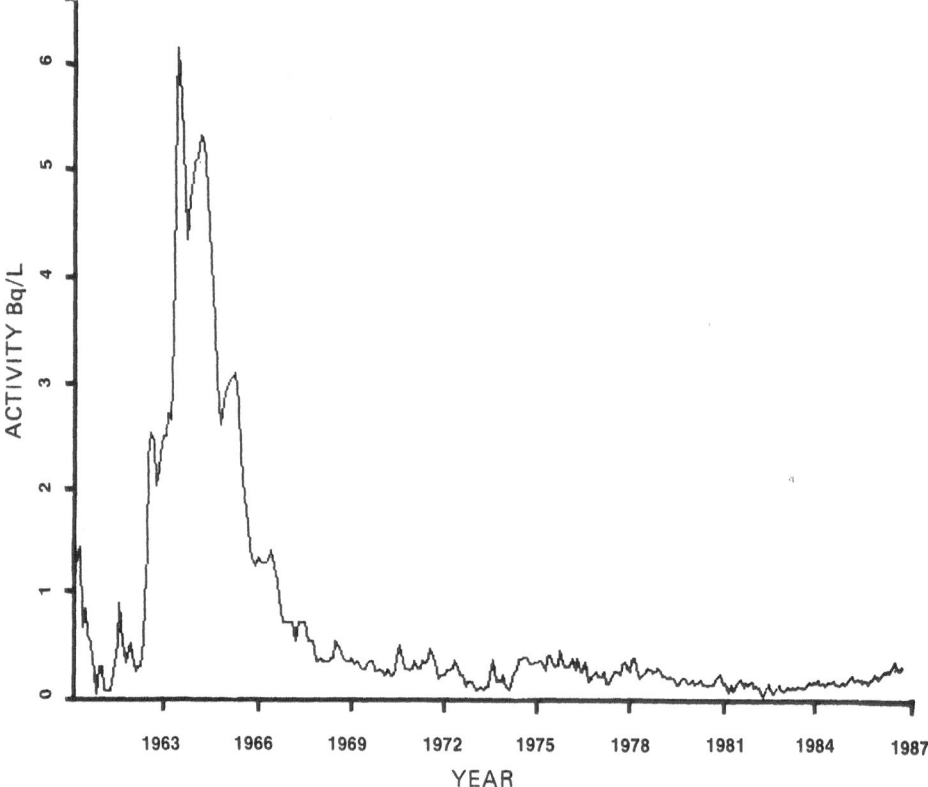

**Fig. 22.6.** Cesium-137 in pasteurized milk: Network averages.

would result from appreciable releases. No increases in radioactivity have been detected in milk or airborne particles from the operation of US nuclear power reactors during the history of operation of the US monitoring program (see discussion of the Three Mile Island accident below). However, elevated concentrations of $^3$H have consistently been recorded in surface water downstream from three nuclear facilities, and $^{60}$Co was found on one occasion (34).

## Chinese Fallout

Fallout from several atmospheric weapons tests by the People's Republic of China was reported in the United States. Following the events in 1976 (35) and 1977 (26) increases were measured both in airborne particulates and in pasteurized milk. For example, heavy rainfall in the midwestern United States during the 1977 episode resulted in relatively high concentrations of radioiodine in regional milk samples (Fig. 22.7). A summary of collective dose and health risk estimates is given in Table 22.4.

### Response to the Accident at Three Mile Island—Unit 2

On the day following the accident at the Three Mile Island Unit-2 nuclear power plant at Middletown, Pennsylvania, the ERAMS networks were fully activated, and analysis and reporting were accelerated to a 24-hour operation. No increased radioactivity was detected in any of the samples. On the basis of no detected activity at the eight milk collection sites within 300 km of the station and a population of 46 million in this region, the population dose was estimated to be less than 50 person-Sv.

In addition to the routine ERAMS stations referred to above, additional stations were established near Three Mile Island. This sampling program continues to the present time. A full report on this additional sampling program is now being prepared (36).

### Response to the Chernobyl Nuclear Plant Accident

Following the reactor accident at Chernobyl on April 26, 1986, the EPA augmented its sample collection by notifying the states and the air particulate station operators on April 29 to increase the sampling frequency from the usual twice weekly to once per day. All airborne particles and precipitation samples were sent by overnight mail to the laboratory for analysis. This increased sampling frequency was continued through the third week of June, when airborne activity returned to near background levels. The EPA also requested through the FDA that all milk sampling stations begin collecting milk samples twice per week. This augmented milk collection began during the first week of May and continued until the last week of June.

The first radioactivity measured by ERAMS sampling stations was at Olympia, Washington, on May 5 and at Bismarck, North Dakota, and Idaho Falls, Idaho, on May 6. These measurements are consistent with those reported by Larsen (37) and Juzdan et al. (38), who reported the first measurements in surface air particles in the United States on May 5. Preliminary estimates of hazard to members of the US population indicated protective action regarding milk and other fresh foodstuff would not be required. Individual doses and collective dose eqivalent commitment and health risks to the US population were estimated to be very low, as shown in Table 22.4 (25).

Because of the potential dose significance of Sr entering dietary components following the Chernobyl accident in April 1986, weekly milk samples were analyzed for Sr at each of 65 collection points within the United States. Presumably because of the relatively intense fallout near the accident, no $^{89}$Sr or $^{90}$Sr was detected at any ERAMS measurement point; therefore, the Chernobyl accident did not add to the present US ambient intake rates. The $^{131}$I milk ingestion

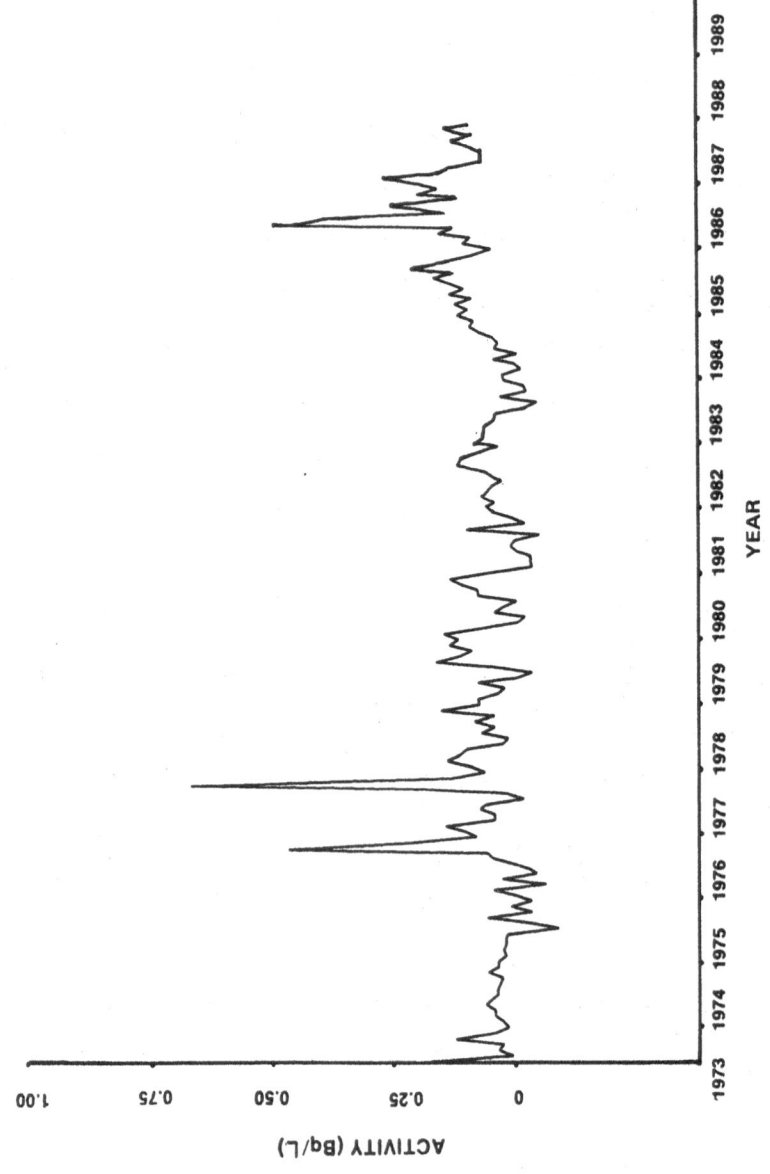

**Fig. 22.7.** Iodine-131 in pasteurized milk: Network averages.

**Bq - days/L**

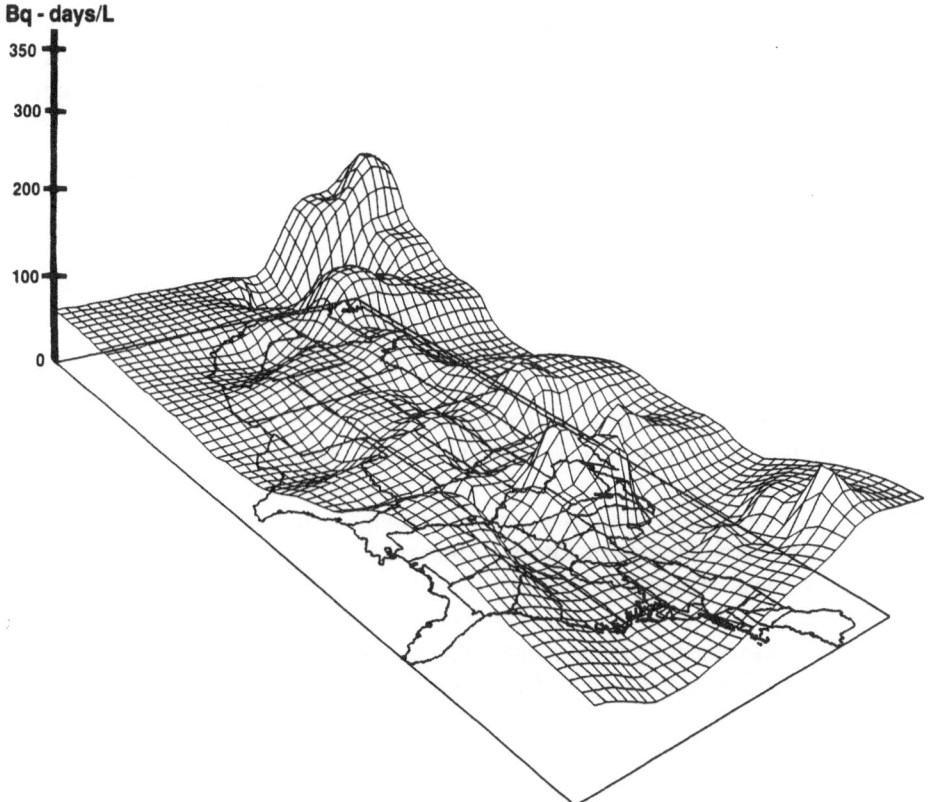

**Fig. 22.8.** Cumulative $^{137}$Cs in pasteurized milk: May 1986 to April 1987.

was estimated to be 2 GBq $^{131}$I during May and June 1986. Similarly, estimates were 1.5 GBq of $^{137}$Cs ingested during the same interval and 7 GBq of $^{137}$Cs ingested in milk for the 12-month interval May 1986 to April 1987. A display of the integrated $^{137}$Cs activity for this interval is shown in Fig. 22.8. The peak concentration of $^{137}$Cs in pasteurized milk was increased by 0.15 Bq/L in nationwide averages following the Chernobyl accident.

We were also able to compare our measured values of $^{131}$I in milk and deposition with the model given by UNSCEAR (16). Deposition of $^{131}$I during May–June 1986 was estimated to be 390 TBq or 51 Bq/m$^2$. For this time interval, the ratio of milk activity to deposition was found to be $8 \times 10^{-4}$ (Bq-a/L)/(Bq/m$^2$). This is quite comparable with the value of $6.3 \times 10^{-4}$ (Bq-a/L)/(Bq/m$^2$) given by UNSCEAR (16).

Similarly for $^{137}$Cs, we estimated total deposition in precipitation from May 1986 to April 1987 as 50 TBq or 6.2 Bq/m$^2$ in the United States. For this time interval the ratio of milk activity to deposition was calculated to be 10 (mBq-a/kg)/(Bq/m$^2$). Assuming the milk intake to be 0.23 of the total dietary intake (the average of values for Denmark and Argentina from UNSCEAR

1982), the total dietary activity is 40 $(mB2q\text{-}a/kg)/(Bq/m^2)$. This is considerably higher than the value of 9 $(mBq\text{-}a/kg)/(Bq/m^2)$ given by UNSCEAR (16). Much of the difference between the two values may be attributable to the possibility that milk may contribute a fraction of activity much larger than 0.23.

## Discussion and Conclusions

The ERAMS has effectively documented the gradual decrease of fallout radioactivity from the contamination in the early 1960s. It identified the major radionuclides and provided input to the calculations for the radiation doses. For public information, it indicated that radiation levels from fallout were well below radiation protection limits and were decreasing.

Possibly an even more important function is preparedness for monitoring radiation accidents that lead to environmental contamination. Only an ongoing program can respond to an emergency on extremely short notice. The ERAMS demonstrated at the time of Chinese Atmospheric nuclear tests and of the two nuclear power station accidents that it could mobilize efforts within a day to increase collection frequency, expand the collection network, and add analytical procedures to monitor specific sources. In these special situations, the monitoring program also provided the technical information about the movement and levels of radionuclides, as well as the public information to reassure that the dose potential was well below limits and was a known quantity, not a mysterious threat.

An extensive program such as ERAMS needs periodic reevaluation to assure that it has kept up with monitoring technology, changes in population and dietary patterns, and developments in understanding the movement of radionuclides through the environment. The experience gained in monitoring the Chernobyl release can be particularly useful for considerations such as assuring sufficient sampling locations around every nuclear facility and matching the desired limit for analytical sensitivity to an acceptable limit for dose determination.

The number of ERAMS stations and their locations have been determined from population data, as influenced by the availability of funds and the cooperation extended by environmental or health protection agencies in the individual states. It is believed, for example, that approximately 40% of the population was represented by the pasteurized milk samples in the early 1970s. A cursory survey now suggests that adding approximately 15 sampling locations would increase coverage significantly; any further expansion of the sampling network would be a balance between the desire for as full coverage as possible and the required cost and effort.

Analytical sensitivity for any monitoring program is strongly correlated with operational costs. Current analytical sensitivity of the ERAMS network has been shown to correspond to collective dose-equivalent commitments in the range of 100 to 1,000 person-Sv, which is well below the background contribution.

Monitoring airborne particles permits quantifying part of the inhalation dose and is generally the most sensitive indicator of radionuclide fallout; hence, it is a desirable component of environmental radioactivity monitoring. The above-cited UNSCEAR dose estimate indicates that inhalation is only a small contributor to the total dose over the long term, but this pathway can be significant during specific accidents.

Improved coverage of the inhalation pathway could be provided by collecting and analyzing shorter-lived gaseous radionuclides and by measuring the gamma rays from airborne radionuclides with continuously operating external radiation detectors. Nuclear power station operators collect gaseous radioiodine on a charcoal cartridge placed behind the filter for airborne particles and analyze it by gamma-ray spectrometry. They measure external radiation either with thermo-luminiscent dosimeters (TLDs) exposed to integrate doses over periods as long as 3 months, or with direct-reading ionization chambers that are extremely sensitive (to $10^{-8}$ Gy/h). The latter are particularly useful for detecting short-lived radioisotopes of Kr and Xe, which usually would be major constitutents of accidental airborne releases at nuclear power stations.

The ERAMS includes an external radiation monitoring program with TLDs. It has only 22 locations at this time, however, and these dosimeters do not permit distinguishing between radiation from radionuclides on the ground, vegetation, and other surfaces and radiation from airborne radionuclides. The detection sensitivity of the TLDs is limited by fluctuations in the natural radiation background to approximately $5 \times 10^{-5}$ Gy for the 3-month monitoring period.

Deposition (i.e., precipitation, together with dry fallout) samples appear to be the broadest predictor of radiation dose from fallout by ingestion. A useful development for ERAMS, therefore, appears to be utilization of the predictive models given in UNSCEAR reports to estimate the average radiation dose in the United States by the food pathway employing the results of gamma spectral and radoiochemical analysis.

The milk monitoring program quantifies an appreciable fraction of the ingestion dose over the long term and the major fraction when short-lived radionuclides are involved. An ongoing activity to confirm this status is needed and could consist of a comparison of the intake of radionuclides by persons who drink milk with the total intake in food determined in existing food monitoring programs. Furthermore, if the potential exists for accidental discharge of radionuclides that do not pass through the lactating system of cows into milk, arrangements need to be made for a standby food monitoring program.

The water supply monitoring program contributes some information concerning ingestion doses due to fallout, especially for tritium, which is a minor but not negligible source of the radiation dose. In case of an accident at a nuclear facility, these samples will be important for evaluating the dose potential of liquid releases and guiding countermeasures such as discontinuing water intake or instituting additional treatment. Sampling and analysis of surface waters support this effort.

Several activities have been directed toward improving the operation of the

ERAMS. A charcoal filter is being added to the particulate air samplers already in use for collection of gaseous iodine. This charcoal filter is to be used only when significant quantities of radioiodine are expected. Second, discussions have been conducted with compilers of data from operating nuclear power plants regarding the feasibility of including these data with the ERAMS. If this proves possible, it would not only extend the geographical coverage but also add measurements of vegetation and soil, which are not now measured by ERAMS. Third, use of sequential water samplers is being expanded at points where contamination of surface water is most likely. These sequential samplers replace quarterly grab samples to collect a more representative sample. If these samplers prove durable in the extremes of weather conditions, then all grab sampling of surface water will eventually be replaced by these units.

The ERAMS is, of course, only one national program among many that should be considered in planning an environmental radiation monitoring network. It has provided useful information, both technically and for public information, but requires ongoing evaluation for optimum efficacy. It functions most effectively when, on the one hand, its operators have available to them prompt information on sources of radioactive releases and their pathways, and, on the other hand, the public health agencies have given the public the needed guidance to relate the reported radionuclide and radiation dose levels to limits, control efforts, and effects.

# References

1. Simpson RE, Shuman FGD, Baratta EJ, Tanner JT (1981) Survey of radionuclides in foods, 1961–1977. Health Phys 40:529–435
2. Stroube WB, Jelinek CF, Baratta EJ (1985) Survey of radionuclides in foods 1978–1982. Health Phys 49:731–735
3. U.S. Department of Energy (1986) Environmental Measurements Laboratory annual report 1985. New York, EML-461
4. Porter CR, Cahill D, Schneider R, Robbins P, Perry W, Kahn B (1961) Determination of strontium-90 in milk by an ion exchange method. Anal Chem 33:1306–1308
5. Porter CR, Kahn B (1964) Improved determination of Sr-90 by an ion exchange method. Anal Chem 36:676–678
6. Porter CR, Carter MW (1965) Field method for rapid collection of iodine-131 from milk. Public Health Rep 80 (5):453–456
7. Porter CR, Carter MW, Kahn B, Pepper EW (1966) Rapid field method for the collection of radionuclides from milk. In: Snyder WS (ed) Proceedings of the First International Congress of Radiation Protection, September 1966. Rome, Italy, pp 339–346
8. Porter CR, Kahn B, Carter MW, Rehnberg GL, Pepper EW (1967) Determination of radiostrontium in food and other environmental samples. Environ Sci Tech 1:745–750
9. Lippman M (ed) (1962) Filter holders and filter media. In Air sampling instruments. American Conference of Government Industrial Hygienists, Cincinnati, Ohio
10. Lockhart LB Jr, Patterson RL Jr (1966) Intercalibration of some systems employed in monitoring fission products in the atmosphere. NRL Report No. 5850

10a. Mercer ER, Burton JD, Bartless BO (1963) Relationships between the deposition of Sr-90 and the contamination of milk in the United Kingdom. Nature 198:662–665

11. Schell WR, Strom D, Rosen J, Yusko J (1986) Fallout from Chernobyl in Western Pennsylvania. Annual Health Conference, Pennsylvania Public Health Association, State College, Pennsylvania, October 9–10

12. Eisenbud M, Harley J (1956) Radiation fallout. Science 124:251–255

13. Robinson PB (1968) A comparison of results between the Public Health Service raw milk and pasteurized milk networks for January 1964 through June 1966. Radiol Health Data 9:475–488

14. Rocklein PD, Smedley CE, Simpson RE (1970) Strontium-90 and cesium-137 in total diet samples: a comparative study of data. Radiol Health Data 11:47–62

15. United Nations Scientific Committee on the Effects of Atomic Radiation (1977) Report to the general assembly, annex C. United Nations, New York, pp 141–146

16. United Nations Scientific Committee on the Effects of Atomic Radiation (1982) Report to the general assembly, annex E. United Nations, New York, pp 220–222

17. Fisher HL, Coleman JR, Grundy RD (1966) Strontium-90 concentrations observed in U.S. pasteurized milk compared with strontiumn-90 levels in precipitation, in total deposition and in soil 1962–1963. Radiol Health Data 7:427–439

18. Grundy RD (1967) Strontium-90 dietary intake estimate based on fractional intakes due to milk. Radiol Health Data Rep 8:73–77

19. Cummings SL, Fendley TT, Jenkins JH, Bankert L, Bedrosian PH, Porter CR (1971) Cs-137 in whitetailed deer as related to vegetation and soils of the southeastern United States. In: Nelson DJ (ed) Proceedings of the Third National Symposium of Radioecology, May 10–12, Oak Ridge, Tennessee CONF-710501-P1, pp 123–128

20. Kahn B, Straub CP, Jones IR (1962) Radioiodine in milk of cows consuming stored feed and of cows on pasture. Science 138:1334–1335

21. Kahn B, Jones IR, Porter CR, Straub CP (1965) Transfer of radiostrontium from cows feed to milk. J Dairy Sci 48:1023

22. Porter CR, Phillips CR, Carter MW, Kahn B (1966) The cause of relatively high Cs-137 concentrations in Tampa, Florida milk. In: Aberg B, Hungate FP (eds) Radioecological concentration processes: proceedings of an international symposium, April 1966. Stockholm, Sweden, pp 95–101

23. U.S. Environmental Protection Agency (1987) ERAMS manual (3 vols). Office of Radiation Programs, EPA 520/5-84-006, 007, and 008

24. U.S. Environmental Protection Agency (1984) Eastern Environmental Radiation Facility Radiochemistry Procedures Manual. Office of Radiation Programs, EPA 520/5-84-006

25. Broadway JA, Smith JM, Norwood DL, Porter CR (1987) Estimates of radiation dose and health risk to the United States population following the Chernobyl nuclear plant accident. Health Phys, in press

26. Smith JM, Norwood DL, Strong AB, Broadway JA (1982) Assessment of fallout in the United States from the atmospheric nuclear test by the People's Republic of China on September 17, 1977. Environmental Protection Agency, EPA 520/5-82-008

27. International Commission on Radiological Protection (1981) Limits for intakes of radionuclides by workers. ICRP Publication 30, Ann ICRP, Vol. 5, Oxford

28. Dunning DE, Leggett RN, Yalcintas MG (1980) A combined methodology for

estimating dose rates and health effects for exposure to radioactive pollutants. Oak Ridge National Laboratory, Oak Ridge, Tennessee, ORNL/TM-7105

29. U.S. Public Health Service Division of Radiological Health (1962) Institutional diet sampling program. Radiol Health Data 2:42–45

30. Setter LR, Smith D, Spector M (1966) Cesium-137 in food—a summary of results on selected foods in the United States, July 1962 to October 1963. Radiol Health Data 7:145–156

31. Porter CR, Pepper EW, Kahn, B (1967) Comparison of the radiostrontium/calcium ratio in milk and in the total diet of children. Presented at American Chemical Society meeting, April 11, 1967

32. Pennington JAT, Young BE, Wilson DB, Johnson RD, Vanderveen JE (1986) Mineral content of foods and total diets: the selected minerals in foods survey, 1982 to 1984. J Am Diet Assoc 86:876–891

33. Cunningham WC, Stroube WB Jr, Baratta EJ (1987) Radionuclides in foods 1983–1986. Health Phys, in press

34. Broadway JA, Mardis HM (1983) Analytical capability of the environmental radiation ambient monitoring system. US Environmental Protection Agency. EPA 520/5-83-024

35. Smith JM, Broadway JA, Strong AB (1978) United States population dose estimates for iodine in the thyroid after the Chinese atmospheric nuclear weapons tests. Science 200:44–46

36. Kirk W (1987) Radiation monitoring in the environment of Three Mile Island Unit 2. US Environmental Protection Agency, in press

37. Larson RJ, Sanderson CG, Rivera W, Zamichelli M (1987) The characterization of radionuclides in North American and Hawaiian surface air and deposition following the Chernobyl accident. Environmental Measurement Laboratory, New York, EML-460, pp 1–104

38. Juzdan ZR, Helfer JK, Miller KM, Rivera W, Sanderson CG, Silvestri S (1987) Deposition of radionuclides in the northern hemisphere following the Chernobyl accident, Environmental Measurements Laboratory, New York, EML-460, pp 105–154

CHAPTER 23

# Identification and Reliability of Parameters for the Assessment of Derived Intervention Levels for Control of Contaminated Foodstuffs

A. Kaul[1]

## Introduction

The contamination of agricultural products from the accidental release of radioactive substances—and therefore the contamination of foodstuff—may result in an unacceptable radiation risk for members of the public from the consumption of such food products. To limit this risk, the International Commission on Radiological Protection (ICRP) set forth the following principles (1):

- avoidance of serious nonstochastic effects,
- introduction of unequivocally beneficial countermeasures to limit stochastic effects, and
- reduction of the collective dose equivalent to limit stochastic effects for members of the public as far as reasonably practicable.

Additionally, intervention levels of dose are to be established, which when exceeded, should or must result in the discontinued consumption of contaminated food products. The lower limit, within the estimable range for the introduction of respective administrative measures, such as the prohibited sale of food products, is an effective dose equivalent of 5 mSv during the first year after an accidental exposure, as recommended by ICRP (1) and internationally recognized by the World Health Organization (WHO), Food and Agriculture Organization (FAO), International Atomic Energy Agency (IAEA), Organization for Economic Cooperation and Development (OECD), and Commission of the European Communities (CEC). The upper limit, beyond which administrative measures are to be unequivocally introduced, is 50 mSv, as recommended by ICRP (1). The related organ dose values for organ-specific radionuclide deposits are 50 and 500 mSv.

To guarantee compliance with these intervention levels of organ or effective dose equivalent, corresponding lower and upper levels must be assessed for the annual rate of activity intake from foods, and upper and lower intervention levels of the specific activity must be derived from these levels by using average

[1] Institute for Radiation Hygiene, Federal Health Office, Neuherberg/Munich, FRG.

consumption rates of individual food products. The intervention levels of annual rates of activity intake from foodstuff are nuclide-specific. The respective derived intervention levels (DILs) of specific activity additionally depend on the proportional contribution of a food product to the annual food basket, and on the assumed mean annual radionuclide contamination of a food product in relation to the respective nuclide-specific DIL of specific activity.

The nuclide-specific ingestion dose factors on which the intervention levels for the annual rates of activity intake are based are age-dependent. Further, the annually consumed amount of food on which the DILs for the specific activity of a food item are based is also quite different in the various population groups.

In the following it has been attempted to quantify the influence of dosimetric and metabolic data on the nuclide-specific ingestion dose factors in general and for some radionuclides in particular. Furthermore, the degree is demonstrated in which the DILs of the specific activity of major foodstuffs are dependent on population age and regional consumption habits.

## The Concept of the DIL for Foodstuff

The DIL (Bq/kg or L) of the specific activity of a foodstuff in the form in which it is consumed is

- proportional to the intervention level of dose for food restrictions (Sv);
- inversely proportional to the committed single organ or effective dose equivalent per unit intake of a radionuclide by ingestion (Sv/Bq);
- inversely proportional to the annual intake of a foodstuff (kg per year); and
- inversely proportional to the ratio of the integral over one year of the specific activity in a foodstuff to the specific activity in that foodstuff at a specified time, or simply to the mean fractional contamination of a foodstuff by a radionuclide over a year in relation to the DIL of the specific activity for that rationuclide.

## The Dosimetric and Metabolic Model

### The Dosimetric Model

The committed dose equivalent to an organ from a single intake of a radionuclide via ingestion is the product of (a) the dose in a target organ due to one nuclear transformation in a source organ, and (b) the number of nuclear transformations in the relevant source organs. Whereas the dose per nuclear transformation in a target can be assessed by applying the dosimetric model of the ICRP for the adult reference man (2), the number of nuclear transformations in the sources has to be derived from appropriate metabolic models (3).

The dosimetric model of the adult reference man consists of skeletal, lung, and other soft tissues. The trunk, legs, and arms, as well as the interior organ phantoms (which are approximately adequate to human organs in size, shape, position in the body, composition and density) are described by simple mathematical models to provide computer calculations by means of a Monte-Carlo-transport code of absorbed dose to targets from nuclear transformations in sources.

## The Metabolic Model

For calculating the dose to a target organ per unit activity intake by ingestion, the biodistribution of the administered radionuclide has to be known. To describe the time-dependent biodistribution of radionuclides in the body, that is, the biokinetics, a linear compartment model was developed by the ICRP (3). According to this model, ingested radionuclides are moving via the stomach, the small intestine, and the large intestine to excretion and are absorbed from the small intestine into the transfer compartment. From the transfer compartment, that is, the body fluids, the radionuclides are moving to various organ and tissue compartments. The rate constants at which material is transferred from the stomach, the small intestine, and the large intestine are assumed to be constant, regardless of the radionuclide or chemical form of the ingested substance. The fractional absorption from the gastrointestinal tract depends on the element, the chemical form, and the matrix in which the radionuclide is ingested. The chemical form of a radionuclide determines the rate constants of the various organ or tissue compartments.

# Parameters Influencing Dose Factors

Using these dosimetric and metabolic models for calculating the committed organ and effective dose equivalent per unit activity intake by ingestion, that is, the dose factors, for members of the public from the intake of radionuclides in the environment, various anatomical and physiological parameters have to be taken into account. These are

- organ and body masses, size, shape, and location in the body;
- model-specific parameters such as residence times in the various body compartments;
- nuclide specific parameters such as fractional distribution and biological half-times in organs and tissues; and
- the radionuclide and matrix-specific fractional absorption from the gastrointestinal tract.

## Organ and Body Masses

Even in the absence of differences in the uptake and retention of a radionuclide, the committed dose equivalent in a particular tissue per unit intake of a radionuclide would be greater in children than in adults because of the smaller mass of their organs and tissues. For a child in the first year of life, whose body mass at age 6 months is about 7 kg (2), the committed dose equivalent in an organ per unit intake of a short-lived radionuclide emitting nonpenetrating radiation, such as alpha-particles or beta-particles, will be about ten times greater than for a 70-kg adult. However, for long-lived radionuclides with longer retention in the body, such as $^{239}$Pu, this factor is only about 2 (4), since the age-dependent increase of the organ or tissue mass decreases proportionally to the specific tissue activity. For radionuclides emitting penetrating radiations, such as gamma-radiation or characteristic x-rays, the committed dose equivalent per unit intake of activity in an organ is inversely proportional to the following factor: (the mass of an organ)$^{2/3}$. Consequently, the modifying factor for considering age-dependence of body mass is smaller than for radionuclides emitting nonpenetrating radiation.

Although the shape and relative position of organs may considerably change with age, these differences are of lesser influence on the dose factors than compared with organ and body mass. Allowing for organ or body size and mass alone, the committed effective dose equivalents for young members of the population will be greater by factors less than 2 and 10 than those for the adult reference man, depending on the type of radiation emitted by the radionuclide. Examples are given in Table 23.1 for radionuclides that might be released accidentally into the environment and for which no relevant age-dependent metabolic data are currently available for dose calculations (5).

## Model-Specific Parameters

The biokinetics of radionuclides may essentially vary between individuals, especially between children and adults, owing to different elimination rates from various compartments of the metabolic model. For intake of radionuclides by

**Table 23.1.** Age-dependent dose factors for selected radionuclides with metabolic data of the adult reference man (5)

| Radionuclide | Effective dose equivalent per unit intake by ingestion (Sv/Bq) | | | Ratio of dose factors: 1 year to adult |
| --- | --- | --- | --- | --- |
| | Adult | 10 years | 1 year | |
| $^{95}$Zr | 9.2 E–10 | 1.9 E–9 | 5.8 E–9 | 6.3 |
| $^{95}$Nb | 6.1 E–10 | 1.3 E–9 | 3.8 E–9 | 6.2 |
| $^{103}$Ru | 7.3 E–10 | 1.7 E–9 | 3.5 E–9 | 4.8 |
| $^{106}$Ru | 5.8 E– 9 | 1.7 E–8 | 5.8 E–8 | 10.0 |
| $^{140}$Ba | 2.3 E– 9 | 5.7 E–9 | 1.9 E–8 | 8.3 |
| $^{141}$Ce | 7.0 E–10 | 1.8 E–9 | 6.2 E–9 | 8.9 |
| $^{144}$Ce | 5.3 E– 9 | 1.3 E–8 | 4.5 E–8 | 8.5 |
| $^{239}$Np | 8.0 E–10 | 1.9 E–9 | 6.7 E–9 | 8.4 |
| $^{239}$Pu | 1.3 E– 6 | 1.4 E–6 | 2.4 E–6 | 1.9 |
| $^{240}$Pu | 1.3 E– 6 | 1.4 E–6 | 2.4 E–6 | 1.9 |
| $^{241}$Am | 1.2 E– 6 | 1.7 E–6 | 3.4 E–6 | 2.8 |

ingestion, the influence of these model parameters was studied by Elsasser et al. (6). According to these authors, the cumulated activity, that is, the total number of transformations as an indicator for the dose, in the lower large intestine increases by a factor of up to 3 for short-lived radionuclides of 0.1 days half-life if the mean residence time of the activity in the upper large intestine decreases by a factor of 5. Assuming a typical age-dependent decrease by a factor of 2, the increase of the cumulated activity in the lower large intestine is between 1.02 and 1.7 for physical half-times of a radionuclide of 5 and 0.1 days.

## The Radionuclide-Specific Parameters for Fractional Distribution and Biological Half-Times in Organs and Tissues

The fractional distribution of a radionuclide in organs and tissues and its biological half-time may differ essentially between children and adults. At present, age-dependent metabolic data are available from the literature for only a few elements, that is, for H, I, Cs, and Sr. For radioisotopes of some of these elements, the influence of fractional distribution and biological half-time on the dose factors may be illustrated as follows (6):

1. Due to the more rapid turnover of Cs in children, the dose equivalent to body tissues from unit intake of $^{137}$Cs is only about 0.7 times greater for the 6-month-old infant than it is for adults, rather than a factor of about 8 from the ratio of the body masses.
2. The biological half-time of I in the thyroid increases with age; the uptake by the thyroid is slightly higher in the first month of life than in the young child, adolescent, and adult. However, because of the comparatively short half-time of $^{131}$I, age-related differences in biological turnover are only of little consequence, so that the 8 times greater thyroid mass in adults is the only factor influencing the ratio of the dose equivalent per unit intake of the young child and the adult.
3. The value for the committed dose equivalent per unit intake of the long-lived $^{90}$Sr for the 6-month-old infant proved to be about 3 times the value for the adult, but for the much shorter-lived $^{89}$Sr the corresponding ratio lies in the range of 10, depending on the bone model used.

For a limited number of radionuclides relevant for environmental contamination in nuclear accidents, Table 23.2 summarizes age-dependent dose factors considering age-dependent organ and body masses (anatomical/dosimetric data) as well as age-dependent data on fractional radionuclide distribution and biological half-times (biokinetic data). According to these results and compared with the data in Table 23.1, the difference in the biokinetics of radionuclides is evident in organ dose factors, which, in view of the age-dependent organ or tissue masses, are for the 1-year-old child in the range of 0.6 to about 10 times the values for the adult.

**Table 23.2.** Age-dependent dose factors for selected radionuclides with age-dependent metabolic and dosimetric (anatomical) data (5)

| Radionuclide | Effective dose equivalent per unit intake by ingestion (Sv/Bq) | | | Ratio of dose factors: |
|---|---|---|---|---|
|  | Adult | 10 years | 1 year | 1 year to adult |
| [89]Sr | 2.2 E– 9 | 6.8 E– 9 | 2.5 E-8 | 11.4 |
| [90]Sr | 3.6 E– 8 | 4.0 E– 8 | 1.1 E-7 | 3.1 |
| [131]I | 1.4 E– 8 | 2.8 E– 8 | 1.1 E-7 | 7.9 |
| [132]I | 1.6 E–10 | 3.6 E–10 | 1.4 E-9 | 8.8 |
| [133]I | 2.7 E– 9 | 5.8 E– 9 | 2.3 E-8 | 8.5 |
| [134]Cs | 2.0 E– 8 | 1.2 E– 8 | 1.2 E-8 | 0.6 |
| [137]Cs | 1.4 E– 8 | 9.3 E– 9 | 9.3 E-9 | 0.7 |

## The Radionuclide and Matrix-Specific Fractional Absorption From the Gastrointestinal Tract

With increased absorption of a radionuclide from the gastrointestinal tract, the cumulated activity—and, consequently, the dose in body organs and tissues—changes proportionally for radionuclides with radioactive half-lifes of more than 1 day (6). For radionuclides with radioactive half-lifes of less than 0.1 days, an underestimation of the value for the gastrointestinal absorption leads to a corresponding nonlinear underestimate of the cumulated activity, for example, by a factor of 20 for an underestimate of the gastrointestinal absorption by a factor of 10 from 0.1 to about 1.

Incorporation of radionuclides into the foodstuff can change the amount absorbed by the gastrointestinal tract. In the case of Pu, for example, animal experiments have shown that biological incorporation can increase the amount absorbed by a factor of 5 to 10, compared with the value found for soluble inorganic forms such as nitrates. Recommended values chosen by an expert group of the Nuclear Energy Agency (NEA) of the OECD on "gut transfer factors" (7) for selected elements into food and drinking water are given in Table 23.3. From comparison with the ICRP values for the adult reference man, the recommended fractional absorption for adult members of the public is higher by a factor of up to 10 in the case of Y. The recommended gut-transfer factors for children up to 10 years of age, however, may be 20 times those of the adult (in case of Y), and 10 for La. For Np, Pu, and Am, the 10 times increased gastrointestinal absorption is only recommended for the first year of life.

Since only radionuclides with radioactive half-times of more than one day are relevant for uptake by ingestion of contaminated foodstuffs, the cumulated activity in organ or tissue compartments of children, and hence the doses, will be higher in children by the ratio of the gut-transfer factors for children and adults if calculated on the basis of the adult gastrointestinal radionuclide absorption.

**Table 23.3.** Recommended values for the fractional absorption from the gastrointestinal tract for selected elements incorporated into food and drinking water for members of the public (adults, children) and comparison to ICRP 30/48 data for the adult reference man (7)

| | Fractional absorption from the gastrointestinal tract | | | Ratio | |
| | ICRP 30/48 | | | | |
| Element | adult ref. man | Adult | Child (0–10 yr) | Adult/ICRP | Child/Adult |
|---|---|---|---|---|---|
| Co | 0.30 | 0.30 | 0.60 | 1 | 2 |
| Sr | 0.30 | 0.30 | 0.60 | 1 | 2 |
| Y | 0.0001 | 0.001 | 0.02 | 10 | 20 |
| Zr | 0.002 | 0.01 | 0.02 | 5 | 2 |
| Nb | 0.01 | 0.01 | 0.02 | 1 | 2 |
| Ru | 0.05 | 0.05 | 0.10 | 1 | 2 |
| Sb | 0.10 | 0.10 | 0.20 | 1 | 2 |
| I | 1.0 | 1.0 | 1.0 | 1 | 1 |
| Cs | 1.0 | 1.0 | 1.0 | 1 | 1 |
| Ba | 0.10 | 0.10 | 0.20 | 1 | 2 |
| La | 0.001 | 0.001 | 0.01 | 1 | 10 |
| Np | 0.001 | 0.001 | 0.01 (0–1 yr) | 1 | 10 |
| Pu | 0.001 | 0.001 | 0.01 (0–1 yr) | 1 | 10 |
| Am | 0.001 | 0.001 | 0.01 (0–1 yr) | 1 | 10 |

# Region-Specific and Age-Specific Foodbaskets and Changed Consumption Habits After Radiation Accidents

As given above, the value of the DILs for foodstuffs (i.e., the DIL for the specific activity of food items) is inversely proportional to the mass of each food item or group of food items consumed per year. Consequently, they are dependent on the specific composition of a foodbasket.

Recently, a WHO Expert Group reviewed food consumption data for about 130 countries (8). On the basis of the pattern of consumption, geographic location, and existing FAO groupings, these countries have been divided into eight regional diet groups. They are African, North African, Eastern Mediterranean, Far Eastern, Central American, South American, Chinese, and European (North American included). Averages have been computed for each region. From the results in Table 23.4, it can be seen that there is a considerably wide spectrum of consumption values for the same food item: if normalized to the European region, the variation may be between 0.01 (milk, Chinese region) and 2.3 (roots and tubers, African region). The mean fractional consumption in the various regions is between 0.3 and 1.3.

Data on the age-dependent European annual food consumption (9) are based on values of budget survey data on adults, a standard unit consumption factor

**Table 23.4.** Non-European food consumption patterns normalized to European data on annual consumption of food items (8)

| Region | | Cereals | Roots and tubers | Vegetables | Fruit | Meat | Fish | Milk |
|---|---|---|---|---|---|---|---|---|
| European | (kg/y) | 121.10 | 72.68 | 86.65 | 81.35 | 75.33 | 20.21 | 154.94 |
| Chinese | (relative) | 1.4 | 1.2 | 1.0 | 0.1 | 0.2 | 0.4 | 0.01 |
| Far Eastern | (relative) | 1.7 | 0.4 | 0.6 | 0.6 | 0.3 | 1.2 | 0.2 |
| Eastern Med. | (relative) | 1.6 | 0.3 | 1.1 | 1.3 | 0.4 | 0.4 | 0.5 |
| South American | (relative) | 1.1 | 0.9 | 0.4 | 1.0 | 0.6 | 0.7 | 0.5 |
| Central American | (relative) | 0.9 | 0.6 | 0.5 | 1.2 | 0.6 | 0.9 | 0.5 |
| African | (relative) | 1.2 | 2.3 | 0.4 | 0.7 | 0.3 | 0.9 | 0.2 |
| North African | (relative) | 1.3 | 0.3 | 0.7 | 0.8 | 0.3 | 0.4 | 0.5 |
| Mean fractional consumption | | 1.3 | 0.9 | 0.7 | 0.8 | 0.4 | 0.7 | 0.3 |

**Table 23.5.** European community food consumption data from budget surveys and as recommended by the Article 31 Group of the EC (5, 9)

| Food item | European community values (kg/y) | | | | | |
| | Budget survey data | | | Recommended dietary data | | |
| | Adult | Child | Infant | Adult | Child | Infant |
|---|---|---|---|---|---|---|
| Cereals | 154 | 108 | 22 | 100 | 50 | 20 |
| Vegetables, roots | 103 | 72 | 18 ⎫ | 100 | 40 | 20 |
| Fruit | 55 | 39 | 7 ⎬ | | | |
| Meat and fish | 56 | 39 | 11 | 80 | 40 | 10 |
| Milk and milk prod. | 105 | 74 | 204 | 120 | 150 | 200 |

for the 10-year-old of 0.7 (the 10-year-old consumes 70% as much as an adult) and on recommendations of the European Society for Pediatric Gastroenterology and Nutrition (10) for infant nutrition. They are summarized in Table 23.5 for major foodstuffs and compared with age-dependent dietary data as recommended by the Expert Group according to Article 31 of the EURATOM Treaty (5). According to the data, the annual consumption of milk and milk products by the infant is about twice that of the adult, while that of other major foodstuffs is about one fifth.

However, even in one and the same region the consumption habits of individual members of the population may be quite different. This becomes particularly clear by comparing vegetarians and nonvegetarians. Furthermore, it is to be taken into account that in case of a nuclear reactor accident, the consumption habits may markedly change not only because of the recommended limitation of foods to be consumed, but also because of a general fear of irradiation by the public. During the months of May to July 1986, 300 randomly selected households in the Federal Republic of Germany and Berlin (West) were interviewed in reference to changes in consumption habits (11). Table 23.6 shows

**Table 23.6.** Changes in nutritional behavior of the public during the months of May to July 1986, following the Chernobyl reactor accident (11)

| Food item | Month | Percent proportion of the public either limiting or discontinuing the consumption of various food items | | |
| | | Men | Women | Total |
|---|---|---|---|---|
| Leafy vegetables | May | 72 | 79 | 76 |
| | June | 65 | 80 | 73 |
| | July | 39 | 43 | 41 |
| Tubers/roots | May | 54 | 63 | 58 |
| | June | 60 | 67 | 64 |
| | July | 30 | 36 | 33 |
| Fruit | May | 48 | 60 | 54 |
| | June | 64 | 63 | 63 |
| | July | 25 | 35 | 30 |

the influence on the consumption habits of the population, according to which in May 1986, up to 75% of the population had either limited or discontinued the consumption of fresh leafy vegetables as a result of the Chernobyl reactor accident. With decreasing contamination, the consumption habit gradually returned to normal.

## Time-Average Specific Activity of Foodstuff

The specific activity of foodstuff can vary strongly with time after deposition, depending on the season of the year in which environmental contamination occurs. This variation is primarily caused by the foliar uptake of leafy vegetables, cereals, and pasture grass. Thus, the DILs for the time-averaged specific activity have to be corrected for this time-dependence of the specific activity of food items to enable a reasonable decision on the control of foodstuff. For a peak concentration of accidentally released long-lived fallout into the environment, a mean contamination of foodstuff may be assumed over the whole year for reasons of practical control.

In the FAO proposal (12) the annual contamination of all foods is assumed to be 100% of the DIL of the specific activity. This is, without doubt, a more than conservative assumption. On the other hand, the European Communities (EC) Expert Group, according to Article 31 of the EURATOM Treaty, assumes the annual contamination of foods from long-lived radionuclides to be only

**Table 23.7.** Maximum and mean values of sepcific[137] Cs activity in foodstuffs in different high-contaminated regions of the Federal Republic of Germany over a period of 10 months after the Chernobyl reactor accident

| Food item | Fed. state (land)[a] | Maximum value, May 1986 (Bq/kg) | Mean value (% of maximum value), May 1986– Feb. 1987 | Mean value (% of intervention level of spec. act.[b]), May 1986– Feb. 1987 |
|---|---|---|---|---|
| Unpasteurized milk | By | 123 | 28 | 9.5 |
| | NRW | 14 | 46 | 1.8 |
| Pasteurized milk | By | 83 | 43 | 9.7 |
| | NRW | 12 | 47 | 1.5 |
| Beef | By | 210 | 30 | 10.7 |
| | NRW | 37 | 46 | 2.8 |
| Leafy vegetables | By | 371 | 14 | 8.7 |
| | NRW | 38 | 24 | 1.5 |

[a] By: Bavaria; NRW: Nothern Rhineland and Westphalia
[b] Milk: 370 Bq/kg      } DILs for radioactive [134]Cs and [137]Cs as instituted by the CEC
  Meat, leafy veg.: 600 Bq/kg } (Directive (CEC) No. 1707 from May 30, 1986, Official Journal of EC, No. L 146/88, p 85).

10% of the DIL (5). Depending on each of these assumptions, the values for the DILs of the specific activity differ by a factor of 10.

The value proposed by the EC Expert Group of a mean annual contamination of foods of 10% of the DIL of the specific activity was substantiated for $^{137}$Cs in milk, beef, and leafy vegetables in the Federal State Bavaria as the highest exposed region of the Federal Republic of Germany after the reactor accident (see Table 23.7). In less exposed regions, the relative contamination was about 2%.

## Reliability of the Parameters Influencing the DILs of the Specific Activity for Foodstuff

The results from discussions on the parameters determining the value of DILs of the specific activity for foodstuff may be summarized as follows:

1. The dose factors for ingestion in infants and children depend on the physical characteristics of a radionuclide owing to the smaller organ masses compared with those of the adult reference man and considering the different anatomical proportions. For infants (1 year) they are higher by factors of 2 to 10.
2. Taking age-dependent values for the fractional distribution and biological half-times for radionuclides into consideration, the dose factors for infants (1 year) are higher than for the adult reference man by factors below 1 and 10.
3. The fractional absorption of radionuclides incorporated into food may be higher in the adult member of the population by a factor up to 10 as compared with that of radionuclides incorporated under normal working conditions. The fractional absorption of radionuclides incorporated into food in children (0 to 10 years of age) may be higher by a factor of up to 20 as compared with the adult member of the population.
4. The mean fractional consumption of single food items in the various regions of the world may be different from that in the European/North American region by factors ranging from 0.3 to 1.3, with values from 0.01 to 1.6.
5. According to surveys in the Federal Republic of Germany after the reactor accident in Chernobyl, in up to 75% of the population the consumption of individual food items may be limited or discontinued either as a result of recommendations or due to public fear of irradiation.
6. The assumed mean annual contamination of foodstuff with long-lived radionuclides in relation to the DILs of specific radionuclide activity depends on the type of nuclear accident and the time of year it occurs. The value of this fractional contamination may vary from 2% to 100%.

*Acknowledgment.* The author wishes to express his gratitude to R. LeMar for assistance in the translation of this text into the English language.

# References

1. ICRP Publication 40 (1984) Protection of the public in the event of major radiation accidents: Principles for planning. A report of a task group of committee 4. Pergamon Press, Oxford
2. ICRP Publication 23 (1975) Report of the task group on reference man. Pergamon Press, Oxford
3. ICRP Publication 30 (1979) Limits for intakes of radionuclides by workers. Ann ICRP 2, No 1, Pergamon Press, Oxford
4. Adams N (1981) Dependence on age at intake of committed dose equivalents from radionuclides. Phys Med Biol 26, 6:1019–1034
5. Derived emergency reference levels in widely distributed foodstuffs. A report of the group of experts set up under Article 31 of the EURATOM Treaty (1986): Henrichs K, Elsasser U, Schotola Ch, Kaul A (1985) Dosisfaktoren für inhalation oder ingestion von radionuklidverbindungen (Altersklasse: 1 Jahr, 10 Jahre). ISH Hefte 78, 80; November 1985, Bundesgesundheitsamt, Institut für Strahlenhygiene, Neuherberg/München. Noβke D, Gerich B, Langner S (1985) Doisisfaktoren für inhalation und ingestion von radionuklidverbindungen (Erwachsene). ISH Hefte 62; April 1985, Bundesgesundheitsamt, Institut für Strahlenhygiene, Neuherberg/München. Greenhalgh JR, Fell TP, Adams N (1985) Doses from intakes of radionuclides by adults and young people. Chilton, NRPB-R162, HMSO, London
6. Elsasser U, Kaul A, Roedler HD (1987) Metabolic and dosimetric characteristics of ingested radionuclides. Int. Scientific Seminar on Foodstuffs Intervention Levels following a Nuclear Accident; Luxembourg, April 27–30, 1987. Commission of the European Communities, Preprint EUR 11232, pp. 223–238
7. Report of NEA/OECD expert group on gut transfer factors. In press. ICRP-Publication 48 (1986) The metabolism of plutonium and related elements. Ann ICRP 16, No 2/3. Pergamon Press, Oxford
8. World Health Organization (1988) Derived intervention levels for radionuclides in food. Guidelines for application after widespread radioactive contamination resulting from a major radiative accident. WHO, Geneva
9. Gouvras G (1987) Dietary consumption patterns in the EC. Int. Scientific Seminar on Foodstuffs Intervention Levels following a Nuclear Accident; Luxembourg, April 27–30, 1987. Commission of the European Communities, Preprint EUR 11232, pp. 155–181
10. European Society for Pediatric Gastroenterology and Nutrition (1981) Guidelines on infant nutrition. Acta Pediatr Scand 287 (1981); 1–36 (1971); 66 (1977)
11. Anders H-J, Rosenbauer J, Matiaske B (1987) Einstellungs- und verhaltensänderungen der bundesdeutschen bevölkerung nach tschernobyl. Ernährungs-Umschau 34, 8:255
12. Report of the expert consultation on recommended limits for radionuclide contamination of foods (1986) Rome, December 1–5, 1986, Food and Agriculture Organization of the United Nations

CHAPTER 24

# Use of Mathematical Models in Risk Assessment and Risk Management

## M. D. Hill[1]

## Introduction

Mathematical models of the transfer of radionuclides through the environment are used extensively in radiological protection. There are two main reasons for this. The first is that although it is relatively easy to detect radionuclides in environmental materials, it is impossible to take enough measurements to enable estimates to be made of the radiation doses to people from all the sources of current exposure. The second reason is that it is necessary for planning purposes to estimate doses to people in the future, both those from radionuclides that are already present in the environment and those from routine and accidental releases that may occur in the short or long term. It must also be remembered that it is not possible to actually measure the radiation dose to a person from the intake of radionuclides. The best that can be achieved is to measure intake, and even this is extremely difficult in the case of members of the public and involves a significant intrusion into people's lives. Thus, all doses from intakes are calculated, using mathematical models of radionuclide metabolism in the human body.

Initially, the models used in radiological protection were very simple and limited in scope. As knowledge of radionuclide transfer through the various parts of the environment increased, more complex and wide-ranging models were developed, and we have now reached the stage where, worldwide, many models are available for predicting radionuclide transfer through the atmosphere, the terrestrial environment, the marine environment, freshwater, and the geosphere (see, e.g., refs. 1–5). It is neither possible, nor necessary, in this chapter to describe all these models; attention here is limited to the types of models available for predicting radionuclide transfer through terrestrial food chains. The ways in which terrestrial food-chain models are used in assessing the risks associated with existing and planned nuclear installations are described, and examples of the results of such assessments are given. The use of assessment

[1] National Radiological Protection Board, Chilton, Didcot, Oxon, OX11 ORQ, UK.

results in planning the management of risks is discussed and illustrated by case studies.

The final sections of the chapter deal with the use of models to estimate the radiological consequences of an accident soon after it has occurred and with quantification of the uncertainties in model predictions.

## Models Available for Predicting Radionuclide Transfer Through Terrestrial Food Chains

The models available for predicting radionuclide transfer through terrestrial food chains can be divided into two categories: "equilibrium" models, which calculate the concentrations of radionuclides in foods under steady-state conditions, and "dynamic" models, which predict radionuclide concentrations in foods as a function of time after the initial input into the environment. Equilibrium models can be used in assessments of the radiological impact of continuous radionuclide releases but are of limited value in the context of accidental releases because in these cases no true equilibrium is reached. Dynamic models, on the other hand, can be applied to both continuous and discrete releases of radionuclides. Since the latter type of release is of most interest, the following discussion focuses on dynamic models.

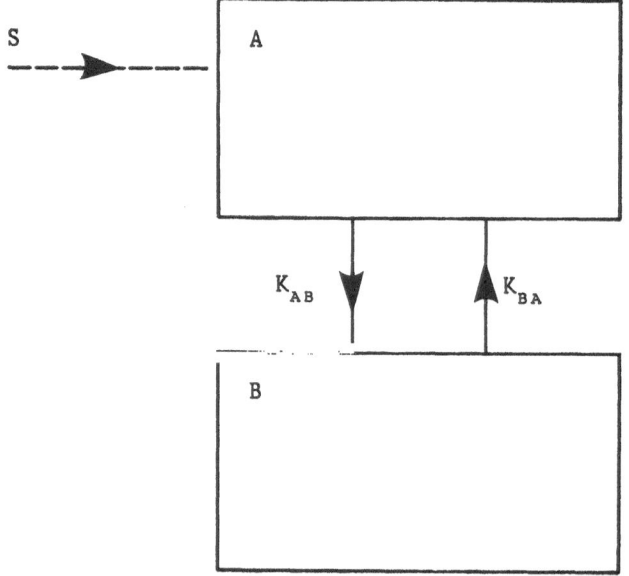

**Fig. 24.1.** Basic structure of compartment models. $\frac{dN_A}{dt} = S - K_{AB}N_A + K_{BA}N_B - \lambda N_A$.

$S$ = input to compartment A, atoms per unit time; $N_A$ = number of atoms in compartment A; $N_B$ = number of atoms in compartment B; $K_{AB}$, $K_{BA}$ = transfer coefficients, per unit time; and $\lambda$ = radioactive decay constant, per unit time.

All the dynamic models currently in use are of the compartment type. In this approach the terrestrial environment, including agricultural animals, is divided into a series of compartments or boxes, with transfer coefficients between them. It is assumed that when a given amount of radionuclide enters each compartment, it immediately becomes uniformly distributed within it. The compartments have to be chosen so that this assumption is reasonable and, therefore, vary in size from "all the grass in a field" to the "udder of a cow." Transfer of radionuclides from one compartment to another is generally assumed to be a first-order process, so that the amount of activity present in a compartment at any time is calculated by solving a series of equations of the type shown in Fig. 24.1. The characteristics of the compartments and the transfer coefficients between them are derived from field observations and experimental data.

The basic features of dynamic compartment models for radionuclide transfer through terrestrial food chains are illustrated in this chapter through a description of the Food Activity from Radionuclide Movement on LAND (FARMLAND) suite of models developed by the National Radiological Protection Board (NRPB) in the United Kingdom. The results of these models have been compared with those of other similar models developed in the United Kingdom (6,7) and in

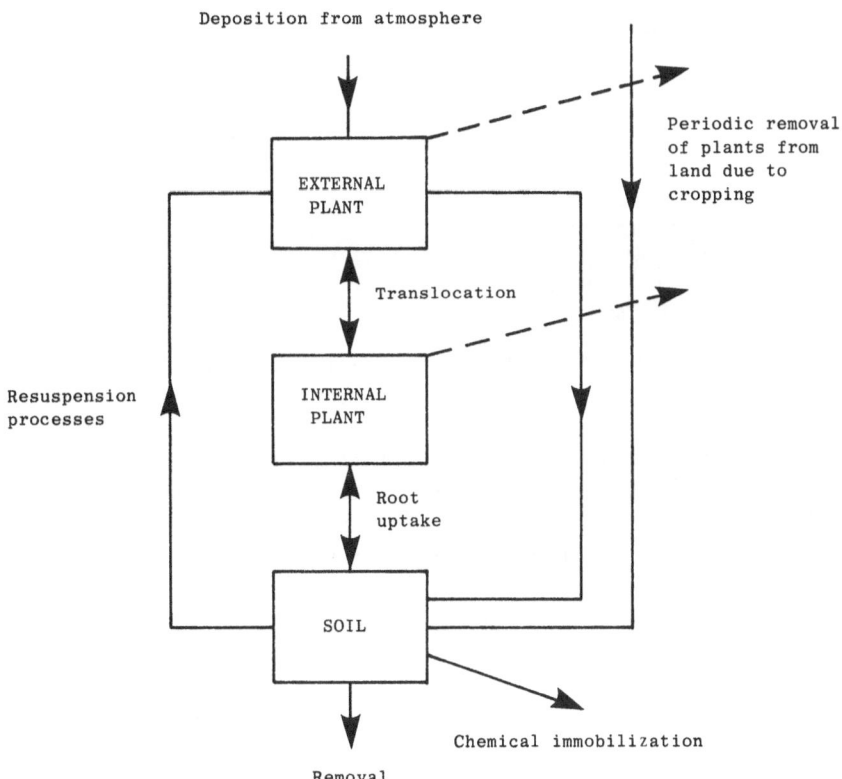

**Fig. 24.2.** Main features of FARMLAND soil-plant models.

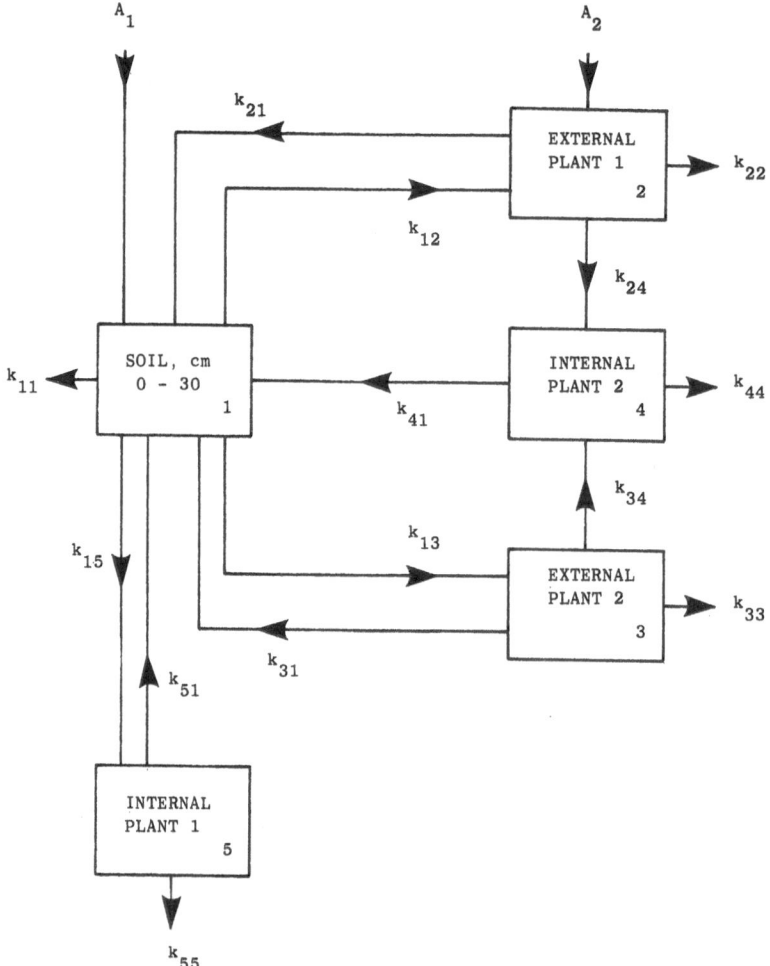

**Fig. 24.3.** FARMLAND green vegetable model.

the Federal Republic of Germany (8). Differences between the results tend to be due to differences in parameter values used and in assumptions about agricultural practices, rather than to differences in the basic structure of the models. Preliminary results of an international model validation and comparison exercise (9) also indicate that there are no fundamental differences of principle between FARMLAND and other comparable models.

The FARMLAND suite of models currently contains separate models for predicting radionuclide transfer into six groups of foods: green vegetables, grain products, root vegetables, milk from cattle, meat from cattle, and meat from sheep.

These groups were chosen because they include the foods that are likely to be most important in the context of the radiological consequences of accidents

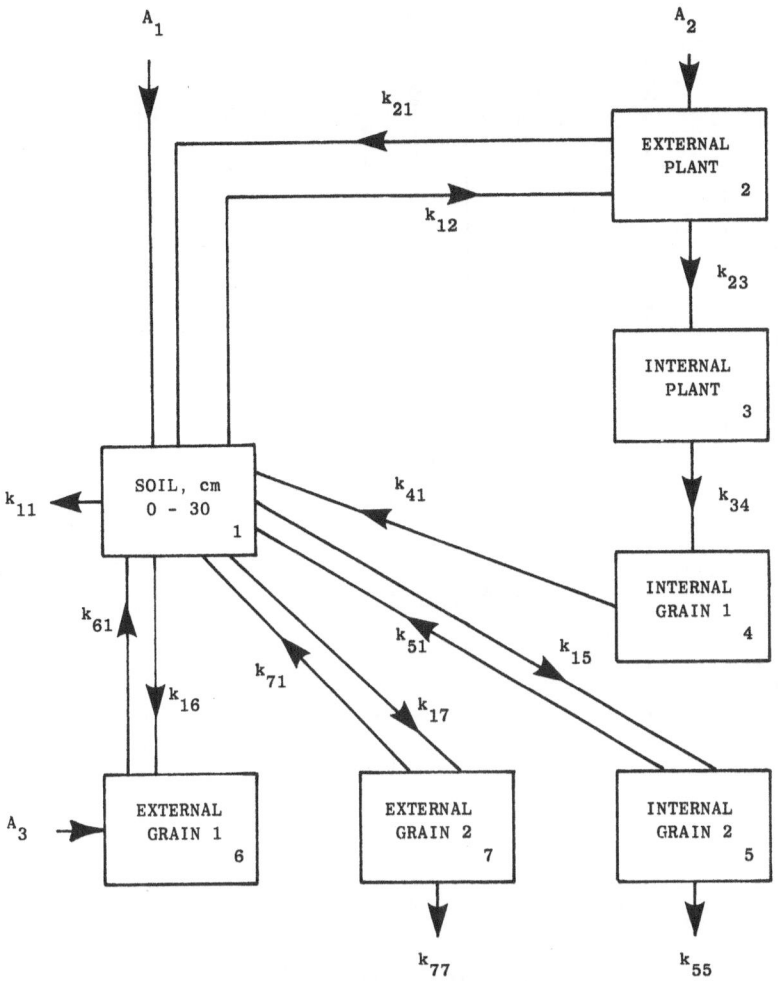

**Fig. 24.4.** FARMLAND grain model.

at UK nuclear facilities. They do not, therefore, encompass foods such as milk and milk products from sheep and goats. At present FARMLAND does not contain a model for predicting radionuclide transfer to meat from pigs because of difficulties in predicting the radionuclide content of pigs' diets.

Figure 24.2 shows the main features of the models for radionuclide transfer to plants, and Figs. 24.3 and 24.4 give further details of the models for green vegetables and grain. The basic processes included in these models are direct deposition on to the plant from the atmosphere and deposition following resuspension of contaminated soil; movement of radionuclides from the surface of plants to the interior (translocation); root uptake; and radionuclide migration down through the soil and out of the rooting zone. In using the models for accident consequence assessments, account is taken of the periodic removal of plants

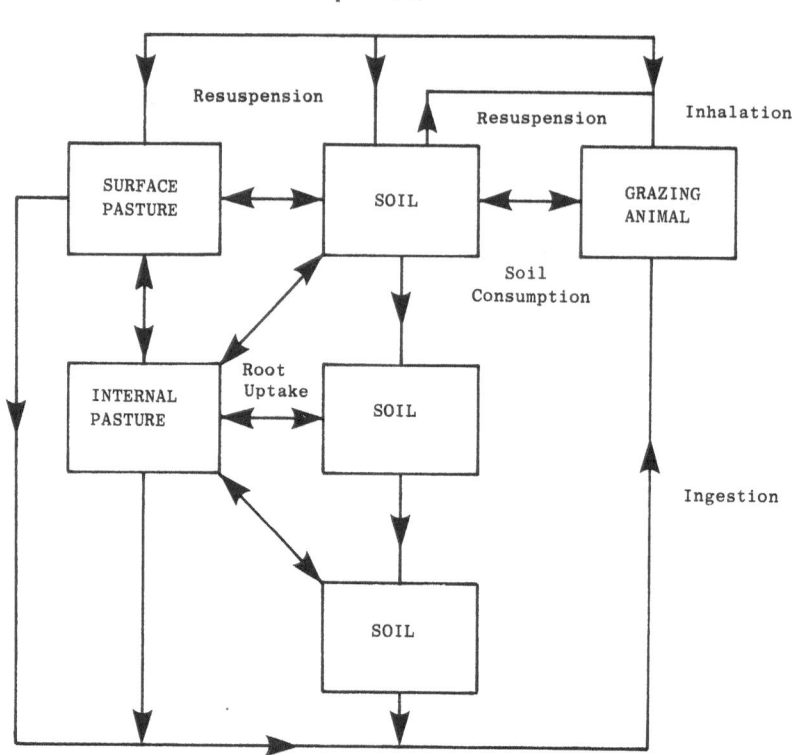

**Fig. 24.5.** Main features of FARMLAND models for radionuclide transfer to cattle and sheep.

from the land due to cropping, of radionuclide losses during preparation and processing, and of delays prior to consumption. It is also possible to take account of the season of the year at which the accident occurs.

Figure 24.5 shows the main features of the models for radionuclide transfer to cattle and sheep. These take into account intake by animals via inhalation and intake via consumption of grass and soil. A separate model is used to calculate radionuclide concentrations on undisturbed pasture and in hay and silage (Fig. 24.6). As an illustration of the structure of the models now used for radionuclide transfer in animals, Fig. 24.7 shows the compartment model for strontium transfer in cows. The main differences between these models and the simpler ones employed in the past (10) are that the newer models include a compartment representing circulating body fluids and also a representation of the recycling of activity from the important organs and tissues. Due to lack of adequate data, it has only been possible to develop models of the type shown in Fig. 24.7 for isotopes of I, Cs, Sr, Pu, and Am. For all other radionuclides, simpler models are still used (Fig. 24.8).

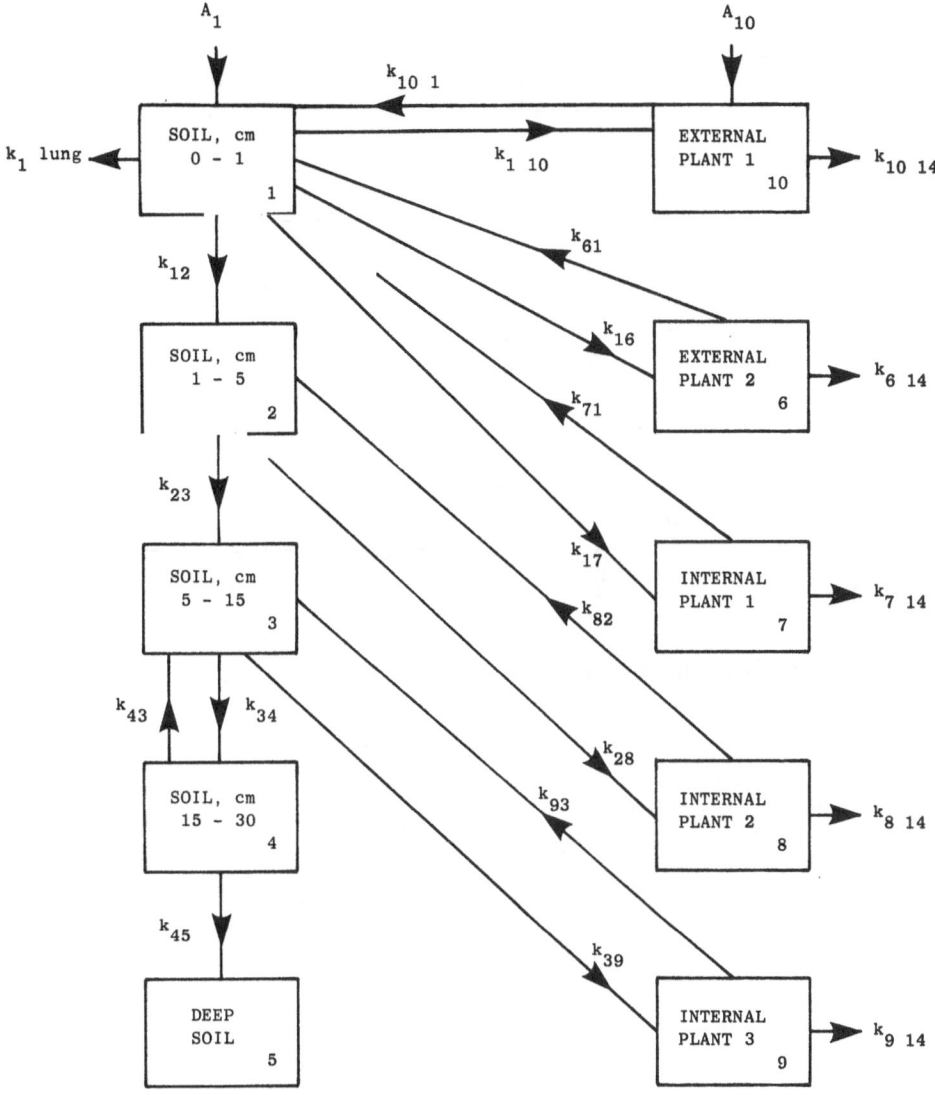

**Fig. 24.6.** FARMLAND models for undisturbed pasture and hay/silage.

## Model Verification and Validation

### Verification

Verification is the process of showing that a mathematical model is a proper representation of the conceptual model on which it is based and of checking that the mathematical equations involved have been solved correctly. For models that are implemented on computers, as all those at NRPB are, verification includes quality assurance of the computer code.

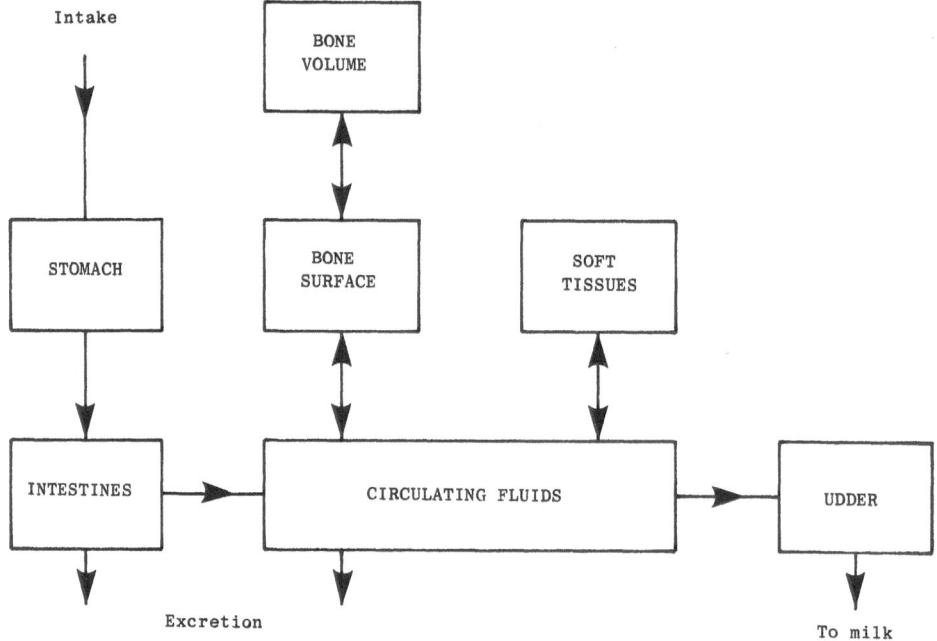

**Fig. 24.7.** Improved compartment model for transfer of Sr in the cow.

Verification procedures vary from one organization to another and also depend on the purpose for which the model is to be used. If the model is primarily a research tool, it may be acceptable to be somewhat less rigorous about verification than if the results of the modeling are to form an input to decisions on authoriza-

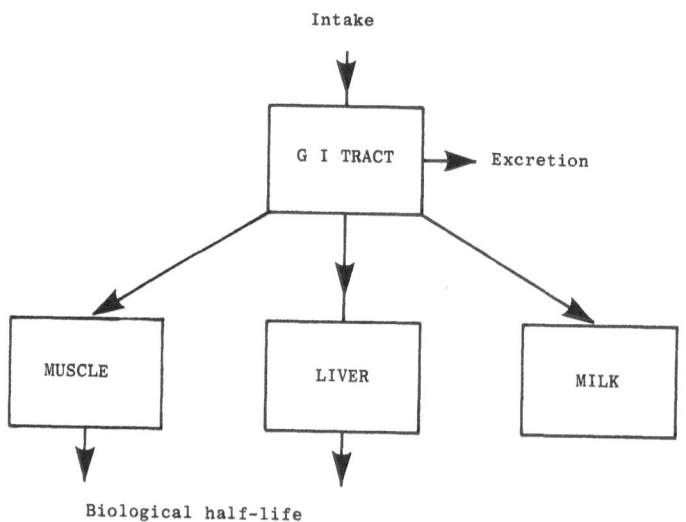

**Fig. 24.8.** Simple compartment model for transfer of radionuclides in the cow.

tions for effluent discharges or to licensing of a nuclear facility. Most of the models used at NRPB are employed for the latter type of work and have therefore been subject to fairly rigorous verification. This is achieved through internal procedures, which include

1. review of model structure and basic equations by staff other than those involved in developing the model,
2. checking computer codes to ensure that programming is correct, and
3. comparing computed results with problem solutions obtained from other models.

These internal procedures are supplemented by taking part in United Kingdom and international model–model comparison exercises (see, for example, refs. 6–9) and by submitting model descriptions and example results to other organizations and experts for external peer review.


## Validation

Validation consists of showing that a conceptual model and the computer code derived from it provide an adequate representation of radionuclide transfer processes in the real environment. The definition of what constitutes an adequate representation necessarily involves some subjective judgment and may depend on whether the model is intended for general or site-specific applications and on the purpose for which model results will be used.

Ideally, validation is carried out by comparing model calculations with sets of field observations and experimental measurements other than those that were used in developing the model. In practice, it is not possible to fully validate any environmental transfer model in this way because independent data sets do not exist for all the radionuclides, time scales, and environmental conditions to which the model is likely to be applied. However, some models, or parts of them, can be partially validated by comparisons of calculations with measurements. This procedure has been applied to parts of the NRPB FARMLAND models for radionuclide transfer through terrestrial food chains (11).

In cases where quantitative validation of the type described above is not feasible, recourse must be made to more qualitative techniques that mainly aim to check that a conceptual model is adequate, rather than that both this and the computer code derived from it are valid. These techniques include the use of data on the environmental behavior of chemically analogous natural or artificial radionuclides or stable elements, use of data from laboratory experiments that simulate environmental conditions, and external peer review of the assumptions used in developing a model. In NRPB work, the tendency has been to use the last of these techniques for all models, to rely primarily on environmental data for models for radionuclide transfer in the aquatic environment, and to use a combination of environmental and laboratory data in validation of terrestrial food-chain models (12).

# Use of Terrestrial Food-Chain Models in Risk Assessments

## Accident-Consequence Assessment Models and Computer Codes

Figure 24.9 shows the general structure of suites of models and computer codes that are designed to estimate the off-site radiological impact of accidental releases of radionuclides into the atmosphere. For reasons of computing efficiency, terrestrial food-chain models are not usually directly included in such computer codes. Instead, the results obtained from such models are used as part of the database that is required to calculate the concentrations of radionuclides in environmental materials as a function of time after deposition on to the ground. Thus FOOD-MARC (Methodology for Assessing the Radiological Consequences), which is the terrestrial food-chain part of the NRPB's MARC computer code for calculating the consequences of potential accidental releases to atmosphere, consists of a database obtained from the FARMLAND models (see Fig. 24.10 and refs. 13 and 14).

As can be seen from Figs. 24.9 and 24.10, estimates of the doses that people might receive through consumption of food are only one of the types of results of an accident-consequence assessment. They are an intermediate step in the calculation of risks to individuals, the potential numbers of health effects in the exposed population, the extent of the countermeasures that would be needed to reduce exposure, and the off-site economic costs of an accident. Examples of assessments are described in the following two sections.

## Example of a Deterministic Accident-Consequence Assessment

A deterministic accident-consequence assessment is one in which it is assumed that the accident occurs in particular weather conditions. The NRPB carried out a large number of such calculations as part of the input to evidence given at the Public Inquiry on the proposal to construct a pressurized water reactor (PWR) at Sizewell in Suffolk. Tables 24.1 and 24.2 show some of the results of a study (15) undertaken for the Political Ecology Research Group, who needed information on the agricultural consequences of potential accidents. Table 24.1 gives the predicted durations of restrictions on food production and consumption following one of the smaller degraded core accidents postulated by the Central Electricity Generating Board (CEGB). The results are given as a function of distance from the reactor and angle from the centerline of the dispersing plume of radioactivity. Table 24.2 shows, for the same accident and weather conditions, the total extent of agricultural restrictions if the wind were blowing in three different directions.

## Example of a Probabilistic Accident-Consequence Assessment

In probabilistic assessments, consequences are calculated for an accident occurring in a large number of sets of different weather conditions; the results are weighted by the probability that those conditions will occur in a year, and

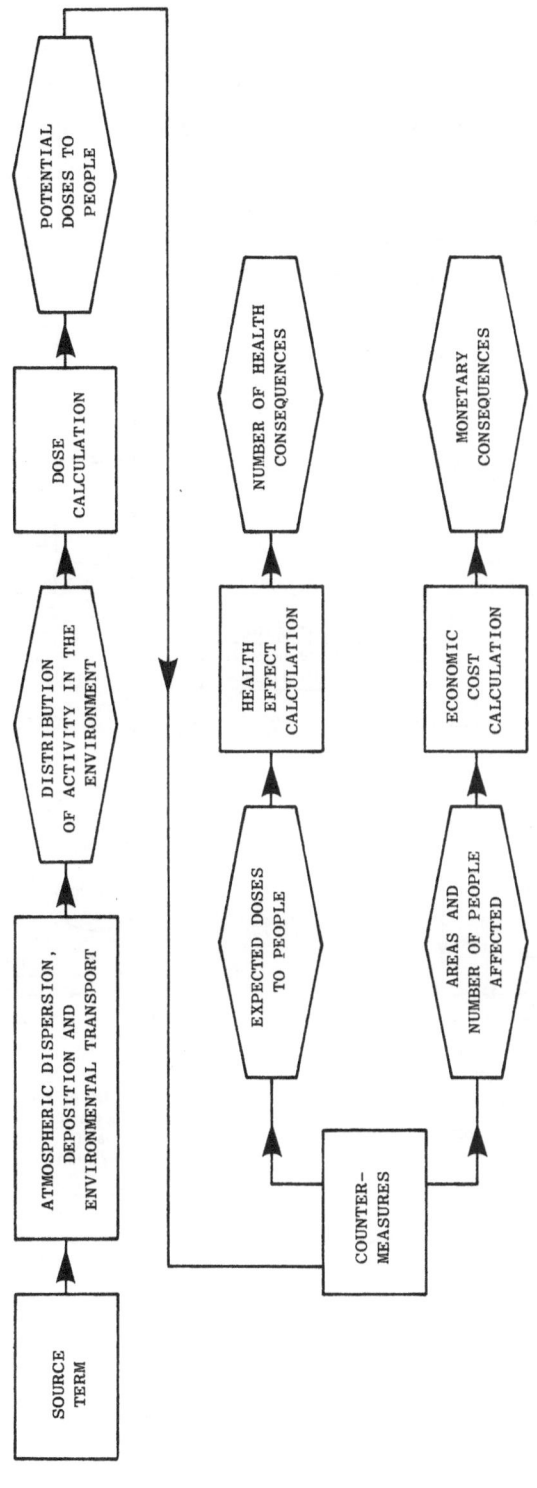

**Fig. 24.9.** Basic features of an accident-consequence assessment.

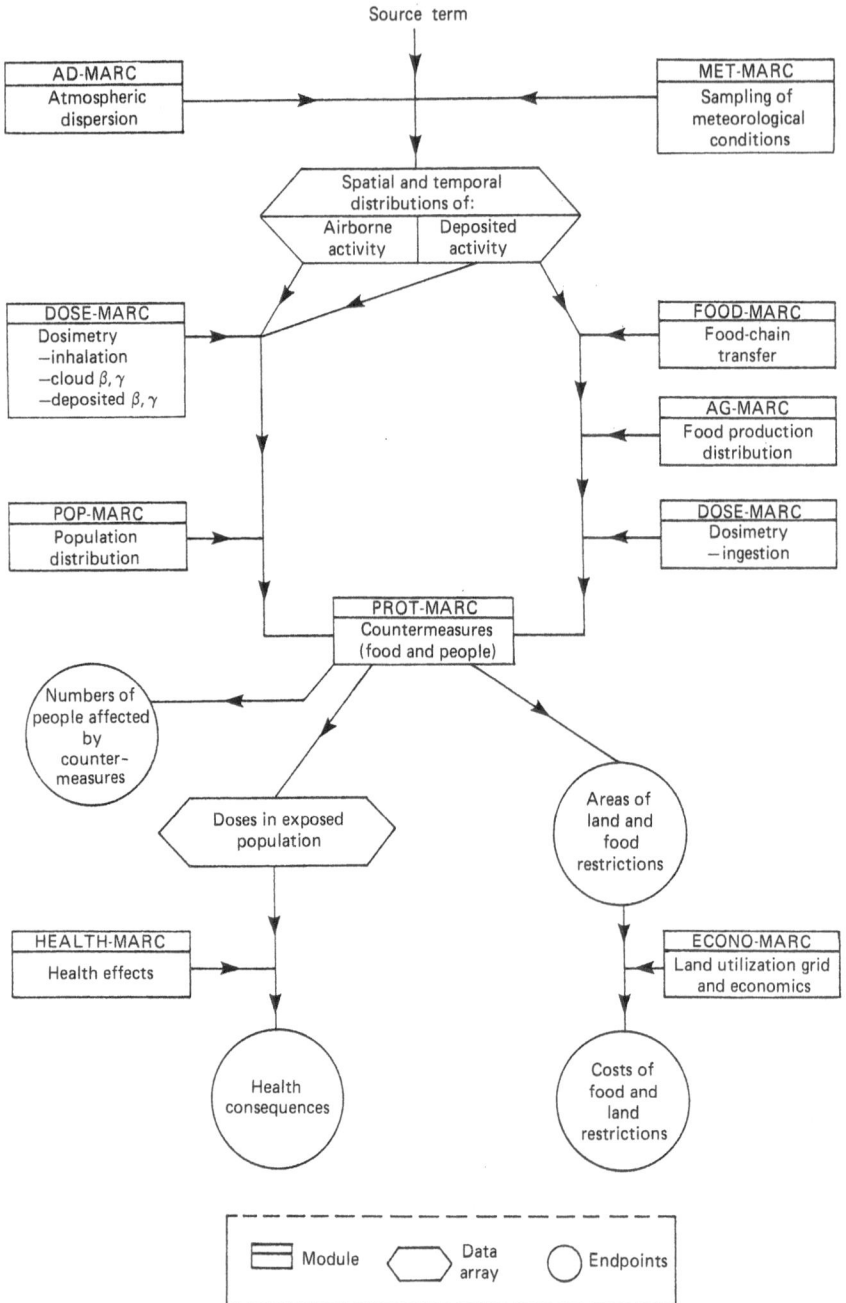

**Fig. 24.10.** Schematic diagram of the major components of MARC.

**Table 24.1.** Predicted duration of food bans for release UK11:[a] neutral weather conditions (Pasquill category D), with rain (1 mm/h), wind speed, 5 m/s

| Distance from release (km) | Crops | | | | | | | | | Milk | | | | | | | | | Meat | | | | | | | | |
|---|---|---|---|---|---|---|---|---|---|---|---|---|---|---|---|---|---|---|---|---|---|---|---|---|---|---|---|
| Angle from plume centerline (deg) | 0 | 5 | 10 | 15 | 20 | 25 | 30 | 35 | 40 | 0 | 5 | 10 | 15 | 20 | 25 | 30 | 35 | 40 | 0 | 5 | 10 | 15 | 20 | 25 | 30 | 35 | 40 |
| 0 –1.5 | 5 | 5 | 5 | 5 | 5 | 5 | 0 | 0 | 0 | 3 | 3 | 2 | 2 | 2 | 2 | 2 | 2 | 1 | 4 | 4 | 4 | 4 | 3 | 1 | 0 | 0 | 0 |
| 1.5–2.5 | 5 | 5 | 5 | 5 | 5 | 0 | 0 | 0 | 0 | 2 | 2 | 2 | 2 | 2 | 2 | 1 | 1 | 0 | 3 | 3 | 3 | 2 | 0 | 0 | 0 | 0 | 0 |
| 2.5–3.5 | 5 | 5 | 5 | 5 | 0 | 0 | 0 | 0 | 0 | 2 | 2 | 2 | 2 | 2 | 1 | 1 | 0 | 0 | 2 | 2 | 2 | 0 | 0 | 0 | 0 | 0 | 0 |
| 3.5–4.5 | 5 | 5 | 5 | 0 | 0 | 0 | 0 | 0 | 0 | 2 | 2 | 2 | 2 | 2 | 1 | 0 | 0 | 0 | 2 | 1 | 0 | 0 | 0 | 0 | 0 | 0 | 0 |
| 4.5–5.5 | 5 | 5 | 5 | 0 | 0 | 0 | 0 | 0 | 0 | 2 | 2 | 2 | 2 | 1 | 1 | 0 | 0 | 0 | 0 | 0 | 0 | 0 | 0 | 0 | 0 | 0 | 0 |
| 5.5–6.5 | 5 | 5 | 0 | 0 | 0 | 0 | 0 | 0 | 0 | 2 | 2 | 2 | 2 | 1 | 1 | 0 | 0 | 0 | 0 | 0 | 0 | 0 | 0 | 0 | 0 | 0 | 0 |
| 6.5–7.5 | 0 | 0 | 0 | 0 | 0 | 0 | 0 | 0 | 0 | 2 | 2 | 2 | 1 | 1 | 0 | 0 | 0 | 0 | 0 | 0 | 0 | 0 | 0 | 0 | 0 | 0 | 0 |
| 7.5–8.5 | 0 | 0 | 0 | 0 | 0 | 0 | 0 | 0 | 0 | 2 | 2 | 2 | 1 | 1 | 0 | 0 | 0 | 0 | 0 | 0 | 0 | 0 | 0 | 0 | 0 | 0 | 0 |
| 8.5–11 | 0 | 0 | 0 | 0 | 0 | 0 | 0 | 0 | 0 | 1 | 1 | 1 | 1 | 0 | 0 | 0 | 0 | 0 | 0 | 0 | 0 | 0 | 0 | 0 | 0 | 0 | 0 |
| 11 –13 | 0 | 0 | 0 | 0 | 0 | 0 | 0 | 0 | 0 | 1 | 1 | 1 | 1 | 0 | 0 | 0 | 0 | 0 | 0 | 0 | 0 | 0 | 0 | 0 | 0 | 0 | 0 |
| 13 –15 | 0 | 0 | 0 | 0 | 0 | 0 | 0 | 0 | 0 | 1 | 1 | 1 | 1 | 0 | 0 | 0 | 0 | 0 | 0 | 0 | 0 | 0 | 0 | 0 | 0 | 0 | 0 |
| 15 –19 | 0 | 0 | 0 | 0 | 0 | 0 | 0 | 0 | 0 | 1 | 1 | 1 | 0 | 0 | 0 | 0 | 0 | 0 | 0 | 0 | 0 | 0 | 0 | 0 | 0 | 0 | 0 |
| 19 –23 | 0 | 0 | 0 | 0 | 0 | 0 | 0 | 0 | 0 | 0 | 0 | 0 | 0 | 0 | 0 | 0 | 0 | 0 | 0 | 0 | 0 | 0 | 0 | 0 | 0 | 0 | 0 |
| 23 –28 | 0 | 0 | 0 | 0 | 0 | 0 | 0 | 0 | 0 | 0 | 0 | 0 | 0 | 0 | 0 | 0 | 0 | 0 | 0 | 0 | 0 | 0 | 0 | 0 | 0 | 0 | 0 |

[a] See ref. 15 of details of release assumptions.

*Key:*

| Symbol | 0 | 1 | 2 | 3 | 4 | 5 |
|---|---|---|---|---|---|---|
| Duration of ban | None | 7 d | 30 d | 100 d | 200 d | 1 yr |

**Table 24.2.** Predicted amounts of agricultural products restricted by countermeasures for release UK11,[a] neutral weather conditions (Pasquill category D), with rain (1 mm/h), wind speed, 5 m/s

| Agricultural product | Produce restricted to distance $d$ (km) |
|---|---|
| | Total |
| Wind direction 240°N | |
| Milk restricted in 7 days (L) | $7.8 \ 10^4$ |
| Total milk restricted (L) | $1.2 \ 10^5$ |
| Initial crop area restricted (km$^2$) | $4.7 \ 10^0$ |
| Time integral of the area of crop restrictions (km$^2$y) | $4.7 \ 10^0$ |
| Initial number of livestock restricted | $1.1 \ 10^3$ |
| Time integral of the number of livestock restricted (livestock-y) | $2.7 \ 10^2$ |
| Wind direction 270°N | |
| Milk restricted in 7 days (L) | $8.8 \ 10^4$ |
| Total milk restricted (L) | $1.5 \ 10^5$ |
| Initial crop area restricted (km$^2$) | $4.4 \ 10^0$ |
| Time integral of the area of crop restrictions (km$^2$-y) | $4.4 \ 10^0$ |
| Initial number of livestock restricted | $6.7 \ 10^2$ |
| Time integral of the number of livestock restricted (livestock-y) | $1.8 \ 10^2$ |
| Wind direction 300°N | |
| Milk restricted in 7 days (L) | $1.0 \ 10^5$ |
| Total milk restricted (L) | $1.6 \ 10^5$ |
| Initial crop area restricted (km$^2$) | $4.9 \ 10^0$ |
| Time integral of the area of crop restrictions (km$^2$-y) | $4.9 \ 10^0$ |
| Initial number of livestock restricted | $7.9 \ 10^2$ |
| Time integral of the number of livestock restricted (livestock-y) | $1.7 \ 10^2$ |

[a] See ref. 15 for details of release assumptions.

summed to produce a distribution. Figure 24.11 shows the various ways in which the output of such an assessment can be expressed. The results given in this figure are conditional upon an assumed release having occurred. Although these are useful, it is often more important to know the probability per unit time of particular consequences occurring or being exceeded. Results in this format can be obtained directly from those shown in Fig. 24.11 by multiplying the probability of exceeding a given number of consequences by the calculated frequency of occurrence of the assumed release. By summing such results for all possible releases, the risks associated with the nuclear installation can be estimated.

In 1987 NRPB carried out a probabilistic consequence assessment for a proposed PWR at Hinkley Point in Somerset (16). In this, 12 hypothetical degraded core accidents were considered; the source terms were provided by the CEGB, who funded the NRPB study. Table 24.3 shows the results for one particular accident, conditional upon it occurring. Figure 24.12 shows the calculated fre-

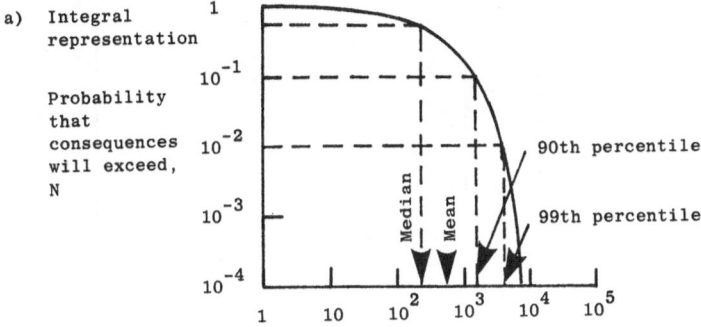

a) Integral representation

b) Differential representation

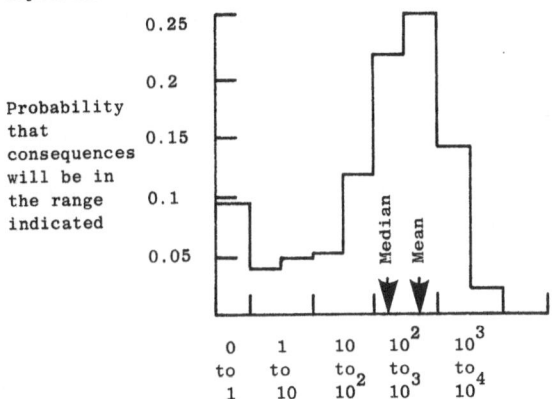

Number of consequences in specified ranges - arbitrary units

c) Characteristic quantities

| Quantity | Number of consequences, N (arbitrary units) | Number of consequences, N (arbitrary units) | Probability of exceeding N % |
|---|---|---|---|
| Mean | 530 | | |
| Median | 210 | 1 | 93 |
| | | 10 | 81 |
| Percentiles | | 100 | 64 |
| 1st | 0 | 1000 | 17 |
| 80th | 820 | 10,000 | 0 |
| 90th | 1400 | | |
| 99th | 3900 | | |

**Fig. 24.11.** Illustrative example of output from probabilistic assessments showing the distribution of consequences (arbitrary units) expressed in different formats.

**Table 24.3.** Characteristics of the distributions of consequences predicted upon occurrence of UK11 (UK11 Frequency: $6.7 \ 10^{-7}$/y)[a]

| Health effect | Expectation value, $E$ | Value at the $p$th percentile | | | | % Probability | |
|---|---|---|---|---|---|---|---|
| | | $p = 1$ | $p = 50$ | $p = 90$ | $p = 99$ | $P(N = 0)$ | $P(N > E)$ |
| **Early** | | | | | | | |
| Death | 0.0 | 0.0 | 0.0 | 0.0 | 0.0 | 100 | 0 |
| Prodromal vomiting | 0.0 | 0.0 | 0.0 | 0.0 | 0.0 | 100 | 0 |
| Lung morbidity | 0.0 | 0.0 | 0.0 | 0.0 | 0.0 | 100 | 0 |
| **Late** | | | | | | | |
| Fatal cancer | 3.2 | $3.0 \ 10^{-1}$ | 2.2 | 7.2 | $1.3 \ 10^1$ | 0 | 34 |
| Nonfatal thyroid cancer | $3.3 \ 10^1$ | 4.4 | $2.6 \ 10^1$ | $6.1 \ 10^1$ | $1.2 \ 10^2$ | 0 | 35 |
| Nonfatal skin cancer | 1.5 | $6.6 \ 10^{-2}$ | $8.6 \ 10^{-1}$ | 3.8 | 6.5 | 0. | 34 |
| Nonfatal breast cancer | $3.4 \ 10^{-1}$ | $1.6 \ 10^{-2}$ | $1.7 \ 10^{-1}$ | $8.8 \ 10^{-1}$ | 1.6 | 0 | 33 |
| Hereditary effects | 1.8 | $8.6 \ 10^{-2}$ | $9.3 \ 10^{-1}$ | 4.8 | 8.5 | 0 | 33 |

Area and number of people evacuated

| Parameter | Expectation value, $E$ | Value at the $p$th percentile | | | | % Probability | |
|---|---|---|---|---|---|---|---|
| | | $p = 1$ | $p = 50$ | $p = 90$ | $p = 99$ | $P(N = 0)$ | $P(N > E)$ |
| Number of people evacuated | $2.9 \ 10^2$ | 0.0 | $8.9 \ 10^1$ | $8.4 \ 10^2$ | $1.4 \ 10^3$ | 18 | 37 |
| Area of land evacuated (km²) | 7.8 | 0.0 | 7.2 | $1.6 \ 10^1$ | $1.7 \ 10^1$ | 18 | 48 |

**Table 24.3.** (*Continued*)

Agricultural products restricted by countermeasures[b]

| Agricultural product | Expectation value, $E$ | Value at the $p$th percentile | | | | % Probability | |
| --- | --- | --- | --- | --- | --- | --- | --- |
| | | $p = 1$ | $p = 50$ | $p = 90$ | $p = 99$ | $P(N = 0)$ | $P(N > E)$ |
| Milk restricted in 7 days (L) | $3.4\ 10^4$ | $7.7\ 10^2$ | $6.2\ 10^3$ | $9.2\ 10^4$ | $4.1\ 10^5$ | 0 | 22 |
| Total milk restricted (L) | $1.2\ 10^5$ | $2.9\ 10^3$ | $1.4\ 10^4$ | $2.8\ 10^5$ | $1.6\ 10^6$ | 0 | 21 |
| Initial crop area restricted (km$^2$) | $1.1\ 10^{-1}$ | 0.0 | $5.7\ 10^{-2}$ | $2.2\ 10^{-1}$ | $9.5\ 10^{-1}$ | 18 | 18 |
| Time integral of the area of crop restrictions (km$^2$ y) | $5.0\ 10^{-2}$ | 0.0 | $1.2\ 10^{-3}$ | $7.2\ 10^{-2}$ | $7.3\ 10^{-1}$ | 18 | 16 |
| Initial number of livestock restricted | $3.5\ 10^2$ | 0.0 | $6.0\ 10^1$ | $5.6\ 10^2$ | $5.1\ 10^3$ | 6.9 | 17 |
| Time integral of the number of livestock restricted (livestock-y) | $8.8\ 10^1$ | 0.0 | 4.7 | $1.5\ 10^2$ | $1.4\ 10^3$ | 6.9 | 16 |

[a] The frequency ($y^{-1}$) with which particular consequences specified in the table are exceeded can be obtained as $0.01\ f(r).(100\text{-}p)$, where $f(r)$ is the frequency of the release category ($y^{-1}$), and $p$ is the percentile appropriate to the value of consequences of interest.

[b] The areas of crop restrictions refer to the actual land area used for crop production.

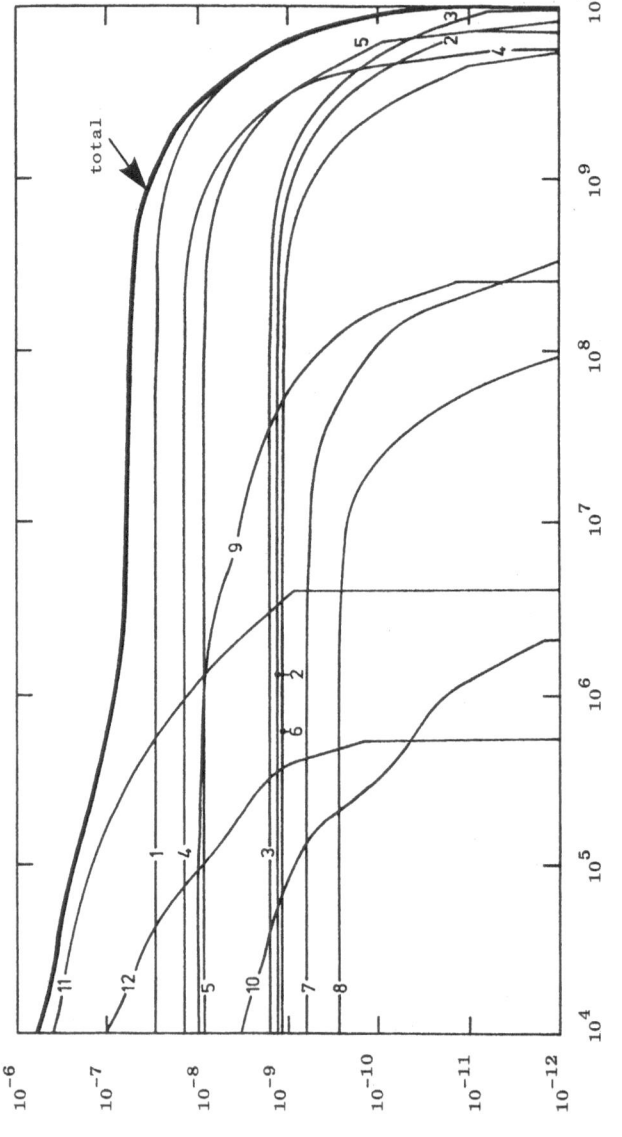

**Fig. 24.12.** The frequency distributions of restrictions on milk from 12 degraded core accidents for one year's operation of the proposed PWR at Hinkley Point.

quency distributions of restrictions on milk for all 12 hypothetical accidents, taking into account their frequencies of occurrence.

## Use of Models in Risk Management

As will be apparent from the previous sections of this chapter, models and calculational techniques for assessing the risks of potential accidents have reached an advanced stage of development. The results of such assessments provide a valuable input to risk management in the sense of taking appropriate decisions on the siting, design and methods of operation of nuclear installations, and of formulating contingency plans to limit the consequences of accidents if they occur (17,18). It is much less clear, however, whether models are useful in providing an input to the decisions that have to be taken in the aftermath of an actual accident because these decisions will rely much more on measurements of radionuclide levels in the environment and predetermined recommendations or regulations about actions to be taken if these levels exceed given values (usually known as "intervention levels" or "derived intervention levels"). Nevertheless, there is a role for the use of models in estimating the total consequences of an actual accident, and this is important because in the current climate of opinion, many of the questions asked by the public and politicians once an accident has happened are likely to relate to long-term effects on health and agriculture. The three possible roles for the use of mathematical models in risk management are addressed, in turn, below.

### Use of Models in Formulating Contingency Plans

To date, models of radionuclide transfer through terrestrial food chains have been used in two ways in providing input to the formulation of contingency plans. The first of these consists of translating selected intervention levels of dose into derived intervention levels of concentrations of radionuclides in food-stuff. Examples of this kind of model application are given in an International Atomic Energy Agency publication (19) and in an NRPB work (20). Figure 24.13 shows the kind of extra information that can be produced from models to aid the interpretation of monitoring data following an accident.

The second type of application of models of radionuclide transfer through the environment in the formulation of contingency plans is in providing an input to decisions on primary intervention levels. One example of such applications is the NRPB work on the choice of the "optimum" dose criterion for the introduction and withdrawal of restrictions on the production and consumption of foodstuff following an accident (21). In this, a method was demonstrated for calculating optimum dose criteria; it involved assigning monetary values to lost food production and to health detriment, expressed as collective effective dose equivalent commitment. It was assumed in the analysis that food-supply restrictions are introduced and withdrawn at the same projected levels of annual

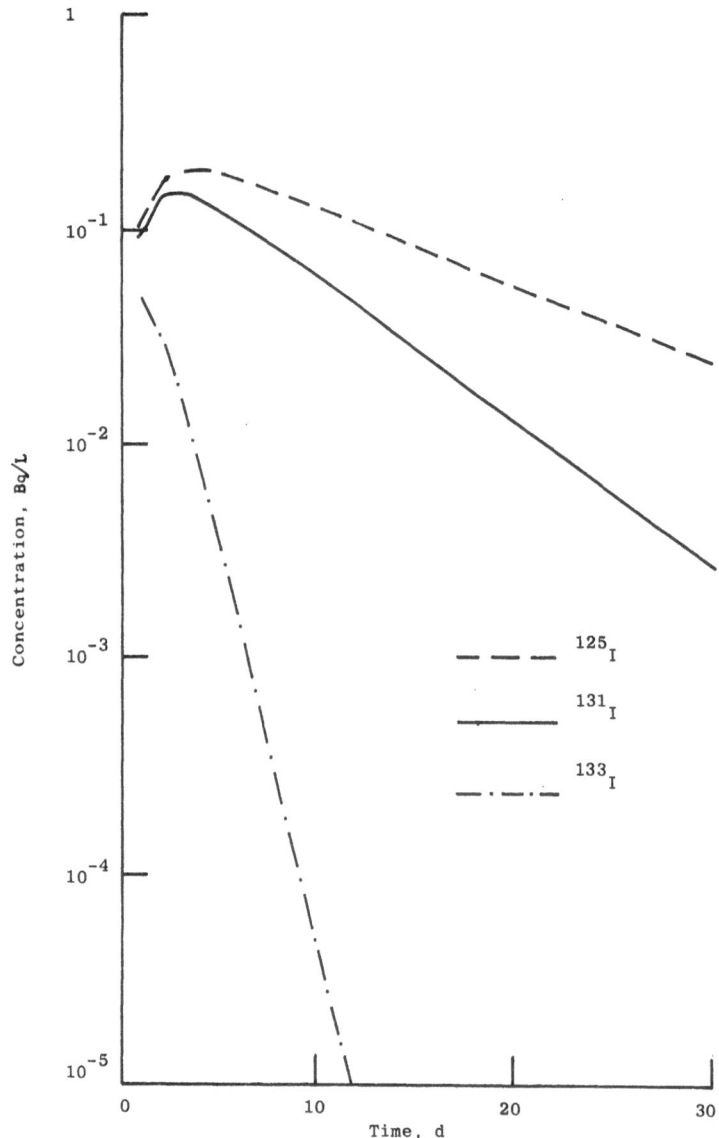

**Fig. 24.13.** Concentration of [125]I, [131]I, and [133]I in milk following a single deposit of 1Bq/m² on pasture. (Reprinted with permission of ref. 20.)

dose to individuals. Using this method, the optimum dose criterion was found to vary over almost the whole range of values considered (0.1 mSv per year to 50 mSv per year). For many of the cases considered, however, the optimum dose criterion lay between 1 and 10 mSv per year. In this study, the most important parameter in determining the calculated optimum was the cost assigned to unit collective dose (essentially the monetary value assigned to averting a radiation health effect). Table 24.4 gives an example of some of the results of

**Table 24.4.** Ratio of the total cost[a] at the optimum to the total cost at 5 mSv/y for four releases[b]

| Value of the person-Sv, £ | Release HRA | | Release HRB | | Release HRC | | Release HRD | |
|---|---|---|---|---|---|---|---|---|
| | Dry | Wet | Dry | Wet | Dry | Wet | Dry | Wet |
| Crops | | | | | | | | |
| 5,000 | 0.96 | 1.00 | 1.00 | 0.94 | 0.97 | 0.83 | 1.00 | 0.94 |
| 20,000 | 0.98 | 0.68 | 0.83 | 0.97 | 1.00 | 1.00 | 1.00 | 1.00 |
| 50,000 | 0.89 | 0.37 | 0.64 | 0.78 | 0.98 | 0.86 | 0.99 | 0.89 |
| Milk | | | | | | | | |
| 5,000 | 0.92 | 0.97 | 1.00 | 0.94 | 1.00 | 0.99 | 1.00 | 1.00 |
| 20,000 | 1.00 | 0.96 | 0.83 | 0.91 | 0.97 | 0.89 | 0.98 | 0.83 |
| 50,000 | 0.75 | 0.75 | 0.59 | 0.69 | 0.90 | 0.70 | 0.96 | 0.60 |
| Livestock | | | | | | | | |
| 5,000 | 0.75 | 0.57 | 0.78 | 0.42 | 0.95 | 0.37 | 0.99 | 0.75 |
| 20,000 | 0.98 | 0.93 | 0.98 | 0.77 | 0.99 | 0.85 | 1.00 | 0.98 |
| 50,000 | 1.00 | 1.00 | 1.00 | 0.97 | 1.00 | 1.00 | 1.00 | 1.00 |

[a] Total cost is the cost of the lost production plus the cost of the collective dose.
[b] See ref. 21 for details.

the NRPB work, which also showed that for a substantial number of the cases considered, 5 mSv per year is the optimum dose criterion. For many other cases use of a 5 mSv per year criterion led to total costs (health detriment plus lost agricultural production) that were similar to those obtained for the optimum dose criterion. The results of the study, therefore, suggested that in many accident situations, a decision to introduce restrictions on the distribution and consumption of foodstuff at a projected annual dose to the most exposed individuals of 5 mSv and to withdraw restrictions when projected doses are below this level would be consistent with the requirements of the radiological protection principle of optimization (i.e., doses should be as low as reasonably achievable, economic and social factors being taken into account). It was also recognized, however, that a more extensive series of analyses would be necessary before firm conclusions could be drawn and that in the event of an actual accident, decisions on countermeasures might well be influenced by factors other than those directly related to radiological protection.

## Inputs to Decisions After an Actual Accident

Once an accident has occurred, there are some decisions about contaminated foodstuff that have to be taken relatively rapidly; for example, whether or not it is necessary to impose bans on the consumption of milk and, depending on the season of the year, fruit and green vegetables. For other decisions more time is available. These include decisions on restrictions on the movement and slaughter of animals for consumption by humans, those on the withdrawal of initial countermeasures, and those on remedial measures such as deep plough-ing of soil to reduce radionuclide concentrations in the rooting zone of arable and pasture land.

Since the initial decisions have to be taken quickly if the countermeasures are to be effective in reducing doses, there is little opportunity to gather all the monitoring data, use models to estimate the radiological impact of the accident, or to weigh the costs and benefits of particular courses of action. Such decisions must be taken on the basis of plans formulated in advance, and there must be enough flexibility in these plans to cater for all reasonably foreseeable situations. As described in the previous section, models of radionu-clide transfer through the food chain can provide part of the basis for such plans.

In the case of later decisions it will be possible to collect all the appropriate monitoring data and to carry out extensive calculations of potential costs and radiological impacts of countermeasures and remedial actions. Computer pro-grams to perform these calculations are being developed in some countries and as part of international projects. Figure 24.14 shows an outline of the logical structure that might be envisaged for such programs. The main point to note about this structure is that it is designed to be flexible enough to allow judgments about the relative importance of the various components of radiological impact and economic cost to be incorporated, as well as views on factors that

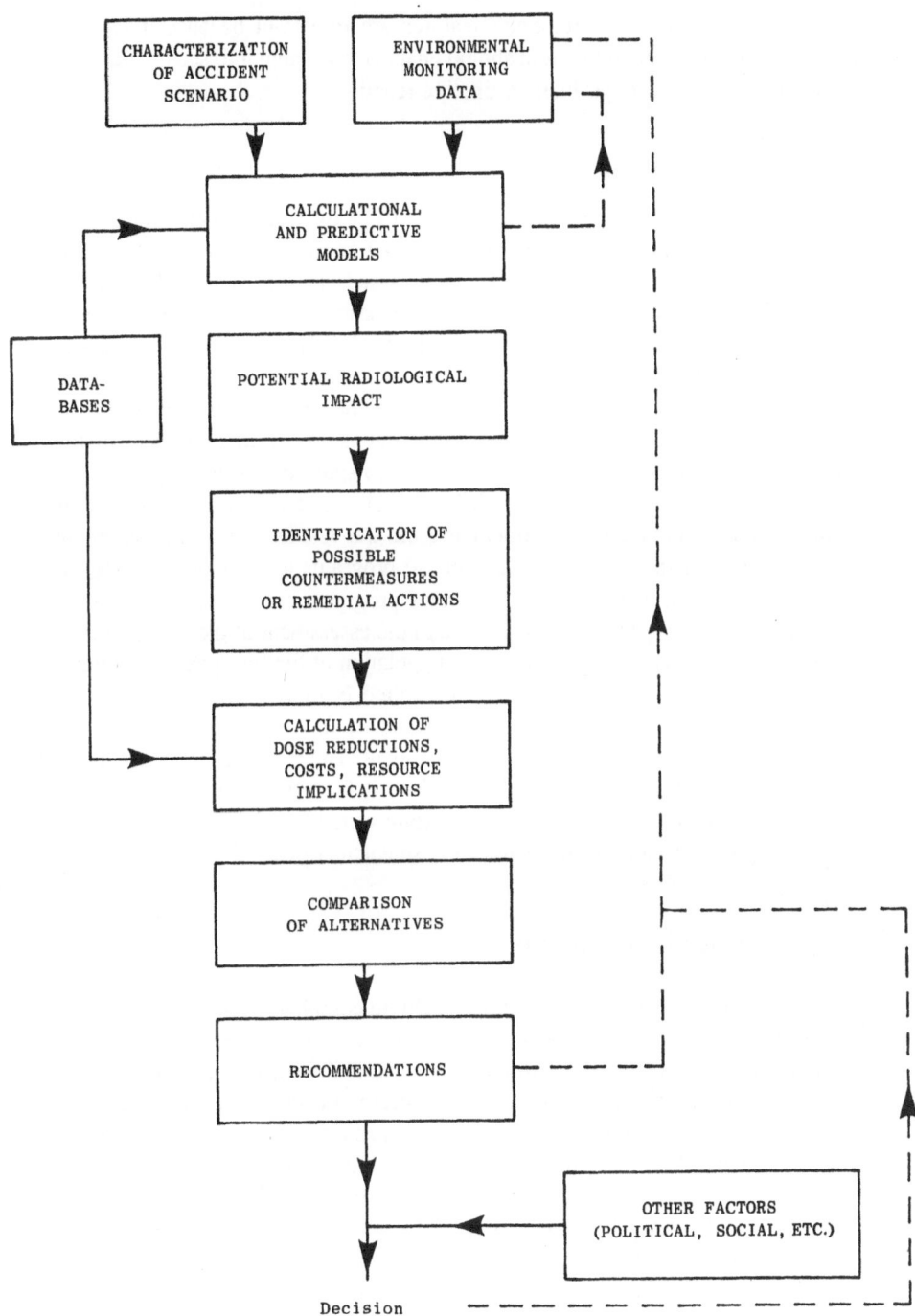

**Fig. 24.14.** Flow chart of possible radiological accident decision-aiding system.

are of a nontechnical nature. Such features are important because it is quite clear that any decision-aiding tool should enable those using it to see the implications of following through their instinctive reactions.

## Consequence Assessments

Following an actual accident there are likely to be two requirements for assessments of its consequences: one to answer initial questions (for example, "how many people will die?," "is it as bad as the Windscale fire, Three Mile Island, or Chernobyl?"), and one to give a detailed description of real and potential effects on individuals and populations. The best solution to fulfill the first requirement is probably to scale the results of precalculations of the consequences of potential accidents; there will be no time available for assimilation of all the results of environmental monitoring data, and using detailed models in the absence of such data would undoubtedly lead to spurious results.

To fulfill the second requirement it is necessary to have computer programs available that can extract and extrapolate from the vast amount of measurement data the numbers that are needed as input to programs to calculate radiological consequences. The latter type of computer programs exists. There are those, for example, developed by NRPB for use in the assessment of the consequences of the Chernobyl reactor accident on the population of the European community (22) and that have been provided to the United Nations Scientific Committee on the Effects of Atomic Radiation (UNSCEAR) for use in its assessment of the worldwide radiological impact of the accident. The former types of computer programs are still being developed; therefore, if faced with an accident today, the scientific community would have to follow traditional practice and select monitoring data by judgment for use in dose calculations.

## Uncertainties in Model Predictions

No chapter on mathematical modeling would be complete without some discussion of the uncertainties in model predictions. These uncertainties can be classified into two categories: those due to stochastic variability in the environment and those due to lack of knowledge of the processes occurring and the values of the parameters determining their patterns and rates. (In the current jargon these are referred to as "type 1" and "type 2" uncertainties, respectively.) In probabilistic accident-consequence assessments, stochastic variability is taken into account explicitly through carrying out calculations for many possible sets of weather conditions. Type 2 uncertainties are dealt with separately using techniques described collectively as "uncertainty analyses." Currently, these techniques are only able to take into account uncertainties about the values of the parameters used in models. This is done by deriving probability distributions of parameter values based on existing knowledge and subjective, expert judgment, together with correlations between the parameters. Models are then run many times, each with a different set of parameter values sampled from their distributions, and the output is a probability distribution of results. From this, quantitative statements can be made about the degree of confidence in model predictions (see, e.g., ref. 23). The advantage of such techniques is that they enable more realistic estimates of uncertainty to be made than is possible by simply examining the range of values that each particular parameter might take. Their disadvantage is that they can be computationally expensive and, more importantly, that significant amounts of effort need to be devoted to agreeing and justifying distributions of parameter values and correlations. This is, however, a small price to pay for the amount of information gained and the degree of consensus that can be reached about the reliability of model predictions. In addition, these techniques allow the major contributions to the uncertainty in model predictions to be identified, and hence their results are extremely useful in indicating priorities for future research on environmental transfer of radionuclides.

## Conclusions

Mathematical models of radionuclide transfer through the environment are essential tools in assessing the risks associated with existing and planned nuclear installations. They also have an important role to play in the formulation of plans to deal with accidental releases of radionuclides and in estimating the consequences of accidents once they have occurred. Although models are continuously being improved, much of the focus of current work is on verification and validation of existing models, on making quantitative estimates of the uncertainties associated with their predictions, and on using models for calculations designed to provide an input to planning the actions to be taken in the event of a future accidental release.

## Summary

This chapter outlined the types of mathematical models that are available for calculating the risks to people from routine and accidental releases of radionuclides into the environment, with particular reference to prediction of radionuclide transfer through terrestrial food chains. The ways in which these models can be used in assessing the risks associated with existing and planned nuclear installations were described. Examples of assessment results were given, and the use of such results in planning the actions to be taken to reduce risk via ingestion of contaminated food following a nuclear accident were illustrated by case studies.

The models described can, in principle, also be used to assess the radiological consequences of an accident soon after it has occurred and hence to provide an input to decisions on mitigating actions. There are, however, practical difficulties in using models in emergency situations. These difficulties have been discussed and conclusions drawn on the most appropriate methods of employing models to aid decisions after an actual accident.

Finally, techniques for quantifying the uncertainties in model predictions were briefly outlined and uses of results of uncertainty analyses were discussed.

## References

1. IAEA (1986) Atmospheric dispersion models for application in relation to radionuclide releases. Vienna, IAEA-TECDOC-379
2. CEC (1986) Proceedings of a workshop on methods for assessing the off-site radiological consequences of nuclear accidents, Luxembourg, April 15–19, 1985, EUR 10397 EN
3. IAEA (1983) An oceanographic model for the dispersion of waste disposed of in the deep sea. IMO/FAO/UNESCO/WHO/IAEA/UN/UNEP Group of Experts on the Scientific Aspects of Marine Pollution. Vienna, GESAMP—Reports and Studies No. 19
4. Hill MD, Cooper JR, Charmasson S, Robeau D (1986) Mathematical models for the transfer of radionuclides in the marine environment and their use in radiological assessments. PARCOM/NEA Workshop on Marine Radioactivity Research and Monitoring in the Paris Convention Area, Paris, February 18–20, 1986, in press
5. Broyd TW, et al. (1984) A directory of computer programs for assessment of radioactive waste disposal in geological formations. EUR 8669 EN
6. Meekings GF, Walters B (1986) Dynamic models for radionuclide transport in agricultural ecosystems: summary of results from a UK code comparison exercise. J Soc Radiol Protect vol 6, No 2
7. Thorne MC, Coughtrey PJ (1984) Validation and refinement of model codes: a comparison of results obtained from three dynamic foodchain models. ANS Report No. 522–2. Associated Nuclear Services, Epsom
8. Brown J, et al. A comparison of two European dynamic foodchain models. CEC, Luxembourg, in press
9. Haegg C, Johansson G (1987) BIOMOVS: An international model validation study. CEC Workshop on Methods for Assessing the Reliability of Environmental Transfer Models Predictions, Athens, October 5–9, 1987, in press
10. Simmonds JR, Linsley GS, Jones JA (1979) A general model for the transfer of radioactive materials in terrestrial food chains. NRPB-R89, HMSO, London
11. Brown J, Haywood SM, Wilkins BT (1987) Validation of the FARMLAND models for radionuclide transfer through terrestrial foodchains. CEC Workshop on Methods for Assessing the Reliability of Environmental Transfer Model Predictions, Athens, October 5–9, 1987, in press
12. Hill MD (ed) Verification and validation of NRPB models for calculating rates of radionuclide transfer through the environment. National Radiological Protection Board, Chilton, in press
13. Clarke RH, Kelly GN (1981) MARC—The NRPB methodology for assessing radiological consequences of accidental releases of activity. NRPB-R127, HMSO, London
14. NRPB. The NRPB methodology for assessing the radiological consequences of accidental releases—MARC-1, in press
15. Haywood SM, Charles D, Kelly GN (1983) Agricultural consequences of postulated accidental releases from the Sizewell PWR in selected meteorological conditions. Chilton NRPB-M105

16. Mills J, Morrey M, Williams J, Jones JA, Simmonds JR (1987) An assessment of the radiological consequences of releases from degraded core accidents for a proposed PWR at Hinkley Point. Chilton NRPB-M141
17. CEC (1986) Methods for assessing the off-site radiological consequences of nuclear accidents. CEC Brussels EUR 10243 EN
18. USNRC (1987) Reactor risk reference document. NUREG-1150 vols 1–3
19. IAEA (1986) Derived intervention levels for application in controlling radiation doses to the public in the event of a nuclear accident or radiological emergency: principles, procedures and data. Vienna, IAEA Safety Series No. 81
20. Linsley GS, Crick MJ, Simmonds JR, Haywood SM (1986) Derived emergency reference levels for the introduction of countermeasures in the early and intermediate phases of emergencies involving the release of radioactive materials to atmosphere. Chilton, NRPB-DL10, HMSO, London
21. Dionian J, Simmonds JR (1985) The choice of individual dose criterion at which to restrict agricultural produce following an unplanned release of radioactive material to atmosphere. NRPB-R183, HMSO, London
22. Morrey M, Brown J, Williams JA, Crick MJ, Simmonds JR, Hill MD (1987) A preliminary assessment of the radiological impact of the Chernobyl reactor accident on the population of the European Community. CEC, Luxembourg
23. Crick MJ, Hofer E, Jones JA, Haywood SM (1988) Uncertainty analysis of the foodchain and atmospheric dispersion modules of MARC. Chilton NRPB-R184, HMSO, London

# Part VII
# Development of Guidelines for Safety Evaluation of Food and Water After Nuclear Accidents

CHAPTER 25

# Development of Guidelines for Safety Evaluation of Food and Water After Nuclear Accidents: Procedures in North America

## G. D. Schmidt[1]

## Introduction

This chapter describes the United States approaches on emergency reference levels for foodstuff. It covers the guidance developed in the 1960s in response to atmospheric fallout, the recommendations for state/local planning for nuclear power plants, and the policy for imported food developed following the Chernobyl accident.

Despite the assigned title, the procedures in Canada and Mexico are not covered; only those in the United States. Somers, Cooper, and Meyerhof will briefly describe Canada's approach in Chapter 32. Mexico has not developed specific guidance. Rather, Mexico has addressed concerns over levels of radioactive materials in food on a case-by-case basis. The basic approach taken by Mexico has been to assure that levels are as low as practicable.

## Atmospheric Fallout Guidance of 1964 to 1965

Guidance on intervention in the event of radiological contamination of food was first given by the Federal Radiation Council (FRC), an interdepartmental group, in the 1960s during the period of atmospheric fallout from nuclear weapons tests. This guidance for federal agencies was defined as Protective Action Guides (PAGs) and was based on the concept of avoiding projected dose (1). The PAG was defined as the projected dose to individuals in the general population that warrants protective action following a contaminating event. Protective action would be warranted if the expected individual dose reduction was not offset by adverse social, economic, or health effects from the action. The PAG applied to the projected or future dose that would be received without any action, and it was expected that the protective action would prevent most of this dose. The PAGs were considered applicable to these types of actions:

---

[1] Food and Drug Administration, Center for Devices and Radiological Health, Office of Health Physics, 5600 Fishers Lane, Rockville, MD 20857 USA. Now retired: 10025 Lloyd Road, Potomac, MD 20854 USA.

1. Altering production, processing, or distribution practices affecting the movemnt of radioactive contamination through the food chain and into the human body. This action included storage of food supplies and animal feeds to allow for radioactive decay.
2. Diverting affected products to uses other than human consumption.
3. Condemning affected foods.

Measures that require an alteration of the normal diet were considered generally less desirable than those listed and would not be undertaken except on the advice of competent medical authorities.

The FRC recommendations addressed $^{131}$I in the initial report. In the second PAG report $^{89}$Sr and $^{90}$Sr and $^{137}$Cs were identified as the radionuclides of particular importance. The PAG recommended doses for $^{131}$I were 300 mSv (30 rem) to the individual and 100 mSv (10 rem) to "the average of a suitable sample of the exposed population group" (1). The concept of "the average of a suitable sample of the exposed population" was developed because it generally is not practical to measure individual doses. As an operational technique, it was considered that conformance to the PAG for the average of the suitable sample would ensure that the individual PAG was met.

The background material to these recommendations further provided data to indicate that if the maximum level in milk was about 2,200 to 2,600 Bq/L (60,000 to 70,000 pCi/L), the infant thyroid dose would be about 100 mSv. It should be noted that this is not a derived intervention level (DIL) as currently used, but an action level for the milk concentration that was equivalent to the PAG. The DIL is the derived concentration in food that would achieve the dose protection level adopted for intervention. As currently used, the DIL is the concentration in food that would be allowed to enter the market. In contrast, the PAG is an action level for taking a specific action. It was expected that if this maximum level was reached, protective action would be taken to avoid most of the projected dose. Thus, the radioactive contamination of food entering the market and actually consumed would be appreciably lower. The specific protective actions considered in selecting these PAGs were (a) change the cattle's diet from pasture to stored feed, and (b) divert contaminated milk to processed dairy products to allow decay of $^{131}$I.

The FRC in a later report (2) provided PAG guidance for $^{89}$Sr and $^{90}$Sr and $^{137}$Cs by three categories covering different time periods and pathways as follows:

Category I: pasture–cow–milk–man pathway during the first 100 days;
Category II: other pathways during the first year; and
Category III: long-term pathway (root uptake from the soil).

The PAGs recommended for the active bone marrow or the whole body decreased from 100 mSv (individual) and 33 mSv (average) for Category I, to 50 mSv (individual) and 20 mSv (average) for Category II, and to 5 mSv (individual) and 2 mSv (average) for Category III. No specific guidance was given for concentrations in food considered equivalent to the PAGs.

# Guidance for Nuclear Power Plant Accidents—1982

In the 1970s the US Atomic Energy Commission initiated a program to develop state and local emergency plans to respond to accidents at nuclear reactor power plants. In the United States the state (and local) agencies have broad general authority for public health protection of their populations from local accidents. Since an accident might involve more than one state, it was felt desirable to have specific federal guidance for such accidents to ensure uniformity of the state (and local) protective actions. The responsibility for developing such guidance for food was assigned to the Food and Drug Administration (FDA).

The FDA published its recommendations for state and local agencies in 1982 (3). Table 25.1 shows the basic applicability of the 1982 FDA recommendations, which covered all causes of accidental contamination of food except war. It should be noted that this was a protective action concept (similar to the earlier guidance) and that the intent was to avoid most of the projected dose if the PAGs were exceeded. It also should be noted that this was short-term guidance (first 2 months) based on the observation that adequate time was available to evaluate and act on problems of longer term contamination.

The recommendations provided two types of graded action levels: a preventive PAG and an emergency PAG, as shown in Table 25.2. The intent was somewhat different than the lower and upper bounds of International Commission on Radiological Protection (ICRP) Publication 40 in that the preventive PAG was based on the concept of taking low-impact actions that would avoid (or reduce) the contamination of food (4). The principal action considered here was placing cows on stored feed to avoid contamination of dairy cows and milk and diverting

**Table 25.1.** General applicability of 1982 FDA protective action guidance for accidental contamination of foods

| | |
|---|---|
| Recommendations to: | State/local agencies |
| For: | Emergency planning and response |
| Covering: | Nuclear facility accidents |
| | Transportation accidents, fallout |
| Intent: | To avoid most of the projected (future) dose that would occur in the absence of protective action |
| Time periods: | Actions over 1–2 months |
| Not applicable to: | Routine situations |

**Table 25.2.** FDA recommended PAGs for accidental contamination of food (3)

Preventive PAG: 5 mSv whole body and other organs; 15 mSv thyroid
  Protective actions having minimal impact
  Prevent or reduce contamination of food

Emergency PAG: 50 mSv whole body and other organs; 150 mSv thyroid
  High impact, protective action justified
  Isolate from commerce, consider condemnation

**Table 25.3.** Rationale for FDA's protective action recommendations (3)

| Judgment of acceptable risk: | |
|---|---|
| Risk of death | $10^{-6}$/y over 70 y |
| Radiation equivalent | 5 mSv whole body |
| Perspective: | |
| Natural disasters | |
| Risk of death | $2.4 \times 10^{-6}$/y |
| Natural background | |
| Mean | 0.53 mSv/y |
| 95% Population | 0.28–0.84 mSv/ y |
| 2 Standard deviations | 0.28 mSv/ y |
| 2 Standard deviations over 70 y | 20 mSv |

fresh milk to milk products to allow radioactive decay in the case of [131]I contamination. The principal concern was contamination of the milk pathway by [131]I and, to a lesser extent, contamination by Cs and Sr.

The recommended preventive PAGs, as shown in Table 25.2, are 15 mSv (1,500 mrem) thyroid and 5 mSv (500 mrem) whole body (and other organs). Levels ten times higher are recommended as the emergency PAGs. The relative organ and whole-body limits reflect those of existing US regulations. This is an area that needs updating per current scientific evidence and the newer ICRP guidance.

The rationale for the PAGs was a judgment of acceptable risk, as shown in Table 25.3. This rationale considered the risk of both natural disasters and the variation in the natural radiation background. The risk level accepted was that of approximately $10^{-4}$ ($0.7 \times 10^{-4}$) lifetime risk of cancer death, which was equated as being a dose of 5 mSv whole body. Further, as noted in Table 25.4, the rationale also considered the feasibility of the protective actions on a cost/risk basis, the risk to subpopulations, and consistency with other PAGs. A discussion of these evaluations is found in Shleien et al. (5).

Since this was a projected dose approach, derived response levels (that corre-

**Table 25.4.** Further considerations to rationale for FDA's protective action recommendations (3)

| |
|---|
| Preventive PAG: 5 mSv/15 mSv thyroid |
| Feasibility |
|     Cost of stored feed |
|     Diversion of stored products |
| Emergency PAG: 50 mSv/150 mSv thyroid |
|     Highly exposed individual $v$ average |
|     Costs condemnations $v$ stored feed |
|     Natural disaster risk to subpopulations |
|     Health impact |
|     Consistent with other PAGs |

**Table 25.5.** Derived response levels[a] for milk pathway equivalent to the preventive PAGs (5 mSv/15 mSv) (3)

|  | $^{90}$Sr | | $^{131}$I | | $^{137}$Cs | |
|---|---|---|---|---|---|---|
|  | Infant | Adult | Infant | Adult | Infant | Adult |
| Initial activity |  |  |  |  |  |  |
| Area (Bq/m$^2$) | 19,000 | 74,000 | 4,800 | 66,600 | 110,000 | 185,000 |
| Forage (Bq/kg) | 6,700 | 30,000 | 1,900 | 26,000 | 48,000 | 70,000 |
| Peak milk (Bq/L) | 330 | 1,500 | 550 | 7,400 | 8,900 | 15,000 |
| Total intake (Bq) | 7,400 | 26,000 | 3,300 | 37,000 | 260,000 | 300,000 |

[a] Derived response levels for $^{134}$Cs and $^{89}$Sr were also given.

sponded to the preventive PAG) were recommended for the peak level in milk, total intake, initial deposition and forage contamination. Derived response levels for $^{90}$Sr, $^{131}$I, and $^{137}$Cs are given in Table 25.5 (values also were recommended for $^{89}$Sr and $^{134}$Cs). Derived levels were specified for both infants and adults as the critical or suitable population. These derived levels were those at which protective actions should be initiated with the intent that most of the dose be avoided. A level at which action should cease was not provided.

The recommendations also included suggested protective actions that should be taken at the preventive PAG as presented in Table 25.6. Emphasis was placed on the pasture–cow–milk pathway, but actions applicable to other foods were also addressed. The emergency PAG guidance addressed the isolation of food from distribution and the possible condemnation of food. It was believed important to consider the availability of other protective actions and the adequate supply of uncontaminated food before food was condemned.

# Import Food Guidance Following Chernobyl

The 1982 recommendation was the existing US guidance concerning accidental contamination of food at the time of the Chernobyl accident. The interagency group established to coordinate agency activities and provide information and

**Table 25.6.** Recommended protective actions at the preventive PAG level (3)

For dairy cows
  Use uncontaminated stored feed
  Use uncontaminated water
For milk
  Withhold fresh milk from market
  Storage for decay of short-lived radionuclides
  Diversion to milk products—powdered milk, cheese, butter, etc
For fruits and vegetables
  Wash, brush, peel to remove contamination
  Storage for decay by canning, freezing, etc

**Table 25.7.** FDA/USDA Chernobyl task-force recommendations for levels on import foods (6)

| | Level of concern (screening values) | |
| --- | --- | --- |
| | *Infant Food* | *Other food* |
| $^{131}I$ | 55 Bq/kg | 300 Bq/kg |
| $^{134/137}Cs$ | 370 Bq/kg | 370 Bq/kg |

Rationale
1. Food not under US control
2. Chronic problem
3. Period of contamination
   $^{131}I$—60 days
   $^{134/137}Cs$—365 days
4. Fraction of diet, 100%
   Infant—Protective
   Adult—Very conservative
5. PAG—15 mSv thryoid
   5 mSv whole body (other organs)

guidance to the public adopted the 5 mSv whole-body/15 mSv thyroid preventive PAGs as appropriate for protection of the public. The derived response levels were not applicable, since they were for the peak levels and for the situation where agricultural practices were under our control, which was not the case for imported foods.

For derivation of levels for imported foods, a task force of FDA and US Department of Agriculture (USDA) representatives assumed 100% contamination of the diet. This assumption was adequately protective for infants' foods and very conservative for adults' foods, as given in Table 25.7. This rationale resulted in the DILs (referred to as a "level of concern" by FDA) shown in Table 25.7. It was expected that the majority of the dose would be contributed by $^{131}I$ during early time periods and by $^{134}Cs$ and $^{137}Cs$ for later periods. Therefore, it was considered appropriate to use $^{131}I$ and the sum of $^{134}Cs$ and $^{137}Cs$ as key indicators. Because of the conservatism in these levels, it was considered unnecessary to establish levels of concern for other radionuclides known to be present (see Appendix A). For example, $^{90}Sr$ was monitored in selected samples to ensure that levels were within expected ranges, but a separate level of concern was not established.

The FDA accepted these levels as appropriate for imported foods and published a notice in the *Federal Register* of the availability of the Compliance Policy Guide for Imported Foods (Appendix A). The USDA, however, felt that it was appropriate to consider the fraction of the diet represented by meat and poultry in adopting levels referred to as screening values. Thus, the USDA derived a level of 2,800 Bq/kg (75,000 pCi/kg) for Cs in meat (Table 25.8) compared with the FDA level for Cs in other foods of 370 Bq/kg (10,000 pCi/kg). A summary of the USDA guidance also is found in Appendix A. Because these values were adopted as initial or short-term levels in accordance

**Table 25.8.** USDA Chernobyl screening values for meat and poultry (6)

Screening values
$^{131}I$                              55 Bq/kg (infants only)
$^{134}Cs + {}^{137}Cs$      2,800 Bq/kg

Assumptions
$^{131}I$—Accepted FDA value
PAG—5 mSv (280,000 Bq intake of Cs)
$^{134}Cs/^{137}Cs$—equal amounts
Exposure time—365 d
Consumption—0.275 kg/d (100 kg/y)

Revised screening value (as of October 1986)
$^{134}Cs + {}^{137}Cs$      370 Bq/kg (consistent with FDA value)

with the 1982 FDA recommendations, the USDA reconsidered its levels for imported meat and poultry in the fall of 1986. Although USDA considered a reduction by a factor of 10 for the long-term as appropriate, in October 1986 the screening value was reduced to 370 Bq/kg for total $^{134}Cs$ plus $^{137}Cs$ in conformity with the FDA levels for other foods.

## Future Revisions to US Guidance

The 1982 FDA PAG recommendations and the levels adopted by FDA and USDA for radionuclides in imported food following Chernobyl represent current US guidance and policy on emergency actions for food. The FDA staff has recognized that certain aspects of this guidance may be in need of revision, updating, or further clarification; however, a formal revision of the US guidance has not been initiated. Thus comments cannot be made on future revisions of US guidance for the ingestion pathway.

During the 1½ years since Chernobyl, individuals from the United States have participated in and carefully followed the activities of the international organizations: Commission of European Communities, Food and Agricultural Organization, International Atomic Energy Agency, Nuclear Energy Agency, and World Health Organization. Those involved have found this experience to be very informative and personally rewarding, and hope such efforts have helped foster the harmonization of DILs for international trade.

## Author's Comments

It is appropriate to comment on areas of possible revisions and my perception of possible approaches. These concluding comments are my personal thoughts and may not reflect agency policy.

## DILs

There seems to be little justification for revising the acceptable risk rationale adopted for the PAG recommendations (3). The resulting DIL of 5 mSv for the whole body is generally consistent with current thinking on the maximum dose level for taking actions for food. Cost/risk analysis (optimization), using cost factors generally accepted in the United States, appears to result in similar levels for total dose from all dietary intakes. Dose factors for the organs need to be updated to be made consistent with national and international guidance. An important question here involves the appropriate weighting factors for the public, and this may require further expert committee recommendations.

## PAGs versus DILs

The PAG concept is one of preventive public health, that is, taking an action (e.g., placing cows on stored feed) to prevent the contamination of the cow and the milk supply. The PAG response level is an action level, not a DIL for the introduction of food into commerce. It is expected that food entering the market would be well below the PAG response level to avoid most of the projected dose. The Chernobyl accident clearly indicates the need for a DIL for market food. It is not at all clear, however, that such a two-tier system can be readily understood by the public and the media.

## Optimization Procedure

The proposed optimization procedure, for implementation of principle (c) of ICRP Publication 40, has various problems (4). These include the following:

- It is not related to reduction of collective dose equivalent and it may even increase collective dose equivalent.
- It is dependent on cost factors that vary by country (or by state) and thus does not foster harmonization.
- It would result in a different level of individual protection based on cost of diet.
- It would base a health protection level solely on cost; a premise that is generally unacceptable to health agencies.

## DILs for Total Diet or Specific Foods

The objective of emergency intervention is to limit or reduce the risk to the public and individuals of the critical group to acceptable levels. The intervention and monitoring system should be designed to accomplish this objective. The question of whether to base DILs on total diet or food groups, or on a specific food intervention, involves aspects of the complexity, flexibility, and conservatism of the intervention system. Experience from Chernobyl has shown the need for flexibility to deal with unique food situations, that is, spices, lamb, powdered milk, beef extract, etc, and population groups, such as Laplanders. Further, Chernobyl has indicated that actual intakes are a small fraction (generally less than 10%) of that predicted by assuming the entire year's diet is contaminated at the DIL.

Thus, it appears that the DILs should be general and flexible. This would be consistent with the experience of those writing regulations that has shown it is not possible to foresee fututre applications and situations. Although the DILs should be protective of individuals in the critical group, they need not be excessively conservative.

## Suggested Approach

The DILs should be based on the annual intake of major food groups such as dairy products, grains, fruit and vegetables, and meat. For US diets, an annual intake of about 100 kg is appropriate for each of these food groups. General DILs would then be derived for the important radionuclides such as $^{131}$I, $^{134}$Cs, $^{137}$Cs, $^{239}$Pu, and perhaps $^{90}$Sr. In addition to these general DILs, there should be a specific DIL for infants' food because of the higher conversion factor from intake to dose.

It is felt that such general DILs can be applied to specific food interventions without the need for an additive rule to consider the dose contribution from other radionuclides or other contaminated foods. The basic conservatism in this calculational approach as compared with actual intake over a year should be protective of individuals in the critical group. *It is considered necessary,* however, that if multiple foods are contaminated, the health agency should conduct appropriate monitoring of the actual intake by critical groups to assure that the risk (dose) objective is met. And, if it appears that the dose objective is being exceeded, appropriate adjustment of DILs and food intervention would be indicated.

With regard to minor food items that constitute a very small fraction of dietary intake, such as spices, it would be reasonable to allow DILs higher by a factor of ten.

A basic concern in offering these comments is that adopted levels should be pragmatic. First, it should be easy to explain that they provide adequate protection for the general public and critical groups. Next, they should be readily understood and implemented by the food and agricultural community, which is not familiar with radiation protection concepts. Finally, the DIL/intervention approach should be flexible in responding to unique and specialized situations.

It is important that there be a consensus on DILs for food in international trade and DILs used by adjacent countries or regions. Such agreement should enhance the credibility of the health and radiation protection community and facilitate governmental action. It is strongly recommended that the international organizations consider the need for harmonization and agreement on the DILs recommended.

# References

1. Federal Radiation Council (1964) Radiation protection guidance for federal agencies. Fed Regist (29FR12056) August 22, 1964, and Report No. 5 Background material for the development of radiation protection standards, July 1964
2. Federal Radiation Council (1965) Radiation protection guidance for federal agencies. Fed Regist (30FR6953) May 22, 1965, and Report No. 7 Background material for the development of radiation protection standards: protective action guides for strontium-89, strontium-90 and cesium-137, May 1965
3. Food and Drug Administration (1982) Accidental radioactive contamination of human food and animal feeds; recommendations for state and local agencies. Fed Regist (47FR47073), October 22, 1982
4. International Commission on Radiological Protection (1984) Protection of the public in the event of major radiation accidents: principles for planning. ICRP Publication 40, Pergamon Press, Oxford
5. Shleien B, Schmidt GD, Chiacchierini RP (1982) Background for protective action recommendations: accidental radioactive contamination of food and animal feeds. HHS Publication FDA 82-8196, August 1982, Food and Drug Administration, Rockville, Maryland 20857
6. Food and Drug Administration (1968) Radionuclides in imported foods: levels of concern: availability of compliance policy guide. Notice Fed Regist (51FR23155) June 25, 1986

# Appendix

DEPARTMENT OF HEALTH AND HUMAN SERVICES

Food and Drug Administration

[Docket No. 86D-0214]

Radionuclides in Imported Foods; Levels of Concern; Availability of Compliance Policy Guide

AGENCY: Food and Drug Administration.
ACTION: Notice.

SUMMARY: Food and Drug Administration (FDA) is announcing the availability of Compliance Policy Guide 7119.14 and "Radionuclide Screening Values for Monitoring Imported Meat Products," which describe the protective action guidelines that FDA and the U.S. Department of Agriculture, Food and Safety and Inspection Service (FSIS), will use for monitoring radionuclides in imported foods sampled under field assignments.

ADDRESSES: Requests for single copies of Compliance Policy Guide 7119.14 and any written comments to the Dockets Management Branch (HFA-305), Food and Drug Administration, Rm. 4-62, 5600 Fishers Lane, Rockville, MD 20857. Requests for single copies of "Radionuclides Screening Values for Monitoring Imported Meat Products" and any written comments to Patricia Stolfa, Deputy Administrator for International Programs, Food Safety Inspection Service, U.S. Department of Agriculture, 14th and Independence Ave., SW., Washington, DC 20250, 202-447-3473.

FOR FURTHER INFORMATION CONTACT: Raymond Gill, Center for Food Safety and Applied Nutrition (HFF-300), Food and Drug Administration, 200 C St. SW, Washington, DC 20204, 202-485-0160.

SUPPLEMENTARY INFORMATION: In the Federal Register of October 22, 1982 (47 FR 47073), FDA issued radionuclide protective action recommendations to State and local officials. These recommendations were for use in the event of a domestic nuclear accident. Both FDA and FSIS have decided to utilize these protective action recommendations with certain dose calculation adjustments as a basis for monitoring radionuclide levels in imported foods sampled as a consequence of the fallout expected from the April 26, 1986, accident in the Soviet Union.

Compliance Policy Guide 7119.14 and "Radionuclide Screening Value for Monitoring Imported Meat Products" define levels of radionuclides that are to be used as indicators of the safety of imported foods and food ingredients and imported meat products. If imported products are found to contain unsafe levels of radionuclides, FDA and FSIS will take appropriate regulatory action to assure that the public health is protected.

Copies of Compliance Policy Guide 7119.14, FSIS's "Radionuclide Screening Values for Monitoring Imported Meat Products," and the **Federal Register** reference mentioned above have been placed on public display with the Dockets Management Branch (address above) and are available for public examination between 9 a.m. and 4 p.m., Monday through Friday. In accordance with 21 CFR 10.85(d)(3) and (i), any person may submit written comments on the guide. Requests for single copies of the filed documents and any comments on Compliance Policy Guide 7119.14 should reference the docket number found in brackets in the heading of this document and should be submitted to the Dockets Management Branch. Two copies of any comments are to be submitted, except that individuals may submit one copy.

Requests for single copies of FSIS's document and any written comments on the radionuclide screening values should be submitted to the Deputy Administrator for International Programs (address above).

Although the agencies will consider comments in deciding whether to revise these guidelines, they will not defer regulatory actions pending any such revisions.

Dated: June 18, 1986.

**James W. Swanson,**
*Acting Associate Commissioner for Regulatory Affairs.*
[FR Doc. 88-14291 Filed 6-24-88; 8:45 am]
**BILLING CODE 4160-61-M**

| FOOD AND DRUG ADMINISTRATION<br>COMPLIANCE POLICY GUIDES | GUIDE | 7119.14 |
|---|---|---|

### Chapter 19—IMPORT FOODS

SUBJECT: Radionuclides in Imported Foods—Levels of Concern

BACKGROUND

In the Federal Register of October 22, 1982 (47 FR 47073), FDA issued protective action guides to State and local officials in the event of a domestic nuclear accident. The basis for the guidance recommendations was that preventive action should be taken whenever the projected dose to the thyroid is 1.5 rem or the projected dose to the whole body, bone marrow or any other critical organ is 0.5 rem. The Agency has decided to utilize these protective action guides as the basis for monitoring foods sampled under field assignments issued on the subject of radionuclides in imported foods.

For foods produced in the United States the protective action guides provide a set of control measures that can be taken to alter the domestic food production process so that contaminated foods do not reach the market place. These control measures are provided in the October 22, 1982 Federal Register document. This document also provides an emergency level of concern above which responsible officials should take appropriate action to remove domestic products from channels of commerce.

FDA's monitoring efforts in the U.S. utilize radioactive iodine in milk as the primary indicator of whether or not a public health risk exists. This approach is used for several reasons. Radioactive iodine is more toxic than many other radionuclides. Radioactive iodine is expected to occur in greater abundance in fallout from a nuclear reactor accident of this type than other radionuclides. Radioactive iodine appears in milk within 3–4 days of deposition and milk reaches consumer markets faster than most other foods with the potential for contamination from fallout. In addition milk can be the sole source of nutrition for infants which is the most sensitive group to radioactive iodine. Therefore, the monitoring of milk serves as an early warning system for judging the health risk to consumers. If the predicted peak of contamination for milk reaches 15,000 picocuries per liter from I-131, responsible officials should take immediate preventive measures to alter the food production process such as removing cattle from pasture. If the milk reaches 150,000 picocuries per liter from iodine-131, responsible officials should take appropriate measures to remove the product itself from channels of commerce. For all other domestic foods the FDA also utilizes the guidance provided in the October 22, 1982 document on protective action guides.

The FDA and state and local health officials cannot influence the production of foods abroad and the Protective Action Guide's are intended to be used for a one time short term incident in the United States. Therefore, certain adjustments in calculating projected dose have been made to take into account what is known about the fallout from the April 26, 1986, accident in the Soviet Union and the expectation that foods offered for import into the United States may contain iodine-131 for as long as sixty days and cesium-134 and cesium-137 for as long as a year. In calculating projected dose the agency must take into account the number of days consumers are likely to be exposed to these radionuclides.

## REGULATORY ACTION GUIDANCE

At this time, it is expected that the majority of the radiation dose is contributed by iodine-131 which may be encountered in food offered for import for about 60 days and cesium-134 and cesium-137 which may be encountered in food offered for import for about one year. The levels of radionuclides that are to be used as indicators of the safety of imported foods are listed in Table I.

Table I
Monitoring Level

|  | Infant Food | Other Foods |
|---|---|---|
| I-131 | 1,500 pCi/kg | 8,000 pCi/kg |
| Cs-134 + Cs-137 | 10,000 pCi/kg | 10,000 pCi/kg |

This monitoring procedure will be modified at any time if information on the radionuclides present in foods indicates changes are needed. There are sufficient conservatisms incorporated into the calculations presented in Table I so that we do not believe that additional analysis is warranted except for spotchecks or where we have indications that other radionuclides are present in significant amounts. In the preparation of these levels of concern, FDA has elected to prepare a single level for all foods under their jurisdiction rather than for individual food categories. This is because reliable data on the relative percent of diet for many food categories is not available.

Furthermore, many imported foods are sold as bulk ingredients to be used in preparation of other foods and there is no way to compute the relative consumption rates for these bulk ingredient foods.

Animal feed and feedstuff are not imported in large volume. There is also a delay factor in the preparation of the feedstuff so that during the bulk shipment and distribution considerable reduction in concentration will occur so that there should be no cause to analyze these products on a routine basis.

Gamma spectroscopy should be used as the monitoring procedure for analyzing imported products regulated by FDA. This recommendation is based on the assumption that a majority of the gamma radiation will stem from iodine-131, cesium-134 and cesium-137.

Where there is a potential for contamination, automatically sample and hold all shipments of products offered for entry which are known to have originated within a 250-mile radius of the Chernobyl plant. This sampling effort should continue until we have information to assure us that there is no cause for concern. Samples from all other areas are to be collected on the basis of assignment.

Samples exceeding the levels of concern in Table I should trigger an appropriate control response to assure that the public health is protected. Standard procedures such as automatic detention, release with comment and refusal of entry should be considered.

The Food Safety and Inspection Service, under their authority, will initiate a control program to prevent the importation of meat and poultry adulterated with radioactivity and will be issuing monitoring levels for meat and poultry using the same protective action guideline criteria as FDA.

## RADIONUCLIDE SCREENING VALUES FOR MONITORING IMPORTED MEAT PRODUCTS[2]

The response level of 75,000 picocuries per kilogram (pCi/kg) for imported meat and poultry was determined in accordance with the Food and Drug Administration's (FDA's) Accidental Radioactive Contamination of Human Food and Animal Feeds; Recommendations for State and Local Agencies which was published in the Federal Register of October 22, 1982 (47 FR 47073), and in consultation with FDA. The response level takes into account the total intake of activity from a radionuclide and the average daily consumption of food in the United States.

This response level value is based, in part, upon the expectation that the major contributors of radiation to imported meat and poultry will be cesium-134 (half-life 2.1 years) and cesium-137 (half-life 30 years). In addition, it is not expected that iodine-131 (half-life 8 days) will contribute radioactivity levels of any practical concern to these types of imported products. Also, iodine-131 has a strong affinity for the thyroid—some 30,000 times that of muscle tissue, which comprises most imported meat and poultry products. Therefore, the value for iodine-131, if present, will be set conservatively at the same level as determined by FDA for infant foods, i.e., 1,500 pCi/kg.

The average daily consumption (0.275 kilograms per day) of meat and poultry is the sum of the individual average values provided in 47 FR 47073 for meat (0.220 kilograms per day) and poultry (0.055 kilograms per day). These figures are the same as those obtained from the U.S. Department of Agriculture Household Food Consumption Survey, 1965–1966. More recent estimates of average daily meat and poultry consumption for individuals are in the approximate range of 0.15 to 0.20 kilograms per day. Because these values are less than 0.275 kilograms per day, the response level of 75,000 pCi/kg should be a conservative value for monitoring purposes, i.e., lower than the response level which would result if the more recent consumption figures were used.

We are confident that meat and poultry can be separated from the food items under FDA regulatory control because of later food consumption survey information than was used for the development of 47 FR 47073. Therefore the monitoring values of the two agencies will differ, however, both are based on a 0.5 rem projected dose commitment to the whole body, bone marrow or any other organ.

The specific calculation of this response level is as follows:

Response Level = Total intake (microcuries)/Consumption (kilograms)
Total Intake = 7 microcuries of cesium-134 + 8 microcuries of cesium-137
                (These values are one-tenth of the respective emergency protective action guide levels for adults which are given in 47 FR 47073).

---

[2] U.S. Department of Agriculture, Food Safety and Inspection Service, Washington, DC 20250 USA.

Consumption = (0.275 kilograms per day) × (365 days of potential exposure)
            = 100.375 kilograms
Therefore, Response Level = 15/100.375
                          = 0.149440 microcuries per kilogram

Because cesium-134 and cesium-137 are expected to be present in equal amounts for nuclear reactor accidents, the response level is reduced by a factor of 2 in accordance with 47 FR 47073. Therefore, the response level is 0.07472 microcuries per kilogram or approximately 75,000 pCi/kg.

Following are the monitoring levels for meat and poultry:

Iodine-131                 :   1,500 pCi/kg or 1.5 pCi/gm
  (infant food only)
Cesium-134 + Cesium-137:  75,000 pCi/kg or 75 pCi/gm

CHAPTER 26

# The Development of WHO's Approach to DILs

P. J. Waight[1]

## Background

Prior to Chernobyl, many international organizations such as the Commission of European Communities (CEC) (1), International Commission on Radiological Protection (ICRP) (2), International Atomic Energy Agency (IAEA) (3), and the World Health Organization (WHO) (4), had produced guidelines for national authorities to apply in the event of an accident at a nuclear facility. Following Chernobyl it was generally accepted that this guidance was applicable close to the accident site, but was inadequate in guiding national authorities far removed from the accident. It was hardly surprising, therefore, that different national philosophies produced widely differing intervention and derived intervention levels. This deficiency was recognized early and led to a number of interagency and international meetings to determine how it could be remedied. One outcome of the interagency discussion was that WHO would endeavor to produce guideline values for derived intervention levels in environmental media, below which actions to reduce or avoid the potential health detriment were not justified. It was also felt that such guideline values would provide a basis upon which countries could implement their own derived intervention levels, and so promote a measure of harmonization.

## Introduction

An Expert Group meeting was convened by WHO in April 1987. In its discussions, the group felt that with the wide variation in food consumption patterns across the world and the apparent lack of consensus on what constituted an "acceptable" risk, it would be fruitless to pursue these avenues to arrive at guidelines for derived intervention levels (DILs). Instead, it was proposed that optimization should be used to determine the DILs. This technique was developed in the experts' report, which was then sent to as many WHO focal points as

[1] World Health Organization, 1211 Geneva 27, Switzerland.

possible for comments. Comments from 24 countries were received, and it was clear that many of the reviewers felt that optimization was not the appropriate approach, as it was based on costs that were difficult to determine and that were extremely variable.

As a result of these comments a revised draft document was produced for consideration by a larger Task Group, which met late in September 1987. A number of crucial questions needed to be addressed. These have been distilled as the wide discussions on approach have progressed in the months since the Chernobyl accident. They can be summarized:

1. What reference level of dose should be used?
2. How can the choice be justified?
3. What role does the population detriment play?
4. How can varied food consumption patterns be accommodated?
5. How should additivity be dealt with?
6. What dose-conversion factors are appropriate?
7. Is there a critical group to consider?

These questions are worth bearing in mind when the approach adopted by the Task Group is discussed. It is my view that the majority of these problems have been resolved in a realistic and pragmatic way.

I should emphasize that the guidance is restricted to contamination in the "far-field," where the food pathway is expected to be the major avenue of exposure. It is, of course, true that ground deposition can be significant, but since the associated protective measures (decontamination and relocation) differ from those for food and since the dose from this pathway is not proportional to the dose from the food pathway, no general guidance could include the deposition pathway. The approach adopted by the group is outlined below.

The first point that was emphasized was that the control of dose to the public in normal operation was very different from the control in an accident. On the one hand, in normal operation the measures were preventive, as the source was under control. On the other hand, in an accident the measures were remedial, as the source was no longer controllable. It had been argued that the dose incurred from the ingestion of accidentally contaminated food was, in fact, controllable and consequently the dose limits for the general public applied. The Task Group rejected this argument, as it was a misinterpretation of the intent of the ICRP recommendations, which were meant to apply to the control of sources during operation at the design and planning stage and did not include radionuclides present in the environment, either natural or accidental. In these circumstances the dose limits for normal operation were not applicable and other reference levels needed to be used following an accident.

## The Choice of the Reference Level of Dose

A lower reference level of committed effective dose equivalent for accidents of 5 mSv, incurred from the first year's exposure, has already been suggested by the international organizations mentioned at the beginning of this chapter.

This indicated some measure of international consensus that could serve as the basis for the development of DILs.

The choice of this level implied the acceptance of a risk of developing a fatal cancer over a lifetime of $10^{-4}$ (5) and was justified by comparison with other naturally occurring radiation risks. For example, the annual dose from background radiation, other than from radon exposure, varies across the world from about 1 to 10 mSv.

The annual dose from radon daughter exposure can be extremely high in certain localities. The WHO (6) has suggested that when the dose exceeds 8 mSv/y, simple remedial measures should be considered. It was felt that a dose level of 5 mSv was acceptable because it was comparable with the annual global variation in effective dose equivalent due to natural radiation sources and because no remedial measures had been recommended for avoiding exposure from other natural sources at doses of 5 mSv or less. The WHO report also emphasized that where a reference dose level is applied, the actual average doses to individuals are very much lower than the reference level.

It was felt that the use of an intervention level of dose of 5 mSv would ensure adequate control of the individual risk. However, this criterion takes no account of the number of people exposed and of the total detriment incurred by this exposure. This was particularly true in the "far-field" where, although the individual dose tended to be low, the population exposed may be quite large. The expert group recommended that cost-benefit techniques could be applied to evaluate this problem. When the individual reference level of dose is not exceeded, cost-benefit techniques can be applied to determine if it is economically feasible and justified to introduce control measures at a dose of less than 5 mSv. Even though the Task Group recognized the limitations of cost/benefit analysis, it still felt that such techniques were a useful aid to decision making, especially when a simple formula could be applied.

It was shown that the level of annual dose, $H$, below which it is not warranted to intervene is given by

$$H = \frac{C}{\alpha}$$

where $C$ is the cost of maintaining the countermeasure per person per unit time, and $\alpha$ is the monetary value of the unit collective dose. When commonly accepted values are substituted for the variables in this equation, the resulting intervention level of dose tends to fall in the range of 1 to 10 mSv.[2] Thus, cost/benefit analysis not only supports the choice of 5 mSv for the primary reference level of dose, but also provides a mechanism by which the "as low as reasonably achievable" (ALARA) principle can be taken into account.

---

[2] Some countries may assign a high value to $\alpha$ and thus adopt a lower reference level of dose at which to intervene compared with a country that assigns a low value to $\alpha$.

## Food Consumption

The value of the DIL for a food varies directly with the reference level of dose and inversely with the mass consumed and the dose per unit intake factor, so that

$$DIL = \frac{RLD}{Vd}$$

Where RLD = reference level of dose (Sv), $V$ = volume or mass of food consumed (kg), $d$ = dose per unit intake (Sv/Bq).

Figure 26.1 shows how, for $^{137}$Cs, the intake of contaminated food to give a dose of 5 mSv varies with the level of contamination. It clearly illustrates that for low food intakes (< 20 kg), very high contamination is required before the reference level of dose is exceeded from consuming that food group alone.

Since the contribution to dose from low consumption foods is so small, for

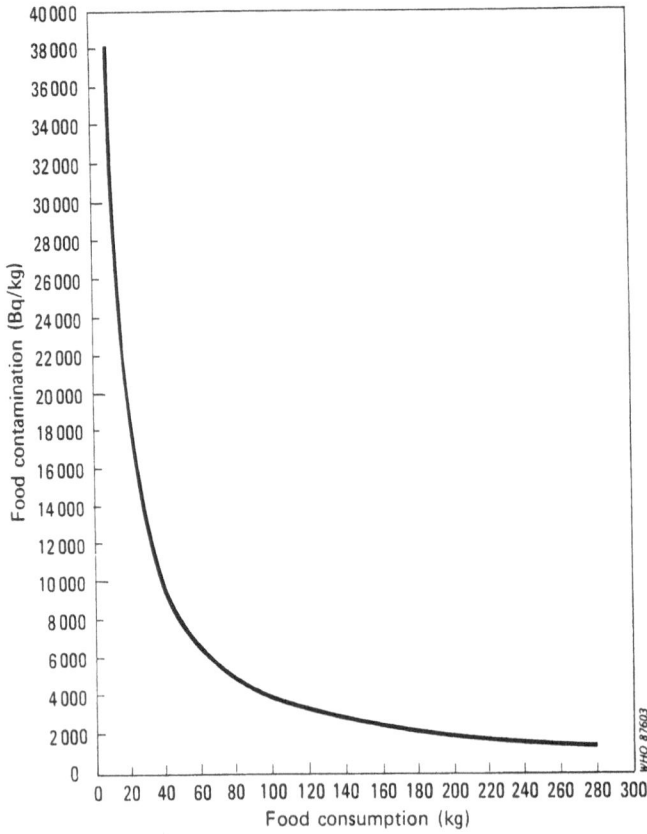

**Fig. 26.1.** Plot of food contamination versus food consumption for $^{137}$Cs to give a dose of 5 mSv, assuming a dose-conversion factor of $1.3 \times ^{-8}$ Sv/Bq.

all practical purposes such foods can be ignored when determining generally applicable guideline values for DILs. The variation in food consumption across the world was tackled in the following way.

Food consumption data were reviewed for about 130 countries, and the national average for food groups that are usually consumed in quantities greater than about 20 kg per year were tabulated. The food groups are cereals, roots and tubers, vegetables, fruit, meat, fish, and milk. On the basis of the pattern of consumption, geographic location, and existing Food and Agriculture Organization (FAO) groupings, the countries have been divided into eight regional diet groups. These diet types are North African, Eastern Mediterranean, Far Eastern, Central American, South American, African, Chinese, and European. Averages have been computed for each region in kilograms per year. The results for each region are tabulated in Table 26.1. It can be seen that there is quite a wide spectrum of the consumption values for the same food item (e.g., milk), but for others, such as cereals, the variation is much smaller. A similar variation exists in the national data between the average and individual values of consumption.

It should be remembered that neither the national nor the regional average consumption data bear much relationship to what an individual actually eats. These are, however, the only data available (no matter how "soft" they may be) that can provide a basis upon which to calculate DILs.

As these consumption rates vary widely, a method is required to establish representative intake rates for the purpose of DIL calculations. The use of global average consumption rates does not adequately reflect the maximum intake rates. On the other hand, assuming the maximum intake rate for each foodstuff leads to too high a total consumption rate in calculating the average individual dose. The usual practice in radiation protection is to assume a group whose habits are representative of the most exposed individuals. The approach adopted is to normalize the maximum intake rates of each food category to an average food consumption rate representative of the regional maximum consumption rates. The FAO figure for total consumption (550 kg/y), which does not include water, is adopted for this normalization calculation.

Table 26.1. Regional food consumption patterns (kg/y/caput)

| Region | Cereals | Roots and tubers | Vege-tables | Fruit | Meat | Fish | Milk |
|---|---|---|---|---|---|---|---|
| European | 121.10 | 72.68 | 86.65 | 81.35 | 75.33 | 20.21 | 154.94 |
| Chinese | 171.65 | 85.76 | 85.17 | 5.48 | 14.98 | 8.87 | 1.90 |
| Far Eastern | 206.63 | 28.38 | 54.25 | 48.32 | 21.51 | 24.42 | 33.56 |
| Eastern Mediterranean | 188.23 | 19.27 | 91.55 | 101.45 | 30.36 | 8.37 | 74.32 |
| South American | 129.50 | 67.64 | 34.18 | 83.19 | 48.38 | 14.35 | 70.67 |
| Central American | 113.17 | 45.95 | 38.82 | 98.56 | 42.30 | 18.65 | 82.20 |
| African | 140.65 | 164.08 | 30.16 | 54.54 | 18.97 | 17.58 | 32.09 |
| North African | 161.90 | 20.00 | 63.44 | 63.92 | 24.01 | 7.26 | 77.19 |

**Table 26.2.** Regional maximum consumption and normalized rates (kg/y/caput)

| | Cereals | Roots and tubers | Vege-tables | Fruit | Meat | Fish | Milk |
|---|---|---|---|---|---|---|---|
| Regional maximum consumption<br>Total   815 | 205 | 165 | 90 | 100 | 75 | 25 | 155 |
| Normalized con-sumption<br>Total   550 | 140 | 110 | 60 | 70 | 50 | 15 | 105 |

The normalized values in Table 26.2 reflect the consumption of a hypothetical reference group and do not represent any single individual's actual consumption. The normalized values compensate for the variation in food consumption among food categories so that the total food intake is not distorted. It is to be expected that contamination of fish is unlikely to be important as a source of exposure, except in some localized areas where the consumption of freshwater fish may be high. A value of 700 L per year was taken as representative of drinking water intake.

## Relevant Radionuclides

This subject is probably the easiest on which to reach a consensus. The radionuclides most likely to be of significance for food pathways were established as $^{90}$Sr, $^{131}$I, $^{134}$Cs, $^{137}$Cs, and $^{239}$Pu.

## Dose-Per-Unit Intake

Although it is recommended that the age-specific dose per unit intake values be used to calculate the DILs for a critical group, these were thought to be inappropriate when calculating guideline values for the population and that more general values could be used. The radionuclides were divided into two classes to reflect their different dose-per-unit intakes. The first class was the actinides, such as $^{239}$Pu, with a dose conversion factor of $10^{-6}$ Sv/Bq, and the second, all others, has a dose conversion factor of $10^{-8}$ Sv/Bq.

**Table 26.3.** General guideline values (Bq/kg)

| Class of radio-nuclide | Cereals | Roots and tubers | Vege-tables | Fruit | Meat | Milk | Fish | Water |
|---|---|---|---|---|---|---|---|---|
| High-dose-per-unit intake factor ($10^{-6}$Sv/Bq) | 35 | 45 | 80 | 70 | 100 | 45 | 350 | 7 |
| Low-dose-per-unit intake factor ($10^{-8}$Sv/Bq) | 3,500 | 4,500 | 8,000 | 7,000 | 10,000 | 4,500 | 35,000 | 700 |

## Guideline Values

The general guideline values that result from the application of the assumptions mentioned are given in Table 26.3.

### Guideline Values for Infants

It was recognized that $^{90}$Sr presented a problem to the infant group, and this was solved by the creation of separate guideline values for infants for the four radionuclides of most concern. The guideline values for infants listed below are based on an assumed infant consumption of 275 L per year of milk and 275 L per year of water and on the doses per unit intake for the radionuclides appropriate to infants.

**Table 26.4.** Guideline values for infants

| Radionuclide | Milk/water |
|---|---|
| $^{90}$Sr | 160 Bq/L |
| $^{131}$I[a] | 1600 Bq/L |
| $^{137}$Cs | 1800 Bq/L |
| $^{239}$Pu | 7 Bq/L |

[a] Based on a mean life of 11.5 d and an organ dose of 50 mSv to the thyroid.

### Additivity

The problem of more than one food type being contaminated by one or more radionuclides was solved by the application of the simple formula:

$$\sum_i \sum_f \frac{C(i,f)}{\mathrm{DIL}(i,f)} \leq 1$$

where $C(i,f)$ is the concentration of the radionuclide $i$ in foodstuff and $f$, and DIL$(i, f)$ is the DIL for that radionuclide in the same foodstuff. Although this additivity formula is simple, its application is more complicated than appears at first sight. A number of examples illustrating its use are given in the WHO report.

It is hoped that the approach adopted in this Expert Group report will serve to guide national authorities in calculating their own DILs and so lead to better harmonization in the future.

*Acknowledgment.* The WHO wishes to thank all the experts who have contributed so much time and effort to arrive at a consensus view, as well as the international organizations, the Nuclear Energy Agency, FAO, IAEA, and the CEC, without whose input and advice the task would have been much more difficult.

# References

1. Commission of the European Communities (1982) Radiological protection criteria for controlling doses to the public in the event of accidental releases of radioactive material, Luxembourg, July 1982
2. International Commission on Radiological Protection (1984) Protection of the public in the event of major radiation accidents: Principles for planning. Vol 14 No. 2, 1984. ICRP Publication 40
3. International Atomic Energy Agency (1985) Principles for establishing intervention levels for the protection of the public in the event of a nuclear accident or radiological emergency, IAEA Safety Series 72, 1985
4. World Health Organization (1984) Nuclear power: Accidental releases—Principles of public health action. Regional Office for Europe, European Series No. 16
5. International Commission on Radiological Protection (1987) Statement from the 1987 Como meeting
6. World Health Organization (1986) Indoor air quality: Radon and formaldehyde. Regional Office for Europe, European Series No. 13

# CHAPTER 27

# FAO Recommended Limits for Radionuclide Contamination of Food

## J. R. Lupien[1] and A. W. Randell[2]

## Summary

The Food and Agriculture Organization (FAO) of the United Nations (UN) is responsible for advising its member governments on a wide range of topics affecting agriculture and food production, processing, storage, and trade, including compliance with food legislation, which controls food quality and safety. This paper discusses appropriate approaches recommended by FAO for assuring orderly trade in foodstuffs in the event that foods are accidentally contaminated with radionuclides or other contaminants.

Accidental releases of radioactive materials into the environment can have a significant effect on trade in agriculture and food commodities. In such situations, national governments must decide under which circumstances their response to radionuclide contamination of food is to be governed by general radiological protection principles and in which circumstances contamination of the food supply requires application of national food legislation and food protection principles, which seek to control contamination before foods reach the consumer. In the case of transboundary releases resulting in contamination of the environment and the food supply, the principles for the protection of the population should not be different to those where the fallout of materials is confined to one country.

Radionuclides are one of several classes of food contaminants that can be carcinogenic or mutagenic. Existing national food legislation requires the use of a number of conservative safety factors in setting regulatory limits for contaminants. This provides the widest possible safety margin between contaminant levels based on public health considerations alone and the levels that are applied to foods in trade. This gives consumers assurance of the safety of the food

[1] Food Quality and Standards Service, Food Policy and Nutrition Division, Food and Agriculture Organization of the United Nations, Via Delle Terme Di Caracalla, 00100 Rome, Italy.

[2] FAO/WHO Food Standards Program Group, Food Quality and Standards Service, Food and Agriculture Organization of the United Nations, Via Delle Terme Di Caracalla, 00100 Rome, Italy.

supply and therefore facilitates national and international food trade. The application of radiological protection principles to the general food supply fails to provide this reassurance. The main approaches used by radiation protection experts for determining derived intervention levels for foods fail to meet the requirement of consumer reassurance and ease of uniform application. The use of monetary derived intervention levels, in particular, runs counter to consumers' expectations of government responsibilities in regard to protection of the food supply.

Action levels established under existing principles for food contamination control, such as the FAO interim International Radionuclide Action Levels for Food (IRALFs), have been shown to provide the necessary consumer assurance and have facilitated trade in foods. Plans for emergency response to an accidental release of radioactive materials should specifically include the use of the interim international radioactivity levels for foods moving in international trade, and in appropriate situations, in domestic trade.

## Introduction

Releases of radioactive materials into the environment can have a significant effect on trade in agricultural and food commodities. This was made clear in 1986 when the fallout of radionuclides from Chernobyl over a wide geographic area of Europe and Asia caused serious disruptions to food production and trade in food products. These disruptions were exacerbated by the lack of uniformity of actions taken by national authorities and the lack of preparedness to respond to such an emergency.

The principles for protection of the population in the case of accidental releases need to be determined as part of any emergency response plan; radiation protection principles should apply for protecting the population from radiation doses in excess of previously agreed levels, and food control legislation and food protection principles should apply to foods moving in trade across national boundaries and within countries in many cases. International agreement on the application of these principles would also assist national governments in deciding on their response to transboundary contamination of the environment and the food supply. Cognizance must also be taken of the perceptions of the national population and the political respone to these perceptions, especially in countries that have not committed themselves to extensive nuclear energy programs, which are the likeliest sources of transboundary contamination with radionuclides. Similarly, existing national food legislation must also be taken into account when proposing levels of contamination by radionuclides in foods since such legislation is normally paramount where food safety is concerned. One of the main decisions to be taken by national governments is to decide in which circumstances its response to contamination is to be governed by radiological protection principles and in which circumstances contamination of the food supply requires application of national food control legislation and food protection principles that seek to reduce contamination to the lowest levels attainable, thereby controlling the contamination before it reaches the consumer.

## Food Control Legislation and Radiation Protection Legislation

International trade in foods, which includes food aid programs, comes within the scope of national food control legislation. Food protection legislation and principles are applied to all food trade in order to reassure the consumer of the safety and quality of foods by providing wide margins of safety to the basic levels derived from public health considerations. Sound and effective national food control laws and regulations provide authorities with simple and uniform action levels that can be applied to all foods moving in trade, whatever their origin and whatever their destination in the distribution chain after clearance. Food control officials are therefore able to use these action levels in a way that is easily understood by the consumers and by food producers, exporters, and importers. Such a system not only gives reassurance to consumers, but also provides quantified action levels that the food industry can use to assess whether or not its products will meet with approval.

There are obvious differences between the radiation protection principles that seek to limit the total exposure to individuals to radiation from all sources, including air, food, and water, and food protection principles that are applied to prevent contaminated foods from freely entering the food chain if the contamination is above a conservatively set limit. Using radiation protection principles, derived intervention levels (DILs) have been estimated based on the current International Commission on Radiological Protection (ICRP) basic recommendation of a dose of 5 mSv per year, the dose limit below which intervention is not warranted from a public health point of view according to radiation protection specialists. Two main radiation protection approaches have been used to estimate DILs for foods, the first one being on the basis of assumed intake rates of foodstuffs for critical groups (1) and the second by the optimization of assumed health detriment costs against the cost of withdrawing the food from use (2). Neither approach is applicable to the control of food moving in international trade.

In the first case, only a few countries have adequate information on national, regional, and local dietary habits and patterns, especially for the most vulnerable groups, such as infants, children, and pregnant or nursing women. Second, this approach assumes that a consignment of food arriving at the point of import in a country will be sold or distributed exclusively to the target group for which the DIL has been established. It also has the effect of establishing a multiplicity of DILs, even for the same foodstuff depending on different levels of intake and, of course, the proportion of contaminated food in the diet. Third, it has the effect of establishing significantly higher DILs for most foods entering countries where imported foods are the main or only source of radionuclide contamination than in their countries of origin, where contamination may be much more widespread. Finally, because of the multiplicity of DILs that would be generated on a worldwide basis, consumers in countries where high DILs were recommended would rightly distrust the competence of their national authorities and cease to have confidence in the safety and quality of the food supply. Notwithstanding the resulting negative effect on international trade in foods, most governments would be unwilling to accept such a reaction by consumers.

The second approach being explored within the context of radiological protection is the development of guideline values, which are monetary DILs based on the optimization of assumed health detriment costs balanced against the cost of withdrawal of the food from use. This is an extension of the optimization procedure established for limiting controlled releases of radionuclides into the environment in accordance with the Basic Safety Standards of the International Atomic Energy Agency (IAEA) (3), using an arbitrary monetary value assigned to the health detriment of the unit collective dose (alpha) to be compared to the theoretical cost of production of various foodstuffs. Through the use of a number of broad assumptions, many of the difficulties associated with the "food intake" approach can be overcome, so that DILs are applied only to two categories of food, namely, high cost and low cost. However, within the Basic Safety Standards of the IAEA, this quantitive method for optimization was expected to be used mainly in facility design and establishing broad radiation protection programs. The Basic Safety Standards admit that optimization becomes more qualitative as the concept is applied to operations in existing facilities (ref. 3: A.IV.216, p 148), and although the IAEA has recommended that risk/benefit considerations in setting intervention levels should be taken into account, no attempt at applying quantitative optimization has been made, nor is one recommended in accident situations.

This optimization approach is inapplicable to the control of radionuclide contamination of foods moving in trade because of its failure to adequately address the costs of adverse effects of radionuclide contamination of foods. This approach is also insensitive to the legitimate concerns of consumers because of its use of cost–benefit analysis for assessing food safety and marketability. As is shown in a recent report by the National Consumer Affairs Advisory Council titled "Consumer Product Safety," made to the Australian government in January 1987, such cost/benefit analysis is not applicable in this case (4). The Council stated that cost/benefit analysis, when used in relation to product safety,

has severe limitations as an appropriate tool in decision making in respect of potentially unsafe products. The apparent objectivity of cost/benefit analysis tends to confuse and/or impede the qualitative debate which is essential in an area inherently concerned with ethical, distributional and political issues. Decisions about the supply of safe consumer products and the recall of unsafe products, often involving matters of life and death, involve essential ethical and political judgements.

Such judgments are provided for in the establishment of national food laws, because such laws are formulated with the consent of consumers, and consumers have made it clear that they do not want their food supply to be a source of exposure to unseen risks.

Since Chernobyl, radiation safety and protection experts have discussed emergency actions to be taken after accidental releases solely on the basis of radiation protection principles, as described previously. The radiation protection experts have tried incorrectly to extend these principles to foods in trade, while at the same time excluding the principles and procedures used under existing food legislation to control contamination of food with any type of contaminant, including radionuclides. Unfortunately, this approach by the radiation protection community does not take into account the expectations of legislators and consumers, which are clearly enunciated in the food control legislation of every country and have been abundantly reinforced since Chernobyl by clearly expressed concerns from countries importing foods.

## Food Contamination Assessment and Control

In previous discussions, it has become clear that radiation protection experts need to appreciate the effect of food legislation on control of radionuclide contamination of foods. There are clear parallels to be drawn between radionuclide contamination in foods and other contaminants, such as aflatoxin in food and feeds. The FAO has been working with member countries to control a number of contaminants in foods, some of which, like radionuclides, are inadvertent and cannot always be avoided. A good example of such a contaminant is aflatoxin, a carcinogenic by-product of mold growth that is formed in varying amounts when the mold *Aspergillus flavus* grows on different crops. Aflatoxin is most commonly found in maize, groundnuts, and certain tree nuts, but it has been found in many other foods that are susceptible to molds. Under certain adverse climatic conditions, aflatoxin can be found in crops throughout a very wide geographic area, such as the entire maize crop from the southeast United States or all groundnuts and groundnut products from a number of countries of Africa and Asia. When such contamination occurs, either on the local level or on a widespread regional level, it can be considered a natural accident and similar to the effects of a nuclear reactor accident, such as Chernobyl, on the food supplies of a region.

Under existing food control legislation, policies, and procedures, aflatoxin contamination has been evaluated by several national governments, the Joint FAO/WHO Expert Committee on Food Additives (which is also responsible for expert evaluation of food contaminants), and the FAO/WHO Codex Alimentarius Commission. Analytical methods have been developed that can routinely detect aflatoxin at the level of 15 parts per billion, and some countries use analytical systems that are legally accurate to 5 or 10 parts per billion. Toxicological testing of aflatoxin has shown that it is carcinogenic at these low levels in several species of test animals, although it does not appear to have a strong carcinogenic effect in short-term feeding periods to steers or swine (5).

In years when climatic conditions are favorable to aflatoxin formation, large volumes of food have been affected and have had to be dealt with under existing food legislation, which does not allow any excessive levels of contaminants, particularly carcinogens, in foods. This has required the establishment of a system of surveillance, sampling, analysis, and compliance decisions based on the level of detection of the analytical method used (5, 10 or 15 ppb) in different countries. Consignments of maize or other crops that exceed this level have to be put through a detoxification process, fed to steers or swine, converted to industrial uses, used as fertilizer, or destroyed since no flexibility exists in the food laws for higher levels to reach consumers. While these remedial actions reduce returns to growers and others involved in the distribution–processing chain, and there have been some countries that have established a payment system to farmers to cover some of the losses, no country has allowed attempts to knowingly market contaminated products because the food laws prohibit this and consumers expect the protection afforded by such laws. Small developing countries that depend heavily on export earnings from groundnuts often suffer serious losses when importing developed countries, particularly in Europe, refuse to accept products contaminated above their detection levels for aflatoxin. Under these circumstances, it is difficult or impossible to accept that one type of accidental food contamination with serious potential for carcinogenesis or muta-genesis can be treated differently under existing food laws than any other contami-nant with similar potential.

It is clear that consumers and governments demand that food legislation require conservative food contamination control policies to be followed in regard to food safety and that these policies be implemented with least cost, that is, efficiently, by national authorities. In most countries, national food law describes contaminated food as "adulterated" and therefore unfit for sale. However, it is recognized that certain contaminants are unavoidably present in food, and maximum levels for their occurrence have to be set. In order to set contaminant levels food control laws require that certain safety information be provided, such as acceptable daily intake recommendations for a pesticide or metallic contaminant based on toxicological considerations. In arriving at a contaminant or tolerance level, toxicological data on test animals are reviewed, and a series of conservative assumptions and safety factors are applied in setting the contami-nation level to be used for food control regulatory purposes. If a no-effect level has been demonstrated in controlled animal feeding tests, that level is the departure point for applying conservative assumptions and safety factors to arrive at a much lower contamination level for foods for human consumption. For contaminants, such as radionuclides or mycotoxins, where a no-effect level cannot be established, additional considerations are applied in setting contaminant levels that acknowledge the impossibility of avoiding all inadvertent contamina-tion of foods with these substances. These considerations must include the same conservative assumptions as before, along with other more conservative approaches that acknowledge the possibility that some very low unavoidable levels of contamination will reach the consumer. This conservative approach is taken in order to reassure consumers and to enable compliance with food laws.

## FAO Recommended Limits for Radionuclide Contamination in Foods

A conservative approach, as described previously, was followed by the FAO Expert Consultation on Recommended Limits for Radionuclide Contamination of Foods, which met in December 1986 (6). Following the April 1986 Chernobyl accident, FAO received many requests from governments for advice on maximum levels of radionuclides in foods. Most of these requests were from countries that were not immediately affected by transboundary fallout, but from which there were concerns about the safety of food originating from contaminated areas. Other countries were concerned that the arbitrary application of limits that were equal or near to background would effectively prevent legitimate trade in foods.

The FAO Expert Consultation (6) proposed the use of IRALFs, below which neither intervention nor constraint would be justified in terms of international movement and trade in food and drink. In deriving the IRALFs from the postaccident radiation dose limits established by radiation protection authorities, the Consultation agreed that

the levels should provide wide margins of safety and be applied as widely as possible to minimize unnecessary interruptions to international trade, and help protect the welfare of otherwise affected agricultural and/or fisheries communities, and that their derivation and application should be simple and easily understood by all food and health authorities.

The value of 5 mSv per year, as recommended by ICRP, was used as a basis for the calculation of the IRALFs. However, a number of conservative assumptions have been applied as safety factors in accordance with normal procedures when dealing with contaminants, such as radionuclides or mycotoxins, which have a linear, nonthreshold dose–response. The conservative assumptions included use of the most senstive population group, most restrictive dose conversion factor, and the supposition that the contaminated food supply would be 100% of the diet. The FAO IRALF approach, because of its conservative nature and the assumption that the entire food supply would be contaminated, is independent of consideration of food consumption and dietary habits and takes into account extreme cases, for example, infants fed entirely or primarily on one food for several months, such as infant formula.

While all of these conditions would not normally be met, the IRALFs resulting from such an approach are entirely consistent with the principles of food protection referred to earlier and should provide a margin of safety that meets the concerns of consumers while not unnecessarily impeding international trade. The IRALFs may also be applied for domestic trade if considered appropriate by national authorities and can be applied immediately after an accident or at any time afterward.

## Effect of Conservatively Estimated Action Levels on Food Trade

At this time, more than two years after the Chernobyl accident, the effect of applying IRALFs or similar limits to radionuclide contamination of foods moving in trade can be assessed qualitatively and, in some cases, quantitatively. Most countries, including those of the European Economic Community (EEC), the United States, Canada, and others, established limits for foods moving in trade that were based on food protection principles in preference to establishing DILs for individual foods. The World Food Program also applied IRALFs to food from non-EEC European donors in its food aid programs (EEC levels were slightly more restrictive). Because the principles were the same, the levels applied were very similar, although some differences occurred when some of the nuclides were grouped together.

Few problems have been encountered, and most of these concern trade in spices and, occasionally, tea. For example, since April 26, 1986, the US Food and Drug Administration analyzed 1,035 samples of imported food for radionuclides using action levels quite similar to the FAO IRALFs. This sampling included all shipments of products offered for entry known to have originated within 400 km of the accident site. A total of 12 shipments was detained, including 2 cheese samples containing excess $^{131}$I. other samples containing excessive $^{134}$Cs or $^{137}$Cs were pasta (5 samples), spices (4 samples), and cheese (1 sample). Calendar-year 1986 imports of foods into the United States from eastern and western Europe, Turkey, and the Soviet Union were valued at about $5,000,000,000 (5 billion US dollars), so the regulatory control levels applied were obviously not too restrictive. Therefore, use of limits for radionuclide contamination, developed along food protection principles and applied under existing food legislation, has not created unwarranted barriers to trade nor has it required national authorities to present to consumers new, unfamiliar, and unacceptable principles for the control of food contamination.

One other alternative, used in some countries following the Chernobyl accident to control radionuclide contamination in foods, was to invoke the as low as reasonably achievable (ALARA) principle. These countries, in general, were not directly affected by an accident and declared that any radionuclide contamination above existing background levels was unacceptable. The confusion that followed the Chernobyl accident shows that this is not a desirable way to regulate or harmonize trade in foods, and this approach also denies the inevitable and unavoidable nature of radionuclide contamination of foods following an event such as Chernobyl.

# Conclusions

In considering emergency responses to the impact of a transboundary release of radioactive materials, the basic principles of radiological protection and of food protection must apply. As to the control of radionuclide contamination of foods moving in international trade, only established food protection procedures will have any validity, since, as has been shown, neither of the approaches for establishing derived intervention levels meet the criteria of reassuring the consumer of the quality and safety of food and facilitating international trade.

Plans for emergency response to accidental releases of radioactive materials should specifically include the use of the FAO approach of using IRALFs moving in international trade and, in appropriate situations, in domestic trade.

# References

1. International Atomic Energy Agency (1986) Derived intervention levels for application in controlling radiation doses to the public in the event of a nuclear accident or radiological emergency: principles, procedures and data. Safety Series No 81, IAEA, Vienna
2. World Health Organization (1988) *Derived Intervention levels for Radionuclides in Food*. WHO, Geneva
3. International Atomic Energy Agency (1982) Basic safety standards for radiation protection, 1982 edition. Safety Series No 9, IAEA, Vienna
4. National Consumers Affairs Advisory Council (1987) Consumer product safety: a report by the National Consumer Affairs Advisory Council. NCAAC, Canberra
5. Food and Agriculture Organization/World Health Organization (1988) Evaluation of certain veterinary drug residues in food. Thirty-second Report of the Joint FAO/WHO Expert Committee on Food Additives, Technical Report Series 763. WHO, Geneva
6. Food and Agriculture Organization of the United Nations (1987) Report of the expert consultation on recommended limits for radionuclide contamination of foods. Document ESN/MISC/87/1, FAO, Rome

CHAPTER 28

# Radionuclides in Food: Radiation Protection Considerations

## D. J. Beninson[1]

## Introduction

Natural radionuclides have always been present in food, with levels depending on the prevailing environmental conditions, such as, for example, concentrations in soil and transfer characteristics of the links in the food chains. Radionuclides have also been released into the environment by nuclear weapons tests and, to a small degree, by nuclear power production and its fuel cycle and by the production, processing, use, and disposal of various radioisotopes. Substantial reviews of the levels, transfer parameters, and resulting radiation doses from all these contributions to exposure of to humans have been published from time to time by the UN Scientific Committee on the Effects of Atomic Radiation (UNSCEAR) (1).

The wide dispersion of radionuclides following the Chernobyl accident and their entry into food chains have brought to attention questions related to the long-term and far-field aspects of nuclear accidents. In particular, problems related to appropriate intervention levels for introducing countermeasures involving restrictions of foodstuffs and the impact on the international movement of such commodities became subjects of obvious importance.

The purpose of this chapter is to review the radiation protection recommendations applicable to the case of radionuclides in food, basically those of the International Commission on Radiological Protection (ICRP), and the developments based on these recommendations. While the author is involved in the work of the ICRP, it should be noted that except when they are straight quotations, the materials presented in this chapter do not necessarily represent the formal position of the ICRP.

[1] Commission Nationale de Energia Atomica, Buenos Aires 8250, Argentina.

# Basic Concepts

## Radiation Effects

The detrimental effects of radiation are classified as stochastic and nonstochastic. The stochastic effects are characterized by the fact that the probability of their occurrence rather than their severity is a function of dose over a substantial range of doses. For nonstochastic effects, on the other hand, the severity depends on the magnitude of the dose and there is a threshold dose below which the effects do not become manifest.

All hereditary effects are stochastic effects. Some somatic effects are stochastic; the main somatic stochastic effect is carcinogenesis, which is of critical importance for radiation protection. Other somatic effects are nonstochastic, such as the acute effects, the cataract of the lens, and the nonmalignant damage of the skin.

Because of the existence of the well-defined threshold dose, the prevention of nonstochastic effects is easily achieved by not allowing doses to exceed limits selected at sufficiently lower levels than the threshold. Protection in this case can be absolute.

For stochastic effects, on the other hand, the situation is different. The mechanism of induction of malignancies by radiation is not fully known. For an individual cancer case, it is impossible to establish a causative relation to radiation exposure, and therefore, the available quantitative information on radiation carcinogenesis in humans stems from epidemiological studies of irradiated population groups. The induction of malignancies is well established after high doses on the order of 1 Gy, but the statistical evidence for most malignancies is still insufficient to infer the shape of a dose-probability curve at low doses. In particular, it is not known whether a threshold exists.

For radiation protection purposes, it is assumed that there is a linear relationship between dose and the probability of a stochastic effect within the relevant range of doses. A consequence of this assumption is that doses are additive in the sense that equal dose increments increase equally the probability of an effect by a value that is independent of the previously accumulated dose and of doses that will be incurred in the future.

The same considerations apply to the hereditary effects, which are also stochastic and are the result of point mutations and chromosomal rearrangements in the germ cells of irradiated individuals. The severity of hereditary effects ranges from inconspicuous to lethal; slight defects will tend to continue in the descendants for many generations, while a severe defect will be eliminated rapidly by the early death of the individual or the zygote carrying the defective gene. There is no relevant quantitative information on radiation-induced hereditary effects in humans. Observations in lower organisms and in some small mammals give information on the rate of induction of hereditary changes, and studies of naturally occurring hereditary diseases in humans suggest which types of diseases might be involved as radiation effects.

## Protection Philosophy

A radiation source always implies some risk to individuals and some detriment (expectation of harmful stochastic effects in the exposed population groups). Present recommendations and standards for radiation protection, therefore, take account of the following considerations:

1. Any level of dose implies a probability of stochastic effects, and therefore, at any level of protection applied to the source, there is an expectation of some harm caused by the source. Nonstochastic effects are not relevant in these considerations because they are prevented by the dose limits for individual organs or tissues.
2. The radiation detriment due to a source is a function of the distribution of all the human radiation exposures, either present or future, caused by the source during its operating lifetime. Under the linear, nonthreshold assumption, the expected number of stochastic effects is proportional to the collective dose commitment from the source.
3. Any level of protection applied to the source implies a cost that, by diverting resources from beneficial uses, can also be considered as economically detrimental to society. Conceptually, therefore, there should be a level of protection for a given source that minimizes the combined detriment to society resulting from the cost of protection applied to the source and the radiation harm from the source.
4. A minimum level of protection is required, irrespective of cost, to restrict individual risks.

The ICRP recognizes two quite different conditions of exposure to radiation (2):

1. one in which the occurrence of the exposure is foreseen and can be limited by control of the sources; and
2. one in which the source of exposure is not subject to control, so that any subsequent exposure can be limited in amount, if at all, only by remedial actions.

For condition 1, the ICRP (2) has recommended the use of a system of dose limitation composed of the following requirements:

1. Justification—no practice resulting in human exposures to radiation should be authorized unless its introduction produces a positive net benefit, even taking into account the resulting radiation detriment.
2. Optimization—all exposures should be kept "as low as reasonably achievable," taking into account the relevant socioeconomical considerations. This requirement implies that the detriment from a practice should be reduced by protective measures to a value such that further reductions become less important than the additional efforts required.
3. Individual dose limitation—the dose equivalent to individuals from all practices should be less than the appropriate dose limits. Applying this requirement,

it must be recognized that many present day practices give rise to dose equivalents that will be received in the future. This should be taken into account to ensure that present or future practice would not be liable to result in a combined undue exposure of any individual.

For condition 2, in which the source of exposure is not under control, any possibility of reducing exposures would have to be based on remedial actions in the environment. While the system of dose limitation does not apply to situations in which the exposures cannot be limited by control of the source, the principles underlying the system can be used to formulate a rationale for intervention planning. This rationale was summarized by the ICRP (3) as follows:

1. Serious nonstochastic effects should be avoided by the introduction of counter-measures to limit individual doses to levels below the thresholds for these effects.
2. The risk from stochastic effects should be limited by introducing countermeasures that achieve a positive net benefit to the individuals involved. This can be accomplished by comparing the reduction in individual dose (and therefore individual risk) that would follow the introduction of a countermeasure with the increase in individual risk resulting from the introduction of that countermeasure.
3. The overall incidence of stochastic effects should be limited, as far as reasonably practicable, by reducing the collective dose equivalent. This source-related assessment may be carried out by cost/benefit analysis techniques and would be similar to a process of optimization in which the cost of a decrease in the health detriment in the affected population is balanced against the cost of further countermeasures.

The three principles apply to all phases of a large accident, but their relative importance changes as the potential for large doses disappears and the individual risk becomes increasingly smaller. In the long-term and far-field situations, the process of deciding whether to introduce a countermeasure or to continue one already established should involve a process similar to those of justification and optimization, as it is stipulated in principle (3) and partially in principle (2).

## Radionuclides in Food

Two different situations can be related to the presence of radionuclides in food and water:

1. one in which radioactive releases into the environment are planned to occur or are occurring from a source under control; the control in this case is based on the application of the system of dose limitation together with satisfactory design and operating procedures; and
2. one in which radionuclides are found in the environment, in the vicinity

after an accident, or are due to widespread dispersion from past accidents, nuclear weapons tests, or, in special circumstances, natural sources; in this case, the source of the environmental contamination is not under control, and exposures can be restricted only by remedial actions in the environment.

The second situation is the one relevant following accidents releasing radioactive materials to the environment. The main issue is the decision whether to initiate a remedial action or not. In principle, it will be appropriate to institute countermeasures only when their economical and social cost and their risk will be less than those resulting from further exposure. If a remedial action is warranted at all, then the proper selection of the intervention level may optimize the situation in the sense of maximizing the net social benefit obtained by the remedial action.

It seems clear from the arguments presented in previous paragraphs that the dose limits recommended for controlled situations do not apply in the situation following accidents. However, several questions have been raised related to the so-called far-field, long-term situations, mainly in the case of the international movement of foodstuffs.

The main problem raised can be summarized as follows: The contaminated foodstuffs are, for those using them, the source of exposure. The situation is therefore controllable by restricting the use of such foodstuffs, and as a consequence, the full system of dose limitation should apply, including the dose limits. This is, however, a misinterpretation of the intent of the recommendations.

The linear nonthreshold assumption for the induction of stochastic effects (the only relevant in this case) implies that a given increment of dose causes the same increment of risk irrespective of previously accumulated doses or of future doses that might be incurred. It is therefore possible to stipulate limits for the dose from any defined combination of sources, corresponding to a risk from that combination of sources that one does not want to exceed. The dose limits recommended by the ICRP in Publication 26 (ref. 2, and additional statements) were meant to apply to the combination of sources stipulated in this document, a combination that does not include radionuclides present in the environment due to a previous accident.

The situation is similar to that of some exposures from natural radiation sources, sources that are also not included in the combination restricted by the dose limits. The ICRP (4) recognized that almost all exposures to natural sources of radiation are controllable to some extent, but the degree of controllability varies very widely, as does the complexity, cost, and inconvenience of the possible control measures. Controllability must therefore be a major factor in any system of limitation: From this point of view, there is a clear difference between existing exposure situations, in which any action would have to be remedial, and future situations, which can be subject to limitation and control at the stages of decision and planning.

In the case of artificial radionuclides in the environment, the same difference applies. For future situations, one deals with decisions and planning regarding the control and limitation of radioactive effluent releases at the source, while in existing situations, exposures can be altered only by taking some remedial action in the environment. As an aid to deciding whether such action should be initiated, the ICRP recommends the use of an intervention level specific to the initiation of the remedial action being considered. An intervention level is not determined by the choice of any limits intended for future situations nor, specifically, by the primary dose limits recommended by the ICRP for members of the public.

## Introduction of a Remedial Action

The decision to introduce a countermeasure should be based on a balance of the radiation risk avoided and the risks and disadvantages caused by the counter-measure itself. It is implicit in the decision that the exposed individuals should be put in a better position by the remedial action. In this sense, better means at a lower risk achieved at a reasonable cost in financial and social terms.

Restricting use of contaminated food could be envisaged to range between two extreme cases: (a) localized situations in which high levels are found at the point of production of given food items and the relevant population group can be readily identified; and (b) general situations in which large amounts of food are involved and the involved population is not readily identifiable.

In both cases, the positive benefit implied by the use of the countermeasure should be demonstrated for the population involved. For general situations mentioned in the previous paragraph, the basic assumption is that all contaminated food, if no countermeasure is applied, is consumed by individuals, even if they cannot be identified.

Ideally the justification of the introduction of a remedial action could be determined by a conceptual cost/benefit analysis, the purpose of which is to ensure that the action is only taken if the net benefit of it is positive. This benefit could be expressed by

$$B = Y_0 - Y_1 - R - X,$$

where $B$ is the net benefit, $Y_0$ is the cost of the radiation detriment if the remedial action is *not* taken, $Y_1$ is the cost of the remaining radiation detriment if the remedial action is carried out, $R$ is the detriment cost caused by risks due to the remedial action itself, and $X$ is the cost of the remedial action.

The cost of radiation detriments $Y_0$ and $Y_1$ can be taken to be proportional to the collective doses involved, that is, to comprise only the so-called $\alpha$ term (5) ($Y = \alpha S$) when the situation is of the generalized type and exposed individuals are not directly identified. For the localized case, individual doses can be assessed for identified groups of individuals, and the detriment cost can be taken to include both the so called $\alpha$ and $\beta$ terms (5).

In practice, it is difficult to quantify all the terms of the previous equation, and it would be necessary to make subjective value judgments similar to those involved in most social and economical decisions.

## Optimization in Selecting Intervention Levels

If intervention is decided at all, then the proper selection of the intervention level may optimize the situation in the sense of maximizing the net social benefit. Assuming that the detriment cost due to the risks from the countermeasure itself is independent of the value of the intervention level, the optimization condition is

$$\frac{dX}{dI} + \frac{dY_1}{dI} = 0,$$

where $I$ is the intervention level or any derived intervention level. It should be noted that the cost of the countermeasure may have a large component independent of the value of the intervention level, which would be relevant for the justification of intervention but does not influence optimization of the selection of the intervention level.

If one assumes that the case is one of far-field, long-term, generalized conditions and that only the so called $\alpha$ term of the detriment cost is relevant for optimization, the optimizing condition can be expressed as

$$\frac{dX(I)}{dI} + \alpha \frac{dS_1}{dI} = 0,$$

where $X(I)$ is the part of the countermeasure cost that is a function of the intervention level, $\alpha$ is the monetary value assigned per unit collective dose, and $S_1$ is the remaining collective dose after applying the countermeasure.

A very simple optimization procedure (6), can be formulated for situations in which the intervention is fully effective while it is applied and both the collective dose and the cost of the countermeasure are proportional to the number of individuals affected by the countermeasure.

If the countermeasure is applied for a time $\tau$, during which individual doses are zero and then removed, the cost and the remaining collective dose can be expressed as

$$X = a N \tau$$

$$S_1 = N \int_\tau^\infty \dot{H}(t)\, dt,$$

where $a$ is the cost of the countermeasure per person and per unit time, and $\dot{H}(t)$ is the individual dose as a function of time if the countermeasure is not applied.

The optimizing condition described previously can be expressed as

$$\frac{dX(\tau)}{d\tau} + \alpha \frac{dS_1}{d\tau} = 0,$$

and therefore,

$$\dot{H}_o(\tau) = a/\alpha ,$$

where $\dot{H}_0(\tau)$ is the optimum value for the individual dose intervention level. It should be noted that the ration $a/\alpha$ is expected to be more insensitive with geographical location than either $a$ or $\alpha$, because richer countries, where $a$ would be higher, are also likely to assign higher values to $\alpha$.

When this optimization procedure is applied to the selection of intervention levels to restrict the use of contaminated food, it can directly produce the derived intervention level expressed in the activity concentration. In this case, the cost of the countermeasure per person and per unit time, $a$, is given by

$$a = bC,$$

where $b$ is the cost per unit mass of the foodstuff and $C$ is the consumption per person and per unit time of that foodstuff.

The individual dose rate (more exactly, the rate of commitment of individual dose), $\dot{H}$, is given by

$$\dot{H} = A C F,$$

where $A$ is the activity concentration and $F$ is a dosimetric factor giving the dose per unit activity ingested of the nuclide in question.

The optimized intervention level $a/\alpha$ can readily be shown to correspond to a derived intervention level of activity concentration, $A_I$:

$$A_I = b/\alpha F,$$

which is likely to be relatively insentive to geographic location for the same reasons indicated previously.

It should be noted that, if derived intervention levels are obtained by this procedure and many types of foodstuffs are contaminated, competent national authorities should keep the situation under review to detect and prevent the possibility of high individual risks resulting from the combined ingestion. This situation, however, is unlikely in the far-field, long-term generalized situations.

## References

1. United Nations Scientific Committee on the Effects of Atomic Radiation (1982) Ionizing radiation: sources and biological effects. Report to the General Assembly and Annexes, Official Record of the General Assembly, 37th Session, Supplement 45 (A/37/45), New York
2. International Commission on Radiological Protection (1977) Recommendations of the International Commission on Radiological Protection, ICRP Publication 26, Pergamon, Oxford

3. International Commission on Radiological Protection (1984) Protection of the public in the event of major radiation accidents: Principles for planning. ICRP Publication 40, Pergamon, Oxford

4. International Commission on Radiological Protection (1984) Principles for limiting exposures of the public to natural sources of radiation. ICRP Publication 39, Pergamon, Oxford

5. International Commission on Radiological Protection (1983) Cost–benefit analysis in the optimization of radiation protection. ICRP Publication 37, Pergamon, Oxford

6. Beninson D (1987) Intervention levels, ICRP policy and developments based on it. International seminar, Commission of the European Communities, Luxembourg April 27–30, 1987

# Part VIII
# Regulatory and Control Programs

CHAPTER 29

# Radionuclides: Regulatory and Control Programs

R. J. Ronk[1] and P. Thompson[1]

## Introduction

### Overview

There are many natural radioactive isotopes present in the environment and a number of ways in which mankind's activities have added others. Normal, peacetime uses of radioactive materials for medical diagnostics, medical research, and other research, as well as power generation, result in ongoing releases of low-level radioactive wastes. There have also been radiological incidents that have resulted in the release of unpredictable contamination. As is the case of environmental contaminants, radionuclides can also become contaminants of food.

As the agency responsible for the safety and wholesomeness of most of the American food supply, the Food and Drug Administration (FDA) maintains a substantial and comprehensive monitoring program for a wide range of contaminants, including radionuclides, in domestic and imported foods. Specifically, the radionuclide monitoring efforts enable the FDA to (a) maintain surveillance of the radionuclide content of the American food supply and (b) maintain the technical proficiency necessary to provide a quick response in the form of in-depth surveillance in the event of a major nuclear incident. The Food, Drug and Cosmetic Act (FD&C Act) provides the FDA with the authority to act against radionuclide contamination, which is defined as unavoidable contamination. Food containing radionuclides, poisonous or deleterious substances that may render the food injurious to health within the meaning of Section 402(a)(1), is considered adulterated.

The activities of the program are coordinated with the radionuclide monitoring efforts of the US Environmental Protection Agency (EPA) and the US Department of Agriculture (USDA). The FDA's Center for Food Safety and Applied Nutrition coordinates the agency's program activities using the FDA's Office of Regional

[1] Center for Food Safety and Applied Nutrition, Food and Drug Administration, 200 C Street, SW, Washington, DC 20204, USA.

Operations for sample collection and analysis and one of its sister centers, the Center for Devices and Radiological Health, for health physics guidance.

Prior to the accident at the Soviet Union's Chernobyl nuclear power plant in April 1986, the United States conducted a modest surveillance program for radionuclide contamination of food. The recent incident at the Chernobyl nuclear power plant, the worst accident of its kind in history, heightened concerns of radioactive contamination. As a result of the accident at Chernobyl, the President of the United States established a series of task forces that mobilized the resources of the federal government. At this time, the EPA increased its monitoring of all air, water, and milk in the United States for radioactivity through its environmental radiation ambient monitoring system, referred to as ERAMS.

International reaction to the Chernobyl accident has had a significant impact on the nuclear power industry. The long-term health effects on exposed populations are still being debated both in the United States and at the international level. The accident will have a global impact for years to come.

## Background

Historically, the FDA has undertaken a continuing effort to monitor radionuclides in foods. The first program was initiated in 1961 in response to concerns about radioactive fallout from nuclear weapons testing in the atmosphere.

In addition to the FDA, at least six other federal, state, and private organizations have conducted monitoring programs for radionuclides in foods in prior years. A statistical study of the data that the FDA had gathered over the years showed that participating organizations were obtaining similar results. The study demonstrated that the radionuclide levels in foods were below levels that would require protective action and were declining as a result of the Limited Test Ban Treaty. Because of this downard trend, and to prevent duplication of effort, the FDA decided in 1969 to discontinue monitoring radionuclides in foods.

Between 1969 and 1973, there was a substantial government-wide reduction in federal monitoring of food products for radioactivity. At the end of 1973, the EPA further reduced its monitoring program by discontinuing its Institutional Total Diet Sampling and Analysis program.

With the decline in food monitoring by other agencies, the projected growth of nuclear power (impacting on food-producing locations around reactor facilities), and increased uses of medical and commercial radioactive materials, the FDA decided again to monitor radioactivity in foods. At the beginning of 1973, the FDA decided that a radiochemical analytical capability should be maintained to analyze and evaluate foods in the event of a radiological incident and to note the development of any upward trends in the radioactive contamination of food (1).

In the *Federal Register* of October 22, 1982 (47 FR 57073), the FDA issued protective action guidelines to state and local officials in the event of a domestic nuclear accident (2). The basis for the guideline recommendations was that preventive action should be taken whenever the projected annual dose to the thyroid is 1.5 rem or the projected dose to the whole body, bone marrow, or any other critical organ is 0.5 rem. For foods produced in the United States, the protective action guides provide a set of control measures that can be taken to alter the domestic food production process so that contaminated foods do not reach the marketplace.

## The FDA's Monitoring Program

The FDA remains vigilant in its efforts to monitor foods for radionuclides since the use of radioactive materials has been a part of modern life and there is a continuing potential for contamination. As mentioned in the introduction, the President of the United States established a number of task forces following the nuclear accident at Chernobyl. One such task force, directed by the EPA, was charged with monitoring the consequences of the accident on health and the environment. The task force later concluded that the radiation released posed no significant increased risks to the American people or to their environment. Deposits of radioactive fallout in the United States were sporadic and in many instances barely measurable using the most sensitive instrumentation.

The FDA's increased monitoring following Chernobyl was precautionary. The FDA did not anticipate a problem with its domestic food supply nor did the agency anticipate that its import monitoring program would find serious radioactive contamination of imported foods. Some initial import samples did have levels above those normally found, but the levels were far below any level thought harmful to the public. In the end, very little food had to be rejected on the basis of radionuclide contamination exceeding the screening action levels that had been established.

The FDA's Radionuclides in Foods program provides ongoing monitoring of the food supply to establish baseline levels and show long-range trends. The program consists of monitoring and analyzing samples from (a) selected domestic foods (the Total Diet Study), (b) foods produced in the areas of selected nuclear reactors, and (c) imported foods. The FDA was immediately able to gear up for monitoring after Chernobyl because the program was already in place and had been operational for years.

### Selected Domestic Foods (Total Diet Study)

The FDA routinely conducts radionuclide analyses of foods through the Total Diet Market Basket Survey program. Foods representative of the total diet of eight subpopulations grouped by age and sex are collected from each of four geographic areas in the United States and analyzed for the presence of radionuclides.

The Total Diet Study (TDS) provides for annual monitoring of a broad range of pesticide residues, contaminants, and nutrient elements in foods in their table-ready form, and it allows for yearly estimates of intake of pesticide residues, contaminants, and nutrient elements of selected age–sex groups based on laboratory analyses. This type of information, which is available from no other source, is essential to monitor the safety and quality of the US food supply and to identify potential public health problems (6). The TDS also provides an invaluable baseline reference for determining the impact on the food supply of environmental contamination accidents. The Radionuclides in Foods program provided this type of information for radionuclide contamination as a result of the Three Mile Island incident in the United States and the Chernobyl incident in the Soviet Union.

Over the years, there have been small changes to the TDS, such as changes in analytical methodologies. The last major change was in 1982. Currently, foods are collected four times per year, once from each of four geographical areas of the United States. Each collection consists of the purchase of identical foods from grocery stores in three cities within a geographical area. The three subsamples of each food (from the three cities) are combined to form a sample for analysis. The current program reflects updated diets, expanded coverage of age–sex groups, and analysis of individual foods (6).

For the Radionuclides in Foods program, each year all TDS homogenates from one market basket are analyzed for $^{90}$Sr (a beta emitter), $^{137}$Cs, $^{106}$Ru, $^{40}$K (naturally occurring), $^{131}$I, and other appropriate gamma emitters if unusually high levels of any of these are noted.

Since domestic foods were not expected to pose a health risk, the FDA did not initiate any specific sample collection and analysis efforts for domestic foods. However, the FDA closely followed the EPA's milk sampling and analysis program. Milk warranted immediate attention after the accident because it serves as an early warning system for radioactive contamination of the food supply. Radioactive I in milk serves as a primary indicator of whether or not a public health risk exists. This approach is used for several reasons. Radioactive I is expected to be relatively high in fallout from a nuclear reactor accident of this type. In addition, cows, goats, and sheep very efficiently collect the radioiodine deposited on grazing areas. Milk is collected daily and radioactive I appears in milk immediately. Milk reaches consumer markets faster than most other foods with the potential for contamination from fallout. Finally, milk is a major source of nutrition for infants and small children who are the most sensitive to radioactive I. Therefore, the monitoring of milk serves as an early warning system for judging health risk to consumers.

The EPA operates the Pasteurized Milk Network. Milk from each of 65 sampling sites throughout the United States is collected each month and analyzed for the presence of radionuclide contamination. These sample stations can assess the status of more than 80% of the milk consumed in major American population centers. The EPA publishes the results of these surveys in its Environmental Radiation Data reports.[2]

---

[2] Eastern Environmental Radiation Facility, Environmental Protection Agency, 1890 Federal Drive, Montgomery, AL 36109, USA.

In May 1986, immediately after the Chernobyl accident, the EPA increased the sampling frequency to 2 collections each week for a total of 8 samples per month from each of the 65 sites. Sampling continued at this rate through June when the frequency returned to one collection per month. Average activity levels of $^{137}$Cs and $^{131}$I were elevated in May and dropped to near-normal levels by August. The average, however, does not adequately describe the results because many results were in the normal (not detected) range with sporadic intermittent elevated concentrations. This type of behavior is typical of much of the post-Chernobyl data generated in radioanalytical laboratories around the world.

## Foods Produced in the Areas of Nuclear Power Reactors

Each year, the FDA collects and analyzes samples of fresh fish, fresh dairy products, and fresh fruits and vegetables near selected nuclear power plants. This monitoring provides baseline data and enables the FDA to perform sample collection and processing procedures and maintain the FDA in a state of technical proficiency should a nuclear incident occur at a nuclear power facility.

Products harvested, produced, or caught downwind (prevailing) or downstream from 8 selected nuclear power plants and as near to the plants (within 10 miles) as feasible were collected. The direction and distance the product was harvested, produced, or caught from each nuclear power plant was recorded in detail. Four commercially significant products of each of the above-mentioned food types were sampled, for a total of 16 samples per plant site. Fish samples were obtained from fish and game commissions or park authorities if they were not available from a commercial source. All samples are analyzed for gamma emitters and tritium and half of them are analyzed for $^{90}$Sr (to monitor for radioactive fission and corrosion products).

## Selected Imported Foods

In May 1986, within two weeks of the Chernobyl accident, the FDA–USDA Food Safety Inspection Service (FSIS) Task Force established by then Director of the FDA's Center for Food Safety and Applied Nutrition, Dr. Sanford A. Miller, recommended the radioactive levels of concern shown in Table 29.1 and instituted the import sampling programs of both organizations to deal with the problem. The FDA issued a very targeted assignment to its field offices to

**Table 29.1.** Monitoring level[a]

| Indicators of concern | Infant food | Other foods |
|---|---|---|
| $^{131}$I | 1,500 pCi/kg | 8,000 pCi/kg |
| $^{134}$Cs + $^{137}$Cs | 10,000 pCi/kg | 10,000 pCi/kg |

[a] Reproduced from ref. 2.

begin collecting specified imported foods from specific geographic regions (for gamma spectroscopy analysis for radionuclides by the FDA's Winchester Engineering and Analytical Center. Within this context, targeted meant focusing on those food commodities most likely to be contaminated, such as broad, leafy vegetables, milk and milk products, and other fresh fruits and vegetables from areas known to have significant levels of fallout, using the most current data. Other samples were collected from aged and long-shelf-life dairy products (e.g., hard cheeses, caseins, and milk chocolates) and manufactured foods likely to contain post-Chernobyl harvested fruits, vegetables, or grains, such as fruit concentrates, jams, preserves, bulk herbs, spices, and Georgian (USSR) tea.

At the same time, the FDA issued a compliance policy guide that established screening action levels for $^{131}I$ and the sum of $^{134}Cs$ and $^{137}Cs$ using gamma spectroscopy for analyzing imported products regulated by the FDA. (See Table 29.1). The screening level for $^{131}I$ in infant food was established at 1,500 pCi/kg and for all other foods it was 8,000 pCi/kg. For the sum of $^{134}Cs$ and $^{137}Cs$, the screening action level was 10,000 pCi/kg for all foods.

Because the 1982 protective action guidelines were intended to be used for a one-time, short-term incident in the United States, certain adjustments in calculating projected doses had to be made to take into account what was known about the fallout from the April 26, 1986, accident in the Soviet Union and what was likely to be seen in the way of imports in the months following the accident.

During the process of determining these screening actions levels, the FDA made several conservative assumptions about the level of contamination and about those radionuclides likely to be the major contributors of doses from an ingestion pathway. These conservative assumptions were necessary because very little usable information was available on which to make public health decisions. One assumption was that a majority of the gamma radiation would stem from $^{131}I$, $^{134}Cs$, and $^{137}Cs$. Since sufficient conservatisms were already incorporated into the calculations presented in Table 29.1, the FDA did not believe that additional analysis was warranted except for spotchecks or where there were indications that other radionuclides were present in significant amounts. In preparing these levels of concern, the FDA elected to prepare a single level for all foods under its jurisdiction rather than for individual food categories. This was done because the ranges of intake are so diverse. Furthermore, many imported foods are sold as bulk ingredients to be used in the preparation of other foods, and there is no way to know the relative consumption rates for these bulk ingredient foods.

Where there was a potential for contamination, FDA personnel automatically sampled and held all shipments of products offered for entry that were known to have originated within a 250-mile radius of the Chernobyl plant. This sampling effort continued until the FDA had sufficient information to be assured that there was no cause for concern. Samples from all other areas were collected on the basis of assignment. Initially, the sampling program to cover imports was limited to products originating or suspected of originating from these countries (7):

| | |
|---|---|
| Austria | Norway |
| Czechoslovakia | Poland |
| Denmark | Soviet Union |
| East Germany | Sweden |
| Finland | West Germany |
| Hungary | |

Products from the Mideast and Asia, transshipped through these countries were not covered initially. Samples were first selected from the Soviet Union. Samples of fresh dairy products (fresh cheese, milk, etc, if available), fresh fruit, other fresh vegetables, fresh fish, and frozen fish were collected from these countries. Samples exceeding the levels of concern triggered an appropriate control response to assure that the public health was protected. Standard procedures, such as automatic detention, release with comment, and refusal of entry, were employed where appropriate.

Since animal feed and feedstuff are not imported into the United States in large volume and since there is also a delay factor in the preparation of the feedstuff so that during the bulk shipment and distribution considerable reduction in concentration will occur, there was no reason to analyze these products on a routine basis. The USDA FSIS initiated a control program to prevent the importation of meat and poultry adulterated with radioactivity and issued similar monitoring levels for meat and poultry using the same criteria as the FDA. The FSIS and FDA worked with the US Customs Bureau to inspect susceptible imported food products.

Although the FDA's screening levels for radionuclides were conservative, they were not as conservative as some other countries that imposed limits on radionuclide contamination of foods offered for import to their respective countries. In the United States, very little imported food had to be rejected on the basis of radionuclide contamination exceeding the established screening action levels. Very early on, the United States made public its screening levels. Furthermore, several countries developed certification programs to assure that there was no radiation contamination problem with their exported products. Products from the certifying countries were sampled at a survey (reduced) level; a maximum of two samples per FDA district per month of any one product–country combination, was analyzed.

Early after the accident, two lots of cheese were detained for excess [131]I. Since then, the United States has detained some cheese, spices, and pasta products for [134]Cs and [137]Cs, but the total dollar value is trivial when compared to the total amount of food imported into the United States (see Table 29.2 for sample analysis and detention by country and Table 29.3 for a listing of import samples detained to date). As of late September 1987, the FDA had analyzed more than 1100 samples of imported food products representing 26 countries for

**Table 29.2.** Report of sample analysis by country[a]

| Country of origin | Number of samples analyzed | Number of samples detained |
|---|---|---|
| Italy | 397 | 2 |
| West Germany | 124 | 0 |
| Norway | 102 | 0 |
| Austria | 95 | 1 |
| Greece | 59 | 6 |
| Denmark | 50 | 0 |
| Turkey | 60 | 4 |
| Switzerland | 43 | 0 |
| Hungary | 40 | 0 |
| Yugoslavia | 31 | 1 |
| Japan | 23 | 0 |
| Sweden | 22 | 1 |
| Poland | 17 | 0 |
| France | 13 | 0 |
| Soviet Union | 13 | 0 |
| Finland | 7 | 0 |
| Romania | 3 | 0 |
| Holland | 4 | 0 |
| United Kingdom | 2 | 0 |
| Belgium | 2 | 0 |
| Thailand | 1 | 0 |
| Morocco | 2 | 0 |
| Iran | 1 | 0 |
| Albania | 1 | 0 |
| Bulgaria | 1 | 0 |
| Egypt | 1 | 0 |

[a] Data compiled through October 15, 1987.

**Table 29.3.** Import samples detained to date

| Date | Country | Product | $^{131}I$[a] (pCi/kg) | $^{134}Cs$[a] (pCi/kg) | $^{137}Cs$[a] (pCi/kg) |
|---|---|---|---|---|---|
| May 16, 1986 | Italy | Cheese | 8256 | 300 | 707 |
| May 23, 1986 | Italy | Cheese | 11666 | 475 | 632 |
| Sept 9, 1986 | Austria | Cheese | ND | 4653 | 6204 |
| Sept 26, 1986 | Turkey | Oregano | ND | 5738 | 7651 |
| Apr 22, 1987 | Yugoslavia | Sage | ND | 3073 | 7032 |
| May 13, 1987 | Turkey | Oregano | ND | 5695 | 14080 |
| May 15, 1987 | Greece | Pasta | ND | 4990 | 11880 |
| May 28, 1987 | Greece | Pasta | ND | 4208 | 11280 |
| May 28, 1987 | Greece | Pasta | ND | 3857 | 10050 |
| May 28, 1987 | Greece | Pasta | ND | 3736 | 9235 |
| June 10, 1987 | Turkey | Oregano | ND | 2784 | 7755 |
| Aug 3, 1987 | Greece | Pasta | ND | 3055 | 7698 |
| Aug 18, 1987 | Greece | Pasta | ND | 2574 | 7640 |
| Aug 20, 1987 | Turkey | Hazelnuts | ND | 3402 | 8641 |
| Oct 5, 1987 | Sweden | Reindeer | ND | 2907 | 10410 |

[a] Data reproduced from ref. 3.

radionuclides. While the incidence and levels of contamination in import samples are expected to decline, systematic surveillance of radionuclide contamination from Chernobyl will continue along with the other food monitoring efforts.

## What We Have Learned and Where We Should Be Headed

### Communication With the Public

It has been the FDA's experience that in the event of a release of radioactive material into the environment, public perception of the risk is so much greater than the actual risk involved that communication with the public becomes critical (4). Much of the misconception could be minimized if the public could understand and recognize the different factors important for chronic versus acute risk. Many of the preventive measures to minimize doses are common sense ideas, such as sheltering, bathing frequently after being outdoors, not drinking rain water, and limiting food consumption to unexposed sources. The nuclear accident at Chernobyl and the many different actions by various countries are good examples of the fear radiation produces in people everywhere.

### Uniform Standards

It has been and will continue to be the FDA's policy to work with other countries and international organizations in an effort to develop international uniformity and consensus. We must agree that standards will be used only for world public health and not as trade barriers.

Many affected countries were hampered in managing the public health risk because they did not have the monitoring and analytical technology in place and did not have any experience with this type of problem. As a result, these countries were often forced to take much more conservative action than was warranted since they could not make discriminating decisions without quick access to real data (4).

Several countries immediately placed a ban on the import of foods from specific areas known to have been affected by radioactive fallout. There was also a great deal of confusion. Since there were no international limits for foodstuffs and no agreement about intervention levels if food contained radioactivity, each country was left to establish its own limits, and the various limits subsequently established placed an additional burden on trade. It is impossible to quantitate the effects these actions have had on the availability of foodstuffs and their impact on trade. Clearly, a number of actions taken by individual governments were not warranted by the balance of the dose avoided and severeness of the action itself.

Several countries, such as the United States, Israel, and Canada, set intervention limits only on foods offered for import since there were no significant amounts of deposition within their boundaries. In general, limits in these countries were established on the basis of feasibility and were much more conservative than limits that would have been established only on the basis of public health protection.

There continues to be a problem with international trade because of the presence of cesium isotopes in food. Part of this problem stems from the lack of uniform requirements among trading partners. Several countries established radionuclide limits on imports at or below background limits and this has led to requirements for certification of foods even from countries that did not experience any significant fallout from the accident. This results in additional burdens on international trade.

The Food and Agriculture Organization (FAO) of the United Nations has established limits for radionuclides in foods and has offered them to member states with the understanding that radionuclide levels occurring below these limits in food do not warrant control actions for public health reasons. Other international organizations are also responsible for providing advice concerning levels of radionuclides in foods—either from a public health or food trade point of view. The World Health Organization (WHO), the International Atomic Energy Agency (IAEA), the European Economic Community, and the Organization for Economic Co-operation and Development all have an interest in providing guidance on the level of control that should be placed on radionuclide contamination of foodstuffs and the environment (5). Not long ago, the WHO convened an expert panel in Geneva that prepared a technical paper to be submitted to the governing officers in the WHO. There are also plans for an FAO–WHO meeting to hopefully achieve a joint agreement on recommendations about the level of radionuclides to be controlled in foods. If agreement can be reached, it is planned that such guidance be formally adopted through the Codex Alimentarius standard setting mechanism. The IAEA, having joint committees with both the WHO and FAO, has also been an active participant in the deliberations and also has other plans focusing on training and adoption of uniform methods of analysis and reporting of data.

## Better Risk Assessment

Because we live in an environment that exposes us constantly to radiation, it has been necessary to develop radiation protection standards both for the general population and for workers who are potentially at greater risk through occupational exposure. The establishment of general standards for maximum doses to individuals, however, is only the first step in the process of determining the risk to individuals. As we all know well, there is a variety of radionuclides that can be involved in fallout from a nuclear reactor. They all make a contribution to the actual calculation of dose, but each does so in its own unique way.

## Summary

Continuing to monitor foods for radionuclides is essential because of the potential for accidental contamination brought about by development of the nuclear power industry, increased peacetime uses of radioactive materials, and other emerging problems. The FDA has in place a compliance program that enables the agency to maintain the technical proficiency necessary to analyze food products (fresh fish, fresh dairy products, and fresh fruits and vegetables near selected nuclear power plants and portions of the total diet samples) for radionuclide contamination.

The April 1986 incident at the Soviet Union's Chernobyl nuclear power plant has heightened concerns about radioactive contamination in FDA regulated products. Both FDA and the US Department of Agriculture's FSIS (meat and poultry products) have routine procedures in place to monitor food from foreign countries. During the Chernobyl incident, both agencies worked with the US Customs Bureau to inspect susceptible imported food products. The FDA's position was precautionary. Results have shown no problem with domestic foods and few in imports.

Because of the whims of nature and the fortunes of geography, not all countries were affected by the fallout from Chernobyl. Meteorological considerations dictated when and where the radioactive material was deposited. The significant deposits were in Scandinavia, eastern Europe, and parts of central and western Europe. The next accident, however, could affect other areas, for no area or country is immune (5).

We must work together through understanding, cooperation, collaboration, and commitment. If agreement can be reached on uniform requirements and standards, if monitoring and analytical techniques and skills can be developed and refined, and if clear, open communication can be initiated and maintained, we as an organization of countries will be far better equipped to deal with the next nuclear accident—if and when it occurs.

*Acknowledgment.* The authors wish to thank Dr. William C. Cunningham, Division of Contaminants Chemistry, US Food and Drug Administration, for his excellent comments and suggestions in the preparation of this manuscript.

## References

1. Cunningham WC, Stroube WB Jr., Baratta EJ (1988) Radionuclides in foods, 1983–1986. US Food and Drug Administration, manuscript in preparation
2. US Food and Drug Administration (1982) Federal Register, October 22, 1982 47:47,021–47,022
3. Food and Drug Administration (1987) (computerized database) Chernobyl Database Office of Regulatory Affairs, Rockville, Maryland, Oct 15
4. Gill RW (1987) Radiation and food supplies—levels of concern. Presented at the Environmental Health Officer Symposium, April 2, 1987, Center for Food Safety and Applied Nutrition, US Food and Drug Administration
5. Gill RW (1987) Radionuclides, US policy and findings after Chernobyl. Presented at the IAEA–NBS International Workshop, October 2, 1987, US Food and Drug Administration
6. Pennington JAT, Gunderson EL (1987) A history of the Food and Drug Administra-

tion's total diet study, 1961 to 1987. JAOAC, September–October, 1987, US Food and Drug Administration
7. US Food and Drug Administration (1986) FDA talk papers, May 9 and May 13, 1986, Foods from 12 countries monitored for radioactivity, and food monitoring update

# Regulation and Control of Radionuclides in Food in a European Socialist Country— Hungary

## L. B. Sztanyik[1]

## Introduction

The reactor accident at Unit 4 of the Chernobyl nuclear power station on April 26, 1986, resulted in an uncontrolled release to the environment of large quantities of radioactive substances. A considerable proportion of these was widely dispersed throughout the whole continent and induced excitement and anxiety among the people. National authorities responsible for the health of the population hastily introduced protective actions differing from each other in both their character and scale. Particularly shocking discrepancies appeared between various countries in their maximum levels of radioactive contamination of foodstuffs declared to be acceptable. This paper presents the thinking followed in Hungary in the early postaccident period for coping with the consequences of the environmental contamination and, in particular, for developing derived intervention levels of radionuclides in foodstuffs.

## Legislation for Radiation Protection

### Normal Exposure Conditions

The fundamental principles of radiation protection given in recommendations by competent international bodies, such as the International Commission on Radiological Protection (ICRP), International Atomic Energy Agency (IAEA), International Labor Organization (ILO), and World Health Organization (WHO), have been widely used in Hungary in the development of national rules and regulations since the establishment of an organized radiation protection system in the country (1). In connection with the introduction of nuclear power and the increasing use of radioactive substances and other radiation sources in industry, medicine, research, and education, the Atomic Energy Act was passed by

---

[1] Frédéric Joliot-Curie National Research Institute for Radiobiology and Radiohygiene, P.O. Box 101, H-1775 Budapest, Hungary.

Parliament, the supreme legislative body of the country, in 1980 (2). Accompanying this, an enacting clause of the Council of Ministers, and several ministerial orders, containing detailed rules and regulations for atomic energy applications were also issued (3).

In preparing this legislation, the new principles and recommendations of the ICRP, issued in Publication No 26 in 1977, were taken into consideration as far as possible (4). The main requirements of the ICRP's dose limitation system were incorporated into the text of the Atomic Energy Act as follows:

(a) In the Hungarian People's Republic, the applications of nuclear energy and related scientific research and development shall serve the interests of the society as a whole. Nuclear energy may only be used in a way that does not endanger human life, the health and living conditions of present and future generations, the man's environment and material goods.

(b) Exposure of workers employed in nuclear energy applications and of the population to all sources of radiation (except normal natural sources and medical applications) must not result in annual doses exceeding the dose-limits authorized by the relevant regulations on the basis of the current level of knowledge and the recommendations of competent national and international advisory bodies.

(c) Within the authorized dose-limits, radiation exposure shall be reduced to a level as low as reasonably achievable.

The enacting clause of the Council of Ministers contains dispositions concerning the duties and responsibilities of various ministries and other national authorities with respect to nuclear energy applications. It specifies that

dose-limits for exposure of workers employed in nuclear energy applications and of the population are to be established by the Minister of Health.

In accordance with this, a ministerial decree covering all essential aspects of radiation protection of people, both workers and the public, has been prepared. The report "Basic Safety Standards for Radiation Protection" by the IAEA, ILO, Nuclear Energy Agency (NEA), and WHO has been of a significant help in formulating this decree (5). Until its coming into force, the national standard, "Protection Against Radioisotope Radiation," issued in 1978 remains valid (6). Acceptable levels of radionuclide concentrations in foodstuffs for normal exposure conditions have not been defined in either the new ministerial decree or the previously used national standards.

## Abnormal Exposure Conditions

It is required by the Atomic Energy Act that

facilities established for the application of atomic energy should be operated in such a manner that abnormal operational occurrences can be avoided and unplanned exposures to radiation and radioactive substances can be prevented;

special dose-levels are to be established for emergency situations;

the management of any establishment or facility used for the application of atomic energy is obliged to report any unusual event to the State Sanitary and Epidemiology Inspectorate of the proper region.

On the basis of these requirements, the enacting clause of the Council of Ministers prescribes that

the operating organization of a facility established for the application of atomic energy is obliged to prepare an *emergency plan*, which will identify; (1) the types of possible abnormal events and means of their prevention; (2) the ways of notification of, and co-operation with the competent public authorities; and (3) the personnel responsible for this notification and for the protective measures.

For the prevention of any dangerous situation in the surroundings of a nuclear facility, the head of the county council has to prepare a complementary emergency plan in cooperation with the competent regional and national authorities and with due consideration to the emergency plan of the facility. This plan shall contain the measures necessary for prevention or alleviation of the danger and the names of persons responsible for taking these measures.

In order to give guidance concerning protection of the public in the event of a radiation accident, temporary emergency reference levels were specified by the Ministry of Health. These were based on the projected whole-body dose in the open air and the projected thyroid dose of children due to inhalation of radioiodine.

As in many other countries, no derived reference or intervention levels of radionuclides in various foodstuffs had been developed before the Chernobyl accident. The emergency plans were only prepared to cope with the consequences of abnormal events that would occur at the country's nuclear facility. No intervention had been foreseen for transboundary effects of abnormal events occuring in another country, either neighboring or at a larger distance.

## Control of Environmental Radioactivity

Studies on contamination of the environment with man-made radioactive substances were initiated in Hungary by a research institute in the early 1950s. These early studies were focused on detection of radioactive fission products of nuclear weapon tests performed in the atmosphere and their deposition on the ground surface. As fallout from testing nuclear weapons continued to appear, some national authorities also established their own monitoring networks. Their control programs, monitoring environmental radioactivity, have analyzed soil, surface waters, agricultural products and foodstuffs, and even levels of radionuclides in human tissues. While in the early period gross beta activity was the general parameter measured, activity concentrations of particular radionuclides, such as $^{85}$Kr, $^{90}$Sr, and $^{137}$Cs were also determined in the subsequent years (7,8).

The emphasis of these programs was shifted to the effects of nuclear power plants on the environment in the mid-1970s, when introduction of nuclear power to the country was decided by the government.

Responsibility for controlling radioactive contamination of the environment is now distributed among four national authorities. A network for monitoring airborne radioactivity is operated by the National Meteorological Service. Surface and drinking waters are monitored by laboratories of the National Water Authority. Concentrations of radioactive substances in soil, agricultural products, and foodstuffs are controlled by institutes of the Ministry of Agriculture and Food. Environmental radiation levels, both indoors and outdoors, and activity concentrations of radionuclides of natural and manmade origin in drinking water, food products, and other substances having direct relevance to the health of the public are subject to control by the Ministry of Health's Radiological Controlling and Data Providing Network. This network operates regional radiation hygiene laboratories in some of the sanitary and epidemiology stations of the country under the general guidance of the National Research Institute for Radiobiology and Radiohygiene (RIRR). In addition to these regulatory, authority-operated institutions, the Central Physical Research Institute and the Polytechnical University in Budapest, as well as the Nuclear Power Plant in Paks, have their own environmental control systems.

Soon after starting with the construction of the nuclear power plant, an extensive preoperational monitoring program was launched by the national authorities in order to establish background levels of environmental radiation and radioactivity and to assess exposure of the population to these radiation sources. The program continued uninterrupted after commissioning the nuclear power plant in order to assess its contribution to the environmental exposure of the population (9).

As soon as information on the reactor accident in Chernobyl was received and radioactive contamination of the domestic environment was expected to occur, all of these monitoring networks of authorities and the environmental control system of research centers were put on alert and requested to switch over to continuous operation. A few days later, when demand in data exceeded the overall capacity of these systems, some other research laboratories and university institutions also became involved in the environmental monitoring and foodstuffs control programs.

# Radioactive Contamination of the Environment Caused by the Chernobyl Accident

## Airborne Radioactivity

Arrival of radioactively contaminated air masses was first detected in the northern and northwestern regions of the country during the night of April 29 to 30. Gross beta activity measurements of aerosol samples taken by the meteorological monitoring network and by our institute showed a first peak of activity between 10 to 15 Bq/m$^3$ in the period of April 30 to May 1. A somewhat less pronounced second peak was observed in the southern and southwestern parts of the country on May 3 and 4. A more significant third peak appeared between May 6 and 8 in the activity of airborne particulate matters of 7 to 20 Bq/m$^3$ all over the country, but particularly in east Hungary (Fig. 30.1). After an intense rainfall

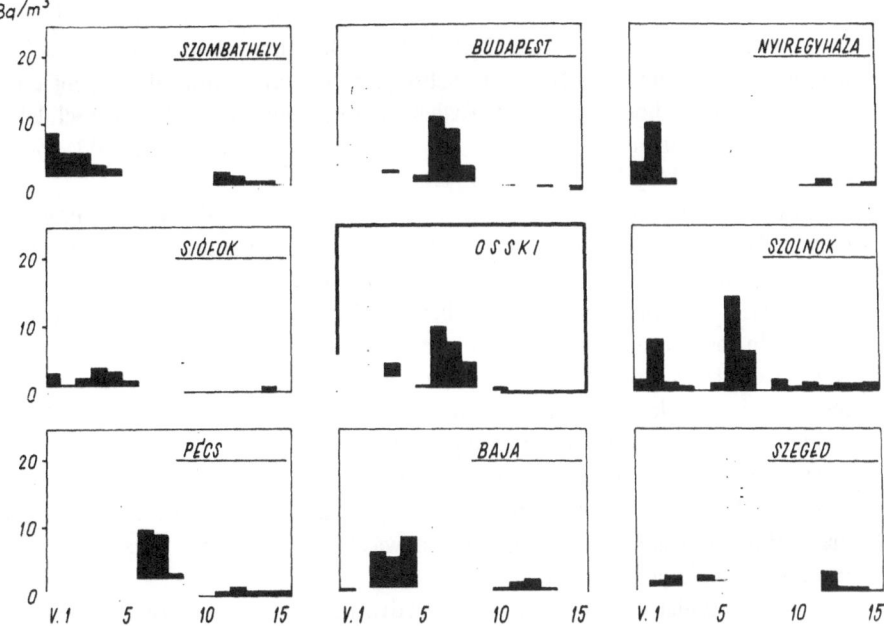

**Fig. 30.1.** Gross beta activity of aerosol samples taken in various regions of Hungary during the first two postaccident weeks (OSSKI is the National Research Institute for Radiobiology and Radiohygiene, Budapest).

during the night of May 9 to 10, radioactivity of the aerosol fell below 1 Bq/m$^3$. In the subsequent days, further decreases occurred down to the range of a few mBq/m$^3$.

With the aid of gamma spectrometry, typical fission products, such as $^{99}$Mo, $^{103}$Ru, $^{131}$I, $^{132}$Te/I, $^{134}$Cs, $^{137}$Cs, and $^{140}$Ba/La, could be found in aerosol samples. On some filters $^{95}$Zr/Nb, $^{141}$Ce, and $^{144}$Ce were also detected. Changes in concentration of these radionuclides showed the same pattern as that of gross beta activity, except that there was an apparent predominance of $^{103}$Ru in the third peak of the environmental contamination. Iodine-131 in vapor form was also found in the atmosphere in concentrations of about 1.5 to 2.5 times higher than that of the particulate $^{131}$I (10).

Follow-up analyses of air filters used during the first days in the postaccident period revealed a number of hot particles containing fission products; such as $^{95}$Zr/Nb, $^{103}$Ru, $^{141}$Ce, and $^{144}$Ce in much higher concentrations than was expected on the basis of a uniform distribution.

## Deposition of Radionuclides

A fraction of airborne radionuclides settled on the ground surface by either dry or wet deposition. Ground surface contamination measured by in situ gamma spectrometry in the region of Budapest at the beginning of May contained the following major radionuclides in decreasing order of concentration: $^{132}$Te/I, $^{131}$I, $^{103}$Ru, $^{137}$Cs, $^{140}$Ba/La, and $^{134}$Cs (Fig. 30.2). Composition of this contamination changed significantly in the subsequent period of time depending on the half-life of radionuclides (11). Rather high concentrations of radionuclides were found in rainwater collected in the area of Budapest April 29 to 30 (about 2.4 kBq/L) and May 8 to 9 (1.2 kBq/L).

All important water bodies (rivers, lakes, and reservoirs) were monitored twice a day in the first half and once a day in the second half of May. Maximum gross beta activity of about 20 Bq/L in Danube water was detected on May 4. This activity concentration decreased below 5 Bq/L within 10 days. Gross beta activity in the second largest river of the country, Tisza, and in Lake Balaton never exceeded a level of 10 Bq/L. Somewhat higher gross beta activities were observed in smaller rivers (Rába, Maros) entering the country from heavily contaminated areas of the neighboring states.

The concentration of each single radionuclide in drinking water was less than 1 Bq/L in the first half of May, except that of $^{132}$Te/I, which was less than 2 Bq/L. No radioactive contamination was found in drinking water obtained from deep, underground sources.

Total accumulated deposition of radioactive I and Cs isotopes on the ground surface in the region of Budapest is given in Table 30.1.

## Geographical Distribution of Radioactive Contamination

Since 1982 systematic measurements of the background radiation levels have been recorded in the country. For this purpose, thermoluminescent dosimeters were used at 123 meterological stations, both indoors and outdoors, and were exposed and evaluated at monthly intervals. On the basis of these measurements, annual averages of background radiation levels and their geographical distribution and seasonal variation were determined.

The dosimeters distributed in the early spring of 1986, just prior to the accident, were left over a longer period of time and exposed to the enhanced level of environmental radiation to the end of July. The dose values measured outdoors during these three months were compared with those that had been established during the same periods in the previous years. The dose increments found at each point are attributable to the accidental contamination and give a reliable picture of its geographical distribution over the whole country (Fig. 30.3).

From a fractional distribution of dose rates measured outdoors during the same period in 1985, a mean value of about 95 nGy/h was received. In April to July 1986, this mean value was about 120 nGy/h, representing approximately a 25% increase (10). It is noteworthy that the mere 1.6% increase in the average dose rate measured indoors in April to July 1986 represents a very efficient shielding provided by houses against gamma rays of the radioactive contaminants (an average shielding factor of about 0.08).

Gamma dose rates measured by a high-pressure ionization chamber in free air at 1 m above ground surface in the area of Budapest reached a maximum

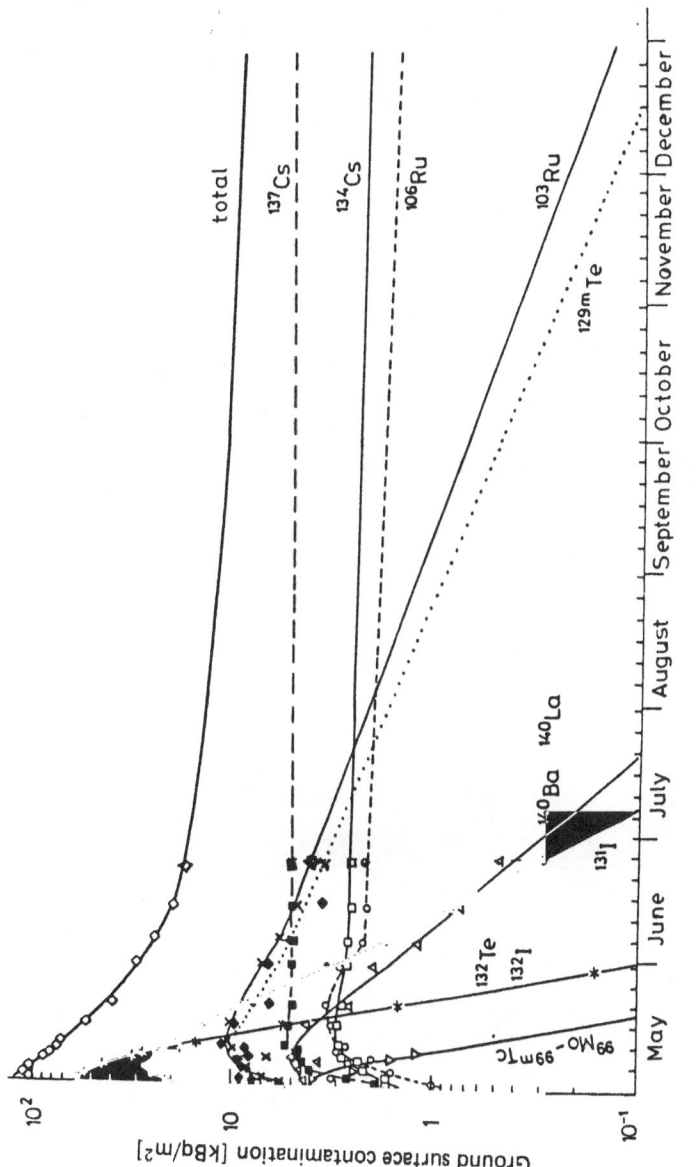

**Fig. 30.2.** Deposition of radionuclides on the ground surface in the region of Budapest, measured by in situ gamma spectrometry and extrapolated to the end of 1986.

427

**Table 30.1.** Total accumulated deposition of radioactive I and Cs isotopes on the ground surface in the region of Budapest

| End point in time | Radionuclide | | |
|---|---|---|---|
| | $^{131}I$ (kBq/m$^2$) | $^{134}Cs$ (kBq/m$^2$) | $^{137}Cs$ (kBq/m$^2$) |
| April 30, 1986 | ~20 | 1.27 | 2.48 |
| May 31, 1986 | ~30 | 2.35 | 4.86 |
| June 30, 1986 | ~30 | 2.41 | 4.99 |

value of 430 nGy/h on May 1. This was about 4.5 times higher than the annual average of background radiation level during the preceding years. The elevated rate decreased rapidly.

## Radioactive Contamination of Foodstuffs

### Contamination of Grass and Vegetables

Gross beta activity well above the natural background level (0.1 to 0.2 kBq/kg wet weight) was found in grass and vegetable samples, such as lettuce, sorrel, and spinach, collected in contaminated areas of the country on the last

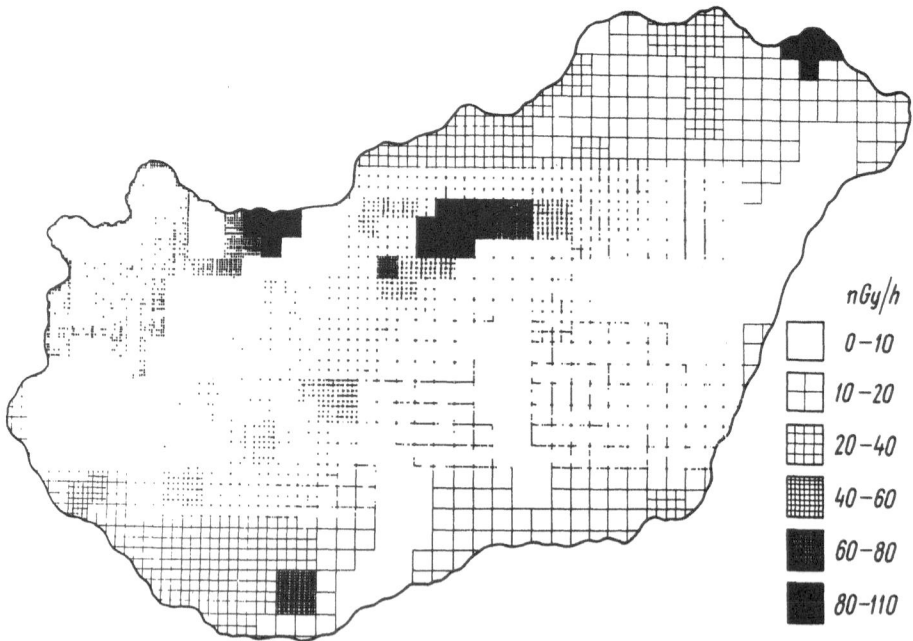

nGy/h
☐ 0–10
⊞ 10–20
▦ 20–40
▩ 40–60
■ 60–80
■ 80–110

**Fig. 30.3.** Increase above normal background of average dose rate values of the environmental radiation in Hungary because of the Chernobyl accident, measured outdoors in free air 1 m above ground surface during April through July 1986.

day of April and the first days of May. With gamma spectrometry, the same radionuclides could be detected in these samples as in air and soil samples. A maximum concentration of $^{131}$I of 10 to 12 kBq/kg wet weight was measured in grass on May 1 and 2. Thereafter, concentration of $^{131}$I decreased rapidly, reaching a value of about 1 kBq/kg at the middle and about 0.1 kBq/kg at the end of May. Concentration of $^{137}$Cs in grass samples was up to 2 to 2.5 kBq/kg wet weight at the beginning of the month. It fell to 0.25 to 0.75 kBq/kg during the first and to 0.05 to 0.15 kBq/kg during the second half of the month.

Radioactive contamination of vegetables was lower by a factor of 2 to 3, at least, and was mainly attributable to surface contamination that could partially be removed by repeated and careful washing in running water. No radioactive contamination was detected on vegetables grown in greenhouses or under plastic tents. Significantly lower and readily removable radioactive contamination was found on the surface of fruit, such as cherries and strawberries.

## Contamination of Milk and Meat

Continued grazing of farm animals during the first days of May resulted in a rapid increase in the concentration of $^{131}$I and $^{137}$Cs in fresh milk samples. A maximum concentration of 1.3 to 1.5 kBq/L of $^{131}$I was found in the first days of May, and about 50 Bq/L of $^{137}$Cs was found between May 5 and 15.

Radioactive $^{131}$I and $^{137}$Cs concentrations in blended milk produced by three plants of the dairy industry in Budapest were significantly lower than those found in fresh farm milk. The maximum concentration of $^{131}$I was less than 200 Bq/L around May 10 and that of $^{137}$Cs was less than 50 Bq/L toward the end of the month (Fig. 30.4).

In all dairies of the country, continuous monitoring of incoming milk was performed throughout the acute postaccident period. Milk with a high contamination level was withdrawn and either converted into milk powder or used for animal feeding.

An extended program was initiated in the middle of May to control radioactive contamination of meat and meat products. Average concentrations of $^{134}$Cs and $^{137}$Cs in pork and beef samples were between 20 to 30 Bq/kg and 30 to 50 Bq/kg, respectively, with great individual variations. Higher radiocesium concentrations were detected in venison, such as deer and stag.

## Emergency Reference Dose Levels in Hungary

The authorities' emergency plan covers all activities that shall be carried out by the public authorities or under their direct responsibility in a serious emergency situation. The protective measures that can be implemented in case of a nuclear accident are defined to reduce the risk of possible radiation exposures to members of the population. Accordingly, decisions to implement any of the protective measures, such as sheltering, stable I prophylaxis, evacuation, etc, are to be

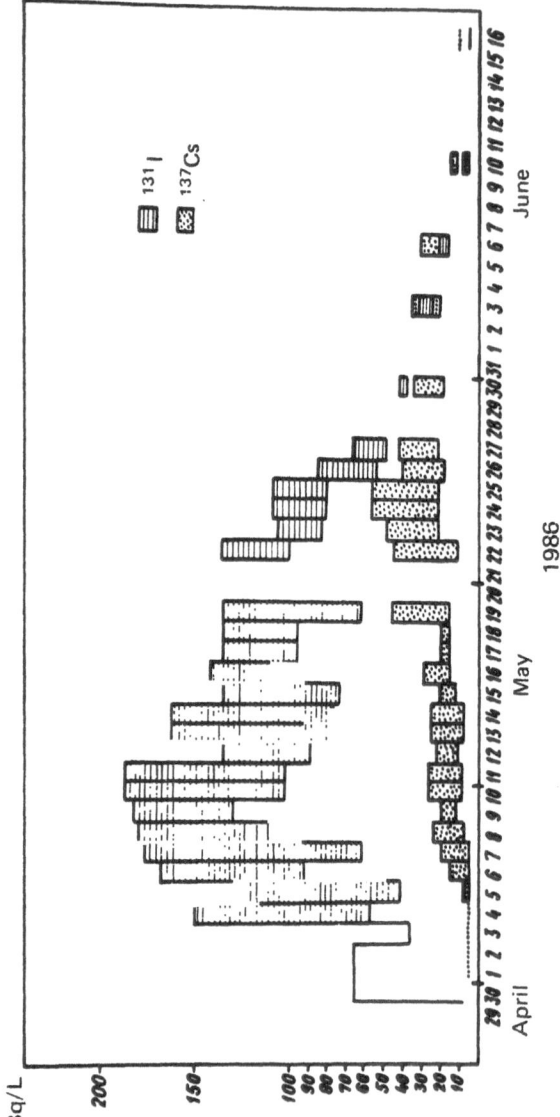

**Fig. 30.4.** Activity concentration of ${}^{131}$I and ${}^{137}$Cs in milk samples marketed by three dairy plants in Budapest. The daily maximum values fell within the shaded areas.

made on the basis of risk to the potentially exposed individuals. This is expressed in dosimetric quantities, that is, projected doses received without taking protective measures.

In accordance with the recommendations of the ICRP (12), the primary objectives of introducing protective measures in an accident situation are (a) the avoidance of serious nonstochastic radiation effects by reducing individual doses to levels below the threshold of these effects and (b) the decrease of risk of stochastic radiation effects by further reducing individual doses, if a positive net benefit can be assured to the individual involved.

In accordance with these principles, the temporary emergency reference levels established by the Ministry of Health before commissioning the first unit of our nuclear power plant were revised (13). The new reference levels were defined in such a way that the level of dose at which implementation of the most disruptive preventive measure (evacuation) should almost certainly be attempted is about half of the dose values at which early mortality is assumed to occur as a consequence of either whole-body or single-organ irradiation, 1 Gy and 10 Gy, respectively (Table 30.2).

Less disruptive protective measures (sheltering and iodine administration) should be implemented if the projected whole-body or single-organ doses are at about half of the threshold doses for nonfatal, but still serious, early consequences of radiation exposure (14).

Lower levels of doses at which the introduction of protective measures may be considered are an order of magnitude lower, about twice the dose limits recommended by the ICRP for members of the public in any one year under normal operating conditions. In an accidental situation, acceptance of a level of dose somewhat higher than the dose limit applicable under normal conditions seems to be justifiable based on this statement of the ICRP:

> The Commission's recommended dose limits are set at a level which is thought to be associated with a low degree of risk; unless a limit were to be exceeded by a considerable amount, the risk would still be sufficiently low as not to warrant such counter-measures as would themselves involve significant risks or undue cost. It is therefore clear that it is not obligatory to take remedial action if a dose-equivalent limit has been or might be exceeded. (4).

**Table 30.2.** Recommended temporary emergency reference levels in Hungary[a]

| | Absorbed dose (Gy) | |
|---|---|---|
| Protective measure | Whole body (bone marrow) | Thyroid or any single organ |
| Sheltering | 0.01–0.1 | 0.1–1 |
| Stable I | | |
|   Adults | | 0.5–5 |
|   Children | | 0.3–3 |
| Evacuation | 0.1–1 | 1–10 |

[a] Reproduced with permission from ref. 13.

No derived intervention levels of environmental dose rates or activity concentrations of radionuclides in air, water, and foodstuffs have been predetermined with the aim of automatically using them to trigger the introduction of protective measures in a particular situation.

Responsibility for coordination of the national efforts in Hungary in the case of a major nuclear accident has been given to a governmental committee made up of representatives of the competent national authorities, including a representative of the Public Health authority. The committee is assisted by an advisory board consisting of experts on nuclear safety, radiation hygiene, environmental protection, meteorology, medical services, etc. The committee can extensively utilize the services, equipment, and facilities of existing research and university institutions, as well as organizations that have permanent emergency functions. This advisory board is available for decision making regarding protective measures to be taken in all phases of an emergency, including control of water and food supplies (13).

## Derived Intervention Levels for Foodstuffs

Usually detailed analyses of all foreseeable types of accidents are requested in the safety report prepared for each facility. From such detailed analyses, the quantity of radioactive material available for release and its radionuclide composition, the most important pathways of population exposure, and a range of possible off-site scenarios can be identified. In addition, there are also generic analyses that are applicable for any given type of nuclear facility (15).

It is known from these analyses that radioactive fission products represent the dominant potential for consequences to the public. These exist in a variety of physical and chemical forms, and their potential for releases to the environment decreases in this order: (a) gaseous materials, (b) volatile substances, and (c) nonvolatile solids. However, consideration of particulate matters should not be completely neglected either. According to these considerations, the key radionuclides that may be expected to be dominant for various exposure pathways at longer distances from the source are $^{131}$I, $^{132}$Te/I, $^{134}$Cs, $^{137}$Cs, $^{106}$Ru, and $^{144}$Ce.

The most important exposure pathways are

1. external exposure to the radioactive substances contained in the passing plume and deposited on the ground;
2. internal exposure caused by inhalation of airborne radioactive materials (volatiles and particulates); and
3. internal exposure caused by ingestion of directly contaminated water and food (e.g., vegetables) or contaminated agricultural and animal products, such as corn, milk, and meat.

Information was received that the uncontrolled release of radioactive substances into the environment from the damaged reactor had not taken place in a sudden "puff" of a relatively short duration, but continued for several days. It was clear from the monitoring results obtained in the early postaccident period that several radionuclides were present in the environment and that exposure of the population might occur simultaneously by more than one pathway. Therefore, application of derived intervention levels developed by competent national and international organizations, such as the IAEA, WHO, and the National Radiological Protection Board (NRPB) of the United Kingdom, for a single radionuclide and a single exposure pathway would not have been appropriate (16,17).

Under such circumstances, it is necessary—in principle—to sum the contributions to the dose in each organ for each population group from all relevant pathways of exposure and from all radionuclides. Then, by comparing the dose in each organ with the appropriate intervention level for the dose, it is possible to decide whether or not countermeasures should be introduced. Given the uncertainties involved in making decisions, in particular those arising from the measurements of the radionuclide concentrations, the duration of radioactive release, and the length of environmental exposure, a complicated procedure for calculating derived intervention levels is not justified and a much simpler approach is recommended (18).

The major pathways in which radionuclides contaminating the domestic environment could contribute to the exposure of the population were (1) external exposure to gamma-emitting radionuclides; (2) internal exposure caused by inhalation of airborne radionuclides; and (3) internal exposure caused by ingestion of contaminated vegetables and fruit, milk and milk products, and meat and meat products. Of the radionuclides present in the environment, not more than 4 to 5 were considered to be critical to the exposure of the population by each pathway (not necessarily the same for each pathway). Furthermore, it was assumed that release of radioactive substances from the damaged reactor might last for about 4 to 5 days with evenly decreasing intensities. On the basis of these considerations, derived intervention levels for some selected radionuclides ($^{90}$Sr, $^{131}$I, $^{134}$Cs, $^{137}$Cs) in foodstuffs were defined for the critical age group of the population (infants). The higher (obligatory) intervention level was either an effective dose equivalent of 100 mSv or a dose equivalent to an organ of 1,000 mSv. Restrictions on food consumption were recommended at derived intervention levels 20/50 times lower than the levels recommended by the NRPB (17) for single radionuclides and particular exposure pathways (i.e., 100 mSv versus 5 mSv for exposure by 4 to 5 critical radionuclides and by multiple pathways over several days).

This was, of course, a rather conservative approach that provided adequate protection of the population. The derived intervention values obtained in this way are given in Table 30.3. In spite of the conservative approach, it was easy to comply with these derived intervention levels, particularly by introducing some relatively inexpensive interventions, such as the temporary prohibition of the grazing of farm animals and the blending of cows' milk collected in more and in less contaminated areas.

Concentrations of $^{131}$I in vegetables, fruit and meat, as well as concentrations

**Table 30.3.** Derived intervention levels for imposing restrictions on consumption of various foodstuffs[a]

| Foodstuffs | Radionuclides | | |
|---|---|---|---|
| | $^{90}Sr$ (Bq/kg) | $^{131}I$ (Bq/kg) | $^{134}Cs$ and $^{137}Cs$ (Bq/kg) |
| Milk and milk products | $5 \times 10^2$ | $8 \times 10^2$ | $1.4 \times 10^3$ |
| Vegetables and fruit | $4 \times 10^4$ | $4 \times 10^4$ | $7.2 \times 10^4$ |
| Meat and meat products | $2 \times 10^4$ | $2 \times 10^4$ | $3.6 \times 10^4$ |

[a] The critical age group of the population is generally constituted of children.

of $^{90}Sr$ and $^{134}Cs$ and $^{137}Cs$ in all foodstuffs were, at least, an order of magnitude lower than the derived intervention levels of these radionuclides in foodstuffs.

Unfortunately, the Commission of the European Communities (CEC) recommended more prohibitive acceptable levels of radionuclides in foodstuffs on May 6 (19). These were not based on strict radiation protection grounds but rather on political issues and were soon extended to imported foodstuffs. Since it was considered by us to be untenable to have two different norms, one stricter for exported foodstuffs and another looser for foodstuffs marketed for internal consumption, there was no other choice than to accept the CEC intervention levels.

In conclusion, one cannot but agree with Eisenbud in that application of protective action guides requires discipline on the part of local government, elected officials, scientists, and the media. The value of advance planning is greatly diminished when protective actions are taken at lower levels of exposure than in the plan announced in advance (20).

# References

1. Sztanyik LB, Bojtor I (1982) Experience in the application of the new ICRP recommendations in Hungary. In: The dose limitation system in the nuclear fuel cycle and in radiation protection. IAEA, Vienna, pp 605–612
2. évi I. törvény az atomenergiáról (Act I of 1980 on atomic energy) (1980) Magyar Közlöny 21:293–301
3. A Minisztertanács (1980) (IV.5.) számú rendelete az atomenergiáról szóló 1980. évi I. számú törvény végrehajtásáról [no 12/1980. (IV.5.) Enacting clause of the Council of Ministers to Act I of 1980 on atomic energy] Magyar Közlöny 21:301–307
4. Recommendations of the International Commission on Radiological Protection (1977) ICRP publication 26, Annals of the ICRP, vol 1(3). Pergamon, Oxford
5. Basic safety standards for radiation protection 1982 edition. Jointly sponsored by the IAEA, ILO, NEA (OECD), WHO. Safety series no 9, IAEA, Vienna
6. Radioizotópok sugárzása elleni védelem (Protection against radioisotope radiation (1979) National standard no MSZ 62–78, Hungarian Bureau of Standards, Budapest
7. Environmental radioactivity in Hungary (1975). Bulletin no 1, Kovács J and Predmerszky T (eds) International Radiation Protection Association Hungarian National Committee, Budapest

8. Environmental radioactivity in Hungary (1979). Bulletin no 2, Kovács J and Pred-merszky T (eds) International Radiation Protection Association Hungarian National Committee, Budapest

9. Sztanyik LB (1983) Experience gained in Hungary on the role and responsibility of the public health authority in the nuclear power program. In: Nuclear power experience, vol 4. IAEA, Vienna, pp 741–750

10. Sztanyik LB, Kanyar B, Koteles GJ, Nikl I, Stur D (1987) Radiological impact of the reactor accident at Chernobyl on the Hungarian population (a report in Hungarian). OSSKI, Budapest

11. Biró T, Fehér I, Sztanyik LB (1986) Radiation consequences in Hungary of the Chernobyl accident. Intern Agrophys 2:291–314

12. Protection of the public in the event of major radiation accidents: principles for planning (1984) ICRP publication 40, Annals of the ICRP, vol 14(2). Pergamon, Oxford

13. Sztanyik LB (1986) Involvement of the public health authority in emergency planning and preparedness for nuclear facilities in Hungary. In: Emergency planning and preparedness for nuclear facilities. IAEA, Vienna, pp 159–167

14. Nonstochastic effects of ionizing radiation (1984) ICRP publication 41, Annals of the ICRP, vol 14(3). Pergamon, Oxford

15. Planning for off-site response to radiation accidents in nuclear facilities (1981) Safety series no 55, IAEA, Vienna

16. Selected radionuclides. Environmental health criteria 25 (1983) WHO, Geneva

17. Derived emergency reference levels for the introduction of countermeasures in the early to intermediate phases of emergencies involving the release of radioactive materials to atmosphere (1986) NRPB-DL10, NRPB, Chilton Didcot Oxon, UK

18. Derived intervention levels for application in controlling radiation doses to the public in the event of a nuclear accident or radiological emergency—principles, procedures and data (1986) Safety series no 81, IAEA, Vienna

19. Commission recommendation of 6 May 1986 addressed to the member states concerning the coordination of national measures taken in respect of agricultural products as a result of radioactive fallout from the Soviet Union (1986) Official Journal of EEC, no L 118, p 28

20. Eisenbud M (1987) Environmental radioactivity from natural, industrial, and military sources, 3rd ed. Academic, San Diego, p 387

CHAPTER 31

# Control of Radioactivity in Foodstuffs in the European Economic Community

P. S. Gray[1] and F. Luyckx[2]

## Relevant Community Law

There are two bodies of European Community (EC) legislation relevant to the control of radioactive contamination in foodstuffs, that concerned with radiation protection requirements based on the Euratom Treaty and that dealing with the free circulation of goods within the EC based on the EEC Treaty. The first has as its keystone the Euratom Basic Safety Standards for the health protection of the general public and workers against the dangers of ionizing radiation (1). These standards take the form of a Council directive, the preparation of which requires the Commission of the European Communities (CEC) to consult a specialized expert group established under Article 31 of the Euratom Treaty and then the Economic and Social Committee before presenting proposals to the Council of Ministers (the Council), which in turn must consult the European Parliament before making a unanimous decision. The present standards cater largely to normal conditions, with responsibility for emergency management placed on the individual member states. Moreover, the procedures described ruled out the possibility of introducing new legislation for communitywide action on the time scale required following the Chernobyl accident.

The free circulation of goods within the Common Market is one of the fundamental principles enshrined in the EEC Treaty. Many of the restrictions on free circulation arise from measures taken to protect health, particularly in the realm of foodstuffs. For nearly two decades, the CEC has been engaged in the difficult task of removing these technical barriers to trade, using Article 30 of the EEC Treaty to attack those measures not justified by essential requirements, such as that of the protection of public health, and Article 100 of the Treaty to harmonize justifiable requirements by means of Council directives. Since the adoption of these directives required unanimity, progress was inevitably

[1] Head of Division, Foodstuffs, Commission of the European Communities, 1049 Brussels, Belgium.
[2] Principal Administrator, Radiation Protection (Euratom), Commission of the European Communities, 1049 Brussels, Belgium.

slow and became more difficult as the number of member states increased. Community legislation on such questions as food additives is still incomplete, and a directive would typically require 4 to 5 years from initiation to adoption, both the European Parliament and the Economic and Social Committee having to be consulted in the process. The adoption of the Single European Act in 1987 should accelerate this process by the introduction of majority voting, but the normal procedure will still be lengthy in terms of taking action in an emergency situation.

## The First Phase After Chernobyl

At the time Chernobyl occurred, therefore, there was neither EEC legislation (nor, in most cases, national legislation) in force relating directly to allowable levels of radioactivity in foodstuffs in the event of an accident. However, under the aegis of the EEC Treaty, the CEC had operated for a number of years a rapid alert system for widespread food contamination problems. This system links named persons in the ministries in each member state that are responsible for food safety with an incident controller in Division III/B/2 (Foodstuffs) in the CEC. Information transmission is effected by an emergency telephone line and a special telex network, which enable information to be received and automatically retransmitted to the contact points in the member states. The system allows contaminated food to be quickly tracked down and the effects of any incident to be limited by the withdrawal of contaminated foodstuffs from the market under national legislation, the CEC playing a coordinating role.

As soon as the CEC was aware that there was likely to be a significant contamination of foodstuffs in the territory of the Communities, it attempted to use Articles 35 and 36 of the Euratom Treaty, whereby environmental radiation monitoring data must be communicated to the CEC. In practice, this met with very limited success since the procedures were not geared to emergency situations. However, at the same time, the EEC food alert system was actuated, and this provided a much better response, although the system had not been designed with nuclear accidents in mind. Divisions III/B/2 and V/E/1, the service responsibile for radiation protection, set up close coordination of the CEC's subsequent activities.

Unfortunately, the initial spread of contamination into western Europe and the EEC occurred during the period April 27 to 28 so that the CEC's initiatives on April 29 to 30 were immediately hampered by the fact that May 1 and 2 were public holidays preceding a weekend. Moreover, at this time, the cloud had not yet reached some member states, and even where contamination of crops had occurred, the first measurements were just coming in, so that data on the extent of contamination was very scanty. Member states' governments, and even regional authorities, began to take uncoordinated action, and trade inside the EEC with the European Free Trade Association (EFTA) countries and with other trading partners became progressively obstructed. It was only possible to bring all the alert system contact points and their advisers on radiation safety together in Brussels on Monday, May 5, but, on the basis of the information and advice they supplied, the CEC took immediate action.

On May 6, the CEC issued a recommendation on the $^{131}$I levels in milk and other products (2). The recommendation also provided for mutual recognition of controls and proposed that no member state would apply stricter standards to products imported from other member states than it applied to its own products. Milk and leafy vegetables were the foodstuffs of immediate interest and while it was possible to get universal agreement on a limit of 500 Bq/L of $^{131}$I for milk, the level applicable to leafy vegetables was more contentious; the recommendation proposed a level of 500 Bq/kg for milk and milk products and 350 Bq/kg for leafy vegetables, both levels to be reduced by a factor of 2 at two successive 10-day intervals.

Following this meeting, and with the increased availability of information, trade was able to resume in most instances. However, there was little information on foodstuffs outside the EEC, and the CEC decided to propose a temporary ban on imports of foodstuffs from certain eastern European countries in order to replace various ad hoc emergency measures taken by member states with a common EEC measure. This proposal was adopted by the Council on May 12, 1986 (3). Had this step not been taken, then products imported into some EEC member states that still allowed imports would not have been able to move throughout the European Communities' territories and the checking necessary to differentiate between these goods and internally produced goods would have disrupted the Common Market. The CEC had, however, the right to act independently in respect to meat under the provisions of Council Directive 72/462/EEC on veterinary questions and, on the basis of an opinion of the Standing Veterinary Committee, temporarily banned on May 7 imports of bovine animals, swine, and fresh meat from the same eastern European countries (4).

Within the framework of EFTA/EEC free-trade relations, a series of informal meetings was held with the EFTA partners, and an active exchange of information was set up, which enabled trade to continue without undue hindrance. The key element in the control of the incident, both inside the EC and with trading partners, was the free flow of information, which was carried out continuously through the rapid alert system and contacts at working level.

## The Second Phase After Chernobyl

By the second week of May, it was evident that, in the medium term, the two Cs isotopes, $^{134}$Cs and $^{137}$Cs, would be the predominant dose-determining sources of radioactivity in the food chain. After several consultations of the EEC food experts and the Article 31 (Euratom) radiation safety experts, the CEC proposed to the Council that the import ban should be replaced by a system whereby imports from all third countries would be subjected to radioactivity limits based on the sum of $^{134}$Cs and $^{137}$Cs. On May 30, the Council adopted a measure in the form of Regulation 1707/86 (5), which set limit values for imported foodstuffs as shown in Table 31.1. These limits are somewhat lower than those recommended to the CEC on a scientific basis (1000 Bq/kg), since the Council took into account other circumstances, such as levels already adopted by some of the EC's trading partners. The scientific limits were intended to ensure compliance with the relevant recommendations of the International Commission on Radiological Protection (ICRP).

**Table 31.1.** Permissible levels of radioactivity in foods imported into the EC[a]

| Products | $^{134}Cs$ and $^{137}Cs$ (Bq/kg) |
|---|---|
| Milk: Liquid, dried, concentrated, condensed, and infant formulas | 370[b] |
| All other foods | 600 |

[a] Reproduced with permission from ref. 5.
[b] The level applicable to concentrated products shall be calculated on the basis of the reconstituted product ready for consumption.

It is important to note that although lower levels were chosen for milk and baby foods, the concentration clause enabled dried products to have radioactivity levels a factor of 4 to 10 above this limit, depending on the extent they were to be diluted for consumption. The exact concentration factors and the methods of analysis were left to the discretion of the control authorities.

It is also interesting that the legal basis of this regulation is quite simply the EEC Treaty as a whole. Had the Regulation been based on the specific articles relating to the internal market, the formal legal procedure of adoption would have been inordinately lengthy. The adoption of Regulation 1707/86 was an example of the EEC's will to act rapidly and effectively.

Two other points are worth noting. At the moment of adoption of Regulation 1707/86, the member states declared that they would not apply stricter limits to trade between member states than those required for import from third countries. This declaration established in one simple operation common standards for intra-EC trade, again avoiding the lengthy procedure referred to at the beginning of this chapter. Although member states were free to adopt whatever levels they wished for foodstuffs produced and consumed in their own territory, in general they adopted limits either identical to those of Regulation 1707/86 or equivalent in their effect. The original regulation was scheduled to expire on September 30, 1986, but was renewed twice—on September 30, 1986 (7), and on February 27, 1987 (8)—and expired on October 31, 1987.

Under the provisions of Regulation 1707/86, the CEC adopted, on June 5, 1986, implementing rules by means of Commission Regulation 1762/86 (9). This described how checks should be carried out by the control authorities and was particularly detailed in respect to live animals, for which sampling posed difficult problems. A list was drawn up of sensitive products to aid control authorities in orientating their checks, and the member states were required to notify the CEC immediately of any case in which the limits were exceeded and of the decisions taken regarding the consignment. They were also required to make monthly reports on imported products, and the rapid alert system was used for transmission of this information to all member states by the CEC. It was therefore possible to build up a picture of the problem areas and to direct control efforts in the most effective way possible.

In the period June 1986 to June 1987 inclusive, 102 cases were reported where consignments exceeded the limits given. A summary of these cases is shown in Table 31.2. In quantity, they represented a very small proportion of the EEC import trade, and most of the cases were products, such as live animals, for which the taking of a representative sample is notoriously difficult.

None of the rejected consignments of meat was derived from farm animals, imports that exceeded the limit were from essentially free-range domestic animals, such as horses or mountain sheep, and game, such as deer. The cases that occurred in fruit were principally from early maturing berries that would have been farmed shortly after the accident; the cases that occurred later in the year were from dried or frozen fruit. Measurements made on fruit that had been contaminated in the early stage of growth showed the expected falloff in activity per unit weight as the fruit increased in size with increasing maturity. The cases of herbs and tea were not surprising in view of the relatively large surface-to-weight ratio and, again, the difficulties in obtaining representative samples of these products. Hazelnuts were at the time of the incident just at the stage

**Table 31.2.** Consignments of food refused by the EC

| Month | Live horses | Game sheep | Fruit | Herbs, tea | Nuts (hazel) | Miscellaneous | Total |
|---|---|---|---|---|---|---|---|
| 1986 | | | | | | | |
| June | 1 | | | | | 2 | 3 |
| July | 5 | 4 | 2 | | | 1 | 12 |
| August | 3 | | | | | | 3 |
| September | | | | 1 | | | 1 |
| October | 2 | 1 | 3 | 1 | 5 | 3 | 15 |
| November | | | | | 3 | | 3 |
| December | | 4 | | 7 | 4 | 10 | 25 |
| 1987 | | | | | | | |
| January | | 2 | 1 | | | | 3 |
| February | | | | 1 | 5 | 2 | 8 |
| March | | | | | 7 | | 7 |
| April | | | | 10 | 8 | | 18 |
| May | | | | 1 | | | 1 |
| June | | | | 1 | 2 | | 3 |
| Total | 11 | 11 | 6 | 22 | 34 | 18 | 102 |

of formation during which contamination could penetrate the kernel. Again, the product is gathered and commercialized in such a way that it is difficult to get representative samples. Snails and fungi figured among the miscellaneous consignments that were registered. There was 1 consignment of milk powder and 10 consignments of a chocolate drink mix (containing milk powder), all from the same source. All 102 consignments of foodstuffs that exceeded the limits were returned to their country of origin.

## EC Exports

European Community exports of foodstuffs are very important, for example, in 1985, exports of liquid, concentrated, and dried milk were valued at 1.888 million ECU. At the time of Chernobyl, many of the EEC trading partners had no prefixed standards for radioactivity in foodstuffs. It was therefore not possible for the EC to fix limits of radioactivity in foodstuffs for export. However, no export refunds were allowed for products exceeding the EC import limits, making their exports uneconomic. Similarly, products taken into intervention, which are frequently exported at a later date, were also required to comply to the EC import limits.

In practice, some countries adopted very stringent standards, some going so far as to require the absence of $^{134}$Cs and $^{137}$Cs. The EC exporters undertook a large program of product control and the CEC was involved in many consultations with the trading partners, a significant number of which adopted permissible levels of radioactivity similar to EC import levels. There is obviously a need for internationally agreed standards for radioactivity in foodstuffs to avoid trade difficulties and provide adequate health protection, and the CEC has welcomed the initiatives of various international agencies in this respect.

## EC Regulatory Proposals

At the time of the adoption of Regulation 1707/86, the need for a permanent system for stipulating maximum permitted levels of radioactivity in foodstuffs was recognized by the EEC, and the CEC undertook to produce a proposal rapidly.

Article 30 of the Euratom Treaty calls for determining basic standards for the protection of the health of workers and the general public against the dangers of ionizing radiation, and this was chosen as the legal basis for the proposal. The expert group set up under Article 31 has the role of advising the CEC to enable the latter to work out these basic standards. This group was asked (initially in May prior to the adoption of Regulation 1707/86 with specific reference to the circumstances resulting from the Chernobyl accident) to advise the CEC on determining levels of radioactivity in foodstuffs following a nuclear accident and, subsequently, to meet any future needs. Again, the starting point for the latter work was the ICRP recommendations on doses (10) and the existing Euratom directive (1).

Although dose objectives can be fairly easily defined, the experts faced an extremely difficult task since, in order to devise permissible limits in food, it is necessary to make assumptions about diet, the degree of radioactivity of each foodstuff, and how this would vary with time. In a localized incident, the ideal control system would be interactive, with decisions made on each foodstuff on the basis of ongoing calculations of accumulated doses.

Such an approach may not, however, be satisfactory even on a local scale and would certainly be impracticable on an EC scale because, in order to ensure continued free circulation of foodstuffs and a situation of legal clarity for those involved in food production and trade, it is necessary to define specific levels for food contamination that are referred to as derived reference levels (DRLs).

In calculating the DRLs, the Article 31 group first considered some 20 radionuclides and a number of different foodstuff categories for several groups. To provide a simpler approach, the foodstuff categories were then reduced to three:

1. milk products,
2. other major foodstuffs, and
3. drinking water.

Similarly, the radionuclides considered were grouped into three classes:

1. isotopes of I and Sr (notably $^{131}$I and $^{90}$Sr),
2. alpha-emitting isotopes of Pu and transplutonium elements (notably, $^{239}$Pu and $^{241}$Am), and
3. all other nuclides of half-life greater than 10 days (notably $^{134}$Cs and $^{137}$Cs).

The derived reference level for each combination of foodstuff category and nuclide class was designed to ensure that the committed effective dose to the thyroid would not exceed 50 mSv in one year and the committed dose equivalent would not exceed 5 mSv for any age group. Typical diet patterns were taken for the EC and it was assumed that the actual intake in one year would amount to no more than 10% of that corresponding to the entire diet being contaminated to 100% of the DRL throughout the year. The Article 31 group estimated that the calculated intakes of radioactivity would generally be higher than those that would occur in practice, even if in certain localized areas the source of foods is not so diverse as that for the greater part of the EEC population.

In addition to consulting the Article 31 group, the CEC consulted a group of senior scientists and also held a seminar at Luxembourg in April 1987. A number of different approaches to modeling the transmission of nuclides in the food chain were discussed, and it became clear that the assumptions made by the Article 31 group gave DRLs that were broadly in line with other approaches. The group took the results of these consultations into account before finalizing its advice. Because of the conservative assumptions already incorporated in the calculation, it was not considered necessary to make further reductions in order to allow for contributions from each of the three nuclide classes finally specified. The values of DRLs given in the opinion of the Article 31 committee are shown in Table 31.3.

**Table 31.3.** Derived reference levels considered as the basis for the control of foodstuffs following an accident[a]

| Isotope | Milk products (Bq/kg) | Other major foodstuffs (Bq/kg) | Drinking water (Bq/L) |
|---|---|---|---|
| $^{131}$I, $^{90}$Sr | 500 | 3,000 | 400 |
| $^{239}$Pu, $^{241}$Am | 20 | 80 | 10 |
| $^{134}$Cs, $^{137}$Cs | 4,000 | 5,000 | 800 |

[a] Reproduced with permission from the CEC.

Finally, the Article 31 committee was of the opinion that foodstuffs with an annual consumption of less than 10 kg could have values 10 times those specified for major foodstuffs.

The CEC, in making its proposal, took the recommendations of the Article 31 committee as a basis, but it also considered that other factors had to be taken into account in setting DRLs, such as the degree of public concern, the implications for EC trade, and the relationship to levels in use elsewhere in the world. As a result, the levels for the Cs nuclide class were further reduced by a factor of 4 for milk products and other major foodstuffs, these being potentially the most significant combinations in the event of a future accident.

The CEC had also carried out some consultations with experts on the transmission of radionuclides in the food chain from animal fodder into animal products, and these showed that less stringent limits could be applied to fodder without the risk that the resultant meat, milk, or other products would exceed the DRLs for food. However, it should be emphasized that the chief consideration here is minimizing economic loss; public health surveillance must rely on monitoring the animal product prior to consumption and not on monitoring animal diets. On the basis of these latter consultations, levels were proposed for fodder in respect of the nuclide class, including the Cs isotopes. For the other classes, the dilution factor was thought to be so high in the feed–food chain that no limits needed be determined in advance.

The limits proposed by the CEC are shown in Table 31.4, and a list of minor foodstuffs is shown in Table 31.5.

To turn now to the legal and administrative provisions of the proposal sent to the Council on June 16, 1987 (11), one of the main problems at the time of Chernobyl was that no preset EEC limits were available to apply immediately. This resulted in the diverse national actions described earlier in this chapter. The proposed permanent system involves a two-stage system that enables rapid action to be taken and to be adjusted later in the light of circumstances. In the first stage, the CEC could adopt a regulation bringing into force the limits set

**Table 31.4.** Maximum permitted levels for foodstuffs, feedstuffs, and drinking water[a]

| | Dairy products[b] (Bq/kg) | Other foodstuffs (except minor foodstuffs)[c] (Bq/kg) | Drinking water and liquid foodstuffs[d] (Bq/kg) | Feedstuffs (Bq/kg) |
|---|---|---|---|---|
| Isotopes of I and Sr, notably, $^{131}$I and $^{90}$Sr | 500 | 3,000 | 400 | e |
| Alpha-emitting isotopes of Pu and transplutonium elements, notably, $^{239}$Pu and $^{241}$Am | 20 | 80 | 10 | e |
| All other nuclides of half-life greater than 10 days, notably, $^{134}$Cs and $^{137}$Cs | 1,000 | 1,250 | 800 | 2,500 |

[a] Reproduced with permission from ref. 11.
[b] Dairy produce is defined as milk falling within headings No. 04.01 and 04.02 of the Common Customs Tariff and those foodstuffs intended for the special feeding of infants during the first 4 to 6 months of life, which meet, in themselves, the nutritional requirements of this category of person and are put up for retail sale in packages that are clearly identified and labeled "food preparation for infants."
[c] Minor foodstuffs are those foodstuffs listed in Annex II of the Regulation. For these, a level of contamination 10 times that quoted in this column may be allowed.
[d] Liquid foodstuffs are defined in chapters 20 and 22 of the Common Customs Tariff.
[e] No value for immediate application.

out in Table 31.4 once it had ascertained that an event had occured of such magnitude as to require the application of emergency measures to foodstuffs and animal fodder. In the second stage, the CEC could, after consulting the Article 31 committee, adapt these limits to the specific circumstances arising from the nuclear accident in question. Both the initial limits and any subsequent changes would apply to foodstuffs and animal fodder which are

imported into the EC,
traded between EC member states,
exported from the EC, or

**Table 31.5.** Minor foodstuffs[a]

| EC common customs tariff heading number | Description |
|---|---|
| 07.01 | Vegetables, fresh or chilled: O. Capers[b] |
| 07.03 | Vegetables provisionally preserved in brine, in sulphur water, or in other preservative solutions but not specially prepared for immediate consumptions: B. Capers[b] |
| Chapter 09 | Coffee, tea, mate, and spices |

[a] Reproduced with permission from ref. 11.
[b] 07.01 O. Capers and 07.03 B. Capers are Customs classifications.

produced within an EC member state and marketed within that EC member state.

The operation of the system is also linked to the operation of a rapid information system currently under consideration by the Council and to the International Atomic Energy Agency Convention on Early Notification of a Nuclear Accident of September 26, 1986.

The EC Council of Ministers did not decide on a permanent system by October 31, 1987, when Regulation 1707/86 lapsed. However, the EC member states agreed not to change their import limits until new levels were decided.

## Postscript

On December 22, 1987, the Council of the European Communities adopted a permanent regulation laying down limits for radioactivity in foodstuffs following a future nuclear incident. The limits to be used are shown in Table 31.6.

**Table 31.6.** Limits for radioactivity in foodstuffs[a]

| Isotopes | Dairy products[b] (Bq/kg) | Other foodstuffs (except minor foods) (Bq/kg) |
|---|---|---|
| Isotopes of Sr, notably, $^{90}$Sr | 125 | 750 |
| Isotopes of I, notably, $^{131}$I | 500 | 2,000 |
| Alpha-emitting isotopes of Pu and transplutonium elements, notably, $^{239}$Pu and $^{241}$Am | 20 | 80 |
| All other nuclides of half-life greater than 10 days, notably, $^{134}$Cs and $^{137}$Cs | 1,000 | 1,250 |

[a] Reproduced with permission from ref. 12.
[b] Levels applicable to concentrates or dried products shall be calculated on the basis of the reconstituted product as ready for consumption. Limits for baby foods, liquid foods, and animal feedstuffs will be decided at a later date.

The Council also adopted a regulation (13) continuing for a further two years the limits for Cs applied by Regulation 1707/86 following Chernobyl, namely, 370 Bq/kg for baby foods and dairy products and 600 Bq/kg for other foods. In addition, the Council adopted a resolution (14) that invites the CEC to bring about international agreement on the basis of EEC permitted levels and considering that the same limits should be applied to exported foodstuffs unless the importing states apply different limits.

## Conclusions

This brief account has described the regulatory actions taken in the EC immediately following Chernobyl, the CEC's experience with the administration of these regulations, and its proposal for a permanent system. When regulating in this field, it is necessary to confront a complex process of the transmission of radionuclides through the food chain with the requirements of trade and public opinion in order to have relatively few and simple limit levels. The consumer is demanding a greater variety in food supplies, and, together with increased specialization, this demand is leading to a growing trade in agricultural products and foodstuffs, both inside the EEC and between the EEC and the world, so that present levels of both internal and external trade are of the order of 300 G ECU (European Currency Unit). There is an imperative need for common regulations not only to avoid the difficulties that were and are being encountered in trade but to resolve the confusion that must arise from diverse measures taken by public authorities.

Good regulatory practice must be founded on a sound scientific approach. The scientific results of Chernobyl in respect to food contamination are still being gathered, sorted, and analyzed, and these will be very valuable in verifying models of radionuclide transfer through food to humans. The CEC, in presenting its proposal for a permanent system, recognizes that these scientific developments are taking place and looks forward to the day when this work will result in a more solid international consensus. It will follow progress closely and will do all it can to contribute to such progress; it will also keep under review the need to adjust EEC regulatory instruments in the light of future developments.

*Acknowledgments.* The authors wish to acknowledge the untiring efforts of the control and surveillance staff in the member states who contributed to the smooth operation of the surveillance and data exchange described in this chapter. They would also like to recognize the valuable contribution made by their colleagues in the CEC, and particularly that of Mr. G. Fraser, in the preparation of this chapter.

## References

1. Council Directive 80/836/Euratom, OJ no L 246 of 17.9.1980, p 1 as amended 84/467/Euratom OJ no L265 of 5.10.84, p 4
2. Commission Recommendation 86/156/EEC, OJ no L 118 of 7.5.1986, p 28
3. Council Regulation (EEC) no 1388/86, OJ no L 127 of 13.5.1986
4. Commission Decision 86/157/EEC, OJ no L 120 of 8.5.1986, pp 66–67
5. Council Regulation (EEC) no 1707/86, OJ no L 146 of 31.5.1986, pp 88–90
6. Council Directive 80/836/Euratom, OJ no L 246 of 17.9.1980, p 1
7. Council Regulation (EEC) no 3020/86, OJ no L 280 of 1.10.86, p 79
8. Council Regulation (EEC) no 624/87, OJ no L 58 of 28.2.87, p 101
9. Commission Regulation (EEC) no 1707/86, OJ no L152 of 6.6.86, p 41
10. ICRP publication 40 (1984) Annals of the ICRP, vol 14 no 32
11. Commission proposal for a Council Regulation COM(87)281 final, OJ no C 174 of 2.7.1987, p 6
12. Council Regulation (Euratom) OJ no L 371 of 30.12.87, p 11
13. Council Regulation (EEC) OJ no L 371 of 30.12.87, p 14
14. Council Resolution OJ no C 352 of 30.12.87, p 1

CHAPTER 32

# Derived Intervention Levels in Food: The Canadian Approach

E. Somers,[1] M. B. Cooper,[1] and D. P. Meyerhof[1]

## Introduction

It was not until the Chernobyl accident that international attention had been given to the control of food contaminated as a result of an accident even though the question of dose intervention levels for countermeasures had been addressed by various international bodies in recent years. In the aftermath of the Chernobyl accident, the emergency response criteria and approaches in the various countries appeared, generally, to be very divergent even taking into account the natural diversity deriving from local situations and requirements. One very significant impact of this diversity in limits was the problems caused to international trade; a situation that persisted for a considerable period of time. There has been a general acknowledgment by international bodies responsible for the establishment of guidelines and recommendations in relation to radiation protection that there is a need for an international consensus in terms of dose intervention levels and the principles for expressing these levels in terms of radionuclide concentrations in various environmental materials. This chapter describes the evolution of criteria for the limitation of radiation exposure to members of the public in Canada from radionuclide contamination in food resulting from a nuclear accident or radiological emergency.

## Basis of Guidelines

The practical implementation of emergency measures in relation to the control of foodstuffs requires the derivation of intervention levels of radionuclide contamination in food. The establishment of Derived Intervention Levels (DILs) and their incorporation in contingency plans are essential to facilitate the decision-making process in emergency situations.

The procedures by which the action levels of the dose are translated into

---

[1] Environmental Health Directorate, Department of National Health and Welfare, Ottawa, Canada.

concentrations of radionuclides present in foodstuffs involve assumptions with respect to a number of important factors, namely,

appropriate dose limits for members of the public;
dietary habits of the population;
the chemical form of the radionuclides;
the most sensitive group in the population;
the period of time during which the release takes place, in comparison to
    the half-life of the contaminant;
what processes, environmental and otherwise, may lead to a reduction or
    increase in the levels of radionuclides in food following the event.

## Canadian Situation Prior to Chernobyl

Emergency planning for reactor accidents within Canada has, as its basis, inter-vention levels of doses at which certain countermeasures would be introduced. These are termed emergency reference levels or protective action levels. Action guidelines are also provided in some instances to translate emergency dose levels into environmental measurements, however, these are mainly in terms of dose rates. The dose levels to be applied for the control of foodstuffs vary between provinces and range from 0.5 to 5 mSv for a whole-body dose and 1.5 to 50 mSv for a thyroid dose.

Prior to the Chernobyl accident, there were no derived limits in Canada for air and food. The only guidelines in existence were for radionuclide contamination of drinking water, although these were not intended to cover emergency situations. They were based on a 0.5-mSv whole-body dose and a 5-mSv dose to critical organs, assuming continuous consumption of 2 L per day for one year.

## Canadian Response Following Chernobyl

Following the Chernobyl reactor accident, interim screening limits were derived to apply to food products suspected of being contaminated. These screening limits were based on the same dose limits set for drinking water contamination. The position was taken that the operation of the Chernobyl nuclear reactor did not result in a benefit to Canadians. Permitting unnecessary exposure to radioactiv-ity through the sale of contaminated foodstuffs, therefore, was unjustified. The objective was to limit the health risks from additional radioactive exposure for the Canadian population from the Chernobyl reactor accident to one in a million.

The approach was to develop a set of guidelines based on the metabolic and dosimetric models of the International Commission on Radiological Protec-tion (ICRP), which relate the limiting radiation dose to a limiting annual intake of radioactivity. Concentrations of radionuclides in a particular food product, which would result in this intake, were derived from an annual consumption

**Table 32.1.** Canadian screening limits for radioactivity in foods from the Chernobyl reactor accident[a]

| Food item | Radionuclide concentration (Bg/kg) | |
|---|---|---|
| | $^{134}Cs$ and $^{137}Cs$ | $^{131}I$ |
| Milk and drinking water (Bq/L) | 50 | 10 |
| Manufactured dairy products | 100 | 40 |
| Other foods | 300 | 70 |
| Spices[b] | 3,000 | |

[a] Reproduced with permission from the International Commission on Radiological Protection.
[b] Limit also includes ruthenium isotopes $^{103}Ru$ and $^{106}Ru$.

rate for the food. The screening limits currently in use are presented in Table 32.1.

In practice, it was found that, even though low levels of fallout from the Chernobyl accident were detected in Canada, the contamination levels in domestically produced foods were well below the screening limits. Application of the screening limits did not result in disruption of the food supply to Canadians, although monitoring of imports from Europe revealed contamination in a number of shipments in excess of the Canadian limits. Action to restrict importation of these products was taken.

Extensive monitoring of $^{137}Cs$ in caribou meat in northern Canada has revealed that, in some areas, the levels of $^{137}Cs$ exceed the interim screening limits. However, the contribution of fallout from the Chernobyl accident is only between 9% and 25% of the activity. The remainder is due to $^{137}Cs$ from nuclear weapons tests in the early 1960s. The interim screening limits were not applied to the caribou as the major part of the activity was not attributable to the Chernobyl accident. The acceptability of the caribou meat was based on a whole-body dose of 5 mSv from an annual consumption of 100 kg of meat. The $^{137}Cs$ concentration based on these criteria is 3,500 Bq/kg (shown in Table 32.2, caribou meat is part of category III, Other major foodstuffs).

**Table 32.2.** Examples of derived intervention levels for selected radionuclides

| Food category[a] | Radionuclide concentration (Bq/kg) | | | |
|---|---|---|---|---|
| | $^{90}Sr$ | $^{131}I$ | $^{137}Cs$ | $^{239}Pu$ |
| I. Milk | 100 | 2,000 | 1,500 | 10 |
| II. Fruit & vegetables | 1,000 | 30,000 | 2,500 | 150 |
| III. Other major foodstuffs | 1,500 | 45,000 | 3,500 | 250 |
| IV. Beverages | 1,000 | 30,000 | 2,500 | 150 |

[a] DILs for tritium ($^3H$) have not been included in this table but are approximately 0.3 MBq/kg for category I and 2.5 MBq/kg for categories II–IV.

In the event of a nuclear accident or radiological emergency, either within Canada or close to its borders, radionuclide contamination of domestically produced food could be higher than that resulting from Chernobyl. The criteria and approach that were used in establishing the interim screening limits after Chernobyl are not compatible with recent international recommendations and provincial emergency plans. At the request of the Province of Ontario, a task force was established to develop Canadian guidelines for DILs in food that would be consistent with provincial emergency plans and that could be applied across all of Canada, taking into account the need for international harmonization.

## Proposed Canadian Guidelines for Food

For food produced and consumed in Canada, the proposed guidelines for the derivation of intervention levels differ somewhat from those recommended by the International Atomic Energy Agency (IAEA), the Food and Agriculture Organization (FAO), and the World Health Organization (WHO). While maintaining the same fundamental intervention levels of doses, Canadian dietary habits have been taken into account in the calculation of the DILs. The general criteria and assumptions used in the Canadian guidelines are outlined in the following.

### Principles

1. The limiting doses for members of the public exposed through the consumption of food contaminated as a result of a nuclear accident or radiological emergency are the lower intervention levels of dose recommended by the IAEA and the ICRP for the intermediate phase of a nuclear accident. For radionuclides that irradiate the whole body (e.g., $^{137}$C, $^{134}$C), the limit is a 5-mSv committed effective dose equivalent. For radionuclides that preferentially irradiate a specific organ (e.g., $^{131}$I, $^{239}$Pu, $^{90}$Sr), the limit is a 50-mSv dose equivalent to the specific organ most at risk.
2. Only two age groups are considered, adults and infants, with the application of the more restrictive DILs where there is a difference in the calculated values between the two groups. Estimates for the annual food intake for the two groups are based on Canadian data. Food items are placed into four categories and DILs have been calculated for each food category on the assumption that the contaminated food item represents the total component of diet within the particular category.
3. To convert the ingestion of radioactivity into doses, dose conversion factors in sieverts per becquerel are selected from literature values for the two groups. The values chosen for a particular radionuclide are those that apply to the nuclide in its most chemically soluble form. Generally, this will yield the most restrictive value for the DIL.
4. No processes, other than radioactive decay, are taken into account in the

calculation of the DIL for a particular radionuclide. For short-lived radionuclides, those with half-lives less than 100 days, an effective food intake, which is related to the half-life of the nuclide, replaces the annual food intake value.

5. The most restrictive approach would be to apply the intervention level of a dose to the total dietary intake of food. However, it was considered unlikely that an individual would consume food, all of which is contaminated, at the intervention level over a whole year. Therefore, in deriving the intervention levels for a particular radionuclide, food consumption was divided into categories with each treated independently. It was assumed that individual contaminated food items represented the total component of a diet within the particular category. Where more than one radionuclide was present in a food item, an additive procedure was applied.

## Calculation of Derived Intervention Levels

### General Expression

For the contamination of a specific food category by a particular radionuclide, the derived intervention level is given by the expression

$$DIL = \frac{\text{Limiting annual radionuclide intake (Bq)}}{\text{Annual or ``effective'' food consumption (kg)}},$$

where the limiting annual radionuclide intake is calculated by

$$\frac{\text{Dose equivalent limit (Sv)}}{\text{Dose conversion factor (Sv/Bq)}}.$$

### Short-Lived Radionuclides

A value for the effective food intake is estimated, and this value replaces the annual food intake for a particular category. The effective food intake is related to the decay constant for the radionuclide. The effective and annual intakes will not differ significantly for radionuclides with half-lives greater than 100 days, and it is recommended that, for these radionuclides, the annual intake figure be used.

### Examples of Derived Intervention Levels for Food

The application of the procedures just described is illustrated by the DILs presented in Table 32.2. These have been calculated for several important radionuclides for each of the food categories. The following qualifications apply to the values:

1. for long-lived radionuclides, they represent concentrations in the foodstuff that persist for one year;

2. for the short-lived radionuclide $^{131}$I, this is a peak concentration, lasting for only one day, followed by radioactive decay;
3. only one radionuclide is present in the foodstuff;
4. only one foodstuff in Category III is contaminated;
5. they would apply to the consumption of food in the first year following an accident or emergency situation;
6. the more restrictive of the DILs, adult or infant, for each radionuclide and category is tabulated; and
7. the DILs are calculated to two significant figures and are also rounded to the nearest value of 5.

## Implementation of the Derived Intervention Levels

The proposed guidelines for food are about to be submitted to a federal provincial committee for approval. Implementation will occur as they are incorporated into provincial plans for nuclear emergencies. Consideration is also being given to incorporating them into the food and drug regulations.

# Derived Intervention Levels for Drinking Water

Because drinking water and foodstuffs are simply components of a person's total diet, there is a need for a coordinated and consistent approach in establishing DILs for both. Water is a major constituent of many manufactured foodstuffs and is the principal component of beer and soft drinks, both of which contribute significantly to the normal diet of many Canadians. The implementation of controls on drinking water supplies is potentially far more complex than for foodstuffs. Proposals for the derived intervention levels in drinking water are currently being developed.

CHAPTER 33

# Perceived Risks of Radionuclides: Understanding Public Understanding

B. Fischhoff[1] and O. Svenson[2]

## Introduction

Although technical experts have the luxury of specializing in the management of particular risks, members of the general public do not. They must approach the risks of radionuclides in the food chain with much the same intellectual and social resources as they approach the other risks in their lives. For example, when evaluating how adequately they understand radionuclide risks, laypeople must, in general, rely on the same critical capacities as they use in assessing their need for additional information about other health risks—or even for information about economic, professional, and interpersonal risks. When laypeople evaluate the trustworthiness of information about radionuclides, they must, in general, ask the same questions as they do about information coming from other sources and about other topics. When laypeople seek confirmation or sympathy for their fears, they must, in general, turn to the same people for advice and support as they do with other problems in their lives.

As a result, the study of how people perceive the risks of radionuclides in the food chain must begin with the study of how they perceive risks in general— before proceeding to consideration of the special properties of these risks and the contexts within which they arise. Moreover, since few actions are taken on the basis of risk perceptions alone, that study must also consider how people perceive the benefits of those actions that may entail exposure to risks and, finally, how people make trade-offs between such risks and benefits.

Fortunately, there exist extensive research literatures regarding the processes of judgment and decision making (1–6). This research provides substantive results that can be tentatively extrapolated to predict (or explain) people's responses to radionuclide risks. For example, many studies have found that people

[1] Department of Engineering and Public Policy, Carnegie Mellon University, Pittsburgh, PA 15213, USA.

[2] University of Lund, Lund, Sweden.

Support for preparation of this report was provided by the US National Science Foundation Grant SES-8715564.

are relatively insensitive to the extent of their own knowledge (7,8). The most common result is overconfidence (e.g, being correct on only 80% of those occasions when one is 100% certain of being correct). The generality of these findings in those settings that have been studied suggests that people might have undue confidence in their beliefs about radionuclide risks (and about how those risks can be controlled). If this seems like a reasonable (and worrisome) hypothesis, then the research literature might be consulted for procedures able to improve people's judgment. For example, telling people that overconfidence is widespread seems to have little effect, whereas presenting people with personalized feedback regarding the appropriateness of their own confidence can make a positive difference (9).

If one wanted to test these hypotheses (or others), then the existing research also provides well-understood methodologies for conducting studies specific to particular risks. Ascertaining people's beliefs and values is a craft having as many nuances as does assessing their physiological functions, or conducting measurements in the natural or biological sciences. For example, two formally equivalent ways of asking people to estimate how large a risk is can produce estimates that vary by several orders of magnitude (10). A study that used one method might make it seem as though people underestimate the risk, whereas a study using the other method would produce apparent overestimates. As in other sciences, such measurement artifacts can sometimes be predicted on the basis of psychological theory (11,12), whereas in other cases they are discovered by trial and error. Exploiting this experience offers the opportunity to avoid the mistaken interpretations, and perhaps even mistaken polices, that such artifacts can produce.

These methodologies have, in fact, been applied to studying laypeople's responses to the risks of technological hazards in general and to those associated with radionuclides in particular (13–16). Although they have seldom challenged the overall conclusions from the general literature on judgment and decision making, these studies have provided important elaborations (e.g., showing just what people believe about these particular risks, just how confident they are in those beliefs, or just how far they trust risk information coming from various sources). They have prompted attention to general issues that have particular significance with radionuclide risks (e.g., the ways in which emotional involvement affects how people process information about risk and how they evaluate the sources of risk information) (17–20).

These detailed, systematic, empirical studies stand in stark contrast to the casual observations that dominate many discussions of the public's behavior. Perhaps surprisingly, even scientists, who would hesitate to make any statements about topics within their own areas of competence without a firm research base, are willing to make strong statements about the public on the basis of anecdotal evidence. Unfortunately, immediate appearances can be deceiving, for example, when salient examples of public behavior are not particularly representative. As mentioned, even systematic observations can be misleading if not undertaken with a full understanding of the relevant methodology. An unfounded belief in having understood the public is the greatest barrier to acquiring a genuine understanding.

The limits to casual observation might be seen in the coexistence of conflicting claims about the public, often associated with conflicting recommendations regarding how to deal with the public. For example, advocates of deregulation frequently describe the public as understanding risks so well that it can readily fend for itself in an unfettered marketplace. This confidence in the public is usually shared by those who advocate extensive public participation in risk management (e.g., hearings and information campaigns). Quite the opposite conclusion about public competence underlies proposals to leave risk management to technical experts and to mandate risk-management practices (e.g., seatbelts, crash helmets, and dietary restrictions). Given the political (and safety) implications of these conflicting perceptions about the public, laypeople's behavior would seem to merit careful study. Good, hard evidence could provide guidance for managing risks, resolving conflicts between the public and technical experts, supplying the information that the public needs for better understanding, and creating technologies whose risks are acceptable to the public.

The following section provides a summary of some conclusions that can be drawn from studies of risk perception, as well as from the general research literature regarding judgment and decison making. As a package, these conclusions point to some fundamental limitations of several common schemes for managing the public's risk perceptions. These limits are discussed in the second section. Acknowledging them allows consideration of some more modest and complex proposals, a common theme of which is taking the details of the public's concerns seriously. These proposals are discussed in the third section.

## What Is Known

### People Simplify

Most substantive decisions require people to deal with more nuances and details than they can readily handle at any one time. People have to juggle a multitude of facts and values when deciding, for example, whether to change jobs, trust merchants, or protest a toxic landfill. To cope with the overload, people simplify. Rather than attempting to think their way through to comprehensive, analytical solutions to decision-making problems, people try to rely on habit, tradition, the advice of neighbors (or the media), and on general rules of thumb (e.g., nothing ventured, nothing gained). Rather than consider the extent to which human behavior varies from situation to situation, people describe other people in terms of encompassing personality traits, such as being honest, happy, or risk seeking (21). Rather than think precisely about the probabilities of future events, people rely on vague quantifiers, such as "likely" or "not worth worrying about"—terms that are also used quite differently by different people (22).

The same desire for simplicity can be observed when people press risk managers to categorize technologies, foods, or drugs as safe or unsafe, rather than to treat safety as a continuous variable. It can be seen when people demand convincing proof from scientists who can provide only tentative findings. It can be seen when people attempt to divide the participants in risk disputes into good guys and bad guys, rather than viewing them as people who, like themselves, have complex and interacting motives. Although such simplifications help people to cope with life's complexities, they can also obscure the fact that most risk decisions involve gambling with people's health, safety, and economic well-being in an arena with diverse actors and shifting alliances.

## Once People's Minds Are Made Up, It Is Hard to Change Them

People are extraordinarily adept at maintaining faith in their current beliefs unless confronted with concentrated and overwhelming evidence to the contrary. Although it is tempting to attribute this steadfastness to pure stubbornness, psychological research suggests that some more complex and benign processes are at work (21).

One psychological process that helps people to maintain their current beliefs is feeling little need to look actively for contrary evidence. Why look if one does not expect that evidence to be very substantial or persuasive? For example, how many environmentalists read the *Wall Street Journal* and how many industrialists read the (US) Sierra Club's *Bulletin* in order to learn something about risks (as opposed to reading these publications to anticipate the tactics of the opposing side)? A second contributing thought process is the tendency to exploit the uncertainty surrounding apparently contradictory information in order to interpret it as being consistent with existing beliefs. In risk debates, a stylized expression of this proficiency is finding just enough problems with contrary evidence to reject it as inconclusive.

A third thought process that contributes to maintaining current beliefs can be found in people's reluctance to recognize when information is ambiguous. For example, the incident at Three Mile Island would have strengthened the resolve of any antinuclear activist who asked only, How likely is such an accident, given a fundamentally unsafe technology?, just as it would have strengthened the resolve of any pronuclear activist who asked only, How likely is the containment of such an incident, given a fundamentally safe technology? Although a very significant event, Three Mile Island may not have revealed very much about the riskiness of nuclear technology as a whole. Nonetheless, it helped the opposing sides polarize their views. Similar polarization has followed the accident at Chernobyl, with opponents pointing to the consequences of a nuclear accident (which come with any commitment to nuclear power) and proponents pointing to the unique features of that particular accident (which are unlikely to be repeated elsewhere, especially considering the precautions instituted in its wake) (23).

## People Remember What They See

Fortunately, given their need to simplify, people are quite good at observing those events that come to their attention (and which they are motivated to understand) (24,25). As a result, if the appropriate facts reach people in a responsible and comprehensible form before their minds are made up, there is a decent chance that their first impression will be the correct one. For example, most people's primary sources of information about risks are what they see in the news media and observe in their everyday lives. Consequently, people's estimates of the principal causes of death are strongly related to the number of people they know who have suffered those misfortunes and the amount of media coverage devoted to them (26).

Unfortunately, it is impossible for most people to gain any firsthand knowledge of hazardous technologies. Rather, what laypeople see are the outward manifestations of the risk-management process, such as hearings before regulatory bodies or statements by scientists to the news media. In many cases, these outward signs are not very reassuring. Often, they reveal acrimonious disputes between supposedly reputable experts, accusations that scientific findings have been distorted to suit their sponsors, and confident assertions that are disproven by subsequent research (27,28).

Although unattractive, these aspects of the risk-management process can provide the public with potentially useful clues as to how well hazardous technologies are understood and managed by industry and regulatory agencies. Presumably, people evaluate these clues just as they evaluate the conflicting claims of advertisers and politicians. It should not be surprising, therefore, that the public sometimes comes to conclusions that differ from what the risk managers hope or expect. For example, it is understandable that the public might conclude that saccharin is an extremely potent carcinogen after seeing the enormous scientific attention that it generated some years back. Indeed, in some cases, the public may have a better overview on the proceedings than the scientists and risk managers mired in them, realizing perhaps that neither side knows as much as it claims.

## People Cannot Readily Detect Omissions in the Evidence That They Receive

Unfortunately, not all problems with information about risk are as readily observable as blatant lies or unreasonable scientific hubris. Often, the information that reaches the public is true, but only part of the truth. Detecting such systematic omissions proves to be quite difficult (29). For example, most young people know relatively few people suffering from the diseases of old age, nor are they likely to see those maladies cited as the cause of death in newspaper obituaries. As a result, young people tend to underestimate the frequency of these causes of death, while overestimating the frequency of vividly reported causes, such as murder, accidents, and tornadoes (26).

Laypeople are even more vulnerable when they have no way of knowing about information that has not been disseminated. In principle, for example, one could always ask physicians if they have neglected to mention any side effects of the drugs they prescribe. Likewise, people could ask merchants whether there are any special precautions for using a new power tool, and they could ask proponents of a hazardous facility if their risk assessments have considered all forms of operator error and sabotage. In practice, however, these questions about omissions are rarely asked. It takes an unusual turn of mind to recognize one's own ignorance and insist that it be addressed.

As a result of this insensitivity to omissions, people's risk perceptions can be manipulated in the short run by selective presentation. Not only will people not know what they have not been told, but they will not even feel how much has been left out (30). What happens in the long run depends on whether the unmentioned risks are revealed by experience or by other sources of information. When deliberate omissions are detected, the responsible party is likely to lose all credibility. Once a shadow of doubt has fallen, it is hard to erase.

## People May Disagree More about What Risk Is Than about How Large It Is

Given this mixture of strengths and weaknesses in the psychological processes that generate people's risk perceptions, there is no simple answer to the question, How much do people know and understand? The answer depends on the risks and the opportunities that people have to learn about them.

One obstacle to determining what people know about specific risks is disagreement about the definition of risk (31–33). The opportunities for disagreement can be seen in the varied definitions used by different risk managers. For some, the natural unit of risk is an increase in probability of death; for others, it is a reduced life expectancy; and for still others, it is the probability of death per unit of exposure (where "exposure" itself may be variously defined).

The choice of definition is often arbitrary, reflecting the way in which a particular group of risk managers habitually collects and analyzes data. The choice, however, is never trivial. Each definition of risk makes a distinct political statement regarding what society should value when it judges the acceptability of risks. For example, "reduced life expectancy" puts a premium on deaths among the young, which would be absent in a measure that simply counted the expected number of premature deaths. A measure of risk could also give special weight to individuals who can make a special contribution to society, individuals who were not consulted (or even born) when a risk-management policy was enacted, or individuals who do not benefit from the technology generating the risk.

If laypeople and risk managers use the term risk differently, then they can agree on the facts about a specific technology but still disagree about its degree of riskiness. Several years ago, the idea circulated in the nuclear power industry that the public cared much more about multiple deaths from large accidents than about equivalent numbers of casualties resulting from a series of small accidents. If this assumption were valid, the industry would be strongly motivated to remove the threat of such large accidents. If removing the threat proved impossible, then the industry could argue that a death is a death, and that in formulating social policy it is totals that matter, not whether deaths occur singly or collectively.

There were never any empirical studies to determine whether this was really how the public defined risk. Subsequent studies, though, have suggested that what bothers people about catastrophic accidents is the perception that a technology capable of producing such accidents cannot be very well understood or controlled (34). From an ethical point of view, worrying about the uncertainties surrounding a new and complex technology, such as nuclear power, is quite a different matter than caring about whether a fixed number of lives is lost in one large accident rather than in many small accidents.

## People Have Difficulty Detecting Inconsistencies in Risk Disputes

Despite their frequent intensity, risk debates are typically conducted at a distance (35,36). The disputing parties operate within self-contained communities and talk principally to themselves. Opponents are seen primarily through their writing or their posturing at public events. Thus, there is little opportunity for the sort of subtle probing needed to discover basic differences in how the protagonists think about important issues, such as the meaning of key terms or the credibility of expert testimony. As a result, it is easy to misdiagnose one another's beliefs and concerns.

The opportunities for misunderstanding increase when the circumstances of the debate restrict candor. For example, some critics of nuclear power actually believe that the technology can be operated with reasonable safety. However, they oppose it because they believe that its costs and benefits are distributed inequitably. Although they might like to discuss these issues, public hearings about risk and safety often provide these critics with their only forum for venting their concern. If they oppose the technology, then they are forced to do so on safety grounds, even if this means misrepresenting their perceptions of the actual risk. Although this may be a reasonable strategy for pursuing their ultimate goals, it makes them look unreasonable to observers who know the facts about nuclear power.

Individuals also have difficulty detecting inconsistencies in their own beliefs or realizing how simple reformulations would change their perspectives on issues. For example, most people would prefer a gamble with a 25% chance of losing $200 (and a 75% chance of losing nothing) to a gamble with a sure loss of $50. Most of the same people would also buy a $50 insurance policy to protect against such a loss. What they will do depends on whether the $50 is described as a sure loss or as an insurance premium. As a result, one cannot predict how people will respond to an issue without knowing how they will perceive it, which depends, in turn, on how it will be presented to them by merchandisers, politicians, or the media (37–39).

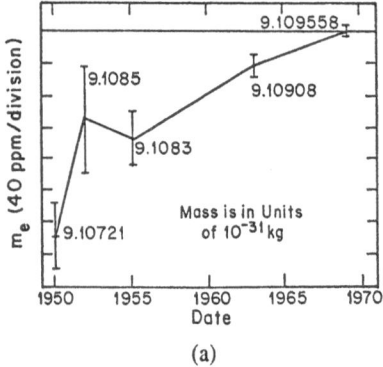

**Fig. 33.1.** Two examples of over-confidence in expert judgment. Overconfidence is represented by the failure of error bars to contain the true value. (a) Estimates of the rest mass of the electron. (b) Estimates of the height at which an embankment would fail. (Reproduced with permission from refs. 42 and 43.)

Thus, people's insensitivity to the nuances of how risk issues are presented exposes them to manipulation. For example, a risk might seem much worse when described in relative terms (e.g., as doubling their risk) than in absolute terms (e.g., as increasing that risk from 1 in a million to 1 in a half million). Although either representation of the risk might be honest, their impacts would be quite different. Perhaps the only fair approach is to present the risk from both perspectives, letting recipients determine which one (or hybrid) best represents their world view.

## Experts Are People, Too

Obviously, experts have more substantive knowledge than laypeople. Often, however, the practical demands of risk management force experts to make educated guesses about critical facts, taking them far beyond the limits of their data. In such situations, debates about risk are often conflicts between competing sets of risk perceptions, those of the public and those of the experts. As a result, one must ask how good those expert judgments are. Do experts, like laypeople, tend to exaggerate the extent of their own knowledge? Are experts more sensitive than others to systematic omissions in the evidence that they receive? Do they, too, tend to oversimplify policy issues?

Available studies suggest that when experts must rely on judgment their thought processes resemble those of laypeople (3,40,41). For example, Fig. 33.1 displays two cases of overconfidence in the judgments of senior scientists in their attempts to determine ranges of possible values (or confidence intervals) for topics within their areas of expertise (42,43). Anecdotal evidence of judgmental limitations can be found in many cases where fail-safe systems have gone seriously awry or where confidently advanced theories have been proved wrong. For example, DDT came into widespread and uncontrolled use before the scientific community had seriously considered the possibility of side effects. The accident sequences at Three Mile Island and Chernobyl seem to have been out of the range of possibility, given the theories of human behavior underlying the design of those reactors (44,45). Of course, science progresses by absorbing lessons that prompt it to discard incorrect theories. However, society cannot make rational decisions about hazardous technologies without knowing how much confidence to place in them and in the scientific theories on which they are based.

In the academic community, vigorous peer review offers technical experts some institutional protection against the limitations of their own judgments. Unfortunately, the exigencies of risk management, including time pressures and resource constraints, often strip away these protections, making risk assessment something of a quasi-science (like much cost/benefit analysis, opinion polling, or evaluation research), bearing more of the rights than the responsibilities of a proper discipline.

On the basis of psychological theory, one would trust experts' opinions most where they have had the conditions to acquire good judgment as a learned skill. These conditions include prompt, unambiguous feedback that rewards them for candid judgment (and not, for example, for exuding confidence). Weather forecasters do have these conditions, and the result is a remarkable ability to assess the extent of their own knowledge (46). It rains almost exactly 70% of the time when they are 70% confident that it will. Unfortunately, such conditions are rare. When feedback is delayed, as with predictions of the carcinogenicity of chemicals having long latencies, certain kinds of learning may become very difficult. Further problems arise when expert predictions are ambiguous and the lessons of history are hard to unravel. Psychological theory also suggests that learning is likely to be fairly local. Thus, one might be as worried about toxicologists making judgments regarding social policy as about social policy makers judging the contents of toxicology reports.

# What Cannot Be Done

The technical and policy issues involved in making risk-management decisions are complex enough in themselves. Dealing with public perceptions of risks creates an additional level of complexity for risk managers. Rather than face that complexity, they may be tempted to look for some "quick fix" that will give the public enough of what it wants to allow industry and government officials to deal with the risk problem free from public interference. Often these simple solutions embody a deep misunderstanding of the role of the public in risk management, reflecting a belief, by risk managers, that the human element in risk management can be engineered in the same way as mechanical and electronic elements. As a result, many attempts to deal with the public and its perceptions have been predictably doomed. The following are some of the more common of these simple strategies for dealing with risk controversies (47–51).

## Give the Public the Facts

The underlying assumption in this strategy is that if people only knew as much as the experts, they would see things the same way. It often results in an incomprehensible deluge of technical details, telling the public more than it needs to know about specific risk research results and much less than it needs to know about the quality of that research. Inundating the public with information also ignores the possibility that there are legitimate differences between the public and the experts regarding either the goals or the facts of risk management. In the short run, it may be comforting for technical experts to assume that risk management is only a matter of facts and that they have a monopoly on relevant wisdom. In the long run, however, such an assumption perpetuates experts' inability to understand what the public really wants from them and from the technologies that they promote.

## Sell the Public the Facts

The premise here is that the public needs persuasion rather than education. It often follows the failure of an information campaign to win public acceptance for a technology. The disrespect for the public underlying this approach frequently leads to heavy-handed implementation—in effect, repeating more loudly (or fancily) messages that the public has already rejected. Here, as elsewhere, obvious attempts at manipulation breed resentment.

## Give the Public More of What It Has Gotten in the Past

The underlying assumption here is that the public will accept in the future the kinds of risks that it has accepted in the past. If true, then what the public wants (and will accept) can be determined simply by examining statistics showing risk/benefit tradeoffs in existing technologies. This "revealed preference" philosophy ignores the fact, consistently revealed by opinion polls showing great public support for environmental regulations, that people are unhappy with how risks have been managed in the past. The risks that people have tolerated are not necessarily acceptable to them. As a result, giving them more of the same means enshrining past inequities in future decisions. In principle, this approach attaches no importance to educating the public, creating a constituency for risk policies, or involving the public in the political process. It seems to respect the public's wishes, while keeping the public itself at arms length.

## Give the Public Clear-Cut, Noncontroversial Statements of Regulatory Philosophy

Examples in the United States are the Delaney clause, prohibiting carcinogenic additives to foods, and the Nuclear Regulatory Commission's (NRC) safety goals for nuclear power. This approach assumes that laypeople are too unsophisticated to understand, in the context of technology management, the sort of routine risk/benefit trade-offs that they routinely make in everyday life, such as when they undergo medical treatments or pursue hazardous occupations. Despite frequent assertions to the contrary, surveys show that people are generally willing to tolerate some risk as long as they receive some compensating benefit. Moreover, such simple statements provide little guidance for many real situations—by denying the complexity of the decisions that need to be made. If perceived as hollow, then they will do little to reassure the public.

## Let the Marketplace Decide

Another frequent hope is that doing away with government regulation will allow people to decide independently what risks they are willing to accept, with the courts addressing any excesses. This approach ignores the fact that there are too many risks around for laypeople to know enough to fend for themselves everywhere. It also ignores people's misperception of some risks and the difficulty of getting risk information to them in a comprehensible form. Even if they knew everything that the experts knew, people might still not want to defend their own welfare when it comes to health and safety, especially where risks have long latencies and it is impossible to prove the source of a health risk and obtain redress.

## Put Risk Managers on the Firing Line

The underlying assumption of this strategy is that what the public needs to deal with risk issues effectively is a coherent story from a credible source. Examples might include the NRC's reliance on a single spokesperson as the Three Mile Island incident wore on and the assumption of center stage by the president of Union Carbide after Bhopal. While fine in principle, this strategy fails if the manager has poor communication skills or is insenstive to listeners' information needs. In addition, focusing on the form of messages may lead to neglect of their content. Oversimplifications, misrepresentations, and unaccept-able policies are just that, even if they come from a nice guy.

## Involve Local Communities in Resolving Their Own Risk-Management Problems

This approach assumes that people will be flexible and realistic about trade-offs when they see the big picture. Such approaches can flounder when the community's authority is more apparent than real or when it lacks the technical capability to understand its alternatives. It may also flounder when those alterna-tives exploit past inequities (e.g., reduce chronic proverty by accepting a hazard-ous waste dump) or are of the jobs-versus-health variety, which people expect government to help them resolve. Ensuring informed consent of the governed for the risks to which they are exposed is a laudable goal. However, its achieve-ment requires that people have tolerable choices, adequate information, and the ability to identify the optimal course of action.

## What Might Be Done

Despite their flaws, each of these simple strategies has some merit. It is important to give people the facts and to employ some persuasion when the facts do not speak for themselves or when existing prejudices must be overcome. It is also important to maintain some consistency with past risk-management decisions, expound clear policies, exploit the wisdom of the marketplace, encourage direct communication between risk managers and the public, and give communities meaningful control over their own destinies. The problem is that each oversimpli-fies the nature of risk issues and the public's involvement with them. When risk managers pin unrealistic hopes on such strategies, then the opportunity to address the public's needs more comprehensively is lost. When these hopes are not met, the frustration that follows is usually directed at the public.

It is both unfair and corrosive for the social fabric to criticize laypeople for failing to respond wisely to risk situations for which they are not adequately prepared to make the appropriate response. It is tragic and dangerous when nuclear engineers, or any other members of our technical elite, feel that they have devoted their lives to creating a useful technology only to have it rejected by a foolish and unsophisticated public. It is equally tragic and dangerous for the public to label those elites as evil and arrogant.

Risk management requires allocating resources and making trade-offs between costs and benefits. Thus, it inherently involves conflicts. Both the substance and the legitimacy of these conflicts are obscured, however, when the participants come to view them as struggles between the forces of good and evil, or of wisdom and stupidity. Effective solutions will have to be respectful solutions, recognizing both the legitimacy and complexity of the public's perspective, giving it no more and no less credit for reasonableness than it deserves.

How can the preceding observations about risk perceptions (and the research literature from which they were drawn) be used to design better procedures for dealing with risk controversies? One necessary starting point is a detailed consideration of the nature of the risk that the public must understand. That consideration must cover not only the best available technical estimates for the magnitude of the risk, but also the best available psychological evidence on how people respond to that kind of risk. Research has shown, for example, that people have special demands for safety—and reassurance—when risks have delayed effect or catastrophic potential; and when they appear to be poorly understood or out of their personal control (52,53). Such risks are likely to grab people's attention and create unrest until they can be put in some acceptable perspective. They demand greater communication resources with particular attention devoted to creating an atmosphere of trust. Perhaps paradoxically, people may need to be treated with the greatest respect in those situations in which they may seem most emotional (or most human).

A second necessary starting point is a detailed description of how information about the risk can reach people. Such information may be the result of accidents at various distances away and attributed to various causes (e.g., malfunctions, human error, sabotage) or of mere incidents, such as newspaper exposés, siting controversies, false alarms, or government inquiries. Proactively, this analysis will show the opportunities for reaching people. For example, is there a chance to educate at least some of the public in advance—or can one but prepare materials for times of crisis? Reactively, this analysis should help one anticipate what people will already know (or believe) when the time comes for systematic communication. It may show people to be buffeted by confusing, contradictory, and erroneous messages—or to have some basic understanding within which they can integrate new information. In any case, communication must build on people's current mental representation for the technology—even if its first step is to challenge inappropriate beliefs and enhance people's ability to examine future information more critically.

Knowing what people do know allows a systematic analysis of what they need to know—the next point of departure in communicating with the public. In some cases, crude estimates of a technology's risks and benefits may be enough; in other cases, it may be important to know how a technology operates. The needs depend on the problems that the public is trying to solve: what to do in an emergency; how to react in a siting controversy; whether to eat vegetables; whether to let their children do so, etc. Perhaps the most efficient description would be in the terms of decision theory (e.g., refs. 6,15,48, and 54), such as the simple decision tree in Fig. 33.2(a), depicting the situation faced by the head of a household deciding whether to test for domestic radon accumulations. Such descriptions allow one to determine how sensitive these decisions are to different kinds of information, so that communication can focus on the things that people really need to know.

Producing comparable descriptions for the different actors in a risk-management episode will help clarify sources of disagreement among them. Are they using terms differently? Are they considering different alternatives? Are different

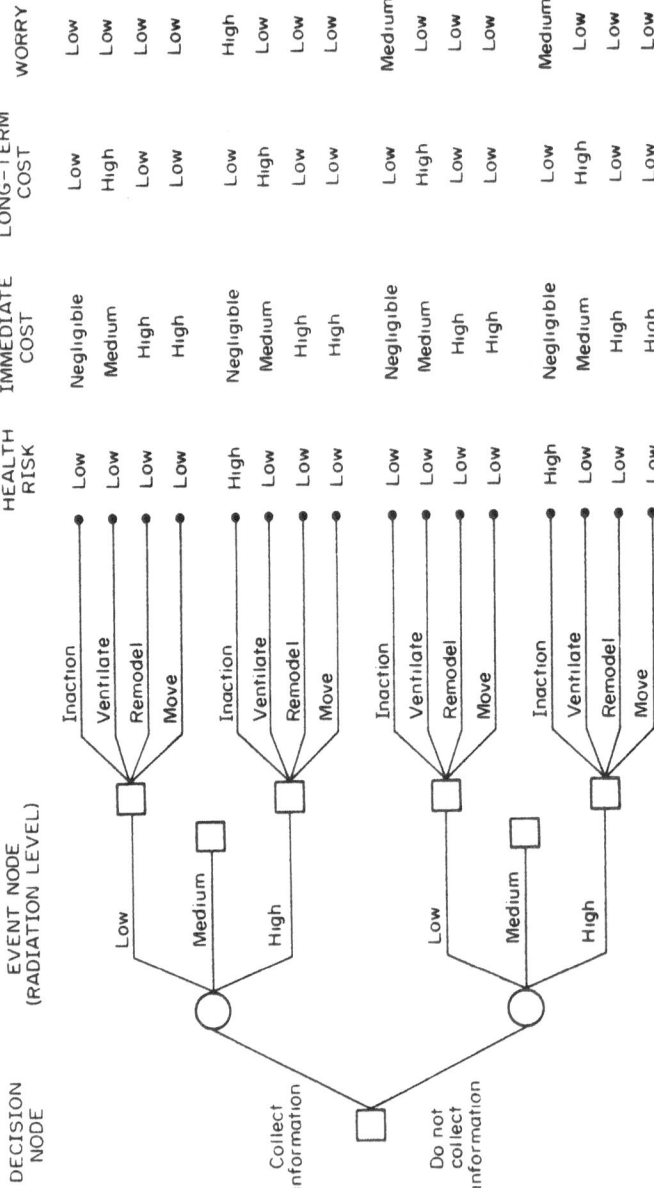

**Fig. 33.2.** A decision tree for a homeowner deciding whether to test for radon. The boxes represent decision points, while the circles are events (namely, the receipt of information). At the right are the consequences of each decision–event–decision sequence from the perspective of a particular homeowner. Such diagrams allow systematic identification of the information needs of potential recipients of risk communications. (Reproduced with permission from ref. 55.)

values important to them? If they really do disagree about the facts, are there defensible grounds for their differing perceptions (e.g., Do at least some experts agree with the public?, and, Is there a history of confident scientific claims about this technology subsequently being overturned?)?

Often, the risk managers' decision problem (e.g., whether to ban ethylene dibromide [EDB]) will be quite different from the public's decision problem (e.g., whether to use blueberry muffin mix). Failure to address the public's information needs is likely to leave it frustrated and hostile. Failure to address the managers' own problems is likely to leave their eventual actions inscrutable. For telling their own story, the managers need a protocol that will ensure that all of the relevant parts get out, including what options they are legally allowed to consider, how they see the facts, and what they consider to be the public interest (56). Such comprehensive accounts are often absent from managers' public pronouncements, preventing the public from responding responsibly and suggesting that the managers failed to consider the issues fully.

After determining what needs to be said, risk managers can start worrying about how to say it. A common worry is that the public will not be able to understand the technical details of how a technology operates. Where those details are really pertinent, the services of good science writers and educators may be needed. Perhaps a more common problem is making the basic concepts of risk management clear. Just what is a one-in-a-million chance? What does it mean to protect wastes for a hundred generations? Must we inevitably set a value on human life when resources are allocated for risk reduction? The psychological research described previously has shown the difficulty of these concepts and is beginning to show ways to communicate them meaningfully. For example, one in a million could be described as requiring each of six one-in-ten events to happen. This might be an effective technique—if the notion of conjunction can be made clear. Here, as elsewhere, detailed empirical research is needed. Often, the health effects of poor risk communications can be considerably worse than the health effects of the risk itself. One should no more expose the public to untested communications than to untested drugs (57).

Taking the details of risk perceptions seriously means reconciling ourselves to a messy process. In managing risks, our societies as a whole are slowly and painfully learning how to make deliberative decisions about very difficult issues, without their citizens ever having been all that good at making analytical decisions about much simpler issues. Avoiding frustration with the failures and with the public that seems responsible for them will help us keep the mental health and mutual respect needed to get through it all.

## Summary

Risk management has always been a major part of everyday life. Laypeople manage risks when they choose their foods, drive their cars, perform their jobs, monitor their health, and allocate their energies to civic causes. In doing so, they cannot have detailed knowledge about more than a few of the risks they face. Nor are they likely to have any specialized training in decision making or risk taking per se. As a result, it should not be surprising that public responses to some, especially new, hazards seem confused, ineffective, frustrated, and emotional—in terms of both laypeople's personal behavior and their demands of government. Empirical study of public responses to specific hazards confirms these general observations. People do, in fact, misunderstand many risks, overestimating some, underestimating others. When they combine evidence, people often violate basic laws of probability and decision theory. Although superficially discouraging, these results do provide some reasons for optimism. One is that there is considerable order in how people cognitively process risk-related information. As a result, there is some opportunity for carefully designed communication programs to help people understand risks (and what to do about them) better. An important component of that design process is testing. Inappropriate risk communications can have worse public health consequences than the risks that they describe. Designed communications have to be particularly effective in order to counteract the unsystematic messages that the public gleans from following the news media, observing public figures, hearing expert pronouncements, and interpreting its own anecdotal experiences. Understanding what people hear (and believe), what decision-making problems they face, and how they think is essential to helping them make decisions in their own, and their society's, best interests.

# References

1. Abelson R, Levi A (1985) Decision making and decision theory. In: Handbook of social psychology. Addison-Wesley, Reading, Massachusetts
2. Fischhoff B (1988) Judgment and decision making. In: Sternberg RJ, Smith EE (eds) The psychology of human thought. Cambridge Univ Press, London and New York
3. Kahneman D, Slovic P, Tversky A. (eds) (1982) Judgments under uncertainty: heuristics and biases. Cambridge Univ Press, London and New York
4. McKean K (June 1985) Decisions, decisions. Discover, 22–31
5. Slovic P, Lichtenstein S, Fischhoff B (1988). Decision making. In: Atkinson (ed) Stevens' handbook of experimental psychology, Wiley, New York
6. von Winterfeldt D, Edwards W (1986) Decision analysis and behavioral research. Cambridge Univ Press, London and New York
7. Fischhoff B, Slovic P, Lichtenstein S (1977) Knowing with certainty: the appropriateness of extreme confidence. Journal of Experimental Psychology: Human Perception and Performance 20:159–183
8. Wallsten T, Budescu D (1983) Encoding subjective probabilities: a psychological and psychometric review. Management Science 29:135–140
9. Fischhoff B (1982) Debiasing. In: Kahneman D, Slovic P, Tversky A (eds) Judgment under uncertainty: heuristics and biases. Cambridge University Press, New York and London, pp 422–444

10. Fischhoff B, MacGregor D (1983) Judged lethality: how much people seem to know depends upon how they are asked. Risk Analysis 3:229–236
11. Poulton EC (1968) The new psychophysics: six models for magnitude estimation. Psychological Bulletin 69:1–19
12. Poulton EC (1982) Biases in quantitative judgments. Applied Ergonomics 13:31–42
13. Cotgrove S (1982) Environment; Catastrophe or cornucopia. Wiley, New York
14. Fischhoff B, Slovic P, S Lichtenstein (1982) Lay foibles and expert fables in judgments about risk. American Statistician 36:240–255
15. Fischhoff B, Svenson O, Slovic P (1987) Active responses to environmental hazards. In: Handbook of environmental psychology. Wiley, New York
16. Freudenberg WR, Rosa EA (eds) (1984) Public reactions to nuclear power. Westview, Boulder, Colorado
17. Baum A, Gatchel R, Schaeffer M (1983) Emotional, behavioral and physiological effects of chronic stress at Three Mile Island. Journal of Consulting and Clinical Psychology 12:349–359
18. Johnson EJ, Tversky A (1983) Affect, generalization and the perception of risk. Journal of Personality and Social Psychology 45:20–31
19. Richardson D, Sorensen J, Soderstrom EJ (1987) Explaining the social and psychological impacts of a nuclear power plant accident. Journal of Applied Social Psychology 17:16–36
20. Weinstein N (ed) (1987) Taking care: understanding and encouraging self-protective behavior. Cambridge Univ Press, London and New York
21. Nisbett RE, Ross L (1980) Human inference: strategies and shortcomings of social judgment
22. Beyth-Marom R (1982) How probable is probable? Journal of Forecasting 1:257–269
23. Krohn W, Weingart P (1987). Commentary: nuclear power as a social experiment— European political "fall out" from the Chernobyl meltdown. Science, Technology, and Human Values 12(2):52–58
24. Peterson CR, Beach LR (1967) Man as an intuitive statistician. Psychological Bulletin 69(1):29–46
25. Hasher L, Zacks RT (1984) Automatic and effortful processes in memory. Journal of Experimental Psychology: General 108:356–388
26. Lichtenstein S, Slovic P, Fischhoff B, Layman M, Combs B (1978) Judged frequency of lethal events. Journal of Experimental Psychology: Human Learning and Memory 4:551–578
27. MacLean D (1987) Understanding the nuclear power controversy. In: Engelhardt HT Jr, Caplan AL, (eds) Scientific controversies: case studies in the resolution and closure of disputes in science and technology. Cambridge Univ Press, London and New York
28. Rothman S, Lichter SR (1987) Elite ideology and risk perception in nuclear energy policy. American Political Science Review 81(2):383–404
29. Tversky A, Kahneman D (1973) Availability: a heuristic for judging frequency and probability. Cognitive Psychology 5:207–232
30. Fischhoff B, Slovic P, Lichtenstein S (1978) Fault trees: sensitivity of assessed failure probabilities to problem presentation. Journal of Experimental Psychology: Human Perception and Performance 4:330–344

31. Fischhoff B, Slovic P, Lichtenstein S (1983) The "public" versus the "experts": perceived versus actual disagreements about the risk of nuclear power. In: Covello V, Flamm G, Roderick J, Tardiff R (eds) Analysis of actual versus perceived risks. Plenum, New York

32. Fischhoff B, Watson S, Hope C (1984) Defining risk. Policy Sciences 17:123–139

33. Slovic P, Fischhoff B, Lichtenstein S (1979) Rating the risks. Environment 24(4):14–20, 36–39

34. Slovic P, Fischhoff B, Lichtenstein S (1984) Modeling the societal impact of fatal accidents. Management Science 30:464–474

35. Mazur A, Disputes between experts. Minerva 11:243–262

36. Nelkin D (ed) (1978) Controversy: politics of technical decisions. Sage, Beverly Hills, California

37. Fischhoff B, Slovic P, Lichtenstein S (1980) Knowing what you want: measuring labile values. In: Wallsten T (ed) Cognitive processes in choice and decision behavior. Erlbaum, Hillsdale, New Jersey

38. Turner CF, Martin E (eds) (1984) Surveying subjective phenomena. Russell Sage Foundation, New York

39. Tversky A, Kahneman D (1981) The framing of decisions and the psychology of choice. Science 211:453–458

40. Fischhoff B (in press) Judgmental aspects of risk assessment. In: Handbook of risk assessment, prepared by the National Science Foundation for the Office of Management and Budget. Plenum, New York

41. Mahoney MJ (1979) Psychology of the scientist: an evaluative review. Social Studies of Science 9:349–375

42. Henrion M, Fischhoff B (1986) Uncertainty assessment in the estimation of physical constants. American Journal of Physics 54:791–798

43. Hynes M, Vanmarcke E (1976) Reliability of embankment performance predictions. Proceedings of the ASCE Engineering Mechanics Division Specialty Conference. Univ of Waterloo Press, Waterloo, Ontario, Canada

44. Aftermath of Chernobyl (1986). Groundswell 9(2):1–7

45. Hohenemser C, Deicher M, Ernst A, Hofsass H, Lindner G, Recknagel E (1986) Chernobyl: an early report. Environment 28(5):6–43

46. Murphy A, Winkler R (1984) Probability of precipitation forecasts: a review. Journal of the American Statistical Association 79:391–400

47. Bingham G (1984) Resolving environmental disputes: a decade of experience. The Conservation Foundation, Washington, DC

48. Fischhoff B, Lichtenstein S, Slovic P, Derby S, Keeney R (1981) Acceptable risk. Cambridge Univ Press, London and New York

49. Lowrance W (1976) Of acceptable risk. Freeman, San Francisco, California

50. MacLean D (ed) (1986) Values at risk. Rowman and Allanheld, Totowa, New Jersey

51. Nelkin D (1977) Technological decisions and democracy. Sage, Beverly Hills, California

52. Fischhoff B, Slovic P, Lichtenstein S, Read S, Combs B (1978) How safe is safe enough? A psychometric study of attitudes towards technological risks and benefits. Policy Sciences 8:127–152

53. Vlek C, Stallen PJ (1981) Judging risks and benefits in the small and in the large. Organizational Behavior and Human Performance 28:235–271

54. Raiffa H (1968) Decision analysis. Addison-Wesley, Reading, Massachusetts

55. Svenson O, Fischhoff B (1985) Levels of environmental decisions: a case study of radiation in Swedish homes. Journal of Environmental Psychology 5:55–68
56. Fischhoff B (winter 1985) Environmental reporting: what to ask the experts. The Journalist, 11–15
57. Fischhoff B (1987) Treating the public with risk communications: a public health perspective. Science Technology and Human Values 12:13–19

# Part IX
## Summary

CHAPTER 34

# Radionuclides in the Food Chain

## W. K. Sinclair[1]

The task of this chapter is to synthesize some of what has been presented in the excellent work preceding this summary, to comment on the present status of our capability to understand and to control radionuclides in the food chain, and to make some suggestions for actions in the future.

## Background

We have known for many decades that natural radionuclides contaminate every aspect of our environment, including our air, food, and water. As a result of careful measurements of the kind described in previous chapters, we know that our annual intake of natural radionuclides ranges between about 2 Bq for $^{232}$Th to about 130 Bq for $^{40}$K (1). The resulting effective dose equivalent to our bodies is about 400 μSv annually, about half of it due to $^{40}$K (2,3). After inhalation of Rn, this contribution is the most important component of the total exposure of members of the US population (and other populations) to natural radiation. The total exposure of the US population to natural radiation is about 3 mSv and in addition there is 0.6 mSv caused by man-made radiation for a total of 3.6 mSv (2). This is within the range of 2.5 to 4 mSv cited to us earlier (4), although it is distributed a little differently.

We have also been aware of some unusually high activity in some portions of our food chain. For example, 30 years ago Brazil nuts were found to be rich in Ra ($^{226}$Ra and $^{228}$Ra) with levels of the order of 200 to 2,000 Bq/kg, some 1,000 times greater than most American foods (5).

We also know that beginning with atmospheric testing of nuclear weapons in the 1950s we have added to our environment a considerable amount of man-made radioactivity in the form of fission product radionuclides (6). These were quite severe contaminants about 20 years ago but have since declined, to deliver less than 10 μSv per year to the total radiation exposure of the average

[1] National Council on Radiation Protection and Measurements, Bethesda, MD 20814, USA.

individual at the present time. We inevitably face the possibility that other radiation sources, components of the nuclear power fuel cycle, nuclear accidents involving power plants, or other less esoteric items, such as discarded and mishandled teletherapy sources, can contaminate the environment and in some circumstances the human food chain.

We also know, perhaps less well, approximately what the risks to human beings are following the ingestion or inhalation of radionuclides. At low dose, the risks of cancer induction and genetic effects are of particular importance, and we know reasonably well what the risks are quantitatively, as indicated in the presentations of Silini (7), Upton and Linsalata (8) and Sankaranarayanan (9). My interpretation of recent results is that the risk of cancer induction for adults is probably between 1% and 5% per Sv, but is age dependent, younger ages being more susceptible. The risk of severe genetic defects is of the same order, about 1% to 2% per Sv, but this has been inferred from animal studies and could be appreciably less for human beings. We have heard furthermore that there will be reevaluations of the risks of cancer following the revisions of the dosimetry and other new features of the Japanese survivor experience, which could raise these risk coefficients. We must remain alert to revisions in our knowledge in this important area.

Most of us in the professions concerned with radiation protection are well aware of these radioactive contaminants and their potential effects, and as a result, many countries took steps years ago to limit the intake of man-made radionuclides based on well-known radiation protection principles.

## The Chernobyl Accident

Most of the operational controls (e.g., monitoring programs) that these limits require were in place in many countries before the accident at Chernobyl (10,11). But nothing that preceded it has focused our attention on the problems of actual and potential contamination of food as has the accident at Chernobyl. Among the many fears that developed in members of the public in the days soon after the Chernobyl accident, the greatest concern of those at a distance, was the possibility of eating contaminated food and drinking contaminated water. Not even the threat of latent cancers was as compelling or caused as much apprehension. My own daughter, an experienced traveler and somewhat familiar with radiation issues, was about to take a trip to Yugoslavia just after Chernobyl; she thought of canceling and was only reassured when I told her to drink only bottled liquids and eat no fresh vegetables on the trip just to be on the safe side. I answered many similar queries, as did a number of other people who contributed to these proceedings. The level of public apprehension about food and water was quite remarkable. It was not helped, as we know, by the fact that different European governments applied very different levels of control and clearly revealed widely differing attitudes to the problem. I had one call from a newspaper in Rome that stated that many of the Italian people were very concerned about what they had heard about radionuclides in their food

**Table 34.1.** Action levels for $^{131}$I in imported foods[a]

| Country | Food | Action level (Bq/kg or Bq/L) |
|---|---|---|
| Canada | Milk | 10 |
| | Dairy products | 40 |
| | Other | 70 |
| European community | Milk | 500 |
| | Vegetables | 350 |
| China | Milk | 1,300 |
| | Fruits and vegetables | 270 |
| | Cereals | 340 |
| | Beverages | 130 |
| Sweden | All foods | 2,000 |
| United States | Infant foods | 55 |
| | All foods | 300 |

[a] Reproduced with permission from ref. 12.

and water. Many felt the government had not paid enough attention to the problem, its regulations were not strict enough, it had failed to inform the people, and so on. Probably in most of the countries where these concerns arose, the harm was more imaginary than real, but a lot of anxiety occurred that might have been avoided.

The differences were not only within Europe but worldwide, as Tables 34.1 and 34.2 on the action levels used for $^{131}$I and for $^{134}$Cs and $^{137}$Cs for a variety of countries and regions indicate (12). The $^{131}$I levels varied by a factor of

**Table 34.2.** Action levels for $^{137}$Cs (or $^{134}$Cs and $^{137}$Cs) in imported foods[a]

| Country | Food | Action level (Bq/kg or Bq/L) |
|---|---|---|
| Brazil | Milk powder | 3,700 |
| | Other foods | 600 |
| Canada | Milk | 50 |
| | Dairy products | 100 |
| | Other foods | 300 |
| | Spices | 3,000 |
| European community | Milk and infant products | 370 |
| | Other foods | 600 |
| China | Milk | 4,600 |
| | Fruits and vegetables | 1,000 |
| | Cereals | 1,200 |
| | Beverages | 460 |
| Sweden | All foods | 300 |
| United States | All foods | 370 |

[a] Reproduced with permission from ref. 12.

200, those for $^{134}$Cs and $^{137}$Cs by a factor of 100. Such variations are not likely to inspire confidence in a concerned and wary public.

Furthermore, the short-term problems are not the only ones of concern. Two recent reports in *Nature* indicate the fact that food contamination problems resulting from Chernobyl are still causing concerns in countries as far apart as the United Kingdom and Japan. The first, in the United Kingdom, concerns the contamination of sheep, which still exceeds the interim limit of 1,000 Bq/kg and where bentonite is being used experimentally to reduce Cs uptake from grazing. According to the report, 500,000 sheep are still under surveillance (13). The second concerns the banning of Italian sphaghetti and macaroni by a private "life club" of 147,000 members in Japan, which has set its own standard of 37 Bq/kg for $^{134}$Cs and $^{137}$Cs in food, much more restrictive than the Commission of the European Communities' (CEC) 600 Bq/kg (14).

Where do we stand now? Granted that we still have many problems from Chernobyl, we seem nevertheless to have entered almost a post-Chernobyl lull. As the result of appraisals that have occurred at meetings such as this one, are we in a better position to deal with a new event if it should occur again? Have we already done what is necessary? Alternatively, what do we have to do to optimize our experience of the recent past in order to deal much more successfully with a future incident?

## Subjects Covered at the Meeting

First of all let us consider what we have learned. We have had a very comprehensive meeting, beginning with a general introduction to the whole subject by Dr. Pochin (15), followed by some fundamental information on biological effects of radiation, sources of radioactive contamination (including natural radionuclides), and general aspects of control via the international recommendations of the International Commission on Radiological Protection (ICRP) from Lindell (16) and Beninson (17). We then heard about environmental pathways critical to humans: air, water, soil, the food chain itself, and the movement of both natural and artificial radionuclides through these pathways. These were embellished with some detailed reports of recent work in the Soviet Union from our Soviet colleagues (see Chapters 8, 11, 12, and 20). We also heard about methods of interfering with these pathways in order to reduce the final exposure. We noted in discussion that getting rid of contaminated food was easier than reducing contamination. But we should not dismiss the important in vivo pathway for reducing radionuclide contamination that Dr. Arnaud described to us (18). We were treated to discussions about experiences following the accidents at Windscale in 1957, Three Mile Island in 1979, and Chernobyl in 1986 and about contamination resulting from weapons tests. The importance of the iodine and Cs radionuclides and to a lesser extent those of Sr and the transuranic alpha emitters has become clear to us. We heard about specific effects that could result from these radionuclides in the food chain. We heard from Dr. Hill on models for risk management of food and water supplies (19) and from Dr. Kaul on the uncertainties in the parameters involved in the assessments (20). From Dr. Rubery, we learned that radionuclides are not the only noxious contaminants of the food chain and that some of the others are just as ubiquitous (21). This was followed by discussions on the development of guidelines for safety evaluations of food and water after nuclear accidents that have taken place in the European community, Hungary, the United States, and elsewhere, leading finally to monitoring and regulatory control programs. We even heard one presentation on the perception of risk by the public and the difficulties of developing a balanced public understanding of the problem, a most important subject that surely deserves more of our attention. These proceedings have been full and informative and have provided us with a comprehensive perspective on the problems associated with the control and management of radionuclides inadvertently arising in our food chain.

## Observations on the Present Status of Food-Chain Control

One thing at least has become clear. It is essential for nations to develop similar ways to treat the problem, especially in regions such as Europe with so many contiguous borders and ready exchanges of food and liquids across national boundaries. Credibility and public acceptance worldwide will be much greater if there is uniformity. Thus, uniform international recommendations are critical as has already been emphasized (22).

It would seem that we still have much to do here, even though the disparity in derived intervention levels (DILs) may seem less than in the early period after Chernobyl. There are still major differences in the approaches used. Most approaches to DILs start from the same place, essentially the lower of the intervention levels proposed by the ICRP in 1984 on a dose basis, that is, no more than 5 mSv to the whole body or 50 mSv to any organ in the first year (23). The Food and Agriculture Organization (FAO) then derives from these a series of interim international radionuclide action levels for food (IRALFs) based on food intake, see Table 34.3 (12). These differ from the recent proposals of the CEC, as shown in Table 34.4 (24); the published DILs of the International Atomic Energy Agency (IAEA), see Table 34.5 (25); guideline values from the World Health Organization (WHO) (26), see Table 34.6; and the levels of concern used by the Food and Drug Administration (FDA) on the import of foods into the United States, see Table 34.7 (27).

The recommendations differ not only in name, which is in itself confusing, but numerically by factors up to 40 or so. In some cases, this is because of a difference in purpose as to how the numbers are to be applied, which we as scientists understand, but such subtle differences will almost certainly not be perceived or understood by the public. Beyond that there are clearly differences on how the basic ICRP standards are applied (e.g., 5 mSv in the first year or 5 mSv in total), in the methods of calculation (the IRALFs of the FAO are

**Table 34.3.** Interim international radionuclide action levels for food[a]

| Radionuclide | Target organ | Food intake (kg) | IRALFs (Bq/kg) |
|---|---|---|---|
| [131]I | Thyroid | | |
| First year | (infant) | 40 | 400 |
| [134]Cs | Whole body | | |
| First year | (adult) | 750 | 350 |
| Following years | Whole body (adult) | 750 | 70 |
| [137]Cs | Whole body | | |
| First year | (adult) | 750 | 500 |
| Following years | Whole body (adult) | 750 | 100 |

[a] Reproduced with permission from ref. 12.

**Table 34.4.** Maximum permitted levels for food, European community proposals[a]

| | Dairy produce (Bq/kg) | Other foodstuffs (except minor) (Bq/kg) | Drinking water and liquid foodstuffs (Bq/L) | Feedstuffs (Bq/kg) |
|---|---|---|---|---|
| Isotopes of I and Sr (e.g., [131]I and [90]Sr) | 500 | 3,000 | 400 | |
| Alpha emitters (e.g., [239]Pu and [241]Am) | 20 | 80 | 10 | |
| Other nuclides T ½ > 10d (e.g., [134]Cs and [137]Cs) | 1,000 | 1,250 | 800 | 2,500 |

[a] Reproduced with permission from ref. 24.

derived differently from the DILs of the IAEA or the guidelines of the WHO), and in the parameters used in the calculations. The ICRP dose-conversion factors for adults seem to be more or less universally used, but the FAO, WHO, and IAEA versions differ somewhat. Furthermore, even such straightforward quantities as annual food intake vary between the FAO, IAEA, CEC, and WHO by a factor of 2, and the period of effective exposure also differs, for [131]I, for

**Table 34.5.** Derived intervention levels for restricting foodstuffs[a]

| | Peak concentrations | | | | |
|---|---|---|---|---|---|
| Isotope | Milk (Bq/L) | Milk products (Bq/L or Bq/kg) | Fruit and vegetables (Bq/kg) | Meat (Bq/kg) | Drinking water (Bq/L) |
| [131]I | 2,000 | 30,000 | 15,000 | 10,000 | 1,500 |
| [239]Pu | 20 | 200 | 40 | 40 | 400 |
| [137]Cs | 20,000 | 100,000 | 3,000 | 10,000 | 700 |

[a] Reproduced with permission from ref. 25.

**Table 34.6.** General guideline values[a]

| Class of radionuclide | Cereals (Bq/kg) | Roots and tubers (Bq/kg) | Vegetables (Bq/kg) | Fruit (Bq/kg) | Meat (Bq/kg) | Milk (Bq/L) | Fish (Bq/kg) | Water (Bq/L) |
|---|---|---|---|---|---|---|---|---|
| High dose per unit intake factor ($10^{-6}$Sv/ Bq) | 35 | 45 | 80 | 70 | 100 | 45 | 350 | 10 |
| Low dose per unit intake factor ($10^{-8}$Sv/ Bq) | 3,500 | 4,500 | 8,000 | 7,000 | 10,000 | 4,500 | 35,000 | 1,000 |

[a] Reproduced with permission from ref. 26.

example, the FAO uses 40 days, the FDA uses 60 days, and the WHO bases their estimates on the mean life or about 11 days.

## Suggestions for the Future

Despite these differences, we, the scientific community, are in a position to solve these problems. We are not really in disagreement about the basic principles or what makes sense in applying them. I would like to make a few points in this regard, so let us consider the following.

First, the basis of all intervention and other control actions derives from the fundamental principles of radiation protection. Therefore, presumably everything should start with the basic considerations developed by the ICRP. Each of the sets of numbers cited above starts with 5 mSv in the first year, the lower level recommended in ICRP Publication 40 for emergency or accident situations (23). (However, the upper level of ICRP of 50 mSv, at which action certainly should be taken, could have been considered for extreme emergency cases.) On the other hand, there has been discussion about whether we have been dealing with emergency or routine situations, and some obvious confusion is apparent. I would point out that the 5 mSv (or the 50 mSv) of ICRP Publication

**Table 34.7.** Recommendations for import of foods, from the FDA, USDA Chernobyl task force[a]

| | Level of concern (screening values) | |
|---|---|---|
| | Infant food (Bq/kg) | Other food (Bq/kg) |
| $^{131}$I | 55 | 300 |
| $^{134}$Cs and $^{137}$Cs | 370 | 370 |

[a] Reproduced with permission from ref. 27.

40 is for emergencies. If the situation is routine, then the normal ICRP dose limits apply (28), and if the exposure is likely to be continuous, the limit of 1 mSv per year for man-made sources applies. In fact, not all of the 1 mSv per year could be assigned to the food chain, so the starting basis would have to be some number substantially smaller than 1 mSv per year, which presumably could be derived by applying as low as reasonably achievable (ALARA) principles to the control of sources in food.

Parenthetically, the ICRP must consider new and developing biological and other pertinent scientific information as it becomes available and possibly modify its levels and approaches accordingly. At the present time, changes in risk estimation may well require some reevaluation. Given stable basic information, the ICRP can then provide revised guidance [it has already provided much guidance in Publication 40 (23), which the ICRP intends to update].

Second, uniform DILs have to be accomplished by some international body or bodies, such as the United Nations Scientific Committee on the Effects of Atomic Radiation (UNSCEAR), FAO, WHO, CEC, and IAEA with adequate input from the United States and other countries (so far this input has not been prominent) working together and agreeing to recommend a single set of DILs, IRALFs, or whatever they are to be called. (It may be that two sets of levels are needed, one for emergency circumstances close to the accidents and the other for the longer term, sometime afterward or further afield, and general use. This might solve some of the problems with numbers. On the other hand, it is important to use the fewest set of numbers that can adequately be used for control purposes.) Even if the numbers generated are relatively complex, such as those of the WHO, which take many facets of the problem into account, methods of simplification might be suggested so that the numbers could be operationally more useful. In any event, concession and cooperation between the international bodies will be essential to obtain agreement. If it would be helpful, perhaps the ICRP should consider extending their basic guidance further into the realm of international control levels, possibly even to developing DILs.

Third, if we have learned anything from Chernobyl, it is that public information and perception are critically important. Given that a satisfactory set of operating controls are in place, internationally and nationally, the public must be made aware of the basis for these controls and the depth of consideration that has taken place in deriving them. Thus, information pamphlets that deal with food and water, the likely contaminants, etc., should be prepared for publication in the newspapers, general distribution by the food industry, and use by the media to quote from and to disseminate. The first of these might be a general pamphlet relating to standardized control measures that uses all our past experiences as guidance. Subsequent pamphlets, which must be prepared at the time, but could be in draft form prior to it, could deal with the specific problems that have arisen in connection with an accident. This could explain why cabbage from X has been banned, but other cabbages on the market are edible.

Still, the most difficult problem for all of us is the development of the right forms of information and the best means to disseminate this information widely. In this respect, scientists are not by any means the only people involved. Indeed, it may be the role of science to put the facts in the hands of others and have them produce material that can be both accurate and readily understood. Perhaps the next conference we need is one on how to do this, such a conference should clearly involve the media and especially those with a talent for purveying technical information to laypersons.

In the meantime, our own scientific house has still to be put in order. This conference has made it clear that we do not have a single set of useful, straightforward recommendations on radionuclides in food to present to the food industry and to a worried world. It is up to us, as a profession, to devote our efforts to accomplishing this with reasonable dispatch or face the risk of letting the public down, when in fact we are in a good position to help them.

# References

1. Harley JH (1988) Naturally occurring sources of radioactive contamination. In: Carter MW (ed) Radionuclides in the food chain. Springer-Verlag, New York, pp 58–71 (see Chap 6 of this monograph)
2. National Council on Radiation Protection and Measurements (1987) Ionizing radiation exposure of the population of the United States. National Council on Radiation Protection and Measurements, NCRP Report 93, Bethesda, Maryland
3. National Council on Radiation Protection and Measurements (1987) Exposure to the population in the United States and Canada from natural background radiation. National Council on Radiation Protection and Measurements, NCRP Report 94, Bethesda, Maryland
4. Jacobi W (1988) Assessment of dose from man-made sources. In: Carter MW (ed) Radionuclides in the food chain. Springer-Verlag, New York, pp 45–57 (see Chap 5 of this monograph)
5. Eisenbud M (1973) Environmental radioactivity. Academic Press, New York
6. Carter MW, Hanley L (1988) Food-chain contamination from testing nuclear devices. In: Carter MW (ed) Radionuclides in the food chain. Springer-Verlag, New York, pp 172–194 (see Chap 15 of this monograph)
7. Silini G (1988) Biological effects of ionizing radiation. In: Carter MW (ed) Radionuclides in the food chain. Springer-Verlag, New York, pp 35–44 (see Chap 4 of this monograph)
8. Upton AC, Linsalata P (1988) Long-term health effects of radionuclides in food and water supplies. In: Carter MW (ed) Radionuclides in the food chain. Springer-Verlag, New York, pp 217–235 (see Chap 17 of this monograph)
9. Sankaranarayanan K (1988) Radionuclides and genetic risks. In: Carter MW (ed) Radionuclides in the food chain. Springer-Verlag, New York, pp 236–263 (see Chap 18 of this monograph)
10. Meekings GF (1988) Radioactivity in food: Surveillance procedures in the United Kingdom. In: Carter MW (ed) Radionuclides in the food chain. Springer-Verlag, New York, pp 291–301 (see Chap 21 of this monograph)
11. Porter CR, Broadway JA, Kahn B (1988) Methodology for surveillance of the food chain as conducted by the United States. In: Carter MW (ed) Radionuclides in the food chain. Springer-Verlag, New York, pp 302–322 (see Chap 22 of this monograph)
12. Food and Agriculture Organization (1987) Recommended limits for radionuclide

contamination of foods. Food and Agriculture Organization of the United Nations, report of a consultation meeting, Rome, 1986

13. Johnston K (1987) British sheep still contaminated by Chernobyl fallout. Nature 328:661

14. Johnston K (1987) Chernobyl takes macaroni off Japan's menu. Nature 329:278

15. Pochin EE (1988) Links in the transmission of radionuclides through food chains. In: Carter MW (ed) Radionuclides in the food chain. Springer-Verlag, New York, pp 22–31 (see Chap 3 of this monograph)

16. Lindell B (1988) International recommendations on radiation protection. In: Carter MW (ed) Radionuclides in the food chain. Springer-Verlag, New York, pp 72–83 (see Chap 7 of this monograph)

17. Beninson DJ (1988) Radionuclides in food: Radiation protection considerations. In: Carter MW (ed) Radionuclides in the food chain. Springer-Verlag, New York, pp 398–406 (see Chap 28 of this monograph)

18. Arnaud MJ (1988) The removal and/or reduction of radionuclides in the food chain. In: Carter MW (ed) Radionuclides in the food chain. Springer-Verlag, New York, pp 195–213 (see Chap 16 of this monograph)

19. Hill MD (1988) Use of mathematical models in risk assessment and risk management. In: Carter MW (ed) Radionuclides in the food chain. Springer-Verlag, New York, pp 335–361 (see Chap 24 of this monograph)

20. Kaul A (1988) Identification and reliability of parameters for the assessment of derived intervention levels for control of contaminated foodstuffs. In: Carter MW (ed) Radionuclides in the food chain. Springer-Verlag, New York, pp 323–334 (see Chap 23 of this monograph)

21. Rubery ED (1988) Evaluation procedures. In: Carter MW (ed) Radionuclides in the food chain. Springer-Verlag, New York, pp 264–281 (see Chap 19 of this monograph)

22. Ronk RJ, Thompson, P (1988) Radionuclides: Regulatory and control programs. In: Carter MW (ed) Radionuclides in the food chain. Springer-Verlag, New York, pp 409–420 (see Chap 29 of this monograph)

23. International Commission on Radiological Protection (1984) Protection of the public in the event of major radiation accidents: principles for planning. International Commission on Radiological Protection, Publication 40, Annals of the ICRP 14:1–22, Pergamon, Oxford

24. Gray PS, Luyckx, F (1988) Control of radioactivity in foodstuffs in the European economic community. In: Carter MW (ed) Radionuclides in the food chain. Springer-Verlag, New York, pp 436–446 (see Chap 31 of this monograph)

25. International Atomic Energy Agency (1986) Derived intervention levels for application in controlling radiation doses to the public in the event of a nuclear accident or radiological emergency. IAEA Safety Series No 81, IAEA, Vienna

26. Waight PJ (1988) The development of WHO's approach to DILs. In: Carter MW (ed) Radionuclides in the food chain. Springer-Verlag, New York, pp 381–388 (see Chap 26 of this monograph)

27. Schmidt GD (1988) Development of guidelines for safety evaluation of food and water after nuclear accidents: Procedures in North America. In: Carter MW (ed) Radionuclides in the food chain. Springer-Verlag, New York, pp 365–380 (see Chap 25 of this monograph)

28. International Commission on Radiological Protection (1977) Recommendations of the International Commission on Radiological Protection. International Commission on Radiological Protection, Publication 26, Annals of the ICRP, Pergamon, Oxford

# Glossary Words

Absorbed dose
Absorption
Actinides
Activation
Activation products
Activity
Airborne
ALARA
Alpha emitter
Alpha particle (alpha radiation)
Anion
Annual limit on intake (ALI)
Atom
Atomic number

Background radiation
Becquerel (Bq)
Beta emitter
Beta particle (beta radiation)
Boiling water reactor (BWR)
Breeder reactor
By-product material

Carcinogenic
Cation
Cladding
Collective dose
Collective dose commitment
Collective effective dose equivalent
Committed dose equivalent ($H_{50}$)
Coolant
Core
Cosmic rays
Coulomb (C)
Critical organ
Curie (Ci)

Daughter
Decay
Decay constant
Decay product
Deoxyribonucleic acid (DNA)
Derived air concentration (DAC)
Discharges
Disintegration
Dose
Dose equivalent (H)
Dose rate

Effective dose equivalent
Electric charge
Electromagnetic radiation
Electron
Electron volt (eV)
Element
Emission
Energy
Excitation
Exposure
External radiation

Fallout
Fast reactor
Fission, nuclear
Fission products
Free radical
Fuel cycle
Fusion, thermonuclear

Gamma emitter
Gamma ray
Genetic effects
Genetically significant dose (GSD)

Gonads
Gray (Gy)

Half-life
High LET radiation

Internal radiation
In vitro studies
In vivo studies
Ion
Ionization
Ionizing radiation
Isotopes

Joule (J)

Leukemia
Linear energy transfer (LET)
Low LET radiation

Mass
Mass number
Mean dose equivalent ($H_T$)
Megawatt (MW)
Moderator
Molecule
Monitoring
Mutation

Natural radiation
Negligible individual risk level (NIRL)
Neutron
Neutron activation
Nonstochastic effects
Nuclear fuel
Nuclear power
Nuclear reactor
Nuclear weapon
Nucleon
Nucleus (atomic)
Nucleus of the cell
Nuclide

Optimization
Order of magnitude
Organ weighting factor

Parent
Permissible dose

Person-Sievert (Man Sievert)
Photographic film
Positron
Pressurized water reactor (PWR)
Probability
Proton

Quality factor (QF)

Rad
Radiation
Radioactive
Radioactive equilibrium
Radioactive series
Radioactive waste
Radioactivity
Radiobiology
Radiological protection
Radionuclide
Radon decay products
RBMK reactor (RBMK)
Reference level
Relative biological effectiveness (RBE)
Rem
Risk
Risk factor
Risk weighting factor
Roentgen (R)

Secondary limit
Sievert (Sv)
SI units (SI)
Somatic effects
Source material
Special nuclear material
Specific activity
Stochastic effects

Teratogenic effects
Terrestrial gamma rays
Thermal neutrons
Thermal reactors
Thermoluminescent dosimeter (TLD)
Tritium ($^3$H)

Whole-body dose equivalent
Working level (WL)
Working level month (WLM)

X-ray

# Selected Glossary for the ILSI Monograph
## *Radionuclides in the Food Chain*

| | |
|---|---|
| Absorbed dose: | Amount of radiation energy absorbed per unit mass of a given tissue. It is measured in grays (Gy), where 1 Gy = 1 J/kg, or rads, where 1 rad = 100 ergs/g. |
| Absorption: | Process by which radiation deposits some or all of its energy in any material through which it passes. |
| Actinides: | Group of 15 elements with atomic numbers 89 to 103, which includes U, Pu, Am, and Cm. |
| Activation: | Process of inducing radioactivity in a stable atom by irradiation (usually with a neutron). |
| Activation products: | Various radionuclides produced by irradiation. Frequently produced by neutron irradiation. |
| Activity: | Amount or quantity of a radionuclide. Describes the rate at which radioactive decay occurs. |
| Airborne: | Discharges of radionuclides, particulates, and gases in a gaseous medium. May also be referred to as effluents or emissions. |
| ALARA: | As low as reasonably achievable. Concept that the effects of radiation and levels of exposure of workers and members of the public should be kept as low as possible with due regard to economic and social factors. |
| Alpha emitter: | Radionuclide that emits alpha particles. |
| Alpha particle (alpha radiation): | Charged particle emitted during the radioactive decay of some radionuclides. It consists |

of two protons and two neutrons and has a net charge of $+2$.

Anion:    Negatively charged ion.

Annual limit on intake (ALI):    Activity of a radionuclide that, taken into the body during a year, would provide a committed effective dose equivalent to a person equal to the annual occupational effective dose equivalent limit (0.05 Sv or 5 rem) or, in some cases, the limit of dose to an individual organ.

Atom:    Smallest portion of an element that is capable of entering into a chemical reaction.

Atomic number:    Number of protons in the nucleus of an atom. Symbol Z.

Background radiation:    Radiation arising from radioactive material other than the one directly under consideration. Normally refers to radiation due to cosmic rays and naturally occurring radioactive substances in earth or building materials.

Becquerel (Bq):    SI unit of radioactivity. 1 Bq = 1 radioactive disintegration per second. 1 Bq = 27 × $10^{-12}$ Ci.

Beta emitter:    Radionuclide that emits beta particles.

Beta particle (beta radiation):    Charged particle emitted during the radioactive decay of some radionuclides. It has a mass and charge $(-1)$ equal to that of an electron.

Boiling water reactor (BWR):    Type of nuclear power plant that has a core cooled by water that is allowed to boil in the pressure vessel. It is a thermal reactor that uses water as both a coolant and a moderator.

Breeder reactor:    Type of nuclear reactor that manufactures more nuclear fuel than it uses.

By-product material:    Radioactive material, except special nuclear material, yielded in or made radioactive by exposure to the radiation incident in the process of producing or utilizing special nuclear material.

Carcinogenic:    Capable of producing cancer.

| | |
|---|---|
| Cation: | Positively charged ion. |
| Cladding: | Covering on nuclear fuel. Designed to resist physical and chemical effects, thus preventing corrosion of the fuel and escape of products of the nuclear reaction. |
| Collective dose: | Frequently used for collective effective dose equivalent. |
| Collective dose commitment: | Sum of the doses to all individuals in a population. |
| Collective effective dose equivalent: | Product of the average effective dose equivalent and the number of persons exposed to a given source of radiation. Expressed in person-sieverts. Frequently abbreviated to collective dose. |
| Committed dose equivalent ($H_{50}$): | Dose equivalent accumulated in the 50 years after intake of a radionuclide, often to age 70. |
| Coolant: | Substance, liquid or gas, used for cooling any part of a reactor in which heat is generated, particularly the core. |
| Core: | Region in a nuclear reactor that contains the fissionable material. |
| Cosmic rays: | High-energy particulate and electromagnetic radiations that originate outside the earth's atmosphere. |
| Coulomb (C): | Unit of electrical charge. |
| Critical organ: | Body organ receiving the radionuclide that results in the greatest overall damage to the body. Usually, but not always, the organ having the greatest concentration. |
| Curie (Ci): | Historical (old) unit of radioactivity. 1 Ci $= 3.7 \times 10^{10}$ disintegrations per second ($27 \text{ Ci} = 10^{12}$ Bq). |
| Daughter: | Product of radioactive decay. |
| Decay: | Process by which radionuclides spontaneously change from atoms of one element to atoms of another element, emitting ionizing radiation as they do so. |
| Decay constant: | Fraction of the number of atoms of a radioactive isotope that decay in unit time. |

| | |
|---|---|
| Decay product: | Nuclide resulting from the radioactive decay of a radionuclide, formed either directly or as the result of successive transformations. |
| Deoxyribonucleic acid (DNA): | Prominent chemical component of nuclei and chromosomes. |
| Derived air concentration (DAC): | Annual limit on intake (ALI) of a radionuclide divided by the volume of air inhaled by reference man in a working year (i.e., $2.4 \times 10^3$ m$^3$). In units of Bq/m$^3$. |
| Discharges: | Release of liquid, gaseous, or airborne effluents from an industrial site. |
| Disintegration: | Spontaneous transformation of an atom of one element into an atom of another element accompanied by the emissions of energy and mass from the nucleus. |
| Dose: | General term denoting the quantity of radiation or energy absorbed. |
| Dose equivalent (H): | Quantity obtained by multiplying the absorbed dose by a quality factor to allow for estimated differences in effectiveness of the various ionizing radiations in causing harm to humans. It is measured in sieverts (Sv), where 1 Sv = 100 rem. |
| Dose rate: | Absorbed dose delivered per unit of time (e.g., Gy/per year). |
| Effective dose equivalent: | Quantity obtained by multiplying the dose equivalents to various tissues and organs by the risk weighting factor appropriate to each and summing the products. An inherent part of the ICRP radiation protection system. |
| Electric charge: | Quantity of electricity. Usually expressed in coulombs (C). |
| Electromagnetic radiation: | Radiation that travels through space in a wave motion. X-rays, gamma rays, and light are examples of this type of radiation. |
| Electron: | Stable elementary particle having a single negative electric charge (equal to $1.6 \times 10^{-19}$ C) and a very small rest mass. |
| Electron volt (eV): | Unit of energy equivalent to the energy gained by an electron in passing through a potential difference of 1 V; 1 eV = $1.6 \times 10^{-19}$ J. |

| | |
|---|---|
| Element: | Substance with atoms all of the same atomic number. |
| Emission: | Release of gaseous effluent from an industrial plant. |
| Energy: | Capacity for doing work. Measured in joules. |
| Excitation: | Process by which radiation imparts energy to an atom or molecule without causing ionization. Dissipated as heat. |
| Exposure: | Measure of the ionization produced in air by X or gamma radiation. The sum of the electrical charges on all ions of one sign produced in air when all electrons liberated by photons in a volume element of air are completely stopped in air divided by the mass of air in the volume element. Measured in units of coulombs per kilogram or in roentgens. |
| External radiation: | Irradiation of the body from radionuclides or other sources of radiation that are external to the body. |
| Fallout: | Radioactive debris from a nuclear detonation, which is airborne or has been deposited on the earth. |
| Fast reactor: | Nuclear reactor where fission is brought about by fast (high-energy) neutrons. |
| Fission, nuclear: | Nuclear transformation characterized by the splitting of a nucleus into at least two other nuclei with the release of a relatively large amount of energy. |
| Fission products: | Radionuclides or other nuclides produced as a result of fission. |
| Free radical: | Fragment of a compound or an element that contains an unpaired electron. Free radicals are highly reactive and chemically toxic. |
| Fuel cycle: | Sequence of steps, such as mining, milling, fabrication, utilization, and reprocessing, through which nuclear fuel passes. |
| Fusion, thermonuclear: | Process in which two or more light nuclei are formed into a single heavier nucleus and energy is released. |

| | |
|---|---|
| Gamma emitter: | Radionuclide that emits gamma radiation. |
| Gamma ray: | Discrete quantity of energy, without mass or charge, that is propagated as a wave. Emitted by some radionuclides. |
| Genetic effects: | Effects occurring in the progeny of those irradiated. |
| Genetically significant dose (GSD): | Dose that, if given to every member of a population, would produce the same genetic harm as the actual doses received by the various individuals. Expressed in sieverts. |
| Gonads: | Ovaries and testes. |
| Gray (Gy): | SI unit for absorbed dose: 1 Gy = 1 J of energy absorbed per kilogram of tissue; 1 Gy = 100 rads. |
| Half-life: | Time for the activity of a radionuclide to lose half its value by decay. |
| High LET radiation: | Type of ionizing radiation that leaves a high average density of energy deposition along the track that it produces in tissue. It is usually more damaging to body tissue than is low LET radiation. Examples of high LET radiations are alpha particles and neutrons. |
| Internal radiation: | Irradiation of the body from radionuclides inside the body. These usually have entered the body by ingestion and inhalation. |
| In vitro studies: | Studies carried out on material from an animal under artificially controlled conditions in the laboratory. |
| In vivo studies: | Studies carried out in the intact animal. |
| Ion: | Atomic particle, atom, or chemical radical bearing an electrical charge, either positive or negative. |
| Ionization: | Process by which a neutral atom or molecule acquires an electric charge. |
| Ionizing radiation: | Radiation that can deliver energy in a form capable of removing electrons from atoms and turning them into ions. |
| Isotopes: | Nuclides with the same atomic number but different mass numbers. |

| | |
|---|---|
| **Joule (J):** | Unit for work and energy equal to 1 newton of force acting through a distance of 1 m; $1 \text{ J} = 10^7$ ergs. |
| **Leukemia:** | Maligant disease of the blood in which the white cells are abnormal in type and number. |
| **Linear energy transfer (LET):** | Measure of the density of energy deposition in the track of ionizing radiation. |
| **Low LET radiation:** | Type of radiation that leaves a low average density of energy deposition along the track that it produces in tissue. It is usually less damaging to biological tissue than high LET radiation. |
| **Mass:** | Material equivalent of energy—different from weight in that it neither increases nor decreases with gravitational force. |
| **Mass number:** | Number of protons plus neutrons in the nucleus of an atom. Symbol A. |
| **Mean dose equivalent ($H_T$):** | Dose equivalent in each organ or tissue to which the ICRP dose limits for nonstochastic effects basically apply. |
| **Megawatt (MW):** | Unit used to measure the power output of nuclear reactors. Frequently expressed as electric output, $MW_e$ (megawatts, electric) or as thermal (heat) output, $MW_t$ (megawatts, thermal). |
| **Moderator:** | Material used in a nuclear reactor to slow down neutrons from high to low or thermal energies. |
| **Molecule:** | Smallest portion of a substance that can exist by itself and retain the properties of the substance. |
| **Monitoring:** | Periodic or continuous determination of the amount of ionizing radiation or radioactive contamination present in an area. |
| **Mutation:** | Chemical change in the DNA in the nucleus of a cell. |
| **Natural radiation:** | Pervades the whole environment. The principal sources of natural radiation are cosmic rays, terrestrial gamma rays, radon decay products, and internal radiation, primarily $^{40}K$. |

Negligible individual risk level (NIRL): Level of risk of death that can be dismissed. Namely, an annual risk of $10^{-7}$. This risk is that associated with an annual effective dose equivalent of 0.01 mSv (1 mrem).

Neutron: Elementary particle with unit atomic mass, approximately, and no electric charge.

Neutron activation: Production of radionuclides by neutron irradiation of stable materials.

Nonstochastic effects: Effects for which the severity of the effect in affected individuals varies with the dose and for which a threshold usually exists.

Nuclear fuel: Usually refers to those nuclear materials that will fission, releasing relatively large amount of energy. Such nuclear materials include $^{235}U$ and $^{233}U$ and $^{239}Pu$.

Nuclear power: Power obtained from the operation of a nuclear reactor. usually measured in $MW_e$ (megawatts, electric) or $MW_t$ (megawatts, thermal).

Nuclear reactor: Device in which nuclear fission may be sustained in a self-supporting chain reaction involving neutrons. Reactors release usable energy.

Nuclear weapon: Explosive device deriving its power from the fission or fusion of nuclei or both.

Nucleon: Constituent particle of the nucleus, a proton or neutron.

Nucleus (atomic): Core of an atom, occupying little of the volume, containing most of the mass, and bearing the total positive electric charge.

Nucleus of the cell: Definitely delineated body within the cell, containing DNA.

Nuclide: Species of atom characterized by the number of protons and neutrons and, in some cases, by the energy state of the nucleus.

Optimization: This has the same general meaning as ALARA, especially in the United States.

Order of magnitude: Quantity given to the nearest power of ten.

Organ weighting factor: Factor that indicates the ratio of the risk of stochastic effects attributable to irradiation of a given organ or tissue to the total risk when the whole body is uniformly irradiated.

Parent:

Radionuclide that upon disintegration yields a specified nuclide (daughter), either directly or as the result of successive transformations.

Permissible dose:

Dose of ionizing radiation that, in the light of present knowledge, is not expected to cause significant bodily injury to a person during his/her lifetime.

Person-Sievert (Man Sievert)

Product of the average effective dose equivalent in sieverts and the number of persons exposed to a given source of radiation. Unit for collective dose.

Photographic film:

Radiation dosimeter using film with emulsion sensitive to ionizing radiation. The degree of blackening is related to dose.

Positron:

Particle emitted from the nucleus that has the exact properties of a beta particle except its unit charge is positive.

Pressurized water reactor (PWR):

Type of nuclear power plant that has a core cooled by water kept under pressure. It is a thermal reactor that uses water as both a coolant and moderator.

Probability:

Mathematical chance that a given event will occur.

Proton:

Elementary nuclear particle with unit atomic mass, approximately, and a single positive electric charge.

Quality factor (QF):

LET factor by which absorbed doses are multiplied to obtain (for radiation protection purposes) a quantity that expresses—on a common scale for all ionizing radiations—the effectiveness of the absorbed dose.

Rad:

Historical (old) unit of absorbed dose. One rad is 0.01 J absorbed per kilogram of any material. (Also defined as 100 ergs per gram).

Radiation:

Energy, emitted from radioactive nuclei, propagated through space or a material medium as waves or particles.

| | |
|---|---|
| Radioactive: | Possessing radioactivity. |
| Radioactive equilibrium: | Among the members of a radioactive series, the state that prevails when the ratios between the amounts of radioactivity of successive members of the series remain constant. |
| Radioactive series: | Succession of nuclides, each of which transforms by radioactive decay into the next until a stable nuclide is formed. The parent is the first member, intermediate members are daughters, and the final stable member is the end product. |
| Radioactive waste: | Useless material containing radionuclides. Frequently categorized in the nuclear power industry according to activity content as low-level, intermediate-level, and high-level waste. |
| Radioactivity: | Property of radionuclides spontaneously emitting ionizing radiation. |
| Radiobiology: | Branch of biology that deals with the effects of radiation on biological systems. |
| Radiological protection: | Science and practice of limiting the harm to humans from radiation. |
| Radionuclide: | Unstable nuclide that emits ionizing radiation. |
| Radon decay products: | All successive radioactive members following $^{222}$Rn in the U decay chain, but often limited to those with short half-lives, excluding $^{210}$Pb, $^{210}$Bi, and $^{210}$Po. |
| RBMK Reactor (RBMK): | Type of nuclear power plant that uses low-enriched U as fuel, is graphite moderated and light water cooled. It is of the channel type wherein the water boils in the channels and has a direct steam cycle to the turbine. These reactors are refueled during operation and usually do not have containment buildings. |
| Reference level: | Predetermined value of a quantity, below a limit, that triggers a specified course of action when the value, usually a dose level, is exceeded or is expected to be exceeded. |
| Relative biological effectiveness (RBE): | Factor used to compare the biological effectiveness of absorbed radiation doses due to different types of ionizing radiation. |

REM:    Historical (old) unit for dose equivalent. The absorbed dose (rads) is multiplied by the quality factor for the particular type of radiation; 100 rem = 1 Sv.

Risk:    Probability of injury, harm, or damage.

Risk factor:    Probability of cancer and leukemia or hereditary damage per unit dose equivalent. Usually refers to fatal malignant diseases and serious hereditary damage. Expressed as the probability per sievert.

Risk weighting factor:    Ratio of the stochastic risk resulting from irradiation of a tissue to the total stochastic risk when the whole body is irradiated uniformly. The concept was developed by the ICRP.

Roentgen (R):    Historical (old) unit of exposure. One roentgen equals $2.58 \times 10^{-4}$ C/kg of air.

Secondary limit:    Limit, derived from a primary limit using conservative assumptions, that assures adherence to the primary limit by methods easier to implement than those required for the primary limit.

Sievert (Sv):    SI unit of dose equivalent. The absorbed dose (in grays) is multiplied by a quality factor for the particular type of radiation; 1 Sv = 100 rem.

SI units (SI):    International system of units.

Somatic effects:    Effects occurring in the individual irradiated, for example, cancer induction.

Source material:    Uranium or Th, or any combination thereof, and ores that contain at least 0.05% U, Th, or any combination thereof.

Special nuclear material:    Plutonium, $^{233}$U, and U enriched in the isotopes $^{233}$U or $^{235}$U.

Specific activity:    Total activity of a given radionuclide per gram of compound, element, or isotope.

Stochastic effects:    Effects, the probability of which, rather than their severity, is a function of radiation dose, without threshold. Stochastic means random in nature.

Teratogenic effects:    Effects occurring in offspring as a result of insults sustained in utero.

Terrestrial gamma rays:

Gamma radiation resulting from the decay of radionucludes in the earth's crust.

Thermal neutrons:

Neutrons that have been slowed to the degree that they have the same average thermal energy as the atoms or molecules through which they are passing. The average energy of neutrons at ordinary temperatures is about 0.025 eV.

Thermal reactors:

Nuclear reactor in which fission is brought about by thermal neutrons.

Thermoluminescent dosimeter (TLD):

Radiation dosimeter using small thermoluminescent crystals that primarily measures X-rays and gamma rays.

Tritium ($^3$H):

Radioactive isotope of hydrogen having 1 proton and 2 neutrons in the nucleus (hydrogen-3 or $^3$H).

Whole-body dose equivalent:

Dose equivalent associated with the uniform irradiation of the whole body.

Working level (WL):

Amount of potential alpha energy in a cubic meter of air that will result in the emission of $2.08 \times 10^{-5}$ J of energy. Formerly defined as the energy emitted by the short-lived daughters of $^{222}$Rn in equilibrium with 100 pCi of $^{222}$Rn per liter of air (3,700 Bq/m$^3$).

Working level month (WLM):

Cumulative exposure equivalent to exposure to one working level for a working month (170 hours), that is, $2.08 \times 10^{-5}$ J/m$^3$ × 170 h = 0.0035 J h/m$^3$.

X-Ray:

Penetrating radiation propagated as an electromagnetic wave having a length shorter than visible light and originating from electron transitions in atoms.

# Index